ELEMENTARY FLUID MECHANICS

ELEMENTARY FLUID MECHANICS
SIXTH EDITION

JOHN K. VENNARD
Late Professor of Fluid Mechanics

ROBERT L. STREET
Professor of Fluid Mechanics
and Applied Mathematics
Stanford University

1807 1982

JOHN WILEY & SONS
New York   Chichester   Brisbane   Toronto   Singapore

Library of Congress Cataloging in Publication Data

Vennard, John King.
  Elementary fluid mechanics.

  Includes bibliographies and index.
  1. Fluid mechanics. I. Street, Robert L.
II. Title.

QA901.V4 1982 532 81-7559
ISBN 0-471-04427-X AACR2

Printed in the United States of America

10 9 8 7 6 5 4 3 2 1

PREFACE

Fluid mechanics is the study of fluids under all conditions of rest and motion. Its approach is analytical and mathematical rather than empirical; it is concerned with basic principles that provide the solution to numerous and diverse problems encountered in many fields of engineering, regardless of the physical properties of the fluids involved.

This textbook is intended for beginners who have completed differential and integral calculus and engineering courses in statics and dynamics, but who have not yet encountered a course in thermodynamics and have no prior experience with fluid phenomena except for that obtained in basic physics courses. The limited experience of the beginner and the broad scope of fluid mechanics have led us to define *elementary fluid mechanics* as that portion of the subject that may be feasibly studied in an introductory course; with more extensive training in mathematics and thermodynamics, the beginner in fluid mechanics could be expected to cover considerably more subject matter, especially in flowfields and compressible fluid motion.

In view of our considerable experience in teaching fluid mechanics to American undergraduates, we believe the difficulties are more of a conceptual than a mathematical nature; for this reason, the emphasis here is on physical concepts rather than mathematical manipulation. Because this book is written for the beginning student, rather than the mature professional, we hope that by itself the text will prove to be an understandable introduction to the various topics, allowing precious classroom time to be spent on amplification and extension of the material. It is our conviction that the use of a textbook in this manner provides vital training for later professional life when new subjects must be assimilated without teachers to introduce them.

Presentation of the subject begins with a discussion of fundamentals and fluid properties, followed by fluid statics, at which point students first encounter a fluid field and the use of partial derivatives. A chapter on kinetics discusses velocity, acceleration, vorticity, circulation, and the conservation of mass (the equation of continuity). Here the key concept of the control volume is first introduced. After this, we move to chapters that discuss the flow of the incompressible and compressible ideal fluid, which include a brief treatment of two-dimensional flowfields; the impulse-momentum principle is then developed in a separate chapter, including applications to incompressible and compressible flow. Discussion of frictional processes and a chapter on similitude and dimensional analysis complete the presentation of the primary tools and lead to applications to pipes, open channels, and measuring devices. Chapter 12, on elementary hydrodynamics, precedes the final chapter on flow about immersed objects.

References to written and visual materials have been provided to guide the inquiring reader to more exhaustive treatment of the various topics. The relevant films listed at the ends of the chapters may be quite useful to the beginner by filling in background and providing visual experience with various fluid phenomena. More than a thousand problems are included at the ends of the chapters, and most quantitative articles in the text are accompanied by Illustrative Problems which show how to achieve useful engineering answers from the derived concepts and procedures.

The traditional American units are derived from the English FSS (foot-slug-second) system. However, most of the world uses the Système International d'Unités (SI). Although SI units are coming into use in the United States, both types of units will be used for many years. To assist in the transition and provide users in the metric countries access to this textbook, this edition uses both SI and FSS unit systems. The emphasis is on SI units, however, and they account for approximately 70% of the usage. In many problems at the ends of the chapters, both sets of units are quoted, and the Solution Manual, which is available to instructors, contains solutions in both sets of units for these problems. Accordingly, instructors may choose the appropriate sets of units to be used at their option.

The revision over the years of a textbook produces an evolution in its content. Such is the case with this book and this edition. Thus, I have maintained the level, style, and basic organization of the original book which made it appealing to many generations of students and their teachers. However, a major effort has been made to improve the clarity of the Illustrative Problems so that they may effectively lead the student from concept to successful solution of engineering problems. In addition, significant changes have been made in several chapters. The role and meaning of control volumes have been emphasized further and they are now more tightly integrated into the text. But there has not been an increase in the complexity or sophistication of presentation. The work-energy principle has been more clearly delineated in Chapter 4. The general quality and depth of the discussion of real fluid flow has been upgraded by a moderate revision of Chapter 7 and by a virtually complete revision and reorganization of Chapter 9, which now leads the student from concrete physical principles and observations on laminar, turbulent, and smooth or rough flows, to their synthesis in the Moody diagram, and then to applications. Finally, the material on similitude and dimensional analysis has been reorganized to stress the key similitude principles, to incorporate the modern approach to the Buckingham Π-theorem, and to raise the practical applicability and scope of the material on turbomachines.

I am deeply indebted to my friend and Stanford colleague, Professor En Yun Hsu, who provided the framework for the revision of Chapters 7 and 9, and who has critically reviewed the entire manuscript. The efforts of Mr. Jeffrey Koseff, who read the manuscript from the student's point of view, are greatly appreciated by me and, without doubt, will be by the readers.

Stanford, California
December 1980

Robert L. Street

CONTENTS

TABLES

1

FUNDAMENTALS

1.1 Development and Scope of Fluid Mechanics

The human desire for knowledge of fluid phenomena began with problems of water supply, irrigation, navigation, and water power. With only a rudimentary appreciation for the physics of fluid flow, we dug wells, constructed canals, operated crude water wheels and pumping devices and, as our cities increased in size, constructed ever larger aqueducts, which reached their greatest size and grandeur in the city of Rome. However, with the exception of the thoughts of Archimedes (287–212 B.C.) on the principles of buoyancy, little of the scant knowledge of the ancients appears in modern fluid mechanics. After the fall of the Roman Empire (A.D. 476) there is no record of progress in fluid mechanics until the time of Leonardo da Vinci (1425–1519). This great genius designed and built the first chambered canal lock near Milan and ushered in a new era in hydraulic engineering; he also studied the flight of birds and developed some ideas on the origin of the forces that support them. However, down through da Vinci's time, concepts of fluid motion must be considered to be more art than science.

After the time of da Vinci, the accumulation of hydraulic knowledge rapidly gained momentum, with the contributions of Galileo, Torricelli, Mariotte, Pascal, Newton, Pitot, Bernoulli, Euler, and d'Alembert to the fundamentals of the science being outstanding. Although the theories proposed by these scientists were in general confirmed by crude experiments, divergences between theory and fact led d'Alembert to observe in 1744, "The theory of fluids must necessarily be based upon experiment." He showed that there is no resistance to motion when a body moves through an ideal (nonviscous or *inviscid*) fluid; yet obviously this conclusion is not valid for bodies moving through real fluids. This discrepancy between theory and experiment, called the d'Alembert paradox, has since been resolved. Yet it demonstrated clearly the limitations of the theory of that day in solving fluid problems. Because of the conflict between theory and experiment, two schools of thought arose in the treatment of fluid mechanics, one dealing with the theoretical and the other with the practical aspects of fluid flow. In a sense, these two schools of thought have persisted to the present day, resulting in the mathematical field of *hydrodynamics* and the practical science of *hydraulics*.

Near the middle of the last century, Navier and Stokes succeeded in modifying the general equations for ideal fluid motion to fit those of a viscous fluid and in so doing showed the possibilities of explaining the differences between hydraulics and hydrodynamics. About the same time, theoretical and experimental work on vortex motion and separated flow by Helmholtz and Kirchhoff was aiding in explaining many of the divergent results of theory and experiment.

1

Meanwhile, hydraulic research went on apace, and large quantities of excellent data were collected. Unfortunately, this research led frequently to empirical formulas obtained by fitting curves to experimental data or by merely presenting the results in tabular form, and in many instances the relationship between physical facts and the resulting formula was not apparent.

Toward the end of the last century, new industries arose that demanded data on the flow of fluids other than water; this fact and many significant advances in our knowledge tended to arrest the empiricism of much hydraulic research. These advances were: (1) the theoretical and experimental work of Reynolds; (2) the development of dimensional analysis by Rayleigh; (3) the use of models by Froude, Reynolds, Vernon-Harcourt, Fargue, and Engels in the solution of fluid problems; and (4) the rapid progress of theoretical and experimental aeronautics in the work of Lanchester, Lilienthal, Kutta, Joukowsky, Betz, and Prandtl. These advances supplied new tools for the solution of fluid problems and gave birth to modern fluid mechanics. The single most important contribution was made by Prandtl in 1904 when he introduced the concept of the boundary layer. In his short, descriptive paper Prandtl, at a stroke, provided an essential link between ideal and real fluid motion for fluids with a small viscosity (e.g., water and air) and provided the basis for much of modern fluid mechanics.

In the twentieth century, fluid problems have been solved by constantly improving rational methods; these methods have produced many fruitful results and have aided in increasing knowledge of the details of fluid phenomena. The trend is certain to continue in large part because of the ever increasing power of the digital computer.

Another continuing trend is that toward greater complexity and challenge in fluid problems. The problems of water supply, irrigation, navigation and water power remain, but on a scale never imagined by the citizens of pre-Christian Rome. The range of new problems added in modern times is virtually infinite, including the sonic boom of the supersonic airplane; dispersion of man's wastes in lakes, rivers, and oceans; blood flow in veins, arteries, kidneys, hearts, and artificial heart and kidney machines; fuel pumping and exhaust flow in moon rockets; the design of 1 millon-ton super oil tankers for speed, cargo-pumping efficiency, and safety; and analysis and simulation of the world's weather and ocean currents. Thus, today fluid mechanics has become an essential part of such diverse fields as medicine, meteorology, astronautics, and oceanography, as well as of the traditional engineering disciplines.

1.2 Physical Characteristics of the Fluid State

Matter exists in two states—the solid and the fluid, the fluid state being commonly divided into the liquid and gaseous states.[1] Solids differ from liquids and liquids from

---

[1] Many would say that matter exists in four states—solid, liquid, gaseous, and plasma, the latter three being classified as fluids. The plasma state is the state of over 99% of the matter of the universe and is distinguished from the others because a significant number of its molecules are ionized. Hence, the plasma contains electrically charged particles and is susceptible to electromagnetic forces. Unfortunately, the intriguing subject of plasma dynamics is beyond the scope of this work.

gases in the spacing and latitude of motion of their molecules, these variables being large in a gas, smaller in a liquid, and extremely small in a solid. Thus it follows that intermolecular cohesive forces are large in a solid, smaller in a liquid, and extremely small in a gas. These fundamental facts account for the familiar compactness and rigidity of form possessed by solids, the ability of liquid molecules to move freely within a liquid mass, and the capacity of gases to fill completely the containers in which they are placed, while a liquid has a definite volume and a well-defined surface. In spite of the mobility and spacing of its molecules, a fluid is considered (for mechanical analysis) to be a *continuum* in which there are no voids or holes; this assumption proves entirely satisfactory for most engineering problems.[2]

A more rigorous mechanical definition of the solid and fluid states can be based on their actions under various types of stress. Application of tension, compression, or shear stresses to a solid results first in elastic deformation and later, if these stresses exceed the elastic limits, in permanent distortion of the material. Fluids possess elastic properties only under direct compression or tension. Application of infinitesimal shear stress to a fluid results in continual and permanent distortion. The inability of fluids to resist shearing stress gives them their characteristic ability to change their shape or to *flow*; their inability to support tension stress is an engineering assumption, but it is a well-justified assumption because such stresses, which depend on intermolecular cohesion, are usually extremely small.[3]

Because fluids cannot *support* shearing stresses, it does not follow that such stresses are nonexistent in fluids. During the flow of real fluids, the shearing stresses assume an important role, and their prediction is a vital part of engineering work. Without flow, however, shearing stresses cannot exist, and compression stress or *pressure* is the only stress to be considered.

Because fluids at rest cannot contain shearing stresses, no component of stress can exist in a static fluid tangent to a solid boundary or tangent to an arbitrary section passed through the fluid. This means that pressures must be transmitted to solid boundaries or across arbitrary sections *normal* to these boundaries or sections at every point. Furthermore, if a *free body* of fluid is isolated as in Fig. 1.1, pressure must be shown as acting *inward* (p_1) and on the free body (according to the usual conventions of mechanics for compression stress). Pressures exerted by the fluid on the container (p_2) will of course act *outward*, but their reactions (p_3) will act inward on the fluid as before. Another property of fluid pressure is that, at a point in a fluid at rest, it has the same magnitude in all directions; this may be proved by considering a convenient two-dimensional[4] free body of fluid (Fig. 1.2) having unit width normal to the plane of the paper. Taking p_1,

---

[2] An important exception is a gas at very low pressure (rarefied gas), where intermolecular spacing is very large and the "fluid" must be treated as an aggregation of widely separated particles. It follows that the continuum assumption is valid only when the smallest physical length scale in a flow is much larger than the average spacing between or than the mean free path of (about 1 μm or $10^{-4} - 10^{-5}$ in.) molecules composing the fluid.

[3] Tests on some very pure liquids have shown them to be capable of supporting tension stresses up to a few thousands of pounds per square inch, but such liquids are seldom encountered in engineering practice.

[4] A more general proof using a three-dimensional element yields the same result.

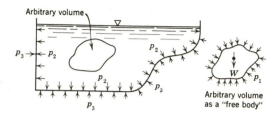

Fig. **1.1**

p_2, and p_3 to be the mean pressures on the respective surfaces of the element, ρ the density[5] of the fluid, g_n the acceleration due to gravity and writing the equations of static equilibrium:

$$\sum F_x = p_1 \, dz - p_3 ds \sin \theta = 0$$

$$\sum F_z = p_2 \, dx - \rho \, g_n \, dx \, dz / 2 - p_3 \, ds \cos \theta = 0$$

From geometry, $dz = ds \sin \theta$ and $dx = ds \cos \theta$. Substituting the first of these relations into the first equation above gives $p_1 = p_3$, whatever the size of dz; substituting the second relation into the second equation yields $p_2 = p_3 + \rho \, g_n \, dz / 2$, whatever the size of dx. From these equations it is seen that p_1 and p_3 approach p_2 as dz approaches zero. Accordingly it may be concluded that at a point $(dx = dz = 0)$ in a static[6] fluid $p_1 = p_2 = p_3$, and the pressure there is the same in all directions.

With the pressure at a point the same in all directions, it follows that the pressure has no vector sense and hence is a *scalar* quantity; however, the differential forces produced by the action of pressures on differential areas are vectors because they have

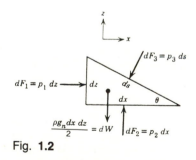

Fig. **1.2**

[5] A summary of symbols and their dimensions is given in Appendix 1.
[6] For the flow of an ideal (inviscid) fluid the same result may be proved, but for viscous fluid motion the presure at a point is generally not the same in all directions owing to the action of shearing stress; however, this is of small consequence in most engineering problems, where shear stress is usually small compared to pressure (or normal stress).

directions normal to the areas. The resultant forces obtained by the integration of the differential forces also are vector quantities.

Another well-known aspect of fluid pressure (which needs no formal proof) is that pressures imposed on a fluid at rest are transmitted undiminished to all other points in the fluid; this follows directly from the static equilibrium of adjacent elements and the fact that the fluid mass is a continuum. This aspect finds practical expression in the hydraulic lift used in automobile service stations.

The reader may be uneasy concerning the treatment of liquids and gases by the same principles in view of their obvious differences of compressibility. In problems where compressibility is of small importance (and there are many of these in engineering), liquids and gases may be treated similarly, but where the effects of compressibility are predominant (as in high speed gas flow) the behavior of liquids and gases is quite dissimilar and is governed by very different physical laws. Usually, when compressibility is unimportant, fluid problems may be solved successfully with the principles of mechanics; when compressibility predominates, thermodynamics and heat transfer concepts must be used as well.

1.3   Units, Density, Specific Weight, Specific Volume, Specific Gravity

The world is presently changing to the use of a single international language of units. The adopted system is the metric *Système International d'Unités* (SI). Many countries have already changed to SI units and the United States is moving toward use of the metric system in lieu of the currently used English (or U.S. customary) system. Unfortunately, both types of units will be used for many years. Accordingly, in this edition of this book, both the metric (SI) system and the English FSS (foot-slug-second) system of units are used. However, usually in illustrative examples and problems only one set of units is used within each example or problem.

For the SI and FSS unit systems, there are four basic dimensions through which fluid properties are expressed. These base dimensions and their associated units (and symbols) are:

Dimension	SI Unit	English FSS Unit
Length (L)	metre (m)	foot (ft)
Mass (M)	kilogram(kg)	slug ($-$)
Time (t)	second (s)	second (s)
Thermodynamic temperature (T)[7]	kelvin (K)	degree Rankine (°R)

In the SI and FSS systems, the above units are known as base units. There are also a number of other units that are derivable from base units and that have special names, for example,

---

[7] Here T°C = T K − 273.15 while T°F = T°R − 459.6. The normal freezing and boiling points of water are (0°C, 32°F) and (100°C, 212°F), respectively.

Quantity	SI Unit	English FSS Unit
Frequency (f)	hertz (Hz = s^{-1})	hertz (Hz = s^{-1})
Force (F)	newton (N = kg $\cdot$ m/s^2)	pound (lb = slug $\cdot$ ft/s^2)
Energy (E),	joule (J = N $\cdot$ m)	British thermal unit
Work (W),		(BTU = 778.2 ft $\cdot$ lb)
Quantity of		
heat (Q)		
Power (P)	watt (W = J/s	horsepower (hp = 550 ft
	= $m^2 \cdot$ kg/s^3)	$\cdot$ lb/s = 550 slug
		$\cdot$ ft^2/s^3)
Pressure (p),	pascal (Pa = N/m^2	pound per square inch
Stress (τ)	= kg/m $\cdot$ s^2)	(psi = (lb/ft^2)/144 =
		(slug/ft $\cdot$ s^2)/144)
Temperature (T;	degree Celsius (°C)	degree Fahrenheit (°F)
practical scales)[7]		

There are several conventions used in the SI systems which should be noted. The names of multiples and submultiples of the basic and derived SI units are related to these units and are formed by means of prefixes which are the same irrespective of the units to which they are applied. For example, the prefixes used herein include mega (M) = 10^6, kilo (k) = 10^3, milli (m) = 10^{-3}, and micro (μ) = 10^{-6}. Thus, 1 megawatt = 1 MW = 10^3 kW = 10^6 W = 10^6 watts and 1 square millimetre = 1 mm^2 = 10^{-6} m^2 = 10^{-6} square metres. In addition, two other conventions are employed with SI units and are extended herein to the FSS system. First, when a compound unit is formed by multiplication of two or more units, a dot is used to avoid confusion, for example, N $\cdot$ m or kg $\cdot$ m/s^2. Second, while the decimal point is employed in the usual way, number spacing is accomplished with spaces in lieu of commas, for example, 14 446 or 0.000 002.

Both the SI and the English FSS systems have the advantage that they distinguish between force and mass and have no ambiguous definitions (such as pounds-mass and pounds-force).

Density is the mass, that is, the amount of matter, contained in a unit volume; specific weight[8] is the weight, that is, gravitational attractive force, acting on the matter in that unit volume. Both these terms are fundamentally dependent on the number of molecules per unit of volume. As molecular activity and spacing increase with temperature, fewer molecules exist in a given volume of fluid as temperature rises; thus density and specific weight decrease with increasing temperature.[9] Because a larger number of molecules can be forced into a given volume by application of pressure, density and specific weight increase with increasing pressure.

Density, ρ, is expressed in the mass-length-time system of dimensions and has the

[8] To the extent that this concept is used in the SI community, the term $\gamma = \rho g_n$ is called the *weight density*, not specific weight.

[9] For example, a variation in the temperature from the freezing point to the boiling point of water will cause the specific weight of water at atmospheric pressure to decrease 4% (Appendix 2) and will cause the density of gases to decrease 37% (assuming no pressure variation).

dimensions of mass units (M) per unit volume (L^3). Thus, $[\rho] = $ slugs/ft^3 or $=$ kg/m^3.

Specific weight,[8] γ, is expressed in the force-length-time system of dimensions and has dimensions of force (F) per unit volume (L^3). Thus, $[\gamma] = $ lb/ft^3 or $=$ N/m^3.

Because the weight (a force), W, is related to its mass, M, by Newton's second law of motion in the form

$$W = Mg_n$$

in which g_n is the acceleration due to the local force of gravity, density and specific weight (the mass and weight of a unit volume of fluid) are related by a similar equation,

$$\gamma = \rho g_n \tag{1.1}$$

Because meaningful physical equations are dimensionally homogeneous, the metre-newton-second dimensions of ρ (which are equivalent to kilograms per cubic metre) may be calculated as follows:

$$\text{Dimensions of } \rho = \frac{\text{Dimensions of } \gamma}{\text{Dimensions of } g_n} = \frac{\text{N/m}^3}{\text{m/s}^2} = \frac{\text{N} \cdot \text{s}^2}{\text{m}^4}$$

Accordingly, the equivalence of dimensions is

$$\text{kg/m}^3 = \text{N} \cdot \text{s}^2/\text{m}^4$$

or

$$1 \text{ N} = 1 \text{ kg} \times 1 \text{ m/s}^2$$

that is, a unit force produces a unit acceleration of a unit mass. Equivalently, in FSS units

$$\text{Dimensions of } \rho = \frac{\text{Dimensions of } \gamma}{\text{Dimensions of } g_n} = \frac{\text{lb/ft}^3}{\text{ft/s}^2} = \frac{\text{lb} \cdot \text{s}^2}{\text{ft}^4}$$

and

$$\text{slugs/ft}^3 = \text{lb} \cdot \text{s}^2/\text{ft}^4$$

or

$$1 \text{ lb} = 1 \text{ slug} \times 1 \text{ ft/s}^2$$

This algebraic use of the dimensions of quantities in the equation expressing physical relationship is employed extensively throughout the text and will prove to be an invaluable check on engineering calculations.[10]

The specific volume, defined as volume per unit of mass, has dimensions of length cubed per unit mass (m^3/kg or ft^3/slug). This definition identifies specific volume as the reciprocal of density.

Specific gravity (s.g.) is the ratio of the density of a substance to the density of

---

[10] A summary of quantities, their units and dimensions, and conversion factors is given in Appendix 1.

water at a specified temperature and pressure. Because these items vary with temperature, temperatures must be quoted when specific gravity is used in precise calculations. Specific gravities of a few common liquids are presented in Table 1[11], from which the specific weights of liquids can be readily calculated by

$$\gamma = (s.g.) \times \gamma_{water} \qquad (1.2)$$

The specific weights of perfect gases can be obtained by a combination of Boyle's and Charles' laws known as the *equation of state*; in terms of specific weight this is

$$\gamma = g_n p / RT \qquad (1.3)$$

in which p is the absolute pressure (force per unit area); T the thermodynamic temperature; and R the engineering gas constant (engery per unit mass per unit of temperature). Naturally, the equation of state can also be written in the form

$$\rho = p / RT \qquad (1.3)$$

The gas constant is constant only if the gas is a *perfect gas*. Common gases in the ordinary engineering range of pressure and temperature may be considered to be perfect for many engineering calculations, but departure from the simple equation of state is to be expected near the point of liquefaction, or at extremely high temperatures or low pressures.

Application of Avogadro's law, "all gases at the same pressure and temperature have the same number of molecules per unit volume," allows the calculation of a universal gas constant. Consider two gases having constants, R_1 and R_2, densities ρ_1 and ρ_2, and existing at the same pressure and temperature, p and T. Dividing their equations of state,

$$\frac{p/\rho_1 T = R_1}{p/\rho_2 T = R_2}$$

results in $\rho_2/\rho_1 = R_1/R_2$. Now, the density is the mass per unit volume so, according to Avogadro's law, the number of molecules per unit volume in each gas must be the same because the temperature and pressure are the same in each gas. Thus, there are the same number of moles n in each unit volume of gas and the actual mass of each gas present is nM_1 and nM_2, where M_1 and M_2 are the molar masses, for example, in units of g/mol.[12] Therefore, $\rho_2/\rho_1 = M_2/M_1$. The molecular weight m of a substance is a dimensionless, relative number which is expressed as the mass of a molecule of the substance relative to the mass of a molecule of a standard substance (usually the mass of a molecule of oxygen, taken as 32.00). Most importantly, the molecular weight is

---

[11] The reader should refer to the International or Smithsonian Physical Tables if precise specific gravities at other temperatures are required. The student may use the values of Table 1 in problem solutions even though the temperatures may not be exactly the same.

[12] In the SI system the mole (mol) is the amount of substance of a system which contains as many elementary entities, for example, atoms, as there are atoms in 0.012 kg of carbon = 12. Consequently, each mole of a gas has the same number of molecules and, for example, one mole of oxygen (O_2) has a molar mass of 0.032 kg. The units of molar mass are, therefore, grams/mole.

numerically equal to the molar mass. It follows that $\rho_2/\rho_1 = m_2/m_1$. Combining this equation with the one involving ρ's and R's above gives $m_2/m_1 = R_1/R_2$, or

$$m_1R_1 = m_2R_2 \qquad (1.4)$$

In other words, the product of molecular weight and engineering gas constant is the same[13] for all gases. This product mR is called the *universal gas constant* $\mathscr{R}$ and is preferred for general use by many engineers. It has the units J/kg · K or ft · lb/slug ·°R, which are the same as those of the engineering gas constant *in this definition*.

_____ **Illustrative Problems** _____

Calculate the specific weight, specific volume, and density of chlorine gas at 80°F and pressure of 100 psia (pounds per square inch, absolute).

Relevant Equations and Given Data

$$\mathscr{R} = mR \qquad (1.4)$$
$$\gamma = \rho g_n = g_n p / RT \qquad (1.1 \ \& \ 1.3)$$
$$T = 80 + 459.6 = 539.6°R$$
$$p = 100 \text{ psia} \qquad \mathscr{R} = 49\ 709 \text{ ft · lb/slug · °R} \qquad \text{(from Table 2)}$$

Solution.   Atomic weight chlorine = 35.45, molecular weight = 2 × 35.45 = 70.9

$$R = 49\ 709/70.9 = 701.1 \text{ ft·lb/slug°R} \qquad (1.4)$$

$$\gamma = 32.2 \times 100 \times 144/701.1 \times 540 = 1.225 \text{ lb/ft}^3 \blacktriangleleft \qquad (1.3)$$

$$\rho = 1.225/32.2 = 0.038\ 1 \text{ slug/ft}^3 \blacktriangleleft \qquad (1.1)$$

$$\text{Specific volume} = 1/\rho = 1/0.038\ 1 = 26.3 \text{ ft}^3/\text{slug} \blacktriangleleft$$

Calculate the density, specific volume, and specific weight of carbon dioxide gas at 100°C and atmospheric pressure (101.3kPa = 101.3 kilonewtons per square metre).

Relevant Equations and Given Data

$$\mathscr{R} = mR \qquad (1.4)$$

$$\rho = p/RT \qquad (1.3)$$

$$\gamma = \rho g_n \qquad (1.1)$$
$$T = 100 + 273.15 \cong 373 \text{ K}$$
$$p = 101.3 \text{ kPa} \qquad \mathscr{R} = 8\ 313 \text{ J/kg · K}$$

_____

[13] The constancy of mR applies particularly to the monatomic and diatomic gases. Gases having more than two atoms per molecule tend to deviate from the law mR = Constant. See Table 2. The nominal values of $\mathscr{R}$ are 49 709 ft · lb/slug ·°R and 8 313 J/kg · K.

Table 1 Approximate Properties of Some Common Liquids at Standard Atmospheric Pressure

English (FSS) Units

	Temperature T, °F	Density ρ, slug/ft³	Specific Gravity s.g., —	Modulus of Elasticity, E, psi	Viscosity $\mu \times 10^5$, lb·s/ft²	Surface Tension[a] σ, lb/ft	Vapor Pressure p_v, psia
Benzene	68	1.70	0.88	150 000	1.37	0.002 0	1.45
Carbon tetrachloride	68	3.08	1.59	160 000	2.035	0.001 8	1.90
Crude oil	68	1.66	0.86	—	15.0	0.002	—
Ethyl alcohol	68	1.53	0.79	175 000	2.51	0.001 5	0.85
Freon-12	60	2.61	1.35	—	3.10	—	—
	−30	2.91	—	—	3.82	—	—
Gasoline	68	1.32	0.68	—	0.61	—	8.0
Glycerin	68	2.44	1.26	630 000	3 120	0.004 3	0.000 002
Hydrogen	−431	0.143	—	—	0.043 5	0.000 2	3.1
Jet fuel (JP-4)	60	1.50	0.77	—	1.82	0.002	1.3
Mercury	60	26.3	13.57	3 800 000	3.26	0.035	0.000 025
	600	24.9	12.8	—	1.88	—	6.85
Oxygen	−320	2.34	—	—	0.58	0.001	3.1
Sodium	600	1.70	—	—	0.690	—	—
	1000	1.60	—	—	0.472	—	—
Water[b]	68	1.936	1.00	318 000	2.10	0.005 0	0.34

SI Units

	T, °C	ρ, kg/m³	s.g., —	E, kPa	$\mu \times 10^4$, Pa·s	σ, N/m	ρ_v, kPa
Benzene	20	876.2	0.88	1 034 250	6.56	0.029	10.0
Carbon tetrachloride	20	1 587.4	1.59	1 103 200	9.74	0.026	13.1
Crude oil	20	855.6	0.86	—	71.8	0.03	—
Ethyl alcohol	20	788.6	0.79	1 206 625	12.0	0.022	5.86
Freon-12	15.6	1 345.2	1.35	—	14.8	—	—
	−34.4	1 499.8	—	—	18.3	—	—
Gasoline	20	680.3	0.68	—	2.9	0.063	55.2
Glycerin	20	1 257.6	1.26	4 343 850	14 939	0.002 9	0.000 014
Hydrogen	−257.2	73.7	—	—	0.21	0.029	21.4
Jet fuel (JP-4)	15.6	773.1	0.77	—	8.7	0.51	8.96
Mercury	15.6	13 555	13.57	26 201 000	15.6	—	0.000 17
	315.6	12 833	12.8	—	9.0	0.015	47.2
Oxygen	−195.6	1 206.0	—	—	2.78	—	21.4
Sodium	315.6	876.2	—	—	3.30	—	—
	537.8	824.6	—	—	2.26	—	—
Water[b]	20	998.2	1.00	2 170 500	10.0	0.073	2.34

[a] In contact with air.
[b] More complete information on properties of water will be found in Appendix 2.

11

Table 2 Approximate Properties of Some Common Gases

	English (FSS) Units				
	Engineering Gas Constant, R, ft·lb/slug·°R	Universal Gas Constant, $\mathscr{R} = mR$ ft·lb/slug·°R	Adiabatic Exponent, k —	Specific Heat at Constant Pressure, c_p, ft·lb/slug·°R	Viscosity at 68°F (20°C), $\mu \times 10^5$ lb·s/ft^2
Carbon dioxide	1 123	49 419	1.28	5 132	0.030 7
Oxygen	1 554	49 741	1.40	5 437	0.041 9
Air	1 715	49 709	1.40	6 000	0.037 7
Nitrogen	1 773	49 644	1.40	6 210	0.036 8
Methane	3 098	49 644	1.31	13 095	0.028
Helium	12 419	49 677	1.66	31 235	0.041 1
Hydrogen	24 677	49 741	1.40	86 387	0.018 9

	SI Units				
	R, J/kg · K	$\mathscr{R} = mR$, J/kg · K	k —	c_p, J/kg · K	$\mu \times 10^5$, Pa · s
Carbon dioxide	187.8	8 264	1.28	858.2	1.47
Oxygen	259.9	8 318	1.40	909.2	2.01
Air	286.8	8 313	1.40	1 003	1.81
Nitrogen	296.5	8 302	1.40	1 038	1.76
Methane	518.1	8 302	1.31	2 190	1.34
Helium	2 076.8	8 307	1.66	5 223	1.97
Hydrogen	4 126.6	8 318	1.40	14 446	0.90

Solution. Atomic weight carbon = 12, atomic weight oxygen = 16, molecular weight CO_2 = 44

$$R = \mathscr{R}/44 = 8\ 313/44 = 189.0 \text{ J/kg · K} \tag{1.4}$$

$$\rho = 101.3 \times 10^3/189.0 \times 373 = 1.44 \text{ kg/m}^3 \ \blacktriangleleft \tag{1.3}$$

$$\text{Specific volume} = 1/1.44 = 0.69 \text{ m}^3/\text{kg} \ \blacktriangleleft$$

$$\gamma = 1.44 \times 9.81 = 14.1 \text{ N/m}^3 \ \blacktriangleleft \tag{1.1}$$

1.4 Compressibility, Elasticity

All fluids can be compressed by the application of pressure, elastic energy being stored in the process; assuming perfect energy conversions, such compressed volumes of fluids will expand to their original volumes when the applied pressure is released. Thus fluids are elastic media, and it is customary in engineering to summarize this property by defining a *modulus of elasticity,* as is done for solid elastic materials such as steel. Since fluids do not possess rigidity of form, however, the modulus of elasticity must be defined on the basis of volume and is termed a *bulk modulus*.

The mechanics of elastic compression of a fluid can be demonstrated by imagining the cylinder and piston of Fig. 1.3 to be perfectly rigid (inelastic) and to contain a volume of elastic fluid V_1. Application of a force, F, to the piston increases the pressure, p, in the fluid and causes the volume to decrease. Plotting p against V/V_1 produces the stress-strain diagram of Fig. 1.3 in which the modulus of elasticity of the fluid (at any point on the curve) is defined as the slope of the curve (at that point); thus

$$E = -\frac{dp}{dV/V_1} \tag{1.5}$$

The steepening of the curve with increasing pressure shows that as fluids are compressed they become increasingly difficult to compress further, a logical consequence of reducing the space between the molecules. The modulus of elasticity of a fluid is not constant but increases with increasing pressure.

Although the schematic curve of Fig. 1.3 applies equally well to liquids and gases, the engineer is usually concerned only with the portion of the curve near $V/V_1 = 1$ for liquids. The slope of the curve in this region is taken as the modulus of elasticity for engineering use; such values of E for common liquids are given in Table 1 and can be used for most engineering problems involving pressures up to a few thousand pounds per square inch or 10 MPa.

Compression and expansion of gases take place according to various laws of thermodynamics. A constant-temperature (*isothermal*) process is characterized by Boyle's law,

$$\frac{p}{\gamma} = \text{Constant} \quad \text{or} \quad \frac{p}{\rho} = \text{Constant} \tag{1.6}$$

whereas a frictionless process in which no heat is exchanged follows the *isentropic* relation

$$\frac{p}{\gamma^k} = \text{Constant} \quad \text{or} \quad \frac{p}{\rho^k} = \text{Constant} \tag{1.7}$$

in which k is the ratio of the two specific heats[14] of the gas, that at constant pressure, c_p, to that at constant volume, c_v. Values of k (frequently called the *adiabatic exponent*) for common gases are given in Table 2.

Expressions for the modulus of elasticity of gases may be easily derived for isothermal and isentropic processes by writing the general form of equation 1.5 in terms of γ or ρ. As the relative increases of γ or ρ are exactly equal to the relative decrease of volume,

$$E = \frac{dp}{d\gamma/\gamma} = \frac{dp}{d\rho/\rho} \tag{1.8}$$

---

[14]Thermodynamics shows, for perfect gases, that $c_p - c_v = R$ which, in combination with $c_p/c_v = k$, yields $c_p = Rk/(k-1)$. These specific heats also are assumed constant (with respect to temperature) for perfect gases.

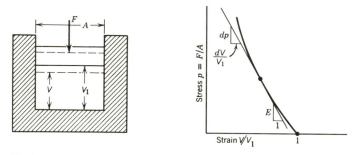

Fig. **1.3**

When this equation is solved simultaneously with the differential forms of equations 1.6 and 1.7, the results are $E = p$ for the isothermal process and $E = kp$ for the isentropic one.

In Appendix 3 it is shown that small pressure disturbances travel through fluids at a finite velocity (or *celerity*) dependent on the modulus of elasticity of the fluid. A small pressure disturbance, that travels as a wave of increased (or decreased) density and pressure, moves at a celerity, a, given by[15]

$$a = \sqrt{\frac{dp}{d\rho}} = \sqrt{\frac{E}{\rho}} \qquad (1.9)$$

The value a is frequently called the *sonic* or *acoustic* velocity because sound, a small pressure disturbance, travels at this velocity. Clearly, pressure disturbances can be transmitted instantaneously between two points in a fluid only if the fluid is inelastic, that is, if $E = \infty$. This never happens, but the assumption of an inelastic or incompressible fluid is often a convenient engineering approximation to the true state of affairs (a is about 4 720 ft/s or 1 440 m/s in water, but only about 1 100 ft/s or 335 m/s in air near sea level).

The disturbance caused by a sound wave moving through a fluid is so small and rapid that heat exchange in the compression and expansion may be neglected and the process considered isentropic. Thus, for a perfect gas, the sonic velocity may be obtained by substituting kp for E in equation 1.9, giving

$$a = \sqrt{kp/\rho} \qquad (1.10)$$

an equation that is accurately confirmed by experiment. This equation can be put into another useful form by substituting RT for p/ρ (obtained from equations 1.1 and 1.3), which results in

$$a = \sqrt{kRT} \qquad (1.11)$$

and shows that the acoustic velocity in a perfect gas depends only on the temperature of the gas.

In this era of high-speed flight, the reader is well aware of the *Mach Number*, **M**,

---

[15]For the derivation see Appendix 3.

the ratio between flow velocity and sonic velocity, and that flow velocities are defined as *subsonic* for **M** < 1 and *supersonic* for **M** > 1. However, **M** is also a useful criterion of relative compressibility of the fluid, which permits decisions on whether or not fluids may be considered incompressible for engineering calculations. For the flow of an incompressible (inelastic) fluid, **M** = 0 because $a = \infty$. Accordingly, as $\mathbf{M} \to 0$, compressibility becomes of decreasing importance; for most engineering calculations, experience has shown that the effects of compressibility may be safely neglected if $\mathbf{M} \gtrsim 0.3$.

Illustrative Problem

Air at 15°C and 101.3kPa is compressed isentropically so that its volume is reduced 50%. Calculate the final pressure and temperature and the sonic velocities before and after compression.

Relevant Equations and Given Data

$$\gamma = g_n p / RT \tag{1.3}$$

$$p_1 / \gamma_1^k = p_2 / \gamma_2^k \tag{1.7}$$

$$a = \sqrt{kRT} \tag{1.11}$$

$$T_1 = 15 + 273.15 \cong 288 \text{ K}$$

$$p_1 = 101.3 \text{ kPa} \qquad R = 286.8 \text{ J/kg} \cdot \text{K}$$

$$k = 1.4$$

Solution. $\quad \gamma_1 = 9.81 \times 101.3 \times 10^3 / 286.8 \times 288 = 12.0 \text{ N/m}^3 \tag{1.3}$

$$\gamma_2 = 2 \times 12.0 = 24.0 \text{ N/m}^3$$

$$\frac{p_2}{(24.0)^{1.4}} = \frac{101.3 \times 10^3}{(12.0)^{1.4}} \qquad p_2 = 267.3 \text{ kPa} \blacktriangleleft \tag{1.7}$$

$$\frac{9.81 \times 267.3 \times 10^3}{286.8 \times T_2} = 24.0 \qquad T_2 = 381 \text{ K}(108°C) \blacktriangleleft \tag{1.3}$$

$$a_1 = \sqrt{1.4 \times 286.8 \times 288} = 20\sqrt{288} = 340 \text{ m/s} \blacktriangleleft \tag{1.11}$$

$$a_2 = 20\sqrt{381} = 391 \text{ m/s} \blacktriangleleft$$

1.5 Viscosity

When various real fluid motions are observed carefully, two fundamentally different types of motion are seen. The first is a smooth, orderly motion in which fluid elements or particles appear to slide over each other in layers or laminae. While there is molecular agitation and diffusion in the fluid, there is virtually no large scale mixing between the layers, and this motion is called *laminar* flow. The second distinct motion which occurs is characterized by random or chaotic motion of individual fluid particles and by rapid macroscopic mixing of these particles through the flow. Eddies of a wide range of sizes are seen, and this motion is called *turbulent* flow. Both types of flow are considered in this book. Turbulent flow is, by far, the most common of the two in nature and engineering applications, but our more complete understanding of the physics of laminar flows makes them a good starting point for analysis, and a number of important laminar flow applications exist. For example, consideration of the laminar flow along a solid boundary is most instructive because it gives some early insight to key general features, such as the importance of the so-called *no-slip* condition at the solid bounding walls of continuum fluid flows and the primary role of viscosity.

The laminar motion of a real fluid along a solid boundary is sketched in Fig. 1.4. Observations show that, while the fluid clearly has a finite velocity, v, at any finite distance from the boundary, there is no velocity at the boundary. Thus, the velocity increases with increasing distance from the boundary. These facts are summarized on the *velocity profile* which indicates relative motion between any two adjacent layers. Two such layers are shown having thickness dy, the lower layer moving with velocity v, the upper with velocity $v + dv$. In Fig. 1.4 two particles 1 and 2, starting on the same vertical line, move different distances $d_1 = (v)\, dt$ and $d_2 = (v + dv)\, dt$ in an infinitesimal time dt. Thus, the fluid is distorted or sheared as a line connecting 1, and 2 acquires an increasing slope and length as t increases. In solids the stress due to shear is proportional to the strain [the relative displacement or strain here is $(d_2 - d_1)/dy = dv\, dt/dy = (dv/dy)\, dt$]. However, a fluid flows under the slightest stress, and the result of the continual application of a constant stress is an infinite strain. In fact, in fluid flow problems, the stress is related to the *rate of strain* (equal to dv/dy here) rather than to total strain.

The frictional or shearing force that must exist between fluid layers can be expressed as a *shearing* or *frictional stress* per unit of contact area and is designated by τ. For laminar (nonturbulent) motion (in which viscosity plays a predominant role), τ is (as noted above) observed to be proportional to the *rate* of relative strain, that is, to the velocity gradient, dv/dy, with a constant of proportionality, μ, defined as the *coefficient of viscosity*.[16] Thus,

$$\tau = \mu \frac{dv}{dy} \tag{1.12}$$

All real fluids possess viscosity and therefore exhibit certain frictional phenomena when motion occurs. Viscosity results fundamentally from cohesion and molecular momentum exchange between fluid layers and, as flow occurs, these effects appear as tangential or shearing stresses between the moving layers.

Equation 1.12 is basic to all problems of fluid resistance. Therefore its implications and restrictions should be emphasized: (1) the nonappearance of pressure in the equation shows that both τ and μ are independent[17] of pressure, and that therefore fluid friction is drastically different from that between moving solids, where pressure plays

---

[16]Also termed *absolute* or *dynamic viscosity*.
[17]Viscosity usually increases very slightly with pressure, but the change is negligible in most engineering problems.

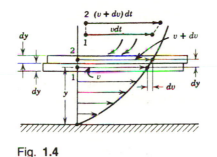

Fig. **1.4**

a large part; (2) any shear stress τ, however small, causes flow because applied tangential forces produce a velocity gradient, that is, relative motion between adjacent fluid layers; (3) where $dv/dy = 0$, $\tau = 0$, regardless of the magnitude of μ; the shearing stress in viscous fluids at rest is zero, and thus its omission in the analysis of the pressure at a point in Fig. 1.2, is confirmed; (4) the velocity profile cannot be tangent to a solid boundary because this would require an infinite velocity gradient there and an infinite shearing stress between fluid and solid; (5) the equation is limited to nonturbulent[18] (laminar) fluid motion, in which viscous action is strong. Also relevant to the use of equation 1.12 is the *observed* fact that the velocity at a solid boundary is zero; that is, there is no "slip" between fluid and solid for all fluids that can be treated as a continuum.

Equation 1.12 may be usefully visualized on the plot of Fig. 1.5 on which μ is the slope of a straight line passing through the origin; as seen above dv is the displacement per unit tin.e and the velocity gradient dv/dy is the time rate of strain. Because of Newton's suggestion, which led to equation 1.12, fluids that follow this law are commonly known as Newtonian fluids. It is these fluids with which this book is concerned. Other fluids are classed as *non-Newtonian* fluids. The science of rheology, which broadly is the study of the deformation and flow of matter, is concerned with plastics, blood, suspensions, paints, and foods, which *flow* but whose resistance is not

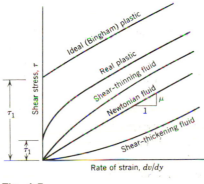

Fig. **1.5**

[18]The criterion defining laminar and turbulent flow is the Reynolds number, discussed in Chapter 7.

characterized by equation 1.12. The relations between τ and dv/dy for two typical *plastics* are sketched[19] on Fig. 1.5, and the essential mechanical difference between fluid and plastic is seen to be the shear, τ_1, manifested by the latter, which must be overcome before flow can begin. Another pair of examples shown in Fig. 1.5 are non-Newtonian, shear-affected fluids that include some suspensions and polymer solutions. The shear versus strain-rate equations corresponding to equation 1.12 are then, for example,

$$\tau - \tau_1 = \mu\frac{dv}{dy} \qquad \tau > \tau_1: \quad \text{Bingham Plastic (oil paint, toothpaste)}$$

$$\tau = k\left(\frac{dv}{dy}\right)^n \qquad n > 1: \quad \text{Shear-thickening fluid}$$

$$n < 1: \quad \text{Shear-thinning fluid}$$

For the latter *power-law* relation, $\mu = k(dv/dy)^{n-1}$. The addition of 4% paper pulp to water reduces n from 1 to about 0.6, while for 33% lime water $n \sim 0.2$.

The dimensions of the viscosity μ are determined from the dimensional homogeneity of equation 1.12 as follows:

$$\text{Dimensions of } \mu = \frac{\text{Dimensions of } \tau}{\text{Dimensions of } dv/dy} = \begin{cases} \dfrac{\text{lb/ft}^2}{\text{ft/s/ft}} = \text{lb} \cdot \text{s/ft}^2 \\[2ex] \dfrac{\text{Pa}}{\text{m/s/m}} = \text{Pa} \cdot \text{s} \end{cases}$$

However, from previous dimensional considerations (Section 1.3), $N = kg \cdot m/s^2$ and $lb = slug \cdot ft/s^2$, and if these are substituted above, viscosity, μ, may be quoted in $kg/m \cdot s$ or $slug/ft \cdot s$. These dimensional combinations are equivalent to $Pa \cdot s$ or $lb \cdot s/ft^2$. The metric combinations times 10^{-1} are given the special name *poises* (after Poiseuille, who did some of the first work on viscosity). Viscosities quoted in *poises* or *centipoises* may be readily converted into the English system from the definitions above and by use of basic conversion factors (e.g., $1 \text{ lb} \cdot \text{s/ft}^2 = 1 \text{ slug/ft} \cdot \text{s} = 478.8$ poises).

Viscosity varies widely with temperature, but temperature variation has an opposite effect on the viscosities of liquids and gases because of their fundamentally different intermolecular characteristics. In *gases*, where intermolecular cohesion is usually negligible, the shear stress, τ, between moving layers of fluid results from an exchange of momentum between these layers brought about by molecular agitation normal to the general direction of motion. The random motion of molecules thus carries them across the direction of the flow from layers at one speed to layers at a different speed. The molecules leaving a low-speed layer collide with molecules in a high-speed layer. In the collision, the exchange of momentum tends to speed up the slower molecules and slow the faster ones. The net effect is an apparent shear force tending to reduce the speed of the higher speed layer. The reverse happens when a molecule

---

[19]The form of the curves is noteworthy, but there is no special significance to their relative positions.

from a fast layer collides with a molecule in a relatively slower layer. Because this molecular activity is known to increase with temperature, the shear stress, and thus the viscosity of gases, will increase with temperature (Fig. 1.6). In liquids, momentum exchange due to molecular activity is small compared to the cohesive forces between

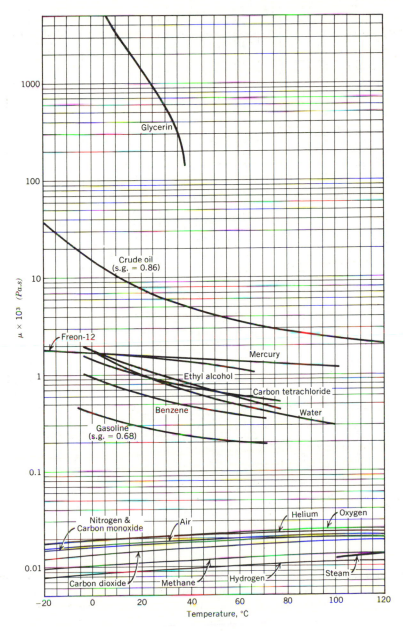

Fig. **1.6a** Viscosities of some common fluids (SI units).

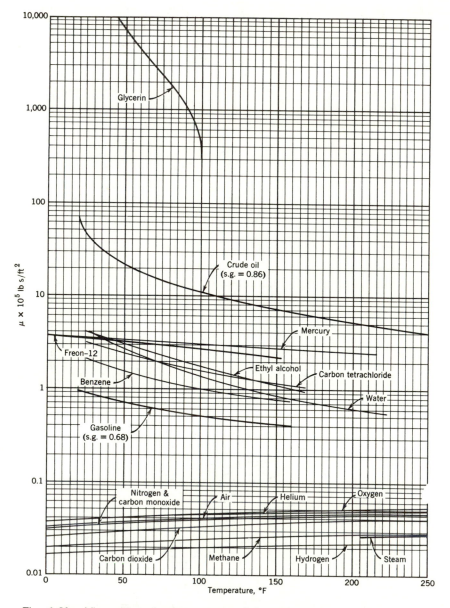

Fig. **1.6b** Viscosities of some common fluids (FSS units).

the molecules, and thus shear stress, τ, and viscosity, μ, are primarily dependent on the magnitude of these cohesive forces that tend to keep adjacent molecules in a fixed position relative to each other and to resist relative motion. Because these forces decrease rapidly with increases of temperature, liquid viscosities decrease as temperature rises (Fig. 1.6).

Owing to the appearance of the ration μ/ρ in many of the equations of fluid flow, this term has been defined by

$$\nu = \frac{\mu}{\rho} \tag{1.13}$$

in which ν is called the *kinematic viscosity*[20]. Dimensional consideration of equation 1.13 shows the dimensions of ν to be square metres per second or square feet per second, a combination of kinematic terms, which explains the name *kinematic* viscosity. In the metric system, this dimensional combination times 10^{-4} is known as *stokes* (after Sir G. G. Stokes). It is fairly common in American practice to quote kinematic viscosities in *centistokes* $= 10^{-6}$ m^2/s $= 1$mm^2/s (1 ft^2/s $= 929$ stokes).

Lamainar flows in which the shear stress is constant (or essentially so) are the simplest ones to examine at this point; such *shear flows* were first studied by Couette and are generally known as *Couette flows*. They can be prodcued by slowly shearing a thin fluid film between two large flat plates or between the surfaces of coaxial cylinders. For constant V in Fig. 1.7a, the force F is the same on both solid surfaces. If these plates have the same surface area, the sheer stress on their surfaces is equal, as are the velocity gradients there. These equalities must also extend through the fluid, yielding a linear velocity profile with $dv/dy = V/h$. In this case the rate of relative strain $(ds/dt)/dy = (dv\,dt/dt)\,dy = dv/dy = V/h$ is constant, where s is the distance a point on the upper plate moves relative to a point on the lower plate.

For the coaxial cylinder of Fig. 1.7b, the situation is somewhat different. If the fluid element in the insert were to rotate *without shearing* and the angular speed were ω, then $v_1 = r_1\omega$ and $v_2 = r_2\omega = (r_1 + dr)\omega$. Thus, in a *solid body rotation* in which

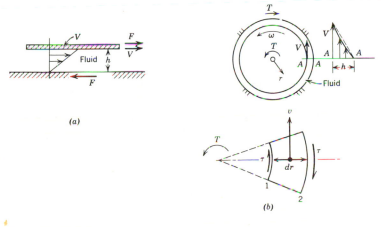

(a)

(b)

Fig. **1.7**

the element is neither sheared nor distorted, a fluid particle at 2 moves farther and faster than one at 1. If the fluid element is to shear, the angular speed of the line 12 must be different from ω at 1. Then, 12 will rotate at a rate $\omega + d\omega$ which is faster (or slower) than the solid body rate ω and the shape of the element will change. The rate of relative strain

$$\frac{ds_{2-1}/dt}{dr} = \frac{\text{Actual velocity of 2} - \text{Rigid body velocity of 2 following 1}}{dr}$$

$$= \frac{[(r_1 + dr)\omega + (r_1 + dr)\,d\omega] - [(r_1 + dr)\omega]}{dr}$$

$$= r_1\frac{d\omega}{dr} + dr\,\frac{d\omega}{dr} = r_1\frac{d\omega}{dr}$$

as dr, $d\omega$, and dt vanish in the limit. In terms of the tangential velocity component v, $\omega = v/r$. Thus,

$$\tau = \mu\frac{ds/dt}{dr} = \mu r\frac{d\omega}{dr} = \mu r\frac{d}{dr}\left(\frac{v}{r}\right)$$

$$= \mu\,dv/dr - \mu v/r \tag{1.14}$$

Only if $v/r \ll dv/dr$, does equation 1.14 reduce to equation 1.12.

For the coaxial cylinders of Fig1.7b, the driving and resisting torques T will be equal for constant velocity V. With cylinders of the same axial length, the surface area of the outer cylinder is larger than that of the inner one, and the former also has the larger radius. Accordingly the shear and the velocity gradient at the outer cylinder will be expected to be less than the respective quanitites on the inner one, and the velocity profile through the fluid to be somewhat as shown. Evidently, $dv/dr \rightarrow -V/h$ as $h \rightarrow 0$; equating dv/dr to $-V/h$ is a common and often satisfactory engineering approximation in such problems if h is small.

—————————— **Illustrative Problem** ——————————

A cylinder 85 mm in radius and 0.6 m in length rotates coaxially inside a fixed cylinder of the same length and 90 mm radius. Glycerin ($\mu = 1.48$ Pa·s) fills the space between the cylinders. A torque of 0.7 N·m is applied to the inner cylinder. After constant velocity is attained, calculate the velocity gradients at the cylinder walls, the resulting r/min, and the power dissipated by fluid resistance. Ignore end effects.

Relevant Equation and Given Data

$$\tau = \mu r\frac{d}{dr}\left(\frac{v}{r}\right) \tag{1.14}$$

$$r_i = 85 \text{ mm} \qquad r_o = 90 \text{ mm} \qquad T = 0.7 \text{ N·m}$$

$$\ell = 0.6 \text{ m} \quad \mu = 1.48 \text{ Pa·s}$$

Solution. Refer to Fig. 1.7*b* for a definition sketch. The torque T is transmitted from inner cylinder to outer cylinder through the fluid layers; therefore (r is the radial distance to any fluid layer and ℓ is the length of the cylinders)

$$T = -\tau(2\pi r \times \ell)\, r \qquad \tau = -\frac{T}{2\pi r^2 \ell} = \frac{\mu r\, d(v/r)}{dr} \tag{1.14}$$

Consequently,

$$\frac{d(v/r)}{dr} = -\frac{T}{2\pi\mu\ell r^3}$$

and

$$\int_{v/r_i}^{0} d\left(\frac{v}{r}\right) = \frac{-T}{2\pi\mu\ell}\int_{r_i}^{r_o}\frac{dr}{r^3} \qquad -\frac{V}{r_i} = \frac{T}{4\pi\mu\ell}\left[\frac{1}{r_o^2} - \frac{1}{r_i^2}\right]$$

$$V = \frac{-0.085 \times 0.7}{4\pi \times 1.48 \times 0.6}\left[\frac{1}{(0.090)^2} - \frac{1}{(0.085)^2}\right] = 0.08 \text{ m/s} = 80 \text{ mm/s}$$

The power P dissipated in fluid friction is equal to the rate of work done at, say, the surface of the inner cylinder (force moving through a distance per unit of time). The force is $2\pi r_i \cdot \ell \cdot \tau_i$; distance per unit time is V! Thus,

$$P = 2\pi r_i \tau_i V \ell = T\omega$$

where $\omega = V/r_i = 0.94\,s^{-1}$ is the rotational speed (rad/s).
Numerically,

$$\left(\frac{dv}{dr}\right)_i = \frac{-T}{2\pi\mu\ell r_i^2} + \frac{v_i}{r_i} = -17.4 + 0.9 = -16.5 \text{ m/s·m} \blacktriangleleft \tag{1.14}$$

$$\left(\frac{dv}{dr}\right)_o = \frac{-T}{2\pi\mu\ell r_o^2} + \frac{v_o}{r_o} = -15.4 + 0 = -15.4 \text{ m/s·m} \blacktriangleleft \tag{1.14}$$

$$\text{r/min} = \left(\frac{\omega}{2\pi}\right) \times 60 = 9.0 \blacktriangleleft$$

$$P = 0.7 \times 0.94 = 0.66 \text{ N·m/s} = 0.66 \text{ W} \blacktriangleleft$$

This power will appear as heat, tending to raise the fluid temperature and decrease its viscosity; evidently a suitable heat exchanger would be needed to preserve the steady-state conditions given.

When $h = r_o - r_i \to 0$, then h/r_i is small and $dv/dr \to -V/h$. As a consequence, v/r is much less than dv/dr, and

$$\tau \approx \mu\frac{dv}{dr} = -\frac{\mu V}{h}$$

Assuming this linear profile case for an approximate calculation gives

$$h = 5 \text{ mm} \qquad \frac{h}{r_i} \approx 0.06$$

$$V = \frac{T}{2\pi\mu\ell}\left(\frac{h}{r_i^2}\right) = 0.125\frac{0.005}{0.0072} = 0.087 \text{ m/s}$$

$$\text{r/min} = 9.8 \blacktriangleleft$$

Because these results differ from the former by almost 9%, the approximation is not satisfactory in this case.

1.6 Surface Tension, Capillarity

The apparent tension effects that occur on the surfaces of liquids, when the surfaces are in contact with another fluid or a solid, depend fundamentally on the relative sizes of intermolecular cohesive and adhesive forces. Although such forces are negligible in many engineering problems, they may be predominant in some, such as the capillary rise of liquids in narrow spaces, the mechanics of bubble formation, the breakup of liquid jets, the formation of liquid drops, and the interpretation of results obtained on small models of larger prototypes.

On a free liquid surface in contact with the atmosphere, there is little force attracting molecules away from the liquid because there are relatively few molecules in the vapor above the surface. Within the liquid bulk, the intermolecular forces of attraction and repulsion are balanced in all directions. However, for liquid molecules at the surface, the cohesive forces of the next layer below are not balanced by an identical layer above. This situation tends to pull the surface molecules tightly to the lower layer and to each other and causes the surface to behave as though it were a membrane; *hence, the name surface tension.* In fact, treating the surface as though it were a membrane capable of supporting tension is an analogy used widely in theoretical treatment of surface tension problems. When the free surface is curved, the surface tension force will support small loads.[21] The surface tension, σ, is thought of as the force in the liquid surface normal to a line of unit length drawn in the surface; thus it will have dimensions of pounds per foot or newtons per metre. Because surface tension is directly dependent on intermolecular cohesive forces, its magnitude will decrease as temperature increases.[22] Surface tension is also dependent on the fluid in contact with the liquid surface; thus surface tensions are usually quoted in contact with air as indicated in the footnote to Table 1.

Consider now (Fig 1.8) the general case of a small element $dx\,dy$ of a surface of double curvature with radii R_1 and R_2. Evidently a pressure difference $(p_i - p_o)$ must accompany the surface tension for static equilibrium of the element. A relation between the pressure difference and the surface tension may be derived from this equilibrium by taking $\Sigma F = 0$ for the force components normal to the element:

$$(p_i - p_o)\, dx\, dy = 2\sigma\, dy\, \sin \alpha + 2\sigma\, dx\, \sin \beta$$

---

[21] A small needle placed gently on a water surface will not sink but will be supported by the tension in the liquid surface. The surface is depressed slightly in the process which causes localized surface curvature and the consequent development of the necessary force.

[22] See Appendix 2.

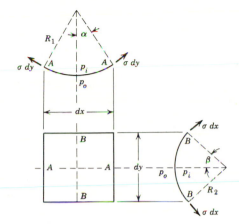

Fig. **1.8**

in which α and β are small angles. However, from the geometry of the element, $\sin \alpha = dx/2R_1$ and $\sin \beta = dy/2R_2$. When these values are substituted above, the basic relation between surface tension and pressure difference is obtained; it is

$$p_i - p_o = \sigma \left(\frac{1}{R_1} + \frac{1}{R_2} \right) \tag{1.15}$$

From this equation the pressure (caused by surface tension) in droplets and tiny jets may be calculated and the rise of liquids in capillary spaces estimated; for a spherical droplet, $R_1 = R_2$; for a cylindrical jet, one R is infinite and the other is the radius of the jet. For the cylindrical capillary tube of Fig. 1.9 (assuming the liquid surface to be a section of a sphere), $p_o = -\gamma h$, $p_i = 0$, and $p_i - p_o = \gamma h$; also $R_1 = R_2 = R$ and $r/R = \cos \theta$. Substituting these in equation 1.15 yields

$$h = \frac{2\sigma \cos \theta}{\gamma r} \tag{1.16}$$

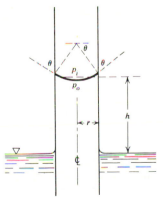

Fig. **1.9**

This result immediately raises several questions: the meaning of the angle θ, the limitations of the equation, and its confirmation by experiment. From the assumption of spherical liquid surface it is clear that the equation is limited to very small tubes; in large tubes the liquid surface is far from the spherical form. The angle θ is known as the *angle of contact*, and it results from surface tension phenonema of complex nature. Figure 1.10 describes the situation when mercury and water surfaces contact a vertical glass surface. Evidently the mercury molecules possess a greater affinity for each other (cohesion) than for the glass (adhesion), whereas the opposite condition obtains for the water and glass. Although the detailed character of these molecular interactions is not completely understood, the contact angles have been measured and found (for pure substances) to be as indicated. Good experimental confirmation of equation 1.16 is obtained for small tubes ($r < 0.1$ in. or 2.5 mm), providing liquids and tube surfaces are extremely clean; in engineering, however, such cleanliness is virtually never encountered and h will be found to be considerably *smaller* than given by equation 1.16. Thus the equation is useful for making conservative estimates of capillary errors. In the sizing of tubes for pressure measurement, the capillarity problem may be avoided entirely by providing tubes large enough to render the capillarity correction negligible for the desired accuracy.

────────────── **Illustrative Problem** ──────────────

Of what diameter must a droplet of water (70°F) be to have the pressure within it 0.1 psi greater than that outside?

Relevant Equation and Given Data

$$p_i - p_o = \sigma\left(\frac{1}{R_1} + \frac{1}{R_2}\right) \tag{1.15}$$

Spherical drop: $R_1 = R_2 = R$

$$p_i - p_o = 0.1 \text{ psi} \qquad T = 70°F$$

Solution. From Appendix 2, $\sigma = 0.004\ 98$ lb/ft

$$p_i - p_o = 0.1 \times 1.44 = 0.004\ 98\left(\frac{2}{R}\right) \tag{1.15}$$

$$R = 0.000\ 69 \text{ ft} \qquad d = 0.016\ 6 \text{ in.} \blacktriangleleft$$

1.7 Vapor Pressure

In the mechanics of *liquids*, the physical property of vapor pressure is frequently important in the analysis of problems. All liquids possess a tendency to vaporize, that is, to change from the liquid to the gaseous phase. Such vaporization occurs because molecules are continually projected through the free liquid surface and lost from the

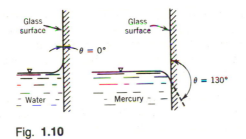

Fig. **1.10**

body of the liquid as a consequence of their natural thermal vibrations. The ejected molecules, being gaseous, then exert their own partial pressure, which is known as the *vapor pressure* (p_v) of the liquid. Because of the increase of molecular activity with temperature, vapor pressure increases with temperature; for water this variation is given in Appendix 2. *Boiling* (formation of vapor bubbles throughout the fluid mass) will occur (whatever the temperature) when the external absolute pressure imposed on the liquid is equal to or less than the vapor pressure of the liquid. This means that the boiling point of a liquid is dependent on the imposed pressure as well as on temperature.[23]

Table 1 offers a comparison of the vapor pressures of a few common liquids at the same temperature. The low vapor pressure of mercury along with its high density makes this liquid well suited for use in barometers and other pressure-measuring devices. The more *volatile* liquids, which vaporize more easily, possess the higher vapor pressures. If a liquid is placed in a sealed container with empty space above the liquid surface, the vaporization of the liquid continues until the vapor exerts the vapor pressure p_v in the once empty space. At this stage the number of molecules escaping from the liquid exactly equals the number that is returning. However, if the space is too large, this equilibrium position is not reached and the liquid continues to vaporize or *evaporate* until the liquid is gone and only vapor remains at a pressure less than or equal to p_v.

--------- **Illustrative Problem** ---------

A vertical cylinder 12 in. (~300 mm) in diameter is fitted (at the top) with a tight but frictionless piston and is completely filled with water at 160°F(~70°C). The outside of the piston is exposed to an atmospheric pressure of 14.52 psia (~100 kPa). Calculate the minimum force applied to the piston that will cause the water to boil.

Given Data

FSS: d_c = 12 in. p_a = 14.52 psia

SI: d_c = 300 mm = 0.3 m p_a = 100 kPa

---

[23] For example, water boils at 212°F (100°C) when exposed to an atmospheric pressure of 14.7 psia (101.3 kPa, absolute), but will boil at 140°F (60°C) if the imposed pressure is reduced to that at an altitude of about 39 000 ft (12 km) in the atmosphere, that is, to 2.89 psia (19.9kPa, absolute). See Appendix 2.

FSS Solution. The force must be applied slowly (to avoid acceleration) in a direction to withdraw the piston from the cylinder. Since the water cannot expand, a space filled with water vapor will be created beneath the piston, whereupon the water will boil. The pressure on the inside of the piston will then be (Appendix 2) 4.74 psia and the force on the piston $(14.52 - 4.74)\pi(12)^2/4 = 1\ 106$ lb. ◄

SI Solution. As noted above, when the piston is slowly withdrawn, the water will boil. The pressure on the inside of the piston will then be (Appendix 2) 31.2 kPa and the force on the piston $(100 - 31.2)\pi(0.3)^2/4 = 4.86$ kN. ◄

References

General and Historical

Busemann, A. 1971. Compressible flow in the thirties. *Ann. Rev. of Fluid Mechanics*. Vol. 3. Palo Alto: Annual Reviews.

Goldstein, S. 1969. Fluid mechanics in the first half of this century. *Ann. Rev. of Fluid Mechanics*. Vol. 1. Palo Alto: Annual Reviews.

McDowell, D. M., and Jackson, J. D., Eds. 1970. *Osborne Reynolds and engineering science today*. New York: Barnes & Noble.

Rouse, H. 1976. Hydraulic's latest golden age. *Ann. Rev. of Fluid Mechanics*. Vol. 8. Palo Alto: Annual Reviews.

Rouse, H., and Ince, S. 1963. *History of hydraulics*. New York: Dover.

Scott Blair, G. W. 1969. *Elementary rheology*. New York: Academic Press.

Tani, I. 1977. History of boundary-layer theory. *Ann. Rev. of Fluid Mechanics*. Vol. 9. Palo Alto: Annual Reviews.

Taylor, G. I. 1974. The interaction between experiment and theory in fluid mechanics. *Ann. Rev. of Fluid Mechanics*. Vol. 6. Palo Alto: Annual Reviews.

Tokaty, G. A. 1971. *A history and philosophy of fluidmechanics*. London: G. T. Foulis.

Wilkinson, W. L. 1960. *Non-Newtonian fluids*. New York: Pergamon Press.

Physical Properties

Adam, N. K. 1958. *Physical chemistry*. New York: Oxford University Press, Chapters 3, 8.

Bridgman, P. W. 1943. Recent work in the field of high pressures. *Amer. Scientist*, 31, 1.

Burdon, R. S. 1949. *Surface tension and the spreading of liquids*. 2nd ed. Cambridge: Cambridge University Press.

Daniels, F., and Alberty, R. A. 1979. *Physical chemistry*. 5th ed. New York: Wiley.

Defay, R., and Prigogine, I. 1966. *Surface tension and absorption*. Trans. by D. H. Everett. New York: Wiley.

Hydraulic Institute. 1979. *Engineering data book*. 1st ed. Cleveland: Hydraulic Institute.

Reid, R. C., Prausnitz, J. M., and Sherwood, T. K. 1977. *The properties of gases and liquids*. 3rd ed. New York: McGraw-Hill.

Weast, R. C., Ed. 1979. *CRC Handbook of chemistry and physics*. 60th ed. Boca Raton: CRC Press.

Units

ASME Orientation and guide for use of SI (metric) units. 8th ed. 1978. Guide SI-1. New York: The Amer. Soc. of Mech. Engin.

ASTM Standard for metric practice. 1980. No. E380-79. Philadelphia: Amer. Soc. for Testing and Mat.

Films

Makovitz, H. Rheological behavior of fluids. NCFMF/EDC Film No. 21613, Encyclopaedia Britannica Educ. Corp.[24]

Rouse, H. Mechanics of fluids: introduction to the study of fluid motion. Film No. U45578, Media Library, Audiovisual Center, Univ. of Iowa.

Trefethen, L. M. Surface tension in fluid mechanics. NCFMF/EDC Film No. 21610, Encyclopaedia Britannica Educ. Corp.[24]

Problems

1.1. A fluid occupying 3.2 m³ has a mass of 4 Mg. Calculate its density and specific volume in SI and FSS units.

1.2. If a power plant is rated at 2 000 MW output and operates (on average) at 75% of rated power, how much energy (in J and ft · lbs) does it put out in a year?

1.3. A rocket payload with a weight on earth of 2 000 lb or 8 900 N is landed on the moon where the acceleration due to the moon's gravity $g_m \approx g_n/6$. Find the mass of the payload on the earth and the moon and the payload's moon weight.

1.4. If a barrel of oil weighs 1.5 kN, calculate the specific weight, density, and specific gravity of this oil. A barrel contains 159 litres or 0.159 m³; the barrel itself weighs 110N.

1.5. Calculate the specific weight of mercury at 60°F and at 600°F.

1.6. Calculate the specific weight, specific gravity, and specific volume of alcohol at 68°F or at 20°C.

1.7. Calculate the density and specific volume of liquid sodium at 538°C.

1.8. The specific gravity of a liquid is 3.0. What is its specific volume?

1.9. A cubic metre of air at 101 kPa and 15°C weighs 12.0 N,. What is its specific volume?

1.10. Calculate the specific weight, specific volume, and density of air at 40°F and 50 psia.

1.11. Calculate the density, specific weight, and specific volume of carbon dioxide at 700 kPa and 90°C.

1.12. Calculate the density of carbon monoxide at 20 psia and 50°F.

1.13. Calculate the temperature of methane gas (CH_4) at 202.6 kPa if its density is 1.05 kg/m³.

1.14. The molecular weight of nitrogen is 28. Use the perfect gas laws to find its density at 7.35 psia (50.6 Pa, absolute) and 212°F (100°C).

1.15. The specific volume of a certain perfect gas at 200 kPa and 40°C is 0.65 m³/kg. Calculate its gas constant and molecular weight.

1.16. Avogadro's number is 6.023×10^{23} molecules/mol. Find the number of molecules in a unit volume of air at standard temperature and pressure (15°C; 101.3 kPa, absolute).

1.17. The derivation of the universal gas constant led to its definition as the product of molecular weight m and engineering gas constant R. Review of the derivation shows, however,

---

[24] Eight millimetre silent film loops selected from NCFMF/EDC films are available as well.

that $M_1R_1 = M_2R_2$. Thus, MR = constant, where M is the molar mass. If $\mathfrak{R}^* = MR$, how is this universal constant different from $\mathfrak{R}$? *Hint*: Stick with SI units.

1.18.   If $h = p/\gamma$, what are the dimensions of h?

1.19.   If $V = \sqrt{2g_nh}$, calculate the dimensions of V.

1.20.   If $F = Q\gamma V/g_n$, what are the dimensions of F?

1.21.   One cubic metre of water is placed under an absolute pressure of 7 000 kPa. Calculate the volume at this pressure.

1.22.   If the volume of a liquid is reduced 0.035% by application of a pressure of 690 kPa or 100 psi, what is its modulus of elasticity?

1.23.   What pressure must be applied to water to reduce its volume 1%?

1.24.   Calculate the specific gravity of carbon tetrachloride at 20°C and 20 000 kPa.

1.25.   When a volume of 1.021 2 ft^3 of alcohol is subjected to a pressure of 7 350 psi, it will contract to 0.978 4 ft^3. Calculate the modulus of elasticity.

1.26.   One cubic metre of nitrogen at 40°C and 340 kPa is compressed isothermally to 0.2 m^3. What is the pressure when the nitrogen is reduced to this volume? What is the modulus of elasticity at the beginning and end of the compression?

1.27.   If the nitrogen in the preceding problem is compressed isentropically to 0.2 m^3, calculate the final pressure and temperature and the modulus of elasticity at the beginning and end of the compression.

1.28.   A gas is compressed isentropically. The measured volume and absolute pressure before and after compression are 0.30 m^3 and 50.7 kPa and 0.111 m^3 and 202.8 kPa, respectively. What are k and E for this gas?

1.29.   Calculate the velocity of sound in air at 0°C (32°F) and absolute pressure of 101.3 kPa (14.7 psia).

1.30.   Calculate the velocity of sound in water at 20°C (68°F).

1.31.   Calculate the Mach numbers for an airplane flying at 1 130 km/h through still air at altitudes 2 and 16 km (see Appendix 4).

1.32.   Calculate the kinematic viscosity of benzene at 20°C (68°F).

1.33.   Calculate the kinematic viscosity of nitrogen at 40°C and 550 kPa.

1.34.   What is the ratio between the viscosities of air and water at 10°C or 50°F? What is the ratio between their kinematic viscosities at this temperature and standard barometric pressure?

1.35.   Using data from Appendix 2, calculate the kinematic viscosity of water at 10°C and 34 475 kPa or at 50°F and 5 000 psi.

1.36.   A certain diatomic gas has a kinematic viscosity of 1.7×10^{-5} m^2/s and viscosity of 2.9×10^{-5} Pa·s at 101 kPa and 40°C. Calculate its molecular weight.

1.37.   The kinematic viscosity and specific weight of a certain liquid are 5.6×10^{-4} m^2/s and 2.00, respectively. Calculate the viscosity of this liquid.

1.38.   Nitrogen at 15 psia and 100°F is compressed adiabatically to 45 psia. Calculate kinematic viscosities and acoustic velocities before and after compression.

1.39. Calculate velocity gradients for $y = 0, 0.2, 0.4,$ and 0.6 m, if the velocity profile is a quarter-circle having its center 0.6 m from the boundary.

10 m/s

0.6 m

Problem 1.39

1.40. Calculate the velocity gradients of the preceding problem, assuming that the velocity profile is a parabola with the vertex 0.6 m from the boundary. Also calculate the shear stresses at these points if the fluid viscosity is 1 poise.

1.41. If the equation of a velocity profile is $v = 4y^{2/3}$ (v fps, y ft), what is the velocity gradient at the boundary and at 0.25 ft and 0.5 ft from it?

1.42. The velocity profile for laminar flow in a pipe is

$$v(r) = v_c(1 - r^2/R^2)$$

where r is measured from the centerline where $v = v_c$ and R is the pipe radius. Find $\tau(r)$ and dv/dr. Plot both versus r.

1.43. The shear stress $\tau = \tau_o(1 - y)$ in a laminar flow in a pipe of radius 1 with τ_o measured at the pipe wall where $y = 0$. If the fluid is a power-law fluid so

$$\tau = k\left(\frac{dv}{dy}\right)^n$$

find the velocity profile in $0 \le y \le 1$. Plot and compare profiles for $n = 0.6, 1.0,$ and 1.4 using $\tau_o/k = 1$. How does the centerline velocity change as a function of n?

1.44. If the viscosity of a liquid is 0.002 lb·s/ft² (0.01 Pa·s), what is its viscosity in poises? in centipoises?

1.45. If the viscosity of a liquid is 3.00 centipoises, what is its viscosity in lb·s/ft²?

1.46. If the kinematic viscosity of an oil is $1\,000$ centistokes, what is its kinematic viscosity in ft²/s (m²/s)? If its specific gravity is 0.92, what is its viscosity in lb·s/ft² (Pa·s)?

1.47. At what temperatures does air have a larger kinematic viscosity than crude oil?

1.48. At a point in a viscous flow the shearing stress is 5.0 psi and the velocity gradient $6\,000$ fps/ft. If the specific gravity of the liquid is 0.93, what is its kinematic viscosity?

1.49. A very large thin plate is centered in a gap of width 0.06 m with different oils of unknown viscosities above and below; one viscosity is twice the other. When the plate is pulled at a velocity of 0.3 m/s, the resulting force on one square metre of plate due to the viscous shear on both sides is 29N. Assuming viscous flow and neglecting all end effects, calculate the viscosities of the oils.

1.50. Through a very narrow gap of height h a thin plate of very large extent is being pulled at constant velocity V. On one side of the plate is oil of viscosity μ and on the other side oil of viscosity $k\mu$. Calculate the position of the plate so that the drag force on it will be a minimum.

1.51. A vertical gap 25 mm wide of infinite extent contains oil of specific gravity 0.95 and viscosity 2.4 Pa · s. A metal plate 1.5 m × 1.6 mm weighing 45 N is to be lifted through the gap at a constant speed of 0.06 m · s. Estimate the force required.

1.52. A cylinder 8 in. in diameter and 3 ft long is concentric with a pipe of 8.25 in. i.d. Between cylinder and pipe there is an oil film. What force is required to move the cylinder along the pipe at a constant velocity of 3 fps? The kinematic viscosity of the oil is 0.006 ft²/s; the specific gravity is 0.92.

1.53. Crude oil at 20°C fills the space between two concentric cylinders 250 mm high and with diameters of 150 mm and 156 mm. What torque is required to rotate the inner cylinder at 12 r/min, the outer cylinder remaining stationary?

1.54. A torque of 4 N · m is required to rotate the intermediate cylinder at 30 r/min. Calculate the viscosity of the oil. All cylinders are 450 mm long. Neglect end effects.

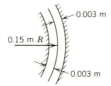

0.003 m

0.15 m R

0.003 m

Problem 1.54

1.55. The viscosity of the oil in the preceding problem is 0.25 Pa · s. What torque is required to rotate the intermediate cylinder at a constant speed of 40 r/min?

1.56. A circular disk of diameter d is slowly rotated in a liquid of large viscosity μ at a small distance h from a fixed surface. Derive an expression for the torque T necessary to maintain an angular velocity ω. Neglect centrifugal effects.

1.57. The fluid drive shown transmits a torque T for steady-state conditions (ω_1 and ω_2 constant). Derive an expression for the slip ($\omega_1 - \omega_2$) in terms of T, μ, d, and h.

1.58 Oil of viscosity μ fills the gap h, which is very small. Calculate the torque T required to rotate the cone at constant speed ω.

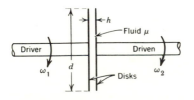

Problem 1.57

Problem 1.58

1.59. A piece of pipe 12 in. long weighing 3 lb and having i.d. of 2.05 in. is slipped over a vertical shaft 2.00 in. in diameter and allowed to fall. Calculate the approximate velocity attained by the pipe if a film of oil of viscosity $0.5 \text{ lb} \cdot \text{s/ft}^2$ is maintained between pipe and shaft.

1.60. The lubricant has a kinematic viscosity of $2.8 \times 10^{-5} \text{ m}^2/\text{s}$ and s.g. of 0.92. If the mean velocity of the piston is 6 m/s, approximately what is the power dissipated in friction?

1.61. Calculate the approximate viscosity of the oil.

Problem **1.60**   Problem **1.61**

1.62. The weight falls at a constant velocity of 46 mm/s. Calculate the approximate viscosity of the oil.

1.63. Calculate the approximate power lost in friction in this bearing.

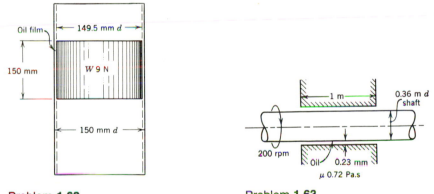

Problem **1.62**   Problem **1.63**

1.64. What excess pressure may be caused within a cylindrical jet of water 5 mm (0.2 in.) in diameter by surface tension?

1.65. Calculate the maximum capillary rise of water (20°C or 68°F) to be expected in a vertical glass tube 1 mm (0.04 in.) in diameter.

1.66. Calculate the maximum capillary rise of water (20°C or 68°F) to be expected between two vertical, clean glass plates spaced 1 mm (0.04 in.) apart.

1.67. Derive an equation for theoretical capillary rise between vertical parallel plates. Plot the rise as a function of separation distance.

1.68. Calculate the maximum capillary depression of mercury to be expected in a vertical glass tube 1 mm (0.04 in.) in diameter at 15.5°C (60°F).

1.69. A soap bubble 50 mm in diameter contains a pressure (in excess of atmospheric) of 20 Pa. Calculate the tension in the soap film.

1.70. What force is necessary to lift a thin wire ring 25 mm in diameter from a water surface at 20°C? Neglect weight of ring.

1.71. Using the assumptions of Section 1.6, derive an expression for capillarity correction h for an interface between liquids in a vertical tube.

1.72. What is the minimum absolute pressure which may be maintained in the space above the liquid in a can of ethyl alcohol at 68°F or 20°C?

1.73. To what value must the absolute pressure over carbon tetrachloride be reduced to make it boil at 68°F or 20°C?

1.74. To what value must the absolute pressure over water be reduced to make it boil at 20°C or 68°F? At 0°C or 32°F?

1.75. At what temperature will water boil at an altitude of 20 000 ft or 6 100 m? See Appendices 2 and 4.

2
Fluid Statics

Fluid statics is the study of fluid problems in which there is no relative motion between fluid elements. With no relative motion between individual elements (and thus no velocity gradients), no shear stress can exist, whatever the viscosity of the fluid.[1] Accordingly, viscosity has no effect in static problems and exact analytical solutions to such problems are relatively easy to obtain.

2.1 Pressure-Density-Height Relationships

The fundamental equation of fluid statics is that relating pressure, density, and vertical distance in a fluid. This equation may be derived readily by considering the static equilibrium of a typical differential element[2] of fluid (Fig. 2.1). The z-axis is in a direction parallel to the gravitational force field (vertical). Applying Newton's first law ($\Sigma F_x = 0$ and $\Sigma F_z = 0$) to the element and using the average pressure on each face to closely approximate the actual pressure distribution on the differential element (recall dx and dz are very small), we obtain

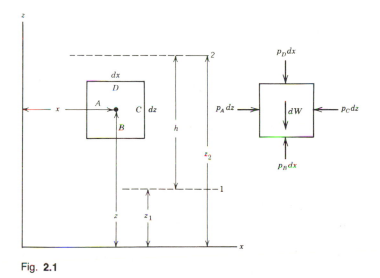

Fig. **2.1**

[1]See equation 1.12.
[2]A two-dimensional element is chosen for simplicity and convenience; a three-dimensional one will yield the same result.

35

$$\sum F_x = p_A \, dz - p_C \, dz = 0 \tag{2.1}$$

$$\sum F_z = p_B \, dx - p_D \, dx - dW = 0 \tag{2.2}$$

in which p and ρ are functions of x and z. In partial derivative notation[3] the pressures on the faces of the element are, in terms of the pressure p in the center

$$p_A = p - \frac{\partial p}{\partial x} \frac{dx}{2} \qquad p_C = p + \frac{\partial p}{\partial x} \frac{dx}{2}$$

$$p_B = p - \frac{\partial p}{\partial z} \frac{dz}{2} \qquad p_D = p + \frac{\partial p}{\partial z} \frac{dz}{2}$$

The weight of the small element is $dW = \rho g_n \, dx \, dz = \gamma \, dx \, dz$ (as dx and dz approach zero in the usual limiting process for partial differentiation, any variations in $\gamma = \rho g_n$ over the element will vanish). Thus, equations 2.1 and 2.2 become

$$\left(p - \frac{\partial p}{\partial x} \frac{dx}{2} \right) dz - \left(p + \frac{\partial p}{\partial x} \frac{dx}{2} \right) dz = -\frac{\partial p}{\partial x} \, dx \, dz = 0$$

and similarly

$$-\frac{\partial p}{\partial z} \, dz \, dx - \gamma \, dx \, dz = 0$$

Cancelling the $dx \, dz$ in both cases gives

$$\frac{\partial p}{\partial x} = 0 \qquad \text{and} \qquad \frac{\partial p}{\partial z} = \frac{dp}{dz} = -\gamma = -\rho g_n \tag{2.3}$$

Because $\partial p / \partial x = 0$, there is no variation of pressure with horizontal distance; that is, pressure is constant in a horizontal plane in a static fluid. Therefore, pressure is a function of z only and it is permissible to replace the partial derivative in the second equation with the total derivative.

The second of equations 2.3 is the basic equation of fluid statics. It can be written in the form $-dz = dp / \gamma$ and so can be integrated directly to find

$$z_2 - z_1 = \int_{p_2}^{p_1} \frac{dp}{\gamma} \tag{2.4}$$

For a *fluid of constant density* (this may be safely assumed for liquids over large vertical distances and for gases over small ones) the integration yields

$$z_2 - z_1 = h = \frac{p_1 - p_2}{\gamma} \qquad \text{or} \qquad p_1 - p_2 = \gamma(z_2 - z_1) = \gamma h \tag{2.5}$$

permitting ready calculation of the increase of pressure with depth in a fluid of constant density. Equation 2.5 also shows that pressure differences ($p_1 - p_2$) may be readily

---

[3]See Appendix 5.

expressed as a *head h* of fluid of specific weight γ. Thus pressures are often quoted as heads in *millimetres of mercury, feet or metres of water*, etc. The relation of pressure to head[4] is illustrated by the open *manometer* and *piezometer columns* of Fig. 2.2.

Equation 2.5 can be rearranged fruitfully to

$$\frac{p_1}{\gamma} + z_1 = \frac{p_2}{\gamma} + z_2 = \text{Constant} \tag{2.6}$$

for later comparison with equations of fluid flow. Takings points 1 and 2 as typical, it is evident from equation 2.6 that the quantity $(z + p/\gamma)$ is the same for all points in a static liquid. This can be visualized geometrically as shown on Fig. 2.3. Frequently in engineering problems the liquid surface is exposed to atmospheric pressure; if the latter is taken to be zero (see Section 2.2), the dashed line of Fig. 2.3 will necessarily coincide with the liquid surface.

For a fluid of variable density, integration of equation 2.4 cannot be accomplished until a relationship between p and γ is known. This problem is encountered in the fields of oceanography and meteorology. In the former, a suitable relationship can be obtained from elasticity considerations (equation 1.8) or empirical relations between pressure and density, which is affected by the temperature and salinity of the sea. In the latter field, certain gas laws provide the $p - \gamma$ relationship. For gases (as in the atmosphere) the polytropic equation[5]

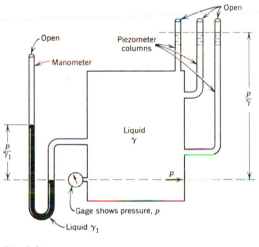

Fig. **2.2**

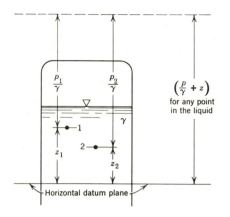

Fig. **2.3**

$$\frac{p}{\gamma^n} = \text{Constant} \tag{2.7}$$

may be employed to develop relations between pressure, density, temperature, and altitude. One of the most important of these is $-dT/dz$, the rate of temperature change with altitude, termed the *temperature lapse rate*, which may be derived as follows: inserting $\gamma RT/g_n$ for p in equation 2.7 and differentiating yields

$$\frac{T \, d\gamma}{\gamma \, dT} = \frac{1}{n-1}$$

Replacing dp in equation 2.3 by $d(\gamma RT/g_n)$ gives

$$\frac{dz}{dT} = -\frac{R}{g_n}\left(\frac{T \, d\gamma}{\gamma \, dT} + 1\right)$$

Substituting the first of these equations into the second results in

$$-\frac{dT}{dz} = \frac{g_n(n-1)}{nR} \tag{2.8}$$

For $n > 1$, $-dT/dz > 0$, which is the familiar situation in the lower portion of the earth's atmosphere (the *troposphere*) where temperature declines with increasing altitude. Between altitudes 11 km (36 000 ft) and 20 km (65 600 ft) in the *stratosphere*, however, the temperature has been observed to be essentially constant at −56.5°C (−69.7°F); here the atmosphere is *isothermal*, $n = 1$, and $-dT/dz = 0$. Through the troposphere the mean lapse rate has been found to be practically constant, and this has led to the definition of a *standard atmosphere* which closely approximates the yearly mean at latitude 40°. The U.S. Standard Atmosphere [6] assumes a sea level pressure of

[6]See Appendix 4.

101.3 kPa (14.70 psia) and a constant lapse rate through the troposphere of 0.006 5°C/m (0.003 56°F/ft) from 15°C (59°F) at sea level to −56.5°C (−69.7°F) at altitude 11 019 m (36 150 ft), at which the stratosphere begins. From the lapse rate above, and taking R to be 286.8 J/kg · K (1 715 ft · lb/slug · °F), n is found (from equation 2.8) to be 1.235, from which pressure and air density may be calculated throughout the (standard) troposphere. Although this standard (and static) atmosphere can be used for design calculations and performance predictions on high altitude aircraft and their appurtenances, the earth's atmosphere with its winds and air currents is not, of course, precisely static, but this is usually a satisfactory approximation for the prediction of pressures and densities. For violent disturbances (e.g., tornadoes and hurricanes) the assumption of a static atmosphere is clearly untenable.

The lapse rate for an *adiabatic atmosphere* is important for purposes of comparison and may be calculated from equation 2.8 using $n = k = 1.40$ and $R = 286.8$ J/kg · K (1 715 ft · lb/slug · °F) providing the air is dry;[7] the result is $−dT/dz =$ 0.009 8°C/m (0.005 35 °F/ft), which is known as the *adiabatic lapse rate* and will be shown to be a criterion of atmospheric stability. Suppose that in an adiabatic atmosphere a mass of fluid is moved from one altitude to another. If it is moved upward, it will expand almost without acceptance or rejection of heat (i.e., adiabatically) because of its poor conduction; accordingly at its new altitude it will have the same density as the surrounding air and thus possess no tendency to move from its new position. The adiabatic atmosphere is thus in a state of *neutral equilibrium* and is inherently *stable*. When this process is imagined for a lapse rate ($−dT/dz$) larger than the adiabatic, the expansion will tend to be adiabatic as before, but in its new position the density of the fluid mass will be smaller than that of its surroundings and its greater buoyancy will cause it to rise further; such an atmosphere is inherently *unstable*—and this fact leads to the expectation that a stable atmosphere can occur only for lapse rates less than the adiabatic. This expectation is confirmed when the foregoing reasoning is applied to an atmosphere with lapse rate less than adiabatic; here displacements of fluid masses produce density changes which tend to restore the air masses to their original position. Thus for diminishing lapse rates atmospheric stability steadily increases, becoming greatest in the case of an *inversion*, when the lapse rate is negative.

---------- **Illustrative Problems** ----------

The liquid oxygen (LOX) tank of a Saturn moon rocket is partially filled to a depth of 10 m with LOX at −196°C. The absolute pressure in the vapor above the liquid surface is maintained at 101.3kPa. Calculate the absolute pressure at the inlet valve at the bottom of the tank.

Relevant Equations and Given Data

$$\gamma = \rho g_n \tag{1.1}$$

[7]Although atmospheric moisture is a critical factor in many meteorology problems, it is disregarded here because it does not change the sense of the development; it would, of course, change the numerical values.

$$p_1 - p_2 = \gamma h \tag{2.5}$$

$$\rho = 1\ 206 \text{ kg/m}^3 \qquad h = 10 \text{ m} \qquad p_2 = 101.3 \text{ kPa, abs.}$$

Solution. $\gamma = \rho \times g_n = 1\ 206 \times 9.81 = 11.8 \text{ kPa/m}$ $\tag{1.1}$

$$p_1(\text{abs.}) = 11.8(10) + 101.3 = 219.3 \text{ kPa} \blacktriangleleft \tag{2.5}$$

Calculate pressure and specific weight of air in the U.S. Standard Atmosphere at altitude 35 000 ft.

Relevant Equations and Given Data

$$\gamma = g_n p / RT \tag{1.3}$$

$$z_2 - z_1 = \int_{p_2}^{p_1} \frac{dp}{\gamma} \tag{2.4}$$

$$\frac{p}{\gamma^n} = \text{Constant} \tag{2.7}$$

$$n = 1.235 \qquad z_2 - z_1 = 35\ 000 \text{ ft}$$

$$R = 1\ 715 \text{ ft} \cdot \text{lb/slug} \cdot °\text{R} \quad dT/dz = -0.003\ 56 \text{ °C/ft}$$

Solution. At sea level (see Appendix 4),

$$p_1 = 14.7 \text{ psia} \qquad T_1 = 519°\text{R} \qquad \gamma_1 = 0.076\ 5 \text{ lb/ft}^3$$

$$\frac{p}{\gamma^{1.235}} = \frac{14.7 \times 144}{(0.076\ 5)^{1.235}} \qquad \frac{1}{\gamma} = \frac{6\ 450}{p^{0.81}} \tag{2.7}$$

$$35\ 000 = 6\ 450 \int_{p_2}^{14.7 \times 144} p^{-0.81}\ dp \qquad p_2 = 504 \text{ psfa} = 3.50 \text{ psia} \blacktriangleleft \tag{2.4}$$

$$T_2 = 519 - 35\ 000 \times 0.003\ 56 = 394°\text{R}$$

$$\gamma_2 = 32.2 \times 504/1\ 715 \times 394 = 0.024 \text{ lb/ft}^3 \blacktriangleleft \tag{1.3}$$

Compare results with the values of Appendix 4.

In the ocean, warm patches of water that are heated by the sun induce thermals (buoyant columns of air) by heating the air above its standard atmosphere value so that the air is buoyant. If this air has a temperature of 17°C and its buoyant rise is adiabatic, at what level will the rising air be in temperature (and density) equilibrium with the standard atmosphere (and stop rising buoyantly)?

Relevant Equation and Given Data

$$T_2 - T_1 = \int_{z_1}^{z_2} \left(\frac{dT}{dz}\right) dz$$

$$T_{1S} = T_{1(\text{Standard})} = 15°\text{C} \qquad -\left(\frac{dT}{dz}\right)_{\text{Standard}} = 0.006\ 5°\text{C/m};$$

$$T_{1T} = T_{1(Thermal)} = 17°C \qquad -\left(\frac{dT}{dz}\right)_{Adiabatic} = 0.009\ 8°C/m$$

$$z_1 = 0$$

Solution.

$$T_{2(Standard)} = 15°C - 0.006\ 5z_2$$

$$T_{2(Thermal)} = 17°C - 0.009\ 8z_2$$

At equilibrium:

$$T_{2S} = T_{2T}$$

so

$$2 = 0.003\ 3z_2$$

$$z_2 = 607\ m \blacktriangleleft$$

2.2 Absolute and Gage Pressures

Pressures, like temperatures, are measured and quoted in two different systems, one relative (gage), and the other absolute; no confusion results if the relation between the systems and the common methods of measurement are completely understood.

The Bourdon pressure gage and the aneroid barometer (shown schematically in Fig. 2.4) are typical mechanical devices for measuring gage and absolute pressures, respectively. In the pressure gage a bent tube (A) of elliptical cross section is held rigidly at B and its free end is connected to a pointer (C) by a link (D). When pressure is admitted to the tube, its cross section tends to become circular, causing the tube to straighten and move the pointer to the right over the graduated scale. If the gage is in proper adjustment, the pointer rests at zero on the scale *when the gage is disconnected*; in this condition the pressures inside and outside of the tube are the same, and thus there is no tendency for the tube to deform. It is apparent that such pressure gages are actuated by the *difference* between the pressure inside and that outside the tube. For example, in the gage system of pressure measurement, if atmospheric pressure exists

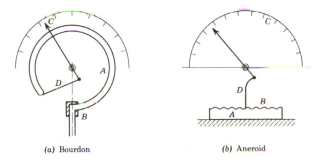

(a) Bourdon   (b) Aneroid

Fig. **2.4**   Mechanical pressure gages.

outside the tube, the local (not standard) atmospheric pressure becomes the zero of pressure. For pressure less than local atmospheric the tube will tend to contract, moving the pointer to the left. The reading for pressure greater than local atmospheric is positive and is called *gage pressure*, or simply *pressure*,[8] and it is usually measured in *pascals* (newtons per square metre) or psi (pounds per square inch); the reading for pressure below local atmospheric is negative, designated as *vacuum*, and usually is measured in millimetres or inches of mercury.

The aneroid gage is a device for measuring absolute pressure. The essential element is a short cylinder (A) with one end an elastic diaphragm (B). The cylinder is evacuated[9] so that the pressure therein is close to absolute zero; pressures imposed on the outside of the diaphragm cause it to deflect inward; these deflections are then a direct measure of the applied pressures, which can be transferred to a suitable scale (C) through appropriate linkages (D); here the pressures recorded are relative to absolute zero and are *absolute pressures*. Although the aneroid cylinder as conventionally used in barometers is capable of measuring only a small range of pressures, the basic idea can be applied to absolute pressure gages for more general use.

Liquid devices that measure gage and absolute pressures are shown on Fig. 2.5; these are the open U-tube and the conventional mercury barometer. With the U-tube open, atmospheric pressure acts on the upper liquid surface; if this pressure is taken to be zero, the applied gage pressure p equals γh and h is thus a direct measure of gage pressure.

The mercury barometer (invented by Torricelli, 1643) is constructed by filling the tube with air-free mercury and inverting it with its open end beneath the mercury surface in the receptacle. Ignoring the small pressure of the mercury vapor,[10] the

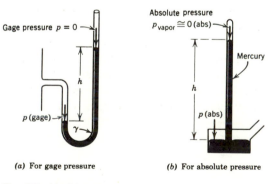

(a) For gage pressure   (b) For absolute pressure

Fig. **2.5** Liquid, pressure gages.

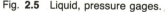

[8]It has been internationally recommended that pressure units themselves should not be modified to indicate whether the pressure is absolute or gage. When the context of use leaves any doubt as to which is meant, the word "pressure" is to be qualified appropriately. Throughout the remainder of this book *pressure* means gage pressure unless the context or the qualified term *absolute pressure* indicates otherwise.
[9]If the cylinder could be completely evacuated, the pressure therein would be the lowest possible (absolute zero) since there would be no fluid molecules to exert pressure.
[10]See Table 1.

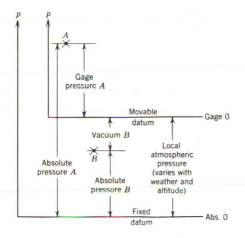

Fig. **2.6**

pressure in the space above the mercury is at absolute zero and again $p = \gamma h$; here the height h is a direct measure of the absolute pressure, p. Although conventional use of the barometer is for the measurement of local atmospheric pressure, the basic scheme is frequently used in industry for the direct measurement of any absolute pressure.

From the foregoing descriptions an equation relating gage and absolute pressures may now be written,

$$\text{Absolute pressure} = \text{Atmospheric pressure} \quad \begin{array}{l} -\text{Vacuum} \\ +\text{Gage pressure} \end{array} \qquad (2.9)$$

which allows easy conversion from one system to the other. Possibly a better picture of these relationships can be gained from a diagram such as Fig. 2.6 in which are shown two typical pressures, A and B, one above, the other below, atmospheric pressure, with all the relationships indicated graphically.

Illustrative Problem

A Bourdon gage registers a vacuum of 12.5 in. (310 mm) of mercury when the atmospheric pressure is 14.50 psia (100 kPa, absolute). Calculate the corresponding absolute pressure.

Relevant Equation and Given Data

$$\text{Absolute pressure} = \text{Atmospheric pressure} \quad \begin{array}{l} -\text{Vacuum} \\ +\text{Gage pressure} \end{array} \qquad (2.9)$$

$$p_G = 12.5 \text{ in.} \quad \text{or} \quad 310 \text{ mm Hg vacuum}$$

$$p_{atm} = 14.50 \text{ psia} \quad \text{or} \quad 100 \text{ kPa, absolute}$$

Note: 29.92 in. Hg is equivalent to 14.70 psi

760 mm Hg is equivalent to 101.3 kPa

Solution.

$$\text{Absolute pressure} = 14.50 - 12.5(14.70/29.92) = 8.35 \text{ psia} \blacktriangleleft \qquad (2.9)$$

$$\text{Absolute pressure} = 100 - 310(101.3/760) = 58.7 \text{ kPa} \blacktriangleleft \qquad (2.9)$$

2.3 Manometry

Bourdon and aneroid pressure gages, owing to their inevitable mechanical limitations, are not usually adequate for precise measurements of pressure; when greater precision is required, *manometers* like those of Fig. 2.7 may be effectively employed (see also Section 11.2 for a brief discussion of electric pressure transducers).

Consider the U-tube manometer of Fig. 2.7a in which all distances and densities are known, and pressure p_x is to be found. Because, over horizontal planes *within continuous columns of the same fluid*, pressures are equal, it is evident at once that $p_1 = p_2$, and from equation 2.5

$$p_1 = p_x + \gamma l \qquad \text{and} \qquad p_2 = 0 + \gamma_1 h$$

Equating p_1 and p_2 gives

$$p_x = \gamma_1 h - \gamma l$$

allowing pressure, p_x, to be calculated.[11]

U-tube manometers are frequently used to measure the difference between two unknown pressures, p_x and p_y, as in Fig. 2.7b. Here

$$p_x + \gamma_1 l_1 = p_4 = p_5 = p_y + \gamma_2 l_2 + \gamma_3 h$$

from which we obtain

$$p_x - p_y = \gamma_2 l_2 + \gamma_3 h - \gamma_1 l_1$$

thus allowing direct calculation of the pressure difference, $p_x - p_y$. Differential manometers of the type above are sometimes made with the U-tube inverted, a liquid of small density existing in the top of the inverted U; the pressure difference measured by manometers of this type may be readily calculated by application of the foregoing principles. When large pressures or pressure differences are to be measured, a pressure gage or a mechanical or electrical transducer (Section 11.2) is usually used.

There are many forms of precise manometers; two of the most common are shown in Fig. 2.7. Figure 2.7 c represents the ordinary *inclined gage* used in measuring the comparatively small pressures in low-velocity gas flows. Its equilibrium position is shown at A, and when it is submitted to a pressure, p_x, a vertical deflection, h, is

---

[11]The use of derived formulas for manometer solutions is not recommended until experience has been gained in their limitations.

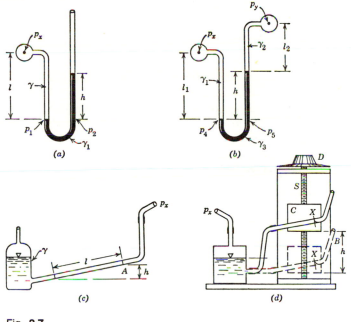

Fig. **2.7**

obtained in which $p_x = \gamma h$. In this case, however, the liquid is forced down a gently inclined tube so that the deflection, l, is much greater than h and, therefore, more accurately read. This type of manometer, when calibrated to read directly in inches of water, is frequently called a *draft gage*.

The principle of the sloping tube is also employed in the alcohol micromanometer of Fig. 2.7d, used in research work. Here the gently sloping glass tube is mounted on a carriage, C, which is moved vertically by turning the dial, D, which actuates the screw, S. When p_x is zero, the carriage is adjusted so that the liquid in the tube is brought to the hairline, X, and the reading on the dial is recorded. When the unknown pressure, p_x, is admitted to the reservoir, the alcohol runs upward in the tube toward B and the carriage is then raised until the liquid surface in the tube rests again at the hairline, X. The difference between the dial reading at this point and the original reading gives the vertical travel of the carriage, h, which is the head of alcohol equivalent to the pressure p_x.

Along with these principles of manometry the following practical considerations should be appreciated: (1) manometer liquids, in changing their relative densities with temperature, will induce errors in pressure measurements if this factor is overlooked; (2) errors due to capillarity may frequently be canceled by selecting manometer tubes of uniform size; (3) although some liquids appear excellent (from density considerations) for use in manometers, their surface-tension effects may give poor menisci and thus inaccurate readings; (4) fluctuations of the manometer liquids will reduce accuracy of pressure measurement, but these fluctuations may be reduced by a throttling device in the manometer line (a short length of small tube is excellent for this purpose); and (5) when fluctuations are negligible, refined optical devices and verniers may be used for extremely precise readings of the liquid surfaces.

--------------------- **Illustrative Problem** ---------------------

This vertical pipeline with attached gage and manometer contains oil and mercury as shown. The manometer is open to the atmosphere. There is no flow in the pipe. What will be the gage reading, p_x?

Relevant Equation and Given Data

$$p_1 - p_2 = \gamma(z_2 - z_1) \tag{2.5}$$

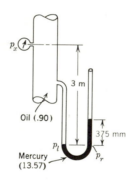

Oil (.90)

3 m

375 mm

p_ℓ

Mercury (13.57)

p_r

p_x

Solution.

$$p_\ell = p_x + (0.90 \times 9.8 \times 10^3)3 \tag{2.5}$$

$$p_r = (13.57 \times 9.8 \times 10^3)0.375 \tag{2.5}$$

Because

$$p_\ell = p_r$$

$$p_x = 23.4 \text{ kPa} \blacktriangleleft$$

2.4 Forces on Submerged Plane Surfaces

The calculation of the magnitude, direction, and location of the total forces on surfaces submerged in a liquid is essential in the design of dams, bulkheads, gates, tanks, ships, and the like.

For a submerged, plane, *horizontal* area the calculation of these force properties is simple because the pressure does not vary over the area; for nonhorizontal planes the problem is complicated by pressure variation. Pressure in constant density liquids,[12] however, has been shown to vary *linearly* with depth (equation 2.5), producing the typical pressure distributions and resultant forces on the walls of the container of Fig. 2.8. The shaded areas, appearing as trapezoids, are really volumes, known as *pressure prisms*. In mechanics it has been shown that the resultant force, F, is equal to the volume of the pressure prism and acts through its centroid.[13]

Now consider the general case[14] of a plane submerged area, A, such as that of Fig. 2.9, located in any inclined plane, $L-L$. Let the centroid of this area be located, as shown, at a depth h_c and at a distance l_c from the line of intersection, $O-O$, of plane $L-L$ and the liquid surface. The force, dF, on the area, dA, is given by $p\ dA = \gamma h\ dA$; because $h = l \sin \alpha$, dF may be expressed as

$$dF = \gamma l\ dA \sin \alpha \qquad (2.10)$$

and the total force on the area A is found by integration of this expression over the area, which gives

$$F = \gamma \sin \alpha \int^A l\ dA \qquad (2.11)$$

Here $\int^A l\ dA$ is the moment of the area A, about the line $O-O$, which is also given by the product of the area, A, and the perpendicular distance, l_c, from $O-O$ to the centroid

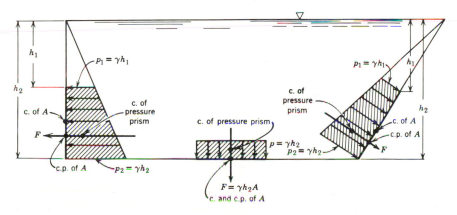

Fig. **2.8** Pressure forces on walls of a container.

[12]In gases, pressure variation with depth is so small that it is usually ignored in the calculation of resultant forces in engineering problems.

[13]The resultant force passes through a point on the plane defined as the *center of pressure* (c.p.). For nonhorizontal areas (with liquid on one side), the center of pressure of an area is always below its centroid. When the same liquid *covers both sides* of the area, examination of the pressure prisms shows that the resultant force of liquid on the area passes through the centroid of the area.

[14]A general solution for the magnitude, direction, and location of the resultant force on this area will allow easy calculation of the forces on areas of more regular shape.

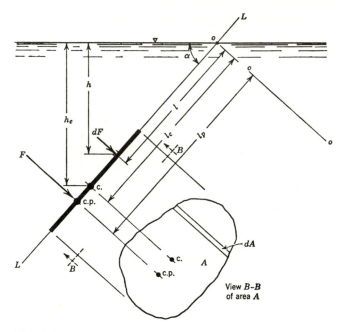

Fig. **2.9**

(c.) of the area. Thus

$$\int^A l\,dA = l_c A$$

and, when this is substituted in equation 2.11,

$$F = \gamma A l_c \sin \alpha$$

However, $h_c = l_c \sin \alpha$, which reduces the foregoing equation to

$$F = \gamma h_c A \qquad (2.12)$$

and indicates that the magnitude of the resultant force on (one side of) any submerged plane area may be calculated by multiplying the area, A, by the pressure at its centroid, γh_c.

The magnitude of the resultant force having been calculated, its direction and location must be considered. Its direction, because of the nonexistence of shear stress, is necessarily normal to the plane, and the distance, l_p, between its line of action and axis O–O may be found by dividing the moment of the force (about this axis) by the magnitude of the force.

From Fig. 2.9, the moment, dM, of the force, dF, about line O–O is equal to $l\,dF$. Substituting the value for dF from equation 2.10 gives

$$dM = \gamma l^2\,dA \sin \alpha$$

and integrating to obtain the total moment, M, yields

$$M = \gamma \sin \alpha \int^A l^2 \, dA$$

in which $\int^A l^2 \, dA$ is the second moment, I_{O-O}, of the area, A, about the line O–O; thus

$$M = \gamma I_{O-O} \sin \alpha$$

From "moment arm = moment divided by force" it is apparent that $l_p = M/F$, and, when the expressions above for M and F are substituted,

$$l_p = I_{O-O}/l_c A \tag{2.13}$$

Thus the resultant force is located with respect to line O–O and the general solution of the problem is completed.

Equation 2.13 may be made more usable by putting it in terms of the second moment,[15] I_c, about an axis lying in the area, parallel to O–O, and through the centroid of the area. Using the transfer equation for second moments of areas produces

$$I_{O-O} = I_c + l_c^2 A$$

which, when substituted in equation 2.13, gives

$$l_p - l_c = I_c/l_c A \tag{2.14}$$

allowing direct calculation of the distance (down the L–L plane) between centroid and center of pressure. This equation also indicates that the center of pressure is always *below* the centroid except for a horizontal area, but that the distance between center of pressure and centroid diminishes as the depth of submergence of the area is increased.[16]

The lateral location of the center of pressure for regular plane areas, such as that of Fig. 2.10, is readily calculated by considering the area to be composed of a large number of rectangles of differential height, dl. The centroid and center of pressure of each of these small rectangles is coincident at the center of the rectangle and, therefore, all the forces on the rectangles act on the median line, AB; thus (from the statics of parallel forces) the line of action of the resultant of these forces must also intersect the median line; this intersection is the center of pressure of the area.

Although a general theorem may be developed to find the centers of pressure of areas of more irregular forms, this is not essential to the solution of the problem. Frequently such areas can be divided into simple areas, the forces located on them, and the location of their resultant found by the methods of statics. The point where the line of action of the resultant force intersects the area is the center of pressure for the composite area. The following illustrative problem demonstrates this method.

---

[15] A summary of I_c's for common areas is found in Appendix 6.
[16] The approximation that the resultant force acts at its centroid can be made for a small area under great submergence or under great pressure.

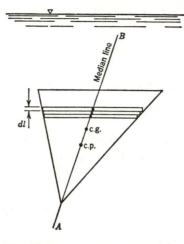

Fig. **2.10**

———————————— **Illustrative Problem** ————————————

Calculate magnitude, direction, and location of the total force exerted by the water on one side of this composite area which lies in a vertical plane.

Relevant Equations

$$F = \gamma h_c A \tag{2.12}$$

$$\ell_p - \ell_c = I_c / \ell_c A \tag{2.14}$$

$$I_{o-o} = I_c + \ell_c^2 A \qquad \text{(Second moment transfer equation)}$$

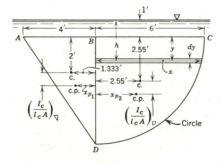

Solution. By inspection the direction of the force is normal to the area. ◀
Magnitude (see Appendix 6):

$$\textit{Force on triangle} = 62.4 \times 3 \times 12 = 2\ 245\ \text{lb} \tag{2.12}$$

$$\text{Force on quadrant} = 62.4 \times 3.55(36\pi/4) = \underline{6\ 275}\ \text{lb} \tag{2.12}$$

$$\text{Total force on composite area} \qquad\qquad = 8\ 520\ \text{lb} ◀$$

Vertical location of resultant force (see Appendix 6):

$$I_c/\ell_c A \text{ for triangle} = \frac{4 \times 6^3}{36 \times 3 \times 12} = 0.667 \text{ ft} \qquad (2.14)$$

$$I_c/\ell_c A \text{ for quadrant} = \frac{70.6}{3.55 \times 28.3} = 0.703 \text{ ft} \qquad (2.14)$$

Taking moments about line AC,

$$2\ 245 \times (2 + 0.667) + 6\ 275 \times (2.55 + 0.703) = 8\ 520(\ell_p - 1) \qquad \ell_p = 4.10 \text{ ft} \blacktriangleleft$$

Lateral location of resultant force: Since the center of pressure of the triangle is on the median line, x_{p_1} is given (from similar triangles) by

$$\frac{(1.333 - x_{p_1})}{2} = \frac{0.667}{6} \qquad x_{p_1} = 1.111 \text{ ft}$$

Dividing the quadrant into horizontal strips of differential height, dy, the moment about BD of the force on any one of them is

$$dM = (dA)p \, (\text{lever arm})$$

$$dM = (x \, dy)\gamma h\left(\frac{x}{2}\right)$$

in which $h = y + 1$ and $x^2 + y^2 = 36$. Substituting and integrating gives the moment about BD of the total force on the quadrant,

$$M = \frac{62.4}{2}\int_0^6 (36 - y^2)(y + 1) \, dy = 14\ 600 \text{ ft·lb}$$

and thus

$$x_{p_2} = \frac{14\ 600}{6\ 275} = 2.33 \text{ ft}$$

Finally, taking moments about line BD,

$$2.33 \times 6\ 275 - 2\ 245 \times 1.111 = 8\ 520 x_p \qquad x_p = 1.425 \text{ ft} \blacktriangleleft$$

Thus the center of pressure of the composite figure is 1.425 ft to the right of BD and 4.10 ft below the water surface. ◄

2.5 Forces on Submerged Curved Surfaces

Resultant forces on submerged curved surfaces cannot be calculated by the foregoing methods but may be readily determined through computation of their horizontal and vertical components. Consider the forces on the curved portion AB in the side of the container of Fig. 2.11. On each small element of surface, the magnitude, direction (normal to the element), and location of the pressure force may be determined by the foregoing principles, and they lead to the indicated pressure distribution which may be reduced to a single resultant force F with components F_H and F_V as shown.

Analysis of the free body of fluid ABC allows calculation of F_H' and F_V', the

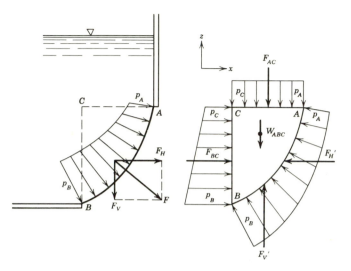

Fig. **2.11**

components of the resultant force exerted *by the surface AB on the fluid*, and the equal-and-opposite of F_H and F_V, respectively. From the static equilibrium of the free body,

$$\sum F_x = F_{BC} - F_H' = 0$$

$$\sum F_z = F_V' - W_{ABC} - F_{AC} = 0$$

and thus $F_H' = F_{BC}$ and $F_V' = W_{ABC} + F_{AC}$. From the inability of the free body of fluid to support shear stress, it follows that F_H' must be collinear with F_{BC} and F_V' collinear with the resultant of W_{ABC} and F_{AC}. The foregoing analysis reduces the problem to one of computation of magnitude and location of F_{BC}, F_{AC}, and W_{ABC}; for F_{AC} and F_{BC} the methods of Section 2.4 may be used, whereas W_{ABC} is merely the weight of the free body of fluid and necessarily acts through its center of gravity.

When the same liquid covers both sides of a curved area but the liquid surfaces are at different levels for the two sides, the net effective pressure distribution is uniform because the effective pressure at any point on the area is dependent only on the difference in surface levels. The resultant force on such an area is obtained by application of the methods above. The horizontal component passes through the centroid of the vertical projection of the area, and the vertical component passes through the centroid of the horizontal projection.

─────────────── **Illustrative Problem** ───────────────

Consider the cross section shown below of the hull of a typical 330 000 tonne (1 tonne = 10^3 kg) *Universe*-class oil tanker. Calculate the magnitude, direction, and location of the resultant force per metre exerted by the seawater ($\gamma = 10$ kN/m³) on the curved surface AB (which is a quarter cylinder) at the corner of the hull.

Relevant Equations and Given Data

$$F = \gamma h_c A \qquad (2.12)$$

$$\ell_p - \ell_c = I_c/\ell_c A \qquad \text{(See Appendix 6)} \qquad (2.14)$$

$$\gamma = 10 \text{ kN/m}^3 = 10^4 \text{ N/m}^3$$

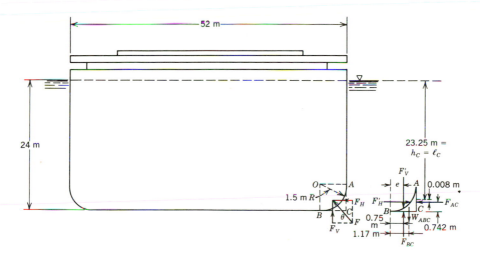

Solution. Isolate the free body of water ABC and show the forces on it. *Horizontal component*: By inspection, this force per metre F_H' (on the free body) has direction to the right and is balanced by F_{AC}:

$$F_H' = F_{AC} = 10^4 \times 23.25 \times (1.5 \times 1) = 348.8 \text{ kN/m} \qquad (2.12)$$

$$(\ell_p - \ell_c)_{AC} = \frac{0.28}{23.25 \times (1.5 \times 1)} = 0.008 \text{ m} \qquad (2.14)$$

Vertical component: By inspection, this force F_V' (on the free body) has a downward direction:

$$\sum F_z = F_{BO} - F_V' - W_{ABC} = 0$$

$$\sum F_z = 10^4 \times 24 \times (1.5 \times 1) - F_V' - 10^4\left(2.25 - \frac{2.25\pi}{4}\right) = 0$$

$$F_V' = 355.2 \text{ kN/m}$$

From statics the center of gravity of ABC is found to be 1.17 m to the right of B, and, taking moments of the forces (on the free body) about 0,

$$F_V' \times e + W_{ABC} \times 1.17 - F_{BC} \times 0.75 = 0$$

$$355.2 \times e + 4.8 \times 1.17 - 360 \times 0.75 = 0 \qquad e = 0.74 \text{ m}$$

Resultant force of water on AB:

Direction: upward to the left, $\theta = \arctan 355.2/348.8 = 45.5°$ ◀

Magnitude:

$$F = \sqrt{(348.8)^2 + (355.2)^2} = 497.8 \text{ kN/m} \blacktriangleleft$$

Location: through a point 0.742 m above and 0.74 m to the right of B. Because the pressure forces on the elements of the cylinder, although different in magnitude, all pass through O, and thus form a concurrent force system, their resultant will also be expected to pass through O. Since $(1.5 - 0.742)/0.74 = 335.2/348.8$, this expectation is confirmed. Note that these tankers are of the order of 350 m long! ◀

2.6 Buoyancy and Flotation

The familiar laws of buoyancy (Archimedes' principle) and flotation are usually stated: (1) *a body immersed in a fluid is buoyed up by a force equal to the weight of fluid displaced*; and (2) *a floating body displaces its own weight of the liquid in which it floats*. These laws are corollaries of the general principles of Section 2.5 and may be readily proved by application of those principles.

A body $ABCD$ suspended in a fluid of specific weight γ is illustrated in Fig. 2. 12. Isolating a free body of fluid with vertical sides tangent to the body allows identification of the vertical forces exerted by the lower (ADC) and upper (ABC) surfaces of the body on the surrounding fluid. These are F_1' and F_2' with ($F_1' - F_2'$) the buoyant force on the body. For the upper portion of the free body

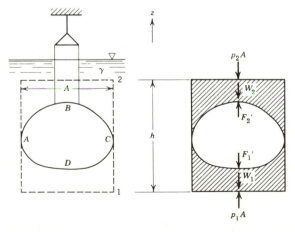

Fig. **2.12**

$$\sum F_z = F'_2 - W_2 - p_2 A = 0$$

and for the lower portion

$$\sum F_z = F'_1 + W_1 - p_1 A = 0$$

with which we obtain (by subtraction of the equations)

$$F_B = F'_1 - F'_2 = (p_1 - p_2)A - (W_1 + W_2)$$

However, $p_1 - p_2 = \gamma h$ and $\gamma h A$ is the weight of a cylinder of fluid extending between horizontal planes 1 and 2, and the right side of the equation for F_B is identified as the weight of a volume of fluid exactly equal to that of the body. Accordingly[17]

$$F_B = \gamma(\text{volume of object}) \qquad (2.15)$$

and the law of buoyancy is proved.

For the floating object of Fig. 2.13 a similar analysis shows that

$$F_B = \gamma(\text{volume displaced}) \qquad (2.16)$$

and, from static equilibrium of the object, its weight must be equal to this buoyant force; thus the object displaces its own weight of the liquid in which it floats.

The principles above find many applications in engineering, for example, in calculations of the draft of surface vessels, the increment in depth of flotation from the increment in weight of the ship's cargo, and the lift of airships and balloons.

The stability of submerged or floating bodies is dependent on the relative location of the buoyant force and the weight of the body. The buoyant force acts upward through the center of gravity of the displaced volume;[18] the weight acts downward at the center of gravity of the body. Stability or instability will be determined by whether a righting

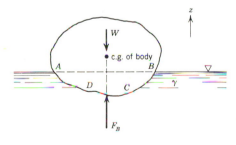

Fig. **2.13**

---

[17]This calculation for buoyant force presumes that a homogeneous fluid completely surrounds the body; if it does not, the concept of the buoyant force must be adjusted accordingly by reapplication of the same principles. A sunken ship embedded in the ocean floor is a classic example of this; here the water does not completely surround the hull.

[18]This is evident from the fact that the buoyant force is the *weight* of the displaced volume; it needs no formal proof.

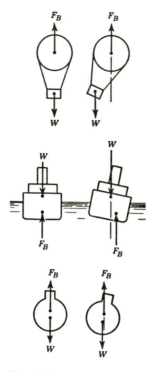

Fig. **2.14**

or overturning moment is developed when the center of gravity and center of buoyancy move out of vertical alignment. Obviously, for the submerged bodies, such as the balloon and submarine of Fig. 2.14, stability requires the center of buoyancy to be above the center of gravity. In surface vessels, however, the center of gravity is usually above the center of buoyancy, and stability exists because of movement of the center of buoyancy to a position outboard of the center of gravity as the ship "heels over," producing a righting moment. An overturning moment, resulting in capsizing, occurs if the center of gravity moves outboard of the center of buoyancy.

Illustrative Problems

A container ship has a cross-sectional area of 32 000 ft² at the waterline when the draft is 29 ft. How many tons of containers can be added before the normal draft of 30 ft is reached. Assume saltwater of specific weight 64.0 lb/ft³.

Relevant Equation and Given Data

$$F_B = \gamma(\text{volume displaced}) \qquad (2.16)$$

$$\gamma = 64.0 \text{ lb/ft}^3 \qquad \Delta_{draft} = 1 \text{ ft}$$

$$A_{c.s.} = 32\ 000 \text{ ft}^2$$

Solution. As the ship floats, the weight of water displaced by the containers equals the weight of the containers. Therefore,

$$\text{Weight of containers} = 32\ 000 \times 1 \times 64.0$$

$$= 2\ 048\ 000 \text{ lb} = 1\ 024 \text{ tons} \blacktriangleleft$$

The solid wooden sphere ($\gamma_S = 8\ 350$ N/m³) of diameter 0.4 m is held in the orifice (0.2 m diameter) by the water. Calculate the force exerted between sphere and orifice plate when the depth is 0.7 m. The sphere will float away if this force becomes zero; is there any depth of water for which this can happen?

Relevant Equation and Given Data

$$F_B = \gamma(\text{volume of object}) \qquad\qquad (2.15)$$

$$\gamma_S = 8\ 350 \text{ N/m}^3 \qquad \gamma_w = 9\ 800 \text{ N/m}^3$$

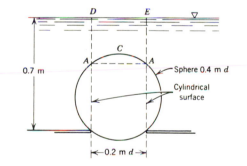

Solution.

$$\text{Volume of sphere} = (\pi/6)(0.4)^3 = 0.033\ 5 \text{ m}^3$$

$$\text{Weight of sphere} = 0.033\ 5 \times 8\ 350 = 280 \text{ N downward}$$

Volume of section sphere outside of cylindrical surface (from solid geometry) is 0.021 8 m³. From this volume the buoyant force F_B may be calculated:

$$F_B = 0.021\ 8 \times 9\ 800 = 214 \text{ N upward}$$

The force downward on surface *ACA* is computed from the volume of *ACADE*, which (from solid geometry) is found to be 0.009 8 m³:

$$F_{ACA} = 0.009\ 8 \times 9\ 800 = 96 \text{ N downward}$$

For static equilibrium of the sphere,

$$\text{Force between sphere and plate} = 280 + 96 - 214 = 162 \text{ N} \blacktriangleleft$$

When the water surface coincides with the horizontal plane *A–A*, the force on surface *ACA* will be zero. Here the force between sphere and plate will be $280 - 214 = 66$ N. Below this point the buoyant force will be less than 214 N; accordingly, the force between sphere and plate can never be zero. ◄

2.7   Fluid Masses Subjected to Acceleration

Fluid masses can be subjected to various types of acceleration without the occurrence of relative motion between fluid particles or between fluid particles and boundaries. Such fluid masses will be found to conform to the laws of fluid statics, modified to allow for the effects of acceleration, and they may often be treated by assuming a change in the magnitude and direction of g_n.

A generalized approach to this problem may be obtained by applying Newton's second law to the fluid element of Fig. 2.15 which is being accelerated in such a way that its components of acceleration are a_x and a_z. The summation of force components on such an element has been indicated in Section 2.1, and is

$$\sum F_x = \left(-\frac{\partial p}{\partial x} \right) dx\ dz \tag{2.17}$$

$$\sum F_z = \left(-\frac{\partial p}{\partial z} - \gamma \right) dx\ dz \tag{2.18}$$

With the mass of the element equal to $(\gamma/g_n)\ dx\ dz$, the component forms of Newton's second law may be written

$$\left(-\frac{\partial p}{\partial x} \right) dx\ dz = \frac{\gamma}{g_n} a_x\ dx\ dz$$

$$\left(-\frac{\partial p}{\partial z} - \gamma \right) dx\ dz = \frac{\gamma}{g_n} a_z\ dx\ dz$$

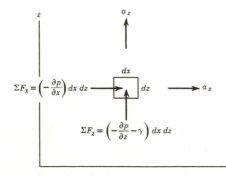

Fig. **2.15**

which reduce to

$$-\frac{\partial p}{\partial x} = \frac{\gamma}{g_n}(a_x) \tag{2.19}$$

$$-\frac{\partial p}{\partial z} = \frac{\gamma}{g_n}(a_z + g_n) \tag{2.20}$$

These equations characterize the pressure variation through an accelerated mass of fluid, and with them specific applications may be studied.

One other useful generalization can be derived from the foregoing equations, namely, a property of a line of constant pressure. By using the chain rule for the total differential for dp in terms of its partial derivations,[19]

$$dp = \frac{\partial p}{\partial x}\, dx + \frac{\partial p}{\partial z}\, dz$$

and by substituting the above expressions for $\partial p/\partial x$ and $\partial p/\partial z$, we obtain

$$dp = -\frac{\gamma}{g_n}(a_x)\, dx - \frac{\gamma}{g_n}(a_z + g_n)\, dz \tag{2.21}$$

However, along a line of constant pressure $dp = 0$ and hence, for such a line,

$$\frac{dz}{dx} = -\left(\frac{a_x}{g_n + a_z}\right) \tag{2.22}$$

Thus the *slope* (dz/dx) of a line of constant pressure is defined; its *position* must be determined from external (boundary) conditions in specific problems.

From the foregoing generalizations some situations of engineering significance may now be examined.

(a) Constant Linear Acceleration with $a_x = 0$

Here a container of liquid is accelerated vertically upward, $\partial p/\partial x = 0$, and with no change of pressure with x, equation 2.20 becomes

$$\frac{dp}{dz} = -\gamma\left(\frac{g_n + a_z}{g_n}\right)$$

For a_z constant, this equation shows that the characteristic linear pressure variation of fluid statics is preserved but that magnitudes of pressure will now depend on a_z. The quantitative aspects of this are shown in Fig. 2.16 for $a_z > 0$ and $a_z < 0$. The latter case is of particular interest when $a_z = -g_n$, yielding $dp/dz = 0$ and showing that the pressure is constant throughout a freely falling mass of fluid; for an *unconfined* mass of freely falling fluid the pressure is therefore equal to that surrounding it—if the surrounding pressure is zero, all pressures within the fluid mass will be zero, a fact which has many applications in subsequent problems.

---

[19]See Appendix 5.

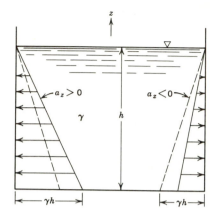

Fig. 2.16

───────────────────   **Illustrative Problem**   ───────────────────

An open tank of water is accelerated vertically upward at 4.5 m/s². Calculate the pressure at a depth of 1.5 m.

Relevant Equation and Given Data

$$\frac{dp}{dz} = -\gamma\left(\frac{g_n + a_z}{g_n}\right)$$

$$\gamma = 9.8 \times 10^3 \text{ N/m}^3 \qquad z = -1.5 \text{ m}$$

$$g_n = 9.81 \text{ m/s}^2 \qquad a_z = 4.5 \text{ m/s}^2$$

Solution.

$$\frac{dp}{dz} = -9.8 \times 10^3\left(\frac{9.81 + 4.5}{9.81}\right) = -14.3 \text{ kN/m}^3$$

$$p = \int_0^p dp = -\int_0^{-1.5} 14.3 \, dz = 21.5 \text{ kPa} \blacktriangleleft$$

(b) Constant Linear Acceleration

Here the slope of a liquid surface (which is a line of constant pressure) is given (from equation 2.22) by $-a_x/(a_z + g_n)$ as shown in Fig. 2.17, and other lines of constant pressure will be parallel to the surface. Along x and z the (linear) pressure variations may be computed from equations 2.19 and 2.20, respectively. In the direction h, normal to the lines of constant pressure, equation 2.21 may be employed. Dividing this equation by dh,

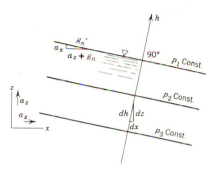

Fig. **2.17**

$$\frac{dp}{dh} = -\gamma\left(\frac{a_x\,dx}{g_n\,dh} + \frac{g_n + a_z\,dz}{g_n\,dh}\right)$$

but, from the similar triangles of Fig. 2.17,

$$dx/dh = a_x/g_n' \qquad \text{and} \qquad dz/dh = (a_z + g_n)/g_n'$$

Substituting these expressions above gives

$$dp/dh = -\gamma(g_n'/g_n)$$

which shows that the pressure variation along h is linear and allows computation of pressures as in statics (equation 2.3), $\gamma g_n'/g_n$ being used for the specific weight of the fluid.

_____ **Illustrative Problem** _____

This open tank moves up the plane with constant acceleration. Calculate the acceleration required for the water surface to move to the position indicated. Calculate the pressure in the corner of the tank at A before and after acceleration.

Relevant Equations and Given Data

$$-\frac{\partial p}{\partial z} = \frac{\gamma}{g_n}(a_z + g_n) \tag{2.20}$$

$$\frac{dz}{dx} = -\left(\frac{a_x}{g_n + a_x}\right) \tag{2.22}$$

$$g_n = 32.2 \text{ ft/s}^2 \qquad \gamma = 62.4 \text{ lb/ft}^3$$

Solution. From geometry, the slope of the water surface during acceleration is -0.229. From the slope of the plane, $a_x = 4a_z$. Using equation 2.22,

$$-0.229 = \frac{-4a_z}{(32.2 + a_z)} \tag{2.22}$$

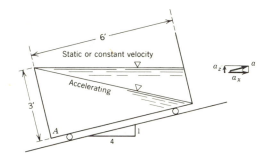

and from the foregoing

$$a_z = 1.96 \text{ ft/s}^2 \qquad a_x = 7.84 \text{ ft/s}^2 \qquad a = 8.08 \text{ ft/s}^2 \blacktriangleleft$$

From geometry, the depth of water vertically above corner A before acceleration is 2.91 ft; hence the pressure there is $2.91 \times 62.4 = 181.5$ psf. After acceleration this depth is 2.75 ft; from equation 2.20,

$$\frac{\partial p}{\partial z} = -\left(\frac{62.4}{32.2}\right)(1.96 + 32.2) = -66.0 \text{ lb/ft}^3$$

Therefore $p = 2.75 \times 66.0 = 181.5$ psf. $\blacktriangleleft$

    The fact that the pressures at A are the same before and after acceleration is no coincidence; general proof may be offered that p_A does not change whatever the acceleration. This means that the force exerted by the end of the tank on the water is constant for all accelerations. However, this is no violation of Newton's second law, since the mass of liquid diminishes with increased acceleration so that the product of mass and acceleration remains constant and equal to the applied force. $\blacktriangleleft$

(c) Centripetal Acceleration with Constant Angular Velocity about a Vertical Axis, and $a_z = 0$

Equations 2.19, 2.20, and 2.22 may be written[20] with radial distance, r, substituted for x and a_r for a_x to give (for $a_z = 0$): $-\partial p/\partial r = \gamma a_r/g_n$, $-\partial p/\partial z = \gamma$, and, for surfaces of constant pressure, $dz/dr = -a_r/g_n$. From kinematics, $a_r = -\omega^2 r$, in which ω is the angular velocity. Substituting this in the foregoing equations,

$$\frac{\partial p}{\partial r} = \frac{\gamma \omega^2 r}{g_n} \qquad (2.23)$$

$$\frac{\partial p}{\partial z} = -\gamma \qquad (2.24)$$

---

[20]For rigorous proof of the validity of this, a complete analysis should be made using the conventional element of polar coordinates; however (after discarding the negligible terms), the same result is obtained.

For surfaces of constant pressure,

$$\frac{dz}{dr} = \frac{\omega^2 r}{g_n} \qquad (2.25)$$

The pressure gradient along r and z can be computed from the first two equations; the third can be easily integrated to

$$z = \frac{\omega^2 r^2}{2g_n} + \text{Constant} \qquad (2.26)$$

showing that lines of constant pressure are parabolas (Fig. 2.18) symmetrical about the axis of rotation. The second equation shows that pressure variation in the vertical is that of fluid statics (equation 2.5) so that $p_3 - p_1 = \gamma h$. The second equation can be integrated (for $z = $ constant) from the axis of rotation (where the pressure is p_c and r is zero) to any radius r where the pressure is p. The result is

$$\frac{p - p_c}{\gamma} = \frac{\omega^2 r^2}{2g_n} \qquad (2.27)$$

which may also be deduced directly from Fig. 2.18.

The foregoing analysis shows the possibility of pressure being created by the rotation of a fluid mass. This principal is utilized in centrifugal pumps and blowers to produce a pressure difference in order to cause fluids to flow.

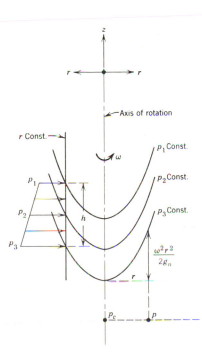

Fig. **2.18**

---------------   **Illustrative Problems**   ---------------

This tank is fitted with three piezometer columns of the same diameter. Calculate the pressure heads at points A and B when the tank is rotating at a constant speed of 100 rpm.

Relevant Equation and Given Data

$$\frac{p - p_c}{\gamma} = \frac{\omega^2 r^2}{2g_n} \qquad (2.27)$$

$$g_n = 32.2 \text{ ft/s}^2$$

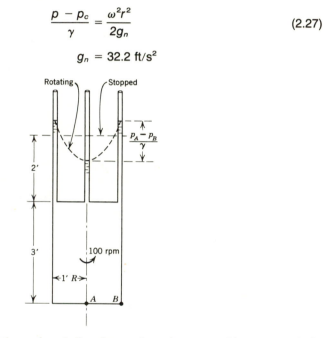

Solution.   During rotation the surface in the piezometer columns must be on a parabola of constant pressure, and liquid lost from the central column must appear in the other two. Accordingly

$$\frac{(p_B - p_A)}{\gamma} = \frac{(2\pi \times 100/60)^2(1)^2}{2g_n} = 1.71 \text{ ft} \qquad (2.27)$$

Since there are two columns at $r = 1$ ft, the level in column "A" must drop twice as far as those in columns "B" rise; therefore,

$$\frac{p_A}{\gamma} = 5 - \tfrac{2}{3}(1.71) = 3.86 \text{ ft} \blacktriangleleft$$

$$\frac{p_B}{\gamma} = 5 + \tfrac{1}{3}(1.71) = 5.57 \text{ ft} \blacktriangleleft$$

A cylinder of radius r_o contains a gas at constant temperature and rotates about its (vertical) axis at constant angular speed. Derive a relation between the pressures on the cylindrical surface and on the axis of rotation.

Relevant Equation

$$\frac{\partial p}{\partial r} = \frac{\gamma \omega^2 r}{g_n} \tag{2.23}$$

Solution.   Ignoring pressure variation in the z-direction (permissible for a gas) and writing $\partial p / \partial r = dp/dr$,

$$dp = \frac{\gamma \omega^2 r \, dr}{g_n} \tag{2.23}$$

Substituting p/RT for γ/g_n (in which T is constant), separating variables, and noting that, for $p = p_c$, $r = 0$ and, for $p = p_o$, $r = r_o$, the equation may be integrated between appropriate limits to yield

$$\ln\left(\frac{p_o}{p_c}\right) = \frac{\omega^2 r_o^2}{2RT}$$

or

$$p_o = p_c \exp\left(\frac{\omega^2 r_o^2}{2RT}\right) \blacktriangleleft$$

Many geophysical problems are concerned with flows on a rotating earth. A simple experiment of some significance involves small perturbations of a fluid in rigid body rotation. In a laboratory experiment, a circular tank of radius $r_o = 2\,\text{m}$ is filled with water to an average depth $d = 75\,\text{mm}$ and rotated with an angular speed $\omega = 5\,\text{r/min}$ about its center. Find the shape of the free surface of the water when it is in rigid body rotation.

Relevant Equation and Given Data

$$z = \frac{\omega^2 r^2}{2g_n} + \text{Constant} \tag{2.26}$$

$$d = 0.075\,\text{m} \qquad g_n = 9.81\,\text{m/s}^2$$

$$r_o = 2\,\text{m} \qquad \omega = 5\,\text{r/min}$$

Solution.   The free surface is a parabola

$$z - z_o = \omega^2 r^2 / 2g_n \tag{2.26}$$

with $z_o = $ the presently unknown depth at $r = 0$. The mass of fluid in the tank must be conserved so

$$\rho \pi r_o^2 d = \rho \int_0^{r_o} 2\pi r z \, dr = \frac{2\pi \rho \omega^2}{2g_n} \int_0^{r_o} r^3 \, dr + 2\pi \rho z_o \int_0^{r_o} r \, dr$$

by use of $z = z_o + \omega^2 r^2 / 2g_n$. Thus,

$$r_o^2 d = \frac{\omega^2 r_o^4}{4g_n} + r_o^2 z_o$$

and

$$z_o = d - \frac{\omega^2 r_o^2}{4g_n} = 0.075 - \left(\frac{2\pi \times 5}{60}\right)^2 \frac{4}{4 \times 9.81}$$

$$z_o = 0.047\,m = 47\,mm$$

Therefore,

$$z = 0.047 + 0.014r^2 \le 0.103 = 103\,mm \blacktriangleleft$$

that is, the water z ranges from 47 mm at the tank center to 103 mm at the tank wall. Compare these differences to those in the confined tank results above. Are the same principles operative?

Problems

2.1. Calculate the pressure in an open tank of crude oil at a point 8 ft or 2.4 m below the liquid surface.

2.2. If the pressure 10 ft (3 m) below the free surface of a liquid is 20 psi (140 kPa), calculate its specific weight and specific gravity.

2.3. If the pressure at a point in the ocean is 140 kPa, what is the pressure 30 m below this point? Specific weight of saltwater is 10 kN/m³.

2.4. An open vessel contains carbon tetrachloride to a depth of 6 ft (2 m) and water on the carbon tetrachloride to a depth of 5 ft (1.5 m). What is the pressure at the bottom of the vessel?

2.5. How many inches of mercury are equivalent to a pressure of 20 psi? How many feet of water?

2.6. How many millimetres of carbon tetrachloride are equivalent to a pressure of 40 kPa? How many metres of alcohol?

2.7. One vertical metre (foot) of air at 15°C (59°F) and 101.3 kPa (14.7 psia) is equivalent to how many pascals (pounds per square inch)? Millimetres (inches) of mercury? Metres (feet) of water?

2.8. The barometric pressure at sea level is 30.00 in. (762 mm) of mercury when that on a mountain top is 29.00 in. (737 mm). If specific weight of air is assumed constant at 0.075 lb/ft³ (11.8 N/m³), calculate the elevation of the mountain top.

2.9. If at the surface of a liquid the specific weight is γ_o, with z and p both zero, show that, if E = constant, the specific weight and pressure are given by

$$\gamma = \frac{E}{(z + E/\gamma_o)} \quad \text{and} \quad p = -E \ln\left(1 + \frac{\gamma_o z}{E}\right)$$

Calculate specific weight and pressure at a depth of 2 miles (2 km) assuming $\gamma_o = 64.0$ lb/ft³ (10.0 kN/m³) and $E = 300\,000$ psi (2 070 MPa).

2.10. In the deep ocean the compressibility of seawater is significant in its effect on ρ and p. If $E = 2.07 \times 10^9$ Pa, find the percentage change in the density and pressure at a depth of 10 000 metres as compared to the values obtained at the same depth under the incompressible assumption. Let $\rho_o = 1\,020$ kg/m³ and the absolute pressure $p_o = 101.3$ kPa.

2.11. The specific weight of water in the ocean may be calculated from the empirical relation $\gamma = \gamma_o + K\sqrt{h}$ (in which h is the depth below the ocean surface). Derive an expression for the pressure at any point h and calculate specific weight and pressure at a depth of 2 miles (3.22 km) assuming $\gamma_o = 64.0$ lb/ft^3 (10 kN/m^3), h in feet (metres), and $K = 0.025$ lb/ft$^{7/2}$ (7.08 N/m$^{7/2}$).

2.12. If the specific weight of a liquid varies linearly with depth below the liquid surface ($\gamma = \gamma_o + Kh$), derive an expression for pressure as a function of depth.

2.13. If atmospheric pressure at the ground is 14.7 psia (101.3 kPa) and temperature is 59°F (15°C), calculate the pressure 25 000 ft (7.62 km) above the ground, assuming (*a*) no density variation, (*b*) isothermal variation of density with pressure, and (*c*) adiabatic variation of density with pressure.

2.14. Calculate pressures and densities of air in the U.S. Standard Atmosphere at 25 000 ft and 50 000 ft or at 8 km and 16 km. Check results with the values in Appendix 4.

2.15. Calculate the depth of an adiabatic atmosphere if temperature and pressure at the ground are, respectively, 59°F (15°C) and 14.7 psia (101.3 kPa).

2.16. Derive a relation between pressure and altitude: (*a*) for an isothermal atmosphere and (*b*) for the U.S. Standard Atmosphere to altitude 35 000 ft (10.7 km).

2.17. If the temperature in the atmosphere is assumed to vary linearly with altitude so $T = T_o - \alpha z$ where T_o is the sea level temperature and $\alpha = -dT/dz$ is the temperature lapse rate, find $p(z)$ when air is taken to be a perfect gas. Give the answer in terms of p_o, α, g_n, R, and z only.

2.18. Show that the temperature lapse rate in an adiabatic atmosphere is the reciprocal of the specific heat at constant pressure.

2.19. Assuming a linear rise of temperature in the U.S. Standard Atmosphere between altitudes of 85 000 and 100 000 ft or 26 and 30 km, calculate the value of the polytropic exponent n over this range of altitude.

2.20. Find the height of a static, perfect gas atmosphere in which the temperature decreases linearly with altitude as $T = T_o - \alpha z$. Take the height to be that altitude where $\rho(z) = 0$.

2.21. A mass of warm (64°F or 17.8°C), moist air is swept by the wind from sea level up the side of a coastal mountain range to the top at an altitude of 2 500 ft (762 m). Assuming the rise is adiabatic in an otherwise standard atmosphere, will the warm air continue to rise, be neutrally buoyant, or move down the leeward slope of the range? Why?

2.22. With atmospheric pressure at 14.5 psia (100 kPa, abs.), what absolute pressure corresponds to a gage pressure of 20 psi (138 kPa)?

2.23. When the barometer reads 30 in. (762 mm) of mercury, what absolute pressure corresponds to a vacuum of 12 in. (305 mm) of mercury?

2.24. If a certain absolute pressure is 85.2 kPa (12.35 psia), what is the corresponding vacuum if the atmospheric pressure is 760 mm (29.92 in.) of mercury?

2.25. A Bourdon pressure gage attached to a closed tank of air reads 20.47 psi (141.1 kPa) with the barometer at 30.50 in. (775 mm) of mercury. If barometric pressure drops to 29.18 in. (741.2 mm) of mercury, what will the gage read?

2.26. A Bourdon gage is connected to a tank in which the pressure is 40.0 psi (276 kPa) above atmospheric at the gage connection. If the pressure in the tank remains unchanged but the gage is placed in a chamber where the air pressure is reduced to a vacuum of 25 in. (635 mm) of mercury, what gage reading will be expected?

2.27. The compartments of these tanks are closed and filled with air. Gage A reads 207 kPa. Gage B registers a vacuum of 254 mm of mercury. What will gage C read if it is connected to compartment 1 but inside compartment 2? Barometric pressure is 101 kPa, absolute.

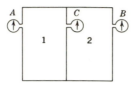

Problem **2.27**

2.28. If the barometer of Fig. 2.5 is filled with a silicon oil of specific gravity 0.86, calculate h if the barometric (absolute) pressure is 101.3 kPa or 14.7 psia. Is this a practical barometer?

2.29. Calculate the height of the column of a water barometer for an atmospheric pressure of 14.70 psia when the water is at 50°F, 150°F, and 212°F.

2.30. Barometric pressure is 29.43 in. (758 mm) of mercury. Calculate h.

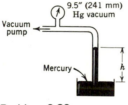

Problem **2.30**

2.31. Calculate the pressure p_x in Fig. 2.7a if $l = 760$ mm, $h = 500$ mm; liquid γ is water, and γ_1 mercury.

2.32. With the manometer reading as shown, calculate p_x.

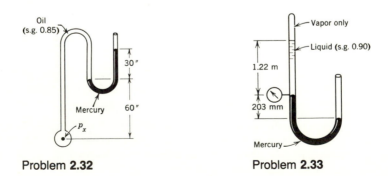

Problem **2.32**   Problem **2.33**

2.33. Barometric (absolute) pressure is 91kPa. Calculate the vapor pressure of the liquid and the gage reading.

2.34. In Fig. 2.7b, $l_1 = 1.27$ m, $h = 0.51$ m, $l_2 = 0.76$ m, liquid γ_1 is water, γ_2 benzene, and γ_3 mercury. Calculate $p_x - p_y$.

2.35. Calculate $p_x - p_y$ for this inverted U-tube manometer.

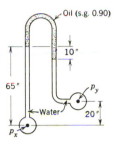

Problem 2.35

2.36. An inclined gage (Fig. 2.7c) having a tube of 3 mm bore, laid on a slope of 1: 20, and a reservoir of 25 mm diameter contains silicon oil (s.g. 0.84). What distance will the oil move along the tube when a pressure of 25mm of water is connected to the gage?

2.37. The meniscus between the oil and water is in the position shown when $p_1 = p_2$. Calculate the pressure difference ($p_1 - p_2$) which will cause the meniscus to rise 2 in.

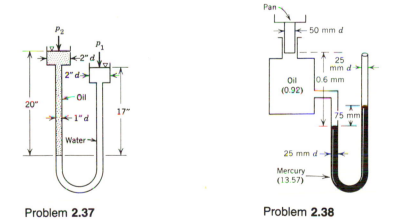

Problem 2.37

Problem 2.38

2.38. Predict the manometer reading after a 1 N weight is placed on the pan. Assume no leakage or friction between piston and cylinder.

2.39. Calculate the gage reading.

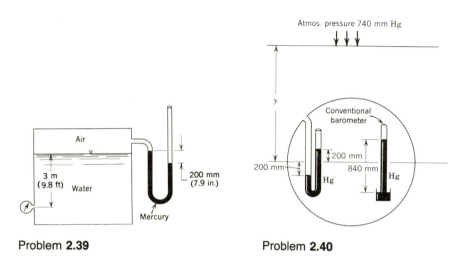

Atmos. pressure 740 mm Hg

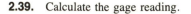

Problem **2.39**   Problem **2.40**

2.40. The sketch shows a sectional view through a submarine. Calculate the depth of submergence, y. Assume the specific weight of seawater is 10.0 kN/m³.

2.41. Calculate the gage reading. Specific gravity of the oil is 0.85. Barometric pressure is 755 mm of mercury.

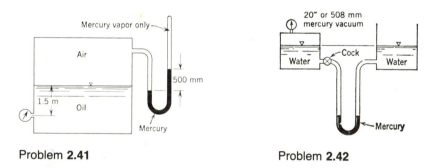

Problem **2.41**   Problem **2.42**

2.42. Calculate magnitude and direction of manometer reading when the cock is opened. The tanks are very large compared to the manometer tubes.

2.43. The manometer reading is 6 in. (150 mm) when the tank is empty (water surface at A). Calculate the manometer reading when the tank is filled with water.

2.44. Barometric pressure is 28 in. or 711 mm of mercury. The cock is opened and the air space pumped out so that the gage reads 20 in. or 508 mm of mercury vacuum. Calculate the absolute pressure in the tank and the manometer reading. Neglect change of water surface in the tank.

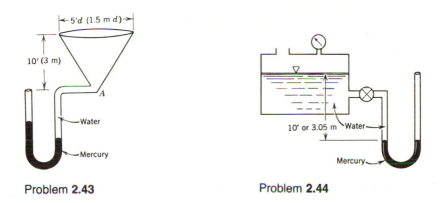

Problem **2.43**　　　　　　　　Problem **2.44**

2.45.　This manometer is used to measure the difference in water level between the two tanks. Calculate this difference.

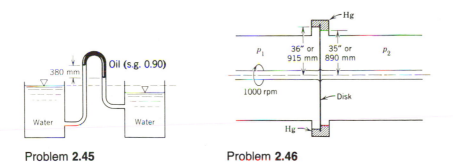

Problem **2.45**　　　　　　　　Problem **2.46**

2.46.　The mercury seal shown is used to support a pressure difference $(p_1 - p_2)$ across the rotating disk. Calculate this pressure difference for a speed of 1 000 rpm (r/min), assuming that the mercury rotates with the disk.

2.47.　Calculate the magnitude of the force on a 24 in. (0.61 m) diameter glass viewing port in a bathyscaphe on the floor of the Pacific Ocean's Marianas Trench (depth = 35 800 feet or 10.9 km).

2.48.　A rectangular gate 6 ft long and 4 ft high lies in a vertical plane with its center 7 ft below a water surface. Calculate magnitude, direction, and location of the total force on the gate.

2.49.　A circular gate 3 m in diameter has its center 2.5 m below a water surface and lies in a plane sloping at 60°. Calculate magnitude, direction, and location of total force on the gate.

2.50.　An isosceles triangle of 12 ft base and 15 ft altitude is located in a vertical plane. Its base is vertical, and its apex is 8 ft below the water surface. Calculate magnitude and location of the force of the water on the triangle.

2.51.　A triangular area of 2 m base and 1.5 m altitude has its base horizontal and lies in a 45° plane with its apex below the base and 2.75 m below a water surface. Calculate magnitude, direction, and location of the resultant force on this area.

2.52. If the specific weight of a liquid varies linearly with depth h according to the equation $\gamma = \gamma_o + Kh$, derive expressions for resultant force per unit width on the rectangular gate and the moment of this force about O.

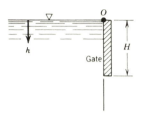

Problem **2.52**

2.53. A square 9 ft by 9 ft (2.75 m × 2.75 m) lies in a vertical plane. Calculate the distance between the center of pressure and centroid, and the total force on the square, when its upper edge is (a) in the water surface, and (b) 50 ft (15 m) below the water surface.

2.54. Calculate the x- and y-coordinates of the center of pressure of this vertical right triangle.

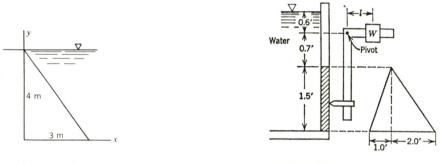

Problem **2.54**   Problem **2.55**

2.55. Calculate ($W \times l$) and the exact position of the pointer to hold this triangular gage in equilibrium.

2.56. Calculate magnitude and location of the resultant force of water on this annular gate.

2.57. Calculate magnitude and location of the resultant force of the liquid on this tunnel plug.

2.58. Calculate the force exerted by water on the largest completely submerged circle (located in a vertical plane) having a distance of 1 ft or 1 m between its centroid and center of pressure.

2.59. A vertical rectangular gate 10 ft (3 m) high and 6 ft (1.8 m) wide has a depth of water on its upper edge of 15 ft (4.5 m). What is the location of a horizontal line which divides this area (a) so that the forces on the upper and lower portions are the same, and (b) so that the moments of the forces about the line are the same?

2.60. A horizontal tunnel of 10 ft (3 m) diameter is closed by a vertical gate. Calculate magnitude, direction, and location of the total force of water on the gate when the tunnel is (*a*) one-half full, (*b*) one-fourth full, and (*c*) three-fourths full.

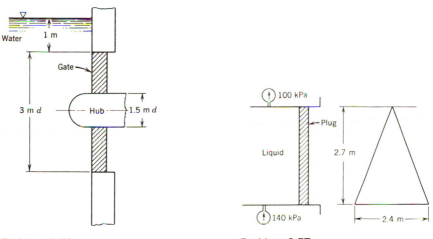

Problem **2.56**   Problem **2.57**

2.61. Calculate the magnitude, direction, and location of the force of the water on one side of this area located in a vertical plane.

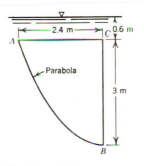

Problem **2.61**

2.62. A vertical rectangular gate 2.4 m wide and 2.7 m high is subjected to water pressure on one side, the water surface being at the top of the gate. The gate is hinged at the bottom and is held by a horizontal chain at the top. What is the tension in the chain?

2.63. A sliding gate 10 ft wide and 5 ft high situated in a vertical plane has a coefficient of friction, between itself and guides, of 0.20. If the gate weighs 2 tons and if its upper edge is at a depth of 30 ft, what vertical force is required to raise it? Neglect buoyancy force on the gate.

2.64. A butterfly valve, consisting essentially of a circular area pivoted on a horizontal axis through its center (and in the plane of the valve), is 2.1 m (6.9 ft) in diameter and lies in a 60° plane with its center 3 m (9.8 ft) below a water surface. What torque must be exerted on the valve's axis to just open it?

2.65. Calculate the h at which this gate will open.

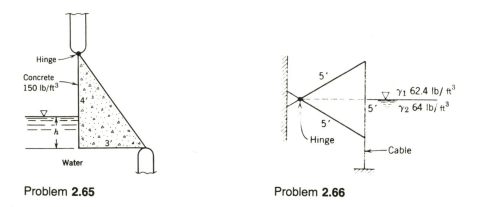

Problem 2.65 **Problem 2.66**

2.66. A solid homogeneous wooden block, with $\gamma = 50$ lb/ft^3, is 5 ft long normal to the paper. The block is anchored by a cable. Calculate the tensile force in the cable.

2.67. These (rectangular) miter gates at the entrance to a canal lock are 4.5 m high. Calculate the reactions at the hinges when the water surface is 0.9 m below the top of the gates.

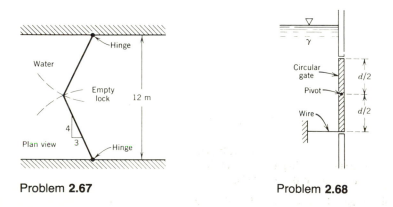

Problem 2.67 **Problem 2.68**

2.68. Derive an algebraic expression for the force in the wire.

2.69. The flashboards on a spillway crest are 1.2 m high and supported on steel posts spaced 0.6 m on centers. The posts are designed to fail under a bending moment of 5 500 N · m. What depth over the flashboards will cause the posts to fail? Assume hydrostatic pressure distribution.

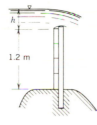

Problem **2.69**

2.70. This rectangular gate will open automatically when the depth of water, d, becomes large enough. What is the minimum depth that will cause the gate to open?

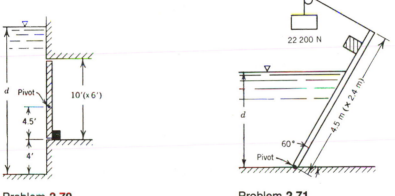

Problem **2.70**   Problem **2.71**

2.71. What depth of water will cause this rectangular gate to fall? Neglect the weight of the gate.

2.72. Calculate magnitude and location of the total force on one side of this vertical plane area.

Problem **2.72**   Problem **2.73**

2.73. Calculate magnitude and location of the total force on one side of this vertical plane area.

2.74. A rectangle 3 m by 4 m lies in a vertical plane with one diagonal horizontal and 7 m below the water surface. Calculate magnitude and location of total force on the rectangle.

2.75. This area lies in a vertical plane beneath a water surface. Calculate magnitude and location of the force of the water on the area. Compare results with those of the preceding problem.

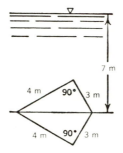

Problem **2.75**

2.76. Calculate the magnitude, direction, and location of the total force on the gate of problem 2.60 when on one side the water surface is 18 ft (5.5 m) above the tunnel invert and on the other side 5 ft (1.5 m) above the tunnel invert. (The "invert" is the low point of the tunnel cross section.)

2.77. A vertical gate in a tunnel 6 m in diameter has water on one side and air on the other. The water surface is 10.5 m above the invert, and the air pressure is 110 kPa. Where could a single support be located to hold this gate in position?

2.78. A rectangular tank 5 ft wide, 6 ft high, and 10 ft long contains water to a depth of 3 ft and oil (s.g. = 0.85) on the water to a depth of 2 ft. Calculate magnitude and location of the force on one end of the tank.

2.79. Using force components, calculate the load in the strut AB if these struts have 1.5 m spacing along the small dam AC? Consider all joints to be pin connected.

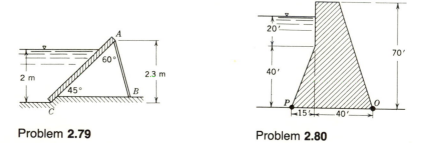

Problem **2.79**   Problem **2.80**

2.80. Using the method of components, calculate the magnitude, direction, and location of the total force on the upstream face of a section of this dam 1 ft wide. What is the moment of this force about O?

2.81. Using hydrostatic principles (not geometry or calculus), calculate the volume $BCDE$ (m³).

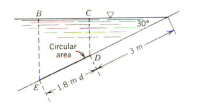

Problem **2.81**

2.82. Calculate the magnitude of the total force in problem 2.51 by the method of components.

2.83. A concrete pedestal, having the shape of the frustum of a right pyramid of lower base 4 ft square, upper base 2 ft square, and height 3 ft, is to be poured. Taking γ for concrete to be 150 lb/ft³, calculate the vertical force of uplift on the forms.

2.84. This tainter gate is pivoted at O and is 10 m long. Calculate the magnitudes of horizontal and vertical components of force on the gate. The pivot is at the same level as the water surface.

Problem **2.84**   Problem **2.85**

2.85. The quarter cylinder AB is 10 ft long. Calculate magnitude, direction, and location of the resultant force of the water on AB.

2.86. Calculate the vertical force exerted by the liquid on this semicylindrical dome AB, which is 1.5 m long.

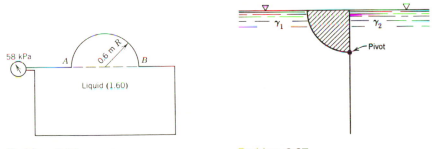

Problem **2.86**   Problem **2.87**

2.87. If this weightless quarter-cylindrical gate is in static equilibrium, what is the ratio between γ_1 and γ_2?

2.88. Calculate the moment about O of the resultant force exerted by the water on this half cylinder, which is 3 m (10 ft) long.

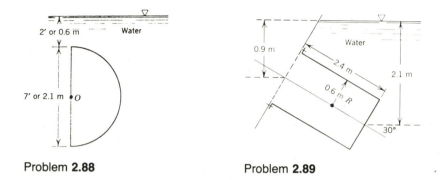

Problem **2.88**   Problem **2.89**

2.89. Calculate magnitude and direction of the resultant forces exerted by the water (a) on the end of the cylinder and (b) on the curved surface of the cylinder.

2.90. Solve problem 2.85, assuming that there is also water to the right of AB with a surface level 3 ft above A.

2.91. The cylinder is 2.4 m long and is pivoted at O. Calculate the moment (about O) required to hold it in position.

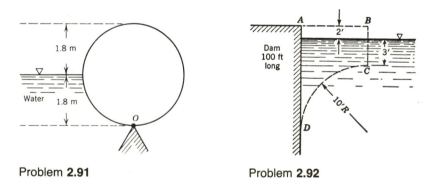

Problem **2.91**   Problem **2.92**

2.92. If this solid concrete (150 lb/ft³) overhang $ABCD$ is added to the dam, what *additional* force (magnitude and direction) will be exerted on the dam?

2.93. If the liquid shown above B in problem 2.85 consists of a 7 ft layer of water on top of a 6 ft layer of carbon tetrachloride, calculate magnitude, direction, and location of horizontal and vertical components of force on AB.

2.94. A hemispherical shell 1.2 m in diameter is connected to the vertical wall of a tank containing water. If the center of the shell is 1.8 m below the water surface, what are the vertical and horizontal force components on the shell? On the top half of the shell?

2.95. This half-conical buttress is used to support a half-cylindrical tower on the upstream face of a dam. Calculate the magnitude, direction, and location of the vertical and horizontal components of force exerted by the water on the buttress (*a*) when the water surface is at the base of the half cylinder, and (*b*) when it is 4 ft (1.2 m) above this point.

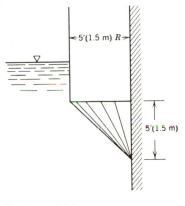

5′(1.5 m) R→

5′(1.5 m)

Problem **2.95**

2.96. A hole 300 mm in diameter in a vertical wall between two water tanks is closed by a sphere 450 mm in diameter in the tank of higher water-surface elevation. The difference in the water-surface elevations in the two tanks is 1.5 m. Calculate the horizontal component of force exerted by the water on the sphere.

2.97. What pressure difference $p_1 - p_2$ is required to open the ball valve if the spring exerts a force of 400 N? The ball is 50 mm in diameter, while the hole in which it rests is 30 mm in diameter. Neglect the weight of the steel ball.

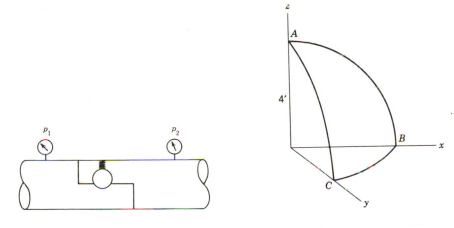

Problem **2.97**

Problem **2.98**

2.98. This one-eighth wooden sphere of specific weight 50 lb/ft³ is placed in the corner of an open rectangular tank, the edges of which coincide with the *x*, *y*, and *z* axes. The joints along the lines *AB, BC,* and *AC* are sealed perfectly so that water cannot enter. Water is then poured into the tank to a depth of 7 ft. Calculate the magnitude, direction, and location of the resultant force (or its components) exerted by the water on the spherical surface.

2.99. If the half cone of problem 2.95 is replaced by a quarter sphere, what are the magnitudes of the vertical and horizontal force components on the buttress?

2.100. This weightless spherical shell with attached *small* piezometer tube is suspended by a cable as shown. Calculate the total tension force in the cable. Also calculate the force by the liquids: (*a*) on the bottom half of the sphere, and (*b*) on the top half of the sphere.

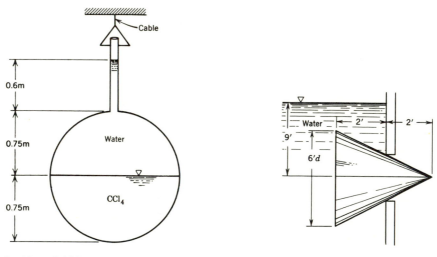

Problem **2.100**   Problem **2.101**

2.101. Calculate magnitude and direction of the resultant force of the water on this solid conical plug.

2.102. The tank is an elliptic cylinder of 2.4 m length. Calculate magnitude and direction of the vertical and horizontal components of force on *AB*.

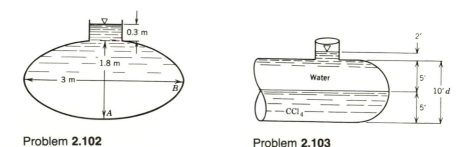

Problem **2.102**   Problem **2.103**

2.103. Calculate the magnitude and location of the resultant force of the liquids on the hemispherical end of this cylindrical tank.

2.104. A stone weighs 60 lb or 267 N in air and 40 lb or 178 N in water. Calculate its volume and specific gravity.

2.105. Ninety-eight millilitres of lead (s.g. 11.4) are suspended from the apex of a conical can having a height of 0.3 m and a base of 0.15 m diameter, and weighing 4 N. When placed in water, to what depth will the can be immersed? The apex is below the base.

2.106. A cylindrical can 76 mm in diameter and 152 mm high, weighing 1.11 N, contains water to a depth of 76 mm. When this can is placed in water, how deep will it sink?

2.107. The timber weighs 40 lb/ft³ and is held in a horizontal position by the concrete (150 lb/ft³) anchor. Calculate the minimum total weight which the anchor may have.

Problem **2.107**　　　　　　　　　　　　　　　　Problem **2.108**

2.108. If the timber weighs 670 N, calculate its angle of inclination when the water surface is 2.1 m above the pivot. Above what depth will the timber stand vertically?

2.109. This homogeneous wooden semicylinder is 3.6 m long and floats on a water surface as shown. What moment, M, would be required to move point A to coincide with the water surface?

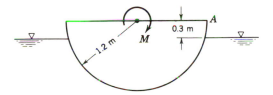

Problem **2.109**

2.110. The barge shown weighs 40 tons and carries a cargo of 40 tons. Calculate its draft in freshwater.

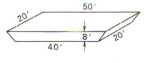

Problem **2.110**

2.111. A modern, half-million tonne (10³ kg) supertanker is roughly rectangular in shape, with a length of 425 m, a width of 67 m, and a draft when fully loaded of 26 m. If the tanker carries 600 Ml (megalitres) of oil (s.g. 0.86) when fully loaded, what is its draft when empty? The specific gravity of seawater is 1.03.

2.112. The weightless sphere of diameter d is in equilibrium in the position shown. Calculate d as a function of γ_1, h_1, γ_2, and h_2.

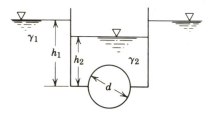

Problem **2.112**

2.113. To what depth will a rigid homogeneous object ($\gamma = 10.4$ kN/m³) sink in the ocean if the weight density therein varies with depth according to the empirical relation $\gamma(\text{kN/m}^3) = 10.0 + 0.008\,1\,\sqrt{\text{depth (m)}}$.

2.114. A balloon having a total solid weight of 800 lb contains 15 000 ft³ of hydrogen. How many pounds of ballast are necessary to hold the balloon on the ground? Barometric pressure is 14.7 psia; temperature of air and hydrogen, 60°F. Assume hydrogen to be at barometric pressure.

2.115. A balloon has a weight (including crew but not gas) of 2.2 kN and a gas-bag capacity of 566 m³. At the ground it is (partially) inflated with 445 N of helium. How high can this balloon rise in the U.S. Standard Atmosphere (Appendix 4) if the helium always assumes the pressure and temperature of the atmosphere?

2.116. An open cylindrical container holding 1.0 m³ of water at a depth of 2 m is accelerated vertically upward at 6 m/s². Calculate the pressure and total force on the bottom of the container. Also calculate this total force by application of Newton's second law.

2.117. Calculate the total forces on the ends and bottom of this container while at rest and when being accelerated vertically upward at 3 m/s². The container is 2 m wide.

Problem **2.117**

2.118. An open conical container 6 ft (1.8 m) high is filled with water and moves vertically downward with a deceleration of 10 ft/s² (3 m/s²). Calculate the pressure at the bottom of the container.

2.119. A rectangular tank 1.5 m wide, 3 m long, and 1.8 m deep contains water to a depth of 1.2 m. When it is accelerated horizontally at 3 m/s² in the direction of its length, calculate the depth of water at each end of the tank and the total force on each end of the tank. Check the difference between these forces by calculating the inertia force of the accelerated mass. Repeat these calculations for an acceleration of 6 m/s².

2.120. A closed rectangular tank 4 ft high, 8 ft long, and 5 ft wide is three-fourths full of gasoline and the pressure in the air space above the gasoline is 20 psi. Calculate the pressures in the corners of this tank when it is accelerated horizontally along the direction of its length at 15 ft/s². Using Newton's second law, calculate the forces on the ends of the tank and check their difference.

2.121. An open container of liquid accelerates down a 30° inclined plane at 16.4 ft/s² or 5 m/s². What is the slope of its free surface? What is the slope for the same acceleration up the plane?

2.122. This U-tube containing water is accelerated horizontally to the right at 3 m/s²; what are the pressures at A, B, and C? Repeat the calculation for mercury. Assume that the tubes are long enough that no liquid is spilled.

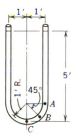

Problem **2.122**

2.123. An open cylindrical tank 1 m in diameter and 1.5 m deep is filled with water and rotated about its axis at 100 r/min. How much liquid is spilled? What are the pressures at the center of the bottom of the tank and at a point on the bottom 0.3 m from the center? What is the resultant force exerted by the water on the bottom of the tank?

2.124. The tank of problem 2.123 contains water to a depth of 0.9 m. What will be the depth at the wall of the tank when the tank is rotated at 60 r/min?

2.125. The tank of problem 2.123 contains water to a depth of 0.3 m. At what speed must it be rotated to uncover a bottom area 0.3 m in diameter?

2.126. A vertical cylindrical tank 1.5 m high and 0.9 m in diameter is filled with water to a depth of 1.2 m. The tank is then closed and the pressure in the space above the water raised to 69 kPa. Calculate the pressure at the intersection of wall and tank bottom when the tank is rotated about a central vertical axis at 150 r/min. Also calculate the resultant force exerted by the water on the bottom of the tank.

2.127. The impeller of a closed filled centrifugal water pump is rotated at 1 750 r/min (or rpm). If the impeller is 1 m (3.3 ft) in diameter, what pressure is developed by rotation?

2.128. When the U-tube of problem 2.122 is rotated at 200 rpm (r/min) about its central axis, what are the pressures at points A, B, and C?

3
KINEMATICS OF FLUID MOTION

The objectives of this chapter are to treat the kinematics of somewhat idealized fluid motion along streamlines and in flowfields and to introduce the important concept of the control volume. As in particle mechanics, kinematics describes motion in terms of displacements, velocities, and accelerations without regard to the forces that cause the motion. However, no attempt is made here to describe the kinematics of turbulence or of the motion of large-scale eddies which the reader has no doubt observed in real fluid flows; these topics are discussed later.

3.1 Steady and Unsteady Flow, Streamlines, and Streamtubes

There are two basic means of describing the motion of a fluid. In the *Lagrangian* view, each fluid particle is labeled (usually by its spatial coordinates at some initial time). Then, the path, density, velocity, and other characteristics of each *individual particle* are traced as time passes. This view is that used in the dynamic analyses of solid particles. If the position of a fluid particle is plotted as a function of time, the result is the trajectory of the particle, called a *path line*. In the *Eulerian* view, attention is focused on *particular points* in the space filled by the fluid. A description is given of the state of fluid motion at each point as a function of time. The motion of individual fluid particles is no longer traced, but the values and variations with time of the velocity, density, and other fluid variables are determined at various spatial points.

The Eulerian view is practical for most engineering problems (indeed, it is used in a great majority of fluid analyses) and is adopted for this introductory text. In the Eulerian view it is easy to determine whether the fluid flow is steady or unsteady. In unsteady flow the fluid variables will vary with time at the spatial points in the flow. In a steady flow, none of the variables at any point in a flow changes with time, although the variables generally are functions of position in the space filled by the fluid. *Thus, in the Eulerian view, a steady flow still can have accelerations.*

For example, in the pipe of Fig. 3.1, leading from an infinite[1] reservoir of fixed surface elevation, unsteady flow exists while the valve A is being opened or closed; with the valve opening fixed, steady flow occurs—under the former condition, pressures, velocities, and the like, vary with time and location; under the latter they may vary only with location. Problems of steady flow are more elementary than those of

---

[1] To preserve steady flow with a finite reservoir the same flowrate must be supplied to the reservoir as flows out of the pipe.

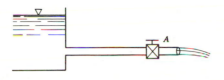

Fig. **3.1**

unsteady flow and have wide engineering application; *therefore, unsteady flow situations are not included in this elementary textbook.*

If curves are drawn in a steady flow in such a way that the tangent at any point is in the direction of the velocity vector at that point, such curves are called *streamlines.* Individual fluid particles must travel on paths whose tangent is always in the direction of the fluid velocity at any point. Thus, Lagrangian path lines are the same as Eulerian streamlines in steady flows. (But not in unsteady flows; why?)

The sketching or plotting of streamlines produces a *streamline picture* or flowfield (Fig. 3.2*a*). Streamline pictures are of both qualitative and quantitative value to the engineer. They allow him to visualize fluid flow through mathematical and experimental determination (Fig. 3.2*b*) of the streamlines and to locate regions of high and low velocity and, from these, zones of high and low pressure.

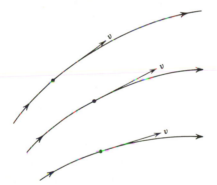

Fig. **3.2a**

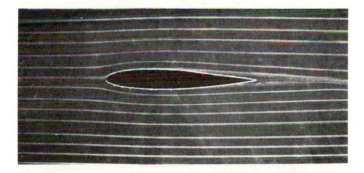

Fig. **3.2b** Steady flow past an airfoil in a smoke tunnel (streamlines are created by introducing small jets of smoke at a number of upstream points in the flow).

When streamlines are drawn through a closed curve (Fig. 3.3) in a steady flow, they form a boundary across which fluid particles cannot pass because the velocity is always tangent to the boundary. Thus the space between the streamlines becomes a tube or passage called a *streamtube,* and such a tube may be treated as if isolated from the adjacent fluid. The use of the streamtube concept broadens the application of fluid-flow principles; for example, it allows treating two apparently different problems such as flow in a passage and flow about an immersed object with the same laws. Also, because a streamtube of differential size essentially coincides with its axis (which is a stream-line), it is to be expected that many of the equations developed for a small streamtube will apply equally well to a streamline.

3.2   One-, Two-, and Three-Dimensional Flows

In a *one-dimensional* flow the change of fluid variables (velocity, pressure, etc.) perpendicular to (across) a streamline is negligible compared to the change along the streamline. In practice for a streamtube of finite cross-sectional area this means all fluid properties are considered uniform over any cross section. Pipe flow is usually taken to be one-dimensional and average fluid properties are used at each section. The flow in a streamtube of differential size is precisely one-dimensional because variations across the tube vanish in this limit (as the area approaches zero). Thus, flow along individual streamlines (however curved) is one-dimensional (the dimension being measured along the streamline). The concept of one-dimensional flow is an extremely powerful and practical one that produces simplicity in analysis and accurate engineering results for a wide range of problems. Later it is shown that some two- and three-dimensional flows may be treated very effectively for engineering purposes as one-dimensional in zones where the streamlines of the flow picture are all essentially straight and parallel.

Two- and three-dimensional flow pictures describe *flowfields,* the former when the flow is completely defined by the streamlines in a single plane, the latter in (three-dimensional) space. Examples of two-dimensional flows are shown over the weir and

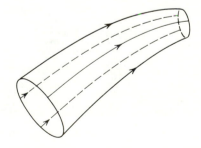

Fig. **3.3**

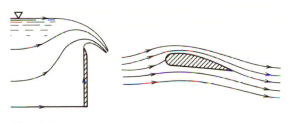

Fig. **3.4**

about the wing of Fig. 3.4. Here the velocities, pressures, and the like, vary throughout the flowfield and thus are functions of position in the field. Such two-dimensional flows are approximations to reality in that they are strictly correct only to the extent that end effects on weir and wing are negligible; this may also be visualized by assuming weir and wing to be infinitely long perpendicular to the plane of the paper. To the extent that such approximations are valid, the flow is completely described by a streamline picture drawn in a single plane. In practice, it is often possible to introduce simple corrections to the two-dimensional results to account for end effects.

Two axisymmetric three-dimensional flows are depicted in Fig. 3.5. Here the streamlines are really stream surfaces and the streamtubes are of annular cross section. On planes passed through the axis of such flows, streamline pictures may be drawn which superficially resemble two-dimensional flows; such pictures may be used fruitfully for streamline visualization, but they do not represent a reduction of a three-dimensional problem to a two-dimensional one because the mathematical descriptions of two-dimensional and axisymmetric flows are *not* the same. Nonaxisymmetric flows such as that over the fuselage and into the air inlets of a single-jet aircraft are *three-dimensional flowfields* of the most general character. These flows are more difficult to visualize and most difficult to predict. To generalize the kinematics and dynamics of the flowfield so that derived equations allow all possible flow configurations leads to mathematical complexities far beyond the scope of an elementary treatment of fluid mechanics and would obscure the physical picture so essential to a real understanding of the problem. Accordingly the treatment hereafter is restricted primarily to one-, two-, and axisymmetric three-dimensional problems.

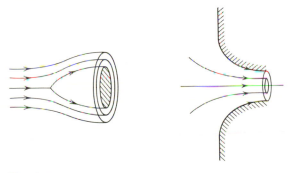

Fig. **3.5**

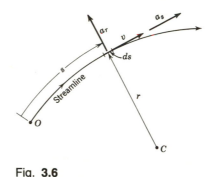

Fig. **3.6**

3.3 Velocity and Acceleration

Velocity and acceleration are vector quantities, having both magnitude and direction. However, often the direction is known or assumed and only the magnitude is to be determined. Then, the scalar components in particular directions are used. In this text most equations and analyses are in terms of scalar components, although vectors are used where they make physical principles more understandable.

For one-dimensional flow along a streamline (Fig. 3.6), velocity and acceleration may be readily defined from past experience in engineering mechanics with the motion of single particles. Select a fixed point O as a reference point and define the displacement s of a fluid particle along the streamline in the direction of motion. In time dt the particle will cover a differential distance ds along the streamline. The velocity magnitude v of this particle over the distance ds is given by $v = ds/dt$; the velocity vector is, of course, tangent to the streamline at s according to the definition of a streamline. Components of acceleration along (tangent to) and across (normal to) the streamline at s may also be written:

$$a_s = \frac{d^2s}{dt^2} = \frac{d}{dt}\left(\frac{ds}{dt}\right) = \frac{dv}{dt} = \frac{ds}{dt}\frac{dv}{ds} = v\frac{dv}{ds} \tag{3.1}$$

and, from particle mechanics,

$$a_r = -\frac{v^2}{r} \tag{3.2}$$

in which r is the radius of curvature of the streamline at s.

————————— **Illustrative Problems** —————————

Along the straight streamline shown, the velocity is given by $v = 3\sqrt{x^2 + y^2}$. Calculate the velocity and acceleration at the point (8, 6).

Relevant Equations and Given Data

$$a_s = \frac{dv}{dt} = \frac{dv}{ds}\frac{ds}{dt} = v\frac{dv}{ds} \qquad (3.1)$$

$$a_r = -v^2/r \qquad (3.2)$$

$r = \infty$ because streamline is straight

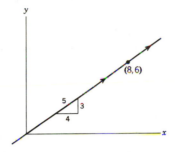

Solution.   Since $s = \sqrt{x^2 + y^2}$, $v = 3s$.
   At (8, 6), $s = 10$; therefore $v = 3 \times 10 = 30$ m/s.

$$a_s = (3s)3 = 9s = 90 \text{ m/s}^2 \blacktriangleleft \qquad (3.1)$$

Obviously a_r is zero because the radius of curvature of the streamline is infinite. ◀

    In the geophysical experiment described in Section 2.7, the fluid at the wall of the tank moves along the circular streamline shown with a constant tangential velocity component of 1.04 m/s. Calculate the tangential and radial components of acceleration at any point on the streamline.

Relevant Equations and Given Data

$$a_r = -v^2/r \qquad (3.2)$$

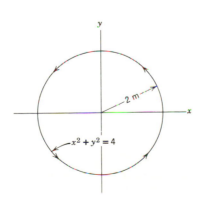

$$a_s = dv/dt \tag{3.1}$$

$$r = 2 \text{ m} \qquad v = 1.04 \text{ m/s}$$

Solution. Because the velocity is constant $a_s = 0$. As the radius of the streamline is 2 m and the velocity along it is 1.04 m/s,

$$a_r = \frac{(1.04)^2}{2} = 0.541 \text{ m/s}^2 \blacktriangleleft \tag{3.2}$$

directed toward the center of the circle.

In flowfields velocity and acceleration are somewhat more difficult to define because a generalization is required which is applicable to the whole flowfield. Consider a steady two-dimensional flow in the $x - y$ plane of Fig. 3.7. In general, the velocities are everywhere different in magnitude and direction at different points in the flowfield. At each point, however, each velocity has components u and v parallel to the x- and y-axes, respectively. If the velocities depend on x and y, their components are also functions of x and y. Written mathematically in the Eulerian view,[2]

$$u = u(x, y) \qquad \text{and} \qquad v = v(x, y)$$

In terms of displacement and time (the Lagrangian view), however,

$$u = \frac{dx}{dt} \qquad \text{and} \qquad v = \frac{dy}{dt}$$

where here x and y are the actual coordinates of a fluid particle. Fortunately, the velocity at a point is the same in both the Eulerian and the Lagrangian view. Acceler-

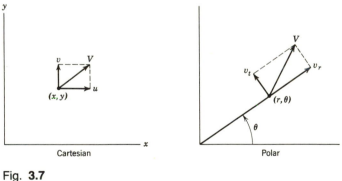

Fig. 3.7

---

[2] In general, the velocities at a point would be functions of time, so $u = u(x, y, t)$ and $v = v(x, y, t)$. However, only *steady* flows are considered herein, so time t does not appear in the definitions of u and v.

ations a_x and a_y are

$$a_x = \frac{du}{dt} \quad \text{and} \quad a_y = \frac{dv}{dt} \tag{3.3}$$

Writing the differentials du and dv in terms of partial derivatives (see Appendix 5),

$$du = \frac{\partial u}{\partial x} dx + \frac{\partial u}{\partial y} dy \quad \text{and} \quad dv = \frac{\partial v}{\partial x} dx + \frac{\partial v}{\partial y} dy$$

Substituting these relations in equations 3.3 and recognizing the velocities u and v as dx/dt and dy/dt, respectively,

$$a_x = u \frac{\partial u}{\partial x} + v \frac{\partial u}{\partial y} \quad \text{and} \quad a_y = u \frac{\partial v}{\partial x} + v \frac{\partial v}{\partial y} \tag{3.4}$$

A similar analysis for polar coordinates, in which v_r and v_t are both functions of r and θ (Fig. 3.7), leads to

$$v_r = \frac{dr}{dt} \quad \text{and} \quad v_t = r \frac{d\theta}{dt} \tag{3.5}$$

and, for the components of acceleration in a steady flow,

$$a_r = v_r \frac{\partial v_r}{\partial r} + v_t \frac{\partial v_r}{r \partial \theta} - \frac{v_t^2}{r} \tag{3.6}$$

$$a_t = v_r \frac{\partial v_t}{\partial r} + v_t \frac{\partial v_t}{r \partial \theta} + \frac{v_r v_t}{r} \tag{3.7}$$

Illustrative Problem

For the circular streamline described in the previous problem along which the velocity is 1.04 m/s, calculate the horizontal, vertical, tangential, and normal components of the velocity and acceleration at the point P (2, 60°).

Relevant Equations and Given Data

$$a_x = u \frac{\partial u}{\partial x} + v \frac{\partial u}{\partial y} \quad \text{and} \quad a_y = u \frac{\partial v}{\partial x} + v \frac{\partial v}{\partial y} \tag{3.4}$$

$$v_r = \frac{dr}{dt} \quad \text{and} \quad v_t = r \frac{d\theta}{dt} \tag{3.5}$$

$$a_r = v_r \frac{\partial v_r}{\partial r} + v_t \frac{\partial v_r}{r \partial \theta} - \frac{v_t^2}{r} \tag{3.6}$$

$$a_t = v_r \frac{\partial v_t}{\partial r} + v_t \frac{\partial v_t}{r \partial \theta} + \frac{v_r v_t}{r} \tag{3.7}$$

$$v_t = 1.04 \text{ m/s} \qquad r = 2$$

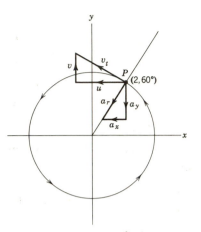

Solution. From similar triangles,

$$u = \frac{-1.04y}{\sqrt{x^2 + y^2}} \quad \text{and} \quad v = \frac{1.04x}{\sqrt{x^2 + y^2}}$$

At P, $x = 1$, $y = \sqrt{3}$, so

$$u = -0.90 \text{ m/s} \blacktriangleleft \qquad v = 0.52 \text{ m/s} \blacktriangleleft$$

Using equations 3.4,

$$a_x = -\frac{1.04y}{\sqrt{x^2 + y^2}} \frac{\partial}{\partial x}\left(\frac{-1.04y}{\sqrt{x^2 + y^2}}\right) + \frac{1.04x}{\sqrt{x^2 + y^2}} \frac{\partial}{\partial y}\left(\frac{-1.04y}{\sqrt{x^2 + y^2}}\right) = -\frac{2.16x}{8}$$

(3.4)

$$a_y = -\frac{1.04y}{\sqrt{x^2 + y^2}} \frac{\partial}{\partial x}\left(\frac{1.04x}{\sqrt{x^2 + y^2}}\right) + \frac{1.04x}{\sqrt{x^2 + y^2}} \frac{\partial}{\partial y}\left(\frac{1.04x}{\sqrt{x^2 + y^2}}\right) = -\frac{2.16y}{8}$$

(3.4)

Substituting $x = 1$, $y = \sqrt{3}$,

$$a_x = -0.27 \text{ m/s}^2 \blacktriangleleft \qquad a_y = -0.47 \text{ m/s}^2 \blacktriangleleft$$

By inspection,

$$v_t = 1.04 \text{ m/s} \blacktriangleleft \qquad v_r = 0 \text{ m/s} \blacktriangleleft \qquad (3.5)$$

Using equations 3.6 and 3.7,

$$a_r = 0\frac{\partial}{\partial r}(0) + 1.04\frac{\partial}{r\partial\theta}(0) - \frac{1.04 \times 1.04}{2} = -0.54 \text{ m/s}^2 \blacktriangleleft \qquad (3.6)$$

$$a_t = 0\frac{\partial}{\partial r}(1.04) + 1.04\frac{\partial}{r\partial\theta}(1.04) + \frac{0 \times 1.04}{2} = 0 \text{ m/s}^2 \blacktriangleleft \qquad (3.7)$$

Note that a_x and a_y might have been obtained more easily (in this problem) by calculating them as the horizontal and vertical components of a_r. Does $a_r^2 = a_x^2 + a_y^2$?

3.4 The Control Volume

A physical *system* is defined as a particular collection of matter and is identified and viewed as being separated from everything external to the system by an imagined or real closed boundary. In particle mechanics the system is a convenient physical entity. Its mass is conserved, and its energy and momentum can be defined precisely and easily analyzed. Indeed, a system-based analysis of fluid flow leads to the Lagrangian equations of motion in which particles of fluid are tracked. Unfortunately, a fluid system is both mobile and very deformable, and in the random and chaotic motion of turbulent flows even its identity may become hard to maintain over time. This suggests the need to define a more convenient object for analysis. This object is a volume which is fixed in space and through whose boundary matter, mass, momentum, energy, and the like, can flow. This volume is called a *control volume* and its boundary is a *control surface*. The fixed control volume can be of any useful size (finite or infinitesimal) and shape, provided only that the bounding control surface is a closed (completely surrounding) boundary. As defined here, neither the control volume nor the control surface change shape or position with time. This approach is consistent with the Eulerian view of fluid motion, in which attention is focused on particular points in the space filled by the fluid rather than on the fluid particles.

The control volume concept is applied in this chapter to derive the *equation of continuity* which expresses the principle of conservation of mass. In later chapters the conservation equations for energy, momentum, and the like, are derived through use of the control volume approach. Because the derivation for each case is repeated as the case arises (as opposed to deriving a single generic equation for all fluid properties, such as mass, momentum, and energy), it is hoped that the reader will build an ability to work from first principles and a better physical understanding of the various processes.

3.5 Conservation of Mass: The Continuity Equation— One-Dimensional Steady Flow

The application of the principle of conservation of mass (in the absence of mass-energy conversions) to a steady flow in a streamtube results in the equation of continuity, which describes the continuity of the flow from section to section of the streamtube. The derivation uses a control volume and the fluid system which just fills the volume at a particular time t. Consider the element of a finite streamtube in Fig. 3.8 through which passes a steady, one-dimensional flow of a compressible fluid. Note the location of the control volume as marked by the control surface that bounds the region between sections 1 and 2 and lies along the inner wall of the streamtube. The velocities at sections 1 and 2 are assumed to be uniform and to be consistent with the assumption of one-dimensional flow. In the tube near section 1 the cross-sectional area and fluid density are A_1 and ρ_1, respectively, and near section 2, A_2 and ρ_2. With the control surface shown coinciding with the streamtube walls and the cross sections at 1 and 2, the control volume comprises volumes I and R. Let a fluid system be defined as the fluid within the control volume $(I + R)$ at time t. The control volume is fixed in space,

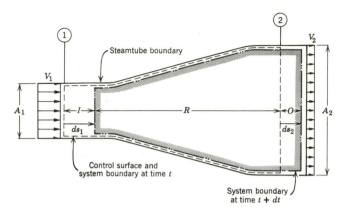

Fig. **3.8**

but in time dt the system moves downstream as shown. From the conservation of system mass

$$(m_1 + m_R)_t = (m_R + m_O)_{t+dt}$$

$$\begin{pmatrix} \text{Mass of fluid in} \\ \text{zones } I \text{ and } R \text{ in} \\ \text{time } t \end{pmatrix} = \begin{pmatrix} \text{Mass of fluid in} \\ \text{zones } O \text{ and } R \text{ at} \\ \text{time } t + dt \end{pmatrix}$$

In a steady flow the fluid properties at points in space are not functions of time so $(m_R)_t = (m_R)_{t+dt}$ and, consequently,

$$(m_I)_t = (m_O)_{t+dt}$$

These two terms are easily expressed in terms of the mass of fluid moving across the control surface in time dt. The volume of I is $A_1 \, ds_1$, and that of O is $A_2 \, ds_2$; accordingly,

$$(m_I)_t = \rho_1 A_1 \, ds_1 \qquad (m_O)_{t+dt} = \rho_2 A_2 \, ds_2$$

and

$$\rho_1 A_1 \, ds_1 = \rho_2 A_2 \, ds_2$$

Dividing by dt,

$$\rho_1 A_1 \frac{ds_1}{dt} = \rho_2 A_2 \frac{ds_2}{dt}$$

However, ds_1/dt and ds_2/dt are recognized as the velocities past sections 1 and 2, respectively; therefore, if $\dot{m} = A\rho V$ is the *mass flowrate,* then

$$\dot{m} = \rho_1 A_1 V_1 = \rho_2 A_2 V_2 \tag{3.8}$$

which is the equation of continuity. In words, it expresses the fact that in steady flow the mass flowrate passing all sections of a streamtube is constant. This equation may also be written

$$\dot{m} = A\rho V = \text{Constant} \qquad d(A\rho V) = 0 \qquad \text{or} \qquad \frac{dA}{A} + \frac{dp}{\rho} + \frac{dV}{V} = 0 \quad (3.9)$$

Multiplication of equation 3.8 by g_n gives the *weight flowrate*

$$G = g_n \dot{m} = A_1 \gamma_1 V_1 = A_2 \gamma_2 V_2 \tag{3.10}$$

while division of equation 3.9 by A gives the *mass velocity*

$$m_v = \rho V$$

which has the dimensions of slugs/ft$^2 \cdot$ s or kg/m$^2 \cdot$ s. The dimensions of $\dot{m}$ are slugs per second or kilograms per second. The dimensions of G are pounds per second or newtons per second; G is often a convenient and concise means of expressing flowrate of a gas in which the specific weight γ may vary along the streamtube with changes of pressure and temperature.

For many flows one is interested in the *volume flowrate*

$$Q = AV \tag{3.11}$$

which has the dimensions of cubic metres or cubic feet per second.[3] However, from equation 3.9 it is clear that Q is *not* a constant along a streamtube in variable density flows. For liquids and for gas flows where variation of density is negligible the equation of continuity reduces to

$$Q = A_1 V_1 = A_2 V_2 \tag{3.12}$$

Thus for fluids of constant density the product of velocity and cross-sectional area is constant along a streamtube. This quantity may, of course, be computed in compressible flow problems, but this is not generally useful because of its variation from section to section along the streamtube.

For two-dimensional flows the flowrate is usually quoted *per unit distance normal to the plane of the flow*. Taking b to be the distance between any two parallel flow planes and h the distance between streamlines, equation 3.10 becomes

$$\frac{G}{b} = \gamma_1 h_1 V_1 = \gamma_2 h_2 V_2 \tag{3.13}$$

in which G/b is known as the two-dimensional (weight) flowrate. The counterpart of this for the incompressible fluid (from equation 3.12) is

$$\frac{Q}{b} = q = h_1 V_1 = h_2 V_2 \tag{3.14}$$

---

[3] There are several common means of expressing volume flowrates. These include the use of litres per second (l/s), cubic feet per second abbreviated as "cusec" or "cfs", cubic feet per minute (cfm), gallons per minute (gpm), and millions of gallons per day (mgd).

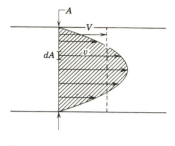

Fig. **3.9**

in which Q/b (hereafter termed q) is the two-dimensional (volume) flowrate, the dimensions of which are evidently m³/s · m or ft³/s · ft.

Frequently in fluid flows the velocity distribution through a flow cross section may be *nonuniform,* as shown in Fig. 3.9. From considerations of conservation of mass, it is evident at once that nonuniformity of velocity distribution does not invalidate the continuity principle as presented above. Thus, for steady flow of the incompressible fluid, equation 3.12 applies as before. Here, however, the velocity V in the equation is the *mean velocity* defined by $V = Q/A$ in which the flowrate Q is obtained from the summation of the differential flowrates, dQ, passing through the differential areas, dA. Thus, V is a fictitious uniform velocity that will transport the same amount of mass through the cross section as will the actual velocity distribution,

$$V = \frac{1}{A} \int^A v \, dA \qquad (3.15)$$

from which the mean velocity can be obtained by performing the indicated integration. With the velocity profile mathematically defined, formal integration may be employed; when the velocity profile is known but not mathematically defined,[4] graphical or numerical methods can be used to evaluate the integral.

The fact that the product AV remains constant along a streamtube (in a fluid of constant density) allows a partial physical interpretation of streamline pictures. As the cross-sectional area of a streamtube increases, the velocity must decrease; hence the conclusion: streamlines widely spaced indicate regions of low velocity, streamlines closely spaced indicate regions of high velocity.

_____ **Illustrative Problems** _____

Three kilonewtons of water per second flow through this pipeline reducer. Calculate the flowrate in cubic metres per second and the mean velocities in the 300 mm and 200 mm pipes.

_____

[4] For example, when obtained from experimental measurements.

Relevant Equations and Given Data

$$G = A_1 \gamma_1 V_1 = A_2 \gamma_2 V_2 \qquad (3.10)$$

$$Q = A_1 V_1 = A_2 V_2 \qquad (3.12)$$

$$G = 3 \text{ kN/s}$$

$$\gamma_1 = \gamma_2 = 9.8 \text{ kN/m}^3$$

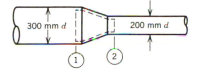

Solution.

$$Q = \frac{3 \times 10^3}{9.8 \times 10^3} = 0.306 \text{ m}^3/\text{s} \blacktriangleleft \qquad (3.10 \text{ and } 3.12)$$

$$V_1 = \frac{0.306}{\dfrac{\pi}{4}(0.3)^2} = 4.33 \text{ m/s} \blacktriangleleft \qquad (3.12)$$

$$V_2 = \frac{0.306}{\dfrac{\pi}{4}(0.2)^2} = 9.74 \text{ m/s} \quad \text{or} \quad V_2 = 4.33 \left(\frac{300}{200}\right)^2 = 9.74 \text{ m/s} \blacktriangleleft \qquad (3.12)$$

Thirty newtons of air per second flow through the reducer of the preceding problem, the air in the 300 mm pipe having a specific weight of 9.8 N/m³. In flowing through the reducer, the pressure and temperature will fall, causing the air to expand and producing a reduction of density; assuming that the specific weight of the air in the 200 mm pipe is 7.85 N/m³, calculate the mass and volume flowrates, the mass velocities, and the velocities in the two pipes.

Relevant Equations and Given Data

$$\dot{m} = \frac{G}{g_n} = \rho_1 A_1 V_1 = \rho_2 A_2 V_2 \qquad (3.8 \text{ and } 3.10)$$

$$Q = AV \qquad (3.11)$$

$$\gamma = \rho g_n \qquad (1.1)$$

$$G = 30 \text{ N/s} \qquad \gamma_1 = 9.8 \text{ N/m}^3$$

$$\gamma_2 = 7.85 \text{ N/m}^3$$

Solution.

$$\dot{m} = \frac{G}{g_n} = \frac{30}{9.81} = 3.06 \text{ kg/s} \blacktriangleleft \qquad (3.8 \text{ and } 3.10)$$

Now, from equations 1.1, 3.8, 3.10, and 3.11

$$G = \gamma_1 Q_1 = \gamma_2 Q_2$$

so

$$Q_1 = \frac{30}{9.8} = 3.06 \text{ m}^3/\text{s} \blacktriangleleft \qquad\qquad Q_2 = \frac{30}{7.85} = 3.82 \text{ m}^3/\text{s} \blacktriangleleft$$

$$V_1 = \frac{3.06}{\frac{\pi}{4}(0.3)^2} = 43.3 \text{ m/s} \blacktriangleleft \qquad V_2 = \frac{3.82}{\frac{\pi}{4}(0.2)^2} = 121.6 \text{ m/s} \blacktriangleleft \qquad (3.11)$$

To check:

$$\frac{\gamma_1 Q_1}{g_n} = 9.8 \times \frac{3.06}{9.81} = 3.06 \text{ kg/s}$$

$$\frac{\gamma_2 Q_2}{g_n} = 7.85 \times \frac{3.82}{9.81} = 3.06 \text{ kg/s}$$

$$m_{v1} = \rho_1 V_1 = 1.0 \times 43.3 = 43.3 \text{ kg/m}^2 \cdot \text{s} \blacktriangleleft$$

$$m_{v2} = \rho_2 V_2 = 0.8 \times 121.6 = 97.3 \text{ kg/m}^2 \cdot \text{s} \blacktriangleleft$$

Compare this problem with the preceding one, noting similarities and differences.

Taking Fig. 3.9 to represent an axisymmetric parabolic velocity distribution in a cylindrical pipe of radius R, calculate the mean velocity in terms of the maximum velocity, v_c.

Relevant Equation

$$V = \frac{1}{A} \int^A v \, dA \qquad (3.15)$$

Solution. Taking r as the radial distance to any local velocity, v, and the element of area dA, $dA = 2\pi r \, dr$, the equation of the parabola is $v = v_c(1 - r^2/R^2)$.

$$V = \frac{1}{\pi R^2} \int_0^R v_c \left(1 - \frac{r^2}{R^2}\right) 2\pi r \, dr \qquad (3.15)$$

Performing the integration shows that for the parabolic axisymmetric profile, the mean velocity is half of the centerline velocity, that is,

$$V = \frac{v_c}{2} \blacktriangleleft$$

3.6 Conservation of Mass: The Continuity Equation—Two-Dimensional Steady Flow

For two-dimensional flow the equation of continuity may be derived by considering a general control volume and repeating the conservation of mass analysis made for

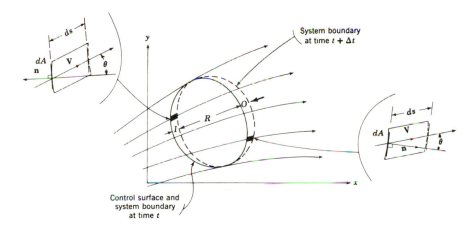

Fig. **3.10**

one-dimensional flow. Consider the flow and control volume in Fig. 3.10. As the system, which is the fluid in the control volume at time t, moves out of the control volume into zone O, new fluid flows into the control volume filling zone I. Part of the system remains in the control volume in zone R. From the conservation of mass

$$(m_I + m_R)_t = (m_R + m_O)_{t+dt}$$

and for a steady flow $(m_R)_t = (m_R)_{t+dt}$. Hence, as before,

$$(m_I)_t = (m_O)_{t+dt}$$

The mass in O is the integral of the masses moving *out* through all the incremental areas dA of the control surface in time dt (see the right-hand insert in Fig. 3.10)

$$(m_O)_{t+dt} = \int_{\text{C.S.OUT}} \rho(ds \cos \theta) \, dA$$

because the volume of the small prism is the product of the area[5] of its base and its height (perpendicular to the base). The distance that the system boundary moves along a streamline $ds = V \, dt$, so that

$$(m_O)_{t+dt} = \int_{\text{C.S.OUT}} \rho(V \cos \theta) \, dA \, dt$$

Note that $V \cos \theta$ is the magnitude of the velocity component *normal* to the control surface at dA; thus, the net flow out of the control volume is determined, *not* by the total velocity, but by the *normal velocity component*. The tangential component does *not* contribute to flow through the surface. In vector terms, if **n** is the *outward unit*

---

[5] For plane (two-dimensional) motion $dA = 1 \cdot dl$ where dl is the length of the differential segment of the control surface and a slice of the flow one-unit thick and perpendicular to the plane of flow is used.

normal vector at dA, from Appendix 5,

$$\mathbf{V} \cdot \mathbf{n} = V \cos \theta$$

Thus,

$$(m_O)_{t+dt} = dt \int_{\text{C.S.}_{\text{OUT}}} \rho \mathbf{V} \cdot \mathbf{n} \, dA = dt \int_{\text{C.S.}_{\text{OUT}}} \rho \mathbf{V} \cdot d\mathbf{A}$$

when $d\mathbf{A} = \mathbf{n} \, dA$ is defined as the *directed area element*. It follows that the mass flow into I is

$$(m_I)_t = \int_{\text{C.S.}_{\text{IN}}} \rho(ds \, \cos \theta) \, dA$$

$$= \int_{\text{C.S.}_{\text{IN}}} \rho(V \cos \theta) \, dA \, dt = dt \int_{\text{C.S.}_{\text{IN}}} \rho \mathbf{V} \cdot (-\mathbf{n}) \, dA$$

$$= dt \left\{ -\int_{\text{C.S.}_{\text{IN}}} \rho \mathbf{V} \cdot \mathbf{n} \, dA \right\} = dt \left\{ -\int_{\text{C.S.}_{\text{IN}}} \rho \mathbf{V} \cdot d\mathbf{A} \right\}$$

because $\mathbf{n}$ is the outward normal and points "against" the IN flow (see the left-hand insert in Fig. 3.10). As the IN and OUT masses are equal, dividing by dt produces

$$-\int_{\text{C.S.}_{\text{IN}}} \rho \mathbf{V} \cdot \mathbf{n} \, dA = \int_{\text{C.S.}_{\text{OUT}}} \rho \mathbf{V} \cdot \mathbf{n} \, dA; \qquad -\int_{\text{C.S.}_{\text{IN}}} \rho \mathbf{V} \cdot d\mathbf{A} = \int_{\text{C.S.}_{\text{OUT}}} \rho \mathbf{V} \cdot d\mathbf{A}$$

Clearly,

$$\int_{\text{C.S.}_{\text{OUT}}} \rho \mathbf{V} \cdot d\mathbf{A} + \int_{\text{C.S.}_{\text{IN}}} \rho \mathbf{V} \cdot d\mathbf{A} = \oint_{\text{C.S.}} \rho \mathbf{V} \cdot d\mathbf{A} = \oint_{\text{C.S.}} \rho \mathbf{V} \cdot \mathbf{n} \, dA = 0 \quad (3.16)$$

in which the circle and arrow symbol on the integral sign indicates that the integral is to be taken once around the control surface in the counterclockwise direction. Thus, over the entire control surface, the sum of all mass flowrates normal to the surface is zero; that is, the mass flowrate IN must exactly equal the mass flowrate OUT in *steady flow*.

The integral continuity equation 3.16 is valid for finite or infinitesimal control volumes. The differential forms are easily obtained from equation 3.16. For the infinitesimal control volume in Fig. 3.11a, the mass flowrates on the sides are[6]

---

[6] In Fig. 3.11a, the density ρ and velocity components u and v are defined at the center (x, y) of the element. Therefore, to first-order accuracy,

$$\rho_{AB} \cong \rho - \frac{\partial \rho}{\partial y} \frac{dy}{2} \qquad \text{and} \qquad v_{AB} \cong v - \frac{\partial v}{\partial y} \frac{dy}{2}$$

while $\mathbf{V} \cdot \mathbf{n} \, dA = -v_{AB} \, dx$, etc.

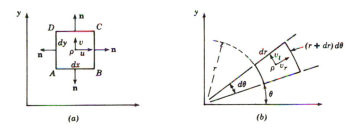

Fig. **3.11**

$$\int_{AB} \rho\mathbf{V} \cdot \mathbf{n}\, dA \cong -\left(\rho - \frac{\partial\rho}{\partial y}\frac{dy}{2}\right)\left(v - \frac{\partial v}{\partial y}\frac{dy}{2}\right) dx$$

$$\int_{BC} \rho\mathbf{V} \cdot \mathbf{n}\, dA \cong \left(\rho + \frac{\partial\rho}{\partial x}\frac{dx}{2}\right)\left(u + \frac{\partial u}{\partial x}\frac{dx}{2}\right) dy$$

$$\int_{CD} \rho\mathbf{V} \cdot \mathbf{n}\, dA \cong \left(\rho + \frac{\partial\rho}{\partial y}\frac{dy}{2}\right)\left(v + \frac{\partial v}{\partial y}\frac{dy}{2}\right) dx$$

$$\int_{DA} \rho\mathbf{V} \cdot \mathbf{n}\, dA \cong -\left(\rho - \frac{\partial\rho}{\partial x}\frac{dx}{2}\right)\left(u - \frac{\partial u}{\partial x}\frac{dx}{2}\right) dy$$

Equation 3.16 now yields

$$\int_{AB} \rho\mathbf{V} \cdot \mathbf{n}\, dA + \int_{BC} \rho\mathbf{V} \cdot \mathbf{n}\, dA + \int_{CD} \rho\mathbf{V} \cdot \mathbf{n}\, dA + \int_{CA} \rho\mathbf{V} \cdot \mathbf{n}\, dA = 0$$

By substituting the flowrate values in differential terms, expanding the products, and retaining only terms of lowest order (largest order of magnitude), we obtain

$$\rho\frac{\partial v}{\partial y} + v\frac{\partial\rho}{\partial y} + \rho\frac{\partial u}{\partial x} + u\frac{\partial\rho}{\partial x} = 0$$

The above continuity equation for compressible, steady, two-dimensional flow can also be written as

$$\frac{\partial}{\partial x}(\rho u) + \frac{\partial}{\partial y}(\rho v) = 0 \qquad (3.17)$$

For steady incompressible flow, ρ is constant and the equation reduces to

$$\frac{\partial u}{\partial x} + \frac{\partial v}{\partial y} = 0 \qquad (3.18)$$

Application of the same principles to the polar element of Fig. 3.11b yields the two-dimensional continuity equation in polar coordinates. For the compressible fluid this is

$$\frac{1}{r}(\rho v_r) + \frac{\partial}{\partial r}(\rho v_r) + \frac{\partial}{r\,\partial\theta}(\rho v_t) = 0 \qquad (3.19)$$

which, for the incompressible fluid, reduces to

$$\frac{v_r}{r} + \frac{\partial v_r}{\partial r} + \frac{\partial v_t}{r\partial\theta} = 0 \qquad (3.20)$$

――――――――――― **Illustrative Problems** ―――――――――――

A mixture of ethanol (grain or ethyl alcohol) and gasoline, called "gasohol," is created here by pumping the two liquids into the "wye" pipe junction shown. The mixture is 10% alcohol and is observed to have a density of 691.1 kg/m³ and a velocity of 1.08 m/s when the gasoline input is 30 l/s. Find the (volume) flowrate and average velocity of the incoming ethanol.

Relevant Equations and Given Data

$$\int_{C.S._{OUT}} \rho\mathbf{V}\cdot d\mathbf{A} + \int_{C.S._{IN}} \rho\mathbf{V}\cdot d\mathbf{A} = \oint_{C.S.} \rho\mathbf{V}\cdot d\mathbf{A} = \oint_{C.S.} \rho\mathbf{V}\cdot \mathbf{n}\, dA = 0 \quad (3.16)$$

$$Q = AV \qquad (3.11)$$

$$Q_{GAS} = 30\ l/s \qquad \rho_{GAS} = 680.3\ kg/m^3 \qquad \rho_{ETH} = 788.6\ kg/m^3$$

$$V_{MIX} = 1.08\ m/s \qquad \rho_{MIX} = 691.1\ kg/m^3$$

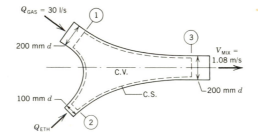

Solution. From the control volume and by assuming that the flow in each pipe is one-dimensional

$$\rho_1 = 680.3\ kg/m^3 \qquad \rho_2 = 788.6\ kg/m^3 \qquad \rho_3 = 691.1\ kg/m^3$$

$$V_3 = 1.08\ m/s$$

$$A_1 = \frac{\pi}{4}(0.2)^2 = 0.031\ m^2 \qquad A_2 = 0.007\ 9\ m^2 \qquad A_3 = 0.031\ m^2$$

$$V_1 = 30 \times 10^{-3}/0.031 = 0.97\ m/s \qquad (3.11)$$

From equation 3.16

$$\int_1 \rho\mathbf{V}\cdot\mathbf{n}\, dA + \int_2 \rho\mathbf{V}\cdot\mathbf{n}\, dA + \int_3 \rho\mathbf{V}\cdot\mathbf{n}\, dA = 0 \qquad (3.16)$$

Using the assumed uniform [a] (i.e., average) velocity at each cross section,

$$\int_1 \rho \mathbf{V} \cdot \mathbf{n}\, dA = -680.3 \times 0.97 \times 0.031 = -20.4 \text{ kg/s}$$

The minus sign arises because the outward normal $\mathbf{n}$ and $\mathbf{V}$ have opposite directions so $(\mathbf{V} \cdot \mathbf{n}\, dA)_1 = -V_1\, dA$. Similarly, for section 2,

$$\int_2 \rho \mathbf{V} \cdot \mathbf{n}\, dA = -788.6 \times V_2 \times 0.007\,9 = -6.23 V_2$$

At section 3 $\mathbf{V}$ and $\mathbf{n}$ have the same direction so

$$\int_3 \rho \mathbf{V} \cdot \mathbf{n}\, dA = 691.1 \times 1.08 \times 0.031 = 23.1 \text{ kg/s}$$

Therefore,

$$\oint_{\text{C.S.}} \rho \mathbf{V} \cdot \mathbf{n}\, dA = -20.4 - 6.23 V_2 + 23.1 = 0 \qquad (3.16)$$

$$V_2 = 0.43 \text{ m/s} \blacktriangleleft$$

$$Q_{\text{ETH}} = Q_2 = A_2 V_2 = 0.007\,9 \times 0.43 = 3.4 \times 10^{-3} \text{ m}^3/\text{s} = 3.4 \text{ l/s} \blacktriangleleft \qquad (3.11)$$

A two-dimensional incompressible flowfield is described by the equations $v_t = \omega r$ and $v_r = 0$, in which ω is a constant. Sketch this flow and show that it satisfies the continuity equation.

Relevant Equation and Given Data

$$\frac{v_r}{r} + \frac{\partial v_r}{\partial r} + \frac{\partial v_t}{r\partial\theta} = 0 \qquad (3.20)$$

$$v_t = \omega r \qquad v_r = 0 \qquad \omega = \text{Constant}$$

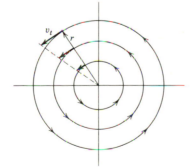

Solution. Evidently the streamlines of this flow must be concentric circles because the radial component of velocity is everywhere zero. This flow is known as a *forced vortex*;

some of its features have been examined in Section 2.7 and others are discussed in Section 12.2. By substituting $v_r = 0$ and $v_t = \omega r$ in equation 3.20, we find

$$\frac{0}{r} + \frac{\partial}{\partial r}(0) + \frac{\partial}{r\partial\theta}(\omega r) = 0 + 0 + 0 = 0 \qquad (3.20)$$

The equation of continuity is satisfied; this flow is physically possible. ◄

3.7   Circulation, Vorticity and Rotation

The integral of the normal component of the velocity over a control surface leads to the continuity equation. In the previous Illustrative Problem tangential components of velocity are seen to give the fluid a swirl. A measure of this swirl is called the *circulation* and is designated by Γ (gamma). Γ is defined as the *line integral* of the tangential component of velocity around a closed curve fixed in the flow, for example, the curve that marks the location of a control surface. In the two-dimensional flowfield of Fig. 3.12 each streamline intersects the control surface at an angle α and the tangential component of velocity at the point of intersection is $V \cos \alpha = V \sin \theta$ (compare Fig. 3.10). An element of circulation $d\Gamma$ is defined as the product of the tangential velocity component and the element dl of the closed curve. Thus

$$d\Gamma = (V \cos \alpha)\, dl$$

The sum of the elements $d\Gamma$ along the curve C marking the control surface defines the circulation:

$$\Gamma = \oint_C d\Gamma = \oint_C (V \cos \alpha)\, dl = \oint_C \mathbf{V} \cdot d\mathbf{l} \qquad (3.21)$$

in which $d\mathbf{l}$ is an elemental vector of size dl and direction tangent to the control surface at each point.

Although the calculation of circulation around an arbitrary curve in a flowfield is generally a tedious step-by-step integration, the principle can be applied easily and fruitfully to specific closed curves such as circles and squares. Calculation of the circulation around the basic differentially sized square element of Fig. 3.11a yields a

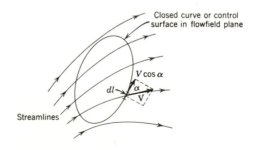

Fig. **3.12**

concept having great general significance because it yields the point value of the circulation in a flow. This value is needed, for example, in calculations of pressure-velocity-head relations (expressed through Bernoulli's equation) in two-dimensional ideal incompressible flows (Section 4.7).

Now to compute the circulation, proceed from A counterclockwise around the boundary of the element, setting down the products of velocity component and distance in order; because the element is of differential size, the resulting circulation is also a differential quantity, $d\Gamma$.

$$
d\Gamma \cong \left[\begin{matrix} \text{Mean velocity} \\ \text{along } AB \end{matrix} \right] dx + \left[\begin{matrix} \text{Mean velocity} \\ \text{along } BC \end{matrix} \right] dy
$$

$$
- \left[\begin{matrix} \text{Mean velocity} \\ \text{along } CD \end{matrix} \right] dx - \left[\begin{matrix} \text{Mean velocity} \\ \text{along } DA \end{matrix} \right] dy
$$

$$
\cong \left[u - \frac{\partial u}{\partial y} \frac{dy}{2} \right] dx + \left[v + \frac{\partial v}{\partial x} \frac{dx}{2} \right] dy
$$

$$
- \left[u + \frac{\partial u}{\partial y} \frac{dy}{2} \right] dx - \left[v - \frac{\partial v}{\partial x} \frac{dx}{2} \right] dy
$$

(Note $\cos \alpha = 1$ on AB and BC, but $\cos \alpha = -1$ on CD and DA.) By expanding the products and retaining only the terms of lowest order (largest magnitude), we obtain

$$
d\Gamma = \left(\frac{\partial v}{\partial x} - \frac{\partial u}{\partial y} \right) dx\, dy
$$

in which $dx\, dy$ is the area inside the control surface. The *vorticity*, ξ (xi), is defined as the differential circulation per unit of area enclosed, which becomes

$$
\xi = \frac{d\Gamma}{dx\, dy} = \frac{\partial v}{\partial x} - \frac{\partial u}{\partial y} \tag{3.22}
$$

For polar coordinates, by the same procedure,

$$
\xi = \frac{\partial v_t}{\partial r} + \frac{v_t}{r} - \frac{\partial v_r}{r\partial \theta} \tag{3.23}
$$

From the definition of circulation, Γ, and the methods used to calculate vorticity, ξ, the reader will sense that the latter quantity is some measure of the rotational aspects of the fluid elements as they move through the flowfield. This can be shown explicitly. Suppose two lines are drawn on the square element in a fluid flow, as shown in Fig. 3.13, so that the lines are parallel to the x and y axes, respectively. If the fluid element tends to rotate, these lines will tend to rotate also and, for the instant at which the lines are drawn on the fluid,[7] their average angular velocity can be calculated. In a small time

---

[7] It is physically possible to carry out such an experiment; see, for example, J.L. Lumley, "Deformation of Continuous Media," NCFMF/EDC Film No. 21608, Encyclopaedia Britannica Educational Corp.

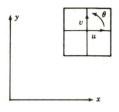

Fig. 3.13

interval dt the vertical line will rotate about the translating center of mass of the element by an amount

$$d\theta_V = -\left[\left(u + \frac{\partial u}{\partial y}\frac{dy}{2}\right) - \left(u - \frac{\partial u}{\partial y}\frac{dy}{2}\right)\right]\frac{dt}{dy}$$

so the angular velocity

$$\omega_V = \frac{d\theta_V}{dt} = -\frac{\partial u}{\partial y}$$

For the horizontal line

$$\omega_H = \frac{d\theta_H}{dt} = \frac{\partial v}{\partial x}$$

The *average rotation* ω of the element is then

$$\omega = \tfrac{1}{2}(\omega_V + \omega_H) = \frac{1}{2}\left(\frac{\partial v}{\partial x} - \frac{\partial u}{\partial y}\right) \tag{3.24}$$

and

$$\xi = \frac{d\Gamma}{dx\,dy} = 2\omega \tag{3.25}$$

The vorticity at a point is twice the rotation there, that is, twice the average angular velocity of the fluid element.

If a flow possesses vorticity (i.e., if $\xi \neq 0$), it is said to be a *rotational* flow; if a flow possesses no vorticity ($\xi = 0$), it is termed *irrotational*.[8] Such definitions are adequate for present purposes but may be misleading because they imply that whole flowfields are either rotational or irrotational. Actually flowfields can possess zones of both irrotational and rotational flows, the latter frequently being concentrated in *singular points*; the free vortex flow (Section 12.2) is a classic example of this.

---

[8] In Section 12.5 it will be shown that *irrotational* flows are also characterized by a *velocity potential* and are therefore known as *potential* flows.

———————————— **Illustrative Problems** ————————————

Calculate the vorticity of the flow described in the second of the Illustrative Problems of Section 3.6.

Relevant Equation

$$\xi = \frac{\partial v_t}{\partial r} + \frac{v_t}{r} - \frac{\partial v_r}{r\partial \theta} \tag{3.23}$$

$$u = U_c(1 - y^2/b^2)$$

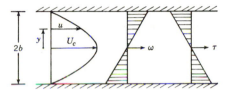

Solution. For the second problem

$$\xi = \frac{\partial}{\partial r}(\omega r) + \frac{\omega r}{r} - \frac{\partial}{r\partial \theta}(0) = \omega + \omega - 0 = 2\omega \blacktriangleleft \tag{3.23}$$

Evidently this is a rotational flow possessing a constant vorticity (over the whole flowfield) of 2ω; the *forced vortex* is thus a *rotational* flowfield.

When a viscous, incompressible fluid flows between two plates and the flow is laminar and two-dimensional, the velocity profile is parabolic:

$$u = U_c(1 - y^2/b^2)$$

Calculate the shear stress τ and rotation ω.

Relevant Equations

$$\tau = \mu \frac{du}{dy} \tag{1.12}$$

$$\omega = \tfrac{1}{2}(\omega_v + \omega_H) = \frac{1}{2}\left(\frac{\partial v}{\partial x} - \frac{\partial u}{\partial y}\right) \tag{3.24}$$

Solution. $\omega = -\dfrac{1}{2}[-2U_c y/b^2] = U_c y/b^2 \blacktriangleleft$ (3.24)

$$\tau = \mu \, du/dy = -2\mu U_c y/b^2 = -2\mu\omega = -\mu\xi \blacktriangleleft \tag{1.12}$$

The rotation and vorticity are large where the shear stress is large.

Films

Lumley, J. L. Deformation of continuous media. NCFMF/EDC Film No. 21608, Encyclopaedia Britannica Educ. Corp.

Lumley, J. L. Eulerian and Lagrangian descriptions in fluid mechanics. NCFMF/EDC Film No. 21621, Encyclopaedia Britannica Educ. Corp.

Rouse, H. Mechanics of Fluids: Fundamental principles of flow. Film No. U45734, Media Library, Audiovisual Center, Univ. of Iowa.

Shapiro, A. H. Vorticity. NCFMF/EDC Film Nos. 21605 and 21606, Encyclopaedia Britannica Educ. Corp.

Problems

3.1. A 4 m diameter tank is filled with water and then rotated at a rate of $\omega = 2\pi(1 - e^{-t})$ rad/s. At the tank walls viscosity prevents slip of fluid particles relative to the wall. What are the speed and the tangential and normal accelerations of those fluid particles next to the tank walls as a function of time?

3.2. The path of a fluid particle is given by the hyperbola $xy = 25$ while at any time t the particle position is $x = 5t^2$. What are the x- and y- components of the particle velocity and acceleration?

3.3. Sketch the following flowfields and derive general expressions for their components of acceleration: (a) $u = 4$, $v = 3$; (b) $u = 4$, $v = 3x$; (c) $u = 4y$, $v = 0$; (d) $u = 4y$, $v = 3$; (e) $u = 4y$, $v = 3x$; (f) $u = 4y$, $v = 4x$; (g) $u = 4y$, $v = -4x$; (h) $u = 4$, $v = 0$; (i) $u = 4$, $v = -4x$; (j) $u = 4x$, $v = 0$; (k) $u = 4xy$, $v = 0$; (l) $v_r = c/r$, $v_t = 0$; (m) $v_r = 0$, $v_t = c/r$.

3.4. When an incompressible, nonviscous (see Chapter 4) fluid flows against a plate in a plane (two-dimensional) flow, an exact solution for the equations of motion for this flow is $u = Ax$, $v = -Ay$, with $A > 0$ for the sketch shown. The coordinate origin is located at the *stagnation point* 0, where the flow divides and the local velocity is zero. Find the velocities and accelerations in the flow.

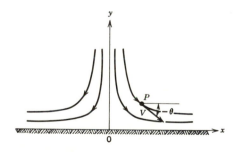

Problem **3.4**

3.5. The mean velocity of water in a 100 mm pipeline is 2 m/s. Calculate the rate of flow in cubic metres per second, newtons per second, and kilograms per second.

3.6. One hundred pounds of water per minute flow through a 6 in. pipeline. Calculate the mean velocity.

3.7. Four hundred litres per minute of glycerin flow in a 75 mm pipeline. Calculate the mean velocity.

3.8. A 0.3 m by 0.5 m rectangular air duct carries a flow of 0.45 m³/s at a density of 2 kg/m³. Calculate the mean velocity in the duct. If the duct tapers to 0.15 m by 0.5 m size, what is the mean velocity in this section if the density is 1.5 kg/m³ there?

3.9. Across a shock wave in a gas flow there is a great change in gas density ρ. If a shock wave occurs in a duct such that $V = 660$ m/s and $\rho = 1.0$ kg/m³ before the shock and $V = 250$ m/s after the shock, what is ρ after the shock?

3.10. Water flows in a pipeline composed of 75 mm and 150 mm pipe. Calculate the mean velocity in the 75 mm pipe when that in the 150 mm pipe is 2.5 m/s. What is its ratio to the mean velocity in the 150 mm pipe?

3.11. A smooth nozzle with a tip diameter of 2 in. terminates a 6 in. waterline. Calculate the mean velocity of efflux from the nozzle when the velocity in the line is 10 ft/s.

3.12. Hydrogen is being pumped through a pipe system whose temperature is held at 273 K. At a section where the pipe diameter is 10 mm, the pressure and average velocity are 200 kPa and 30 m/s. Find all possible velocities and pressures at a downstream section whose diameter is 20 mm.

3.13. At a point in a two-dimensional fluid flow, two streamlines are parallel and 75 mm (3 in.) apart. At another point these streamlines are parallel but only 25 mm (1 in.) apart. If the velocity at the first point is 3 m/s (9.8 fps), calculate the velocity at the second.

3.14. A 300 mm pipeline leaves a large tank through a square-edged hole in its vertical wall. The mean velocity in the pipe is 4.5 m/s. Assuming that the fluid in the tank approaches the center of the pipe entrance radially, what is the velocity of the fluid 0.5, 1, and 2 m from the pipe entrance?

3.15. Five hundred cubic feet per second of water flow in a rectangular open channel 20 ft wide and 8 ft deep. After passing through a transition structure into a trapezoidal canal of 5 ft base width with sides sloping at 30°, the velocity is 6 fps. Calculate the depth of the water in the canal.

3.16. Calculate the mean velocities for these two-dimensional velocity profiles if $v_c = 3$ m/s or 10 ft/s.

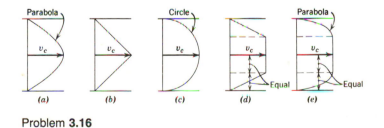

Problem **3.16**

3.17. Calculate the mean velocities in the preceding problem, assuming the velocity profiles to be axisymmetric in a cylindrical passage.

3.18. If the velocity profile in a passage of width $2R$ is given by the equation $v/v_c = (y/R)^{1/n}$, derive an expression for V/v_c in terms of n: (a) for a two-dimensional passage, and (b) for a cylindrical passage.

3.19.   Calculate the mean velocities at C and D in problem 4.68 assuming them to be radial.

3.20.   Fluid passes through this set of thin closely spaced blades. What flowrate q is required for the velocity V to be 10 ft/s (or equivalently, 3.0 m/s)?

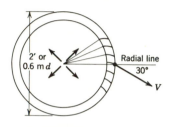

Problem **3.20**

3.21.   A pipeline 0.3 m in diameter divides at a Y into two branches 200 mm and 150 mm in diameter. If the flowrate in the main line is 0.3 m³/s and the mean velocity in the 200 mm pipe is 2.5 m/s, what is the flowrate in the 150 mm pipe?

3.22.   A *manifold pipe* of 3 in. diameter has four openings in its walls spaced equally along the pipe and is closed at the downstream end. If the discharge from each opening is 0.50 cfs, what are the mean velocities in the pipe between the openings?

3.23.   Find the average efflux velocity V if the flow exits from a hole of area 1 m² in the side of the duct as shown.

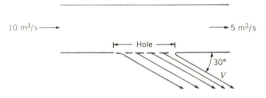

Problem **3.23**

3.24.   Find V for this mushroom cap on a pipeline.

3.25.   Using the wye and control volume of the first Illustrative Problem of Section 3.6, find the mixture velocity and density if freshwater ($\rho_1 = 1$ Mg/m³) enters section 1 at 50 l/s, while saltwater ($\rho_2 = 1.03$ Mg/m³) enters section 2 at 25 l/s.

3.26.   Derive an expression for the flowrate q between two streamlines of radii r_1 and r_2 for the flowfields (a) $v_r = 0$, $v_t = c/r$, and (b) $v_r = 0$, $v_t = \omega r$.

3.27.   For the flowfield $v_r = c/r$, $v_t = 0$, derive an expression for the flowrate q between any two streamlines.

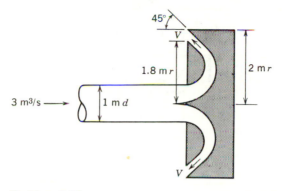

1.8 m r

45°

V

2 m r

3 m³/s ⟶

1 m d

V

Problem **3.24**

3.28. The flow of a uniform, incompressible fluid is described by the equations $v_r = (2\pi r)^{-1}$ and $v_t = -(2\pi r)^{-1}$. Sketch this flow, show that it satisfies the differential continuity equation, find the magnitude and direction of the velocity, and determine if the continuity equation for a finite control volume $r \leq r_0$ is satisfied.

3.29. Derive the equation of continuity in polar coordinates for the incompressible fluid.

3.30. Investigate the flowfields of problem 3.3 to see if they are physically possible (i.e., satisfy the equation of continuity).

3.31. Suppose the two-dimensional flowfield for flow against a plate (problem 3.4) is given by the equations: $u = 4x$, $v = -4y$. Show that this flow satisfies the equation of continuity.

3.32. Calculate the circulation for the flow of problem 3.20.

3.33. Derive the equation for vorticity in polar coordinates.

3.34. For the flowfields of problem 3.3 derive expressions for vorticity, and state whether the flowfield is rotational or irrotational.

3.35. For the velocity profiles of problem 3.16 derive expressions for the vorticity.

3.36. For the velocity profile of problem 3.18 derive an expression for the vorticity.

3.37. Calculate the vorticity for the flow of problem 3.31.

4
FLOW OF AN INCOMPRESSIBLE IDEAL FLUID

Significant insight into the basic laws of fluid flow can be obtained from a study of the flow of a hypothetical *ideal fluid*. An ideal fluid is a fluid assumed to be inviscid, or devoid of viscosity.[1] In such a fluid there are no frictional effects between moving fluid layers or between these layers and boundary walls, and thus no cause for eddy formation or energy dissipation due to friction. The assumption that a fluid is ideal allows it to be treated as an aggregation of small particles that will support pressure forces normal to their surfaces but will slide over one another without resistance. Thus the motion of these ideal fluid particles is analogous to the motion of a solid body on a frictionless plane, and the unbalanced forces existing on them cause the acceleration of these particles according to Newton's second law.

Under the assumption of frictionless motion, equations are considerably simplified and more easily assimilated by the beginner in the field. In many cases these simplified equations allow solution of engineering problems to an accuracy entirely adequate for practical use. The beginner should not jump to the conclusion that the assumption of frictionless flow leads to a useless abstraction which is always far from reality. In those real situations, where the actual effects of friction are small, the frictionless assumption will give good results; where friction is large, it obviously will not. The identification of these situations is part of the art of fluid mechanics; for example, the lift on a wing can often be predicted accurately by an inviscid analysis while the drag rarely can.

The further assumption of an incompressible (i.e., constant-density) fluid restricts the present chapter to the flow of liquids and of gases that undergo negligibly small changes of pressure and temperature. The flow of gases with large density changes is discussed in Chapter 5, but there are numerous practical engineering problems that involve fluids whose densities may safely be considered constant; thus, this assumption proves to be not only a practical one but also a useful simplification in the introduction to fluid flow, because it usually permits thermodynamic effects to be disregarded.

This chapter introduces the important Bernoulli and work-energy equations, which permit us to relate and to predict pressures and velocities in a flowfield. The concepts

---

[1]Compare with the definition of a perfect gas, Section 1.3.

are introduced from several points of view. Thus, the presentation begins with a simple one-dimensional analysis of an elemental fluid system moving along a streamline, moves next to a work-energy relation that allows inclusion of pumps and turbines in the computations, and concludes with an introduction to two-dimensional flows. Significantly, while the simple analysis of a fluid system yields the important Bernoulli equation (Section 4.2), precisely the same equation is obtained *as a simplification* of the more general result of a *control volume analysis* which employs the work-energy principle (Section 4.5).

ONE-DIMENSIONAL FLOW

4.1   Euler's Equation

In 1750, Leonhard Euler first applied Newton's second law to the motion of fluid particles and thus laid the groundwork for an analytical approach to fluid dynamics.

    Consider a streamline and select a small cylindrical fluid system for analysis, as shown in Fig. 4.1. The forces tending to accelerate the cylindrical fluid system are: pressure forces on the ends of the system, $p\,dA - (p + dp)\,dA = -dp\,dA$ (the pressures on the sides of the system have no effect on its acceleration), and the component of weight in the direction of motion, $-\rho g_n\,ds\,dA\,(dz/ds) = -\rho g_n\,dA\,dz$. The differential mass being accelerated by the action of these differential forces is $dM = \rho\,ds\,dA$. Applying Newton's second law $dF = (dM)a$ along the streamline and using the one-dimensional expression for acceleration (equation 3.1) gives (recall only

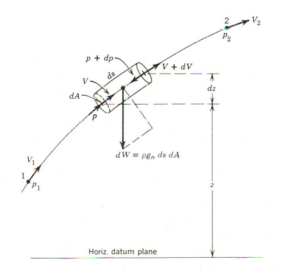

Fig. **4.1**

steady flow is being considered)

$$-dp \, dA - \rho g_n \, dA \, dz = (\rho \, ds \, dA) V \frac{dV}{ds}$$

Dividing by $\rho \, dA$ produces the one-dimensional Euler equation

$$\frac{dp}{\rho} + V \, dV + g_n \, dz = 0$$

For incompressible flow this equation is usually divided by g_n and written

$$\frac{dp}{\gamma} + d\left(\frac{V^2}{2g_n}\right) + dz = 0$$

or for uniform density flows

$$d\left(\frac{P}{\gamma} + \frac{V^2}{2g_n} + z\right) = 0$$

4.2 Bernoulli's Equation

For incompressible flow of uniform density fluid, the one-dimensional Euler equation can be easily integrated between any two points (because γ and g_n are both constant) to obtain

$$\frac{p_1}{\gamma} + \frac{V_1^2}{2g_n} + z_1 = \frac{p_2}{\gamma} + \frac{V_2^2}{2g_n} + z_2$$

As points 1 and 2 are any two arbitrary points on the streamline, the quantity

$$\frac{p}{\gamma} + \frac{V^2}{2g_n} + z = H = \text{Constant} \qquad (4.1)$$

applies to all points on the streamline and thus provides a useful relationship between pressure p, the magnitude V of the velocity, and the height z above datum. Equation 4.1 is known as the *Bernoulli equation* and the *Bernoulli constant H* is also termed the *total head*.

   Examination of the *Bernoulli terms* of equation 4.1 revels that p/γ and z are, respectively, the pressure (either gage or absolute) and potential heads encountered in Section 2.1 and hence may be visualized as vertical distances. Pitot's experiments (1732) showed that the sum of velocity head $V^2/2g_n$ and pressure head p/γ could be measured by placing a tiny open tube (now known as a *pitot tube*) in the flow with its open end upstream. Thus the Bernoulli equation may be visualized for liquids, as in Fig. 4.2, the sum of the terms (total head) being the constant distance between the horizontal datum plane and the *total head line* or *energy line* (**E.L.**). The *piezometric head line* or *hydraulic grade line* (**H.G.L.**) drawn through the tops of the piezometer columns gives a picture of the pressure variation in the flow; evidently (1) its distance

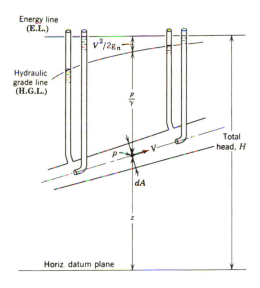

Fig. **4.2**

from the streamtube is a direct measure of the static pressure in the flow, and (2) its distance below the energy line is proportional to the square of the velocity. Complete familiarity with these lines is essential because of their wide use in engineering practice and their great utility in problem solutions.

4.3 The One-Dimensional Assumption for Streamtubes of Finite Cross Section

The foregoing development of the Bernoulli equation has been carried out for a single streamline or infinitesimal streamtube across which the variation of p, V, and z is negligible because of the differential size of the cross section of the element. However, the engineer may apply this equation easily and fruitfully to large streamtubes such as pipes and canals once its limitations are understood. Consider a cross section of a large flow (open or closed but not a free jet) through which *all streamlines are precisely straight and parallel* (Fig. 4.3). The forces, normal to the streamlines, on the element of fluid are $(p_1 - p_2)\,ds$ and the component of the weight of the element, $\gamma h\,ds\,\cos\alpha$, in which $\cos\alpha = (z_2 - z_1)/h$. It is apparent that if (and only if) the streamlines are straight and parallel the acceleration toward the boundary is zero. This means that the forces defined above are in equilibrium:

$$(p_1 - p_2)\,ds = \gamma(z_2 - z_1)\,ds$$

yielding

$$\frac{p_1}{\gamma} + z_1 = \frac{p_2}{\gamma} + z_2 \tag{2.6}$$

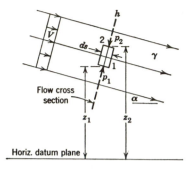

Fig. **4.3**

and demonstrating that the quantity $(z + p/\gamma)$ is constant over the flow cross section normal to the streamlines when they are straight and parallel. This is often called a hydrostatic pressure distribution because the relation between z and p/γ is the same as that from a fluid at rest. This means that the Bernoulli equation of the single streamline may be extended to apply to two- and three-dimensional flows because at a given flow cross section $z + p/\gamma$ is the same for all streamlines as it is for the central streamline; stated in another way, it means that a single hydraulic grade line applies to all the streamlines of a flow, providing these streamlines are straight and parallel. In practice the engineer seldom if ever encounters flows containing precisely straight and parallel streamlines; however, because in pipes, ducts, and prismatic channels such lines are *essentially* straight and parallel, the approximation may be used for many practical calculations.

In ideal fluid flows, because of the absence of friction, the distribution of velocity over a cross section of a flow containing straight and parallel streamlines is uniform; that is, all fluid particles pass a given cross section at the same velocity, which is equal to the average velocity V. Accordingly, no adjustment of the $V^2/2g_n$ term of the Bernoulli equation is to be expected in extending the equation from the infinitesimal to the finite streamtube because $V^2/2g_n$ is constant across the streamtube and H is the same for every streamline in the streamtube. Later when friction is considered it will be shown that there usually is a nonuniform profile of velocity across a finite streamtube. A correction must be applied then to the value of $V^2/2g_n$ in the streamtube before the Bernoulli equation may be used to represent flow along a streamtube of finite cross section.

———————————— **Illustrative Problem** ————————————

Water flows through this section of cylindrical pipe. If the static pressure at point C is 35 kPa, what are the static pressures at A and B, and where is the hydraulic grade line at this flow cross section?

Relevant Equation and Given Data

$$\frac{p_1}{\gamma} + z_1 = \frac{p_2}{\gamma} + z_2 \tag{2.6}$$

$$\gamma = 9.8 \times 10^3 \, \text{N/m}^3 \qquad p_C = 35 \text{ kPa}$$

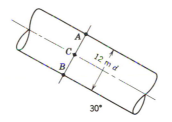

Solution.   Using equation 2.6,

$$p_A = 35.0 \times 10^3 - (9.8 \times 10^3)(0.866)0.6 = 29.9 \text{ kPa} \blacktriangleleft$$

$$p_B = 35.0 \times 10^3 + (9.8 \times 10^3)(0.866)0.6 = 40.1 \text{ kPa} \blacktriangleleft$$

The hydraulic grade line is $(35.0 \times 10^3)/9.8 \times 10^3 = 3.57$ m (vertically) above C. ◄

4.4   Applications of Bernoulli's Equation

Before proceeding to some engineering applications of Bernoulli's equation, it should first be noted that this equation gives further aid in the interpretation of streamline pictures, equation 4.1 indicating that, when velocity increases, the sum $(p/\gamma + z)$ of pressure and potential head must decrease. In many flow problems, the potential head z varies little, allowing the approximate general statement: where velocity is high, pressure is low. Regions of closely spaced streamlines have been shown (Section 3.4) to indicate regions of relatively high velocity, and now from the Bernoulli equation these are seen also to be regions of relatively low pressure.

In 1643, Torricelli showed that the velocity of efflux of an ideal fluid from a small orifice under a static head varies with the square root of the head. Today Torricelli's theorem is written

$$V = \sqrt{2g_n h}$$

the velocity being (ideally) equal to that attained by a solid body falling from rest through a height h. Torricelli's theorem is now recognized as a special case of the Bernoulli equation involving certain conditions appearing in many engineering problems. Torricelli's equation can be easily derived by applying Bernoulli's equation from the reservoir to the tip of the nozzle in Fig. 4.4. The reservoir is assumed to be very large (compared to the nozzle). Thus the small flow from the nozzle produces negligible

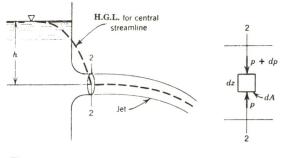

Fig. **4.4**

velocities in the reservoir except near the nozzle. Taking the datum plane at the center of the nozzle and choosing (arbitrarily) the center streamline give $h = z + p/\gamma$ in the reservoir where velocities are negligible. Writing Bernoulli's equation for a streamline between the reservoir and the tip of the nozzle,

$$\frac{p_1}{\gamma} + \frac{V_1^2}{2g_n} + z_1 = h = \frac{p_2}{\gamma} + \frac{V_2^2}{2g_n} + 0$$

Torricelli's equation results if $p_2 = 0$. From the validity of the Bernoulli and Torricelli equations it may be deduced that the pressure throughout the jet in the plane of the nozzle must be zero; however, analytical proof of this from Newton's law may be more convincing.

Consider the streamlines through section 2 of Fig. 4.4 to be essentially straight and parallel.[2] The pressure surrounding the jet is zero (gage) and the vertical acceleration of an elemental fluid mass $p\,dA\,dz$ is equal to g_n. The force causing the acceleration can result only from the pressure difference between top and bottom of the element, and the weight of the element. By writing Newton's second law in the vertical direction, we obtain

$$-(p + dp)\,dA + p\,dA - \gamma\,dA\,dz = -(\rho\,dA\,dz)g_n$$

from which it may be concluded that $dp = 0$. Thus there can be no pressure gradient across the jet at section 2, and with the pressures zero at the jet boundaries it follows that the pressure throughout the jet must be zero. This also shows that V cannot be constant across the jet because $p = 0$ and z varies. Downstream from section 2 it is customary to *assume* that the pressures throughout the free jet are also zero; this is equivalent to assuming that each fluid element follows a free trajectory streamline unaffected by adjacent fluid elements. This is an adequate approximation in many engineering problems, but it is not exact because the curvature and convergence of streamlines and effects such as surface tension were neglected.

[2]Actually the streamlines are curved and convergent (recall the trajectory theory of particle dynamics), but such convergence and curvature may be neglected for large h and high jet velocity.

--------- **Illustrative Problem** ---------

Calculate the flowrate through this pipeline and nozzle. Calculate the pressures at points 1, 2, 3, and 4 in the pipe and the elevation of the top of the water jet's trajectory.

Relevant Equations and Given Data

$$Q = AV = \text{Constant} \qquad\qquad (3.11 \text{ or } 3.12)$$

$$\frac{p}{\gamma} + \frac{V^2}{2g_n} + z = H = \text{Constant} \qquad\qquad (4.1)$$

$$\gamma = 9\,800 \text{ N/m}^3$$

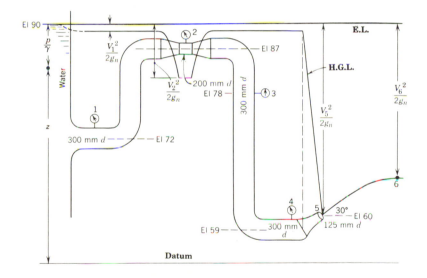

Solution. First sketch the energy line; at all points in the reservoir where the velocity is negligible, $(z + p/\gamma)$ is the same. Thus the energy line has the same elevation as the water surface. Next sketch the hydraulic grade line; this is coincident with the energy line in the reservoir where velocity is negligible but drops below the energy line over the pipe entrance where velocity is gained. The velocity in the 300 mm pipe is everywhere the same, so the hydraulic grade line must be horizontal until the flow encounters the constriction upstream from section 2. Here, as velocity increases, the hydraulic grade line must fall (possibly to a level below the constriction). Downstream from the constriction the hydraulic grade line must rise to the original level over the 300 mm pipe and continue at this level to a point over the *base of the nozzle* at section 4. Over the nozzle the hydraulic grade line must fall to the nozzle tip and after that follow the jet, because the pressure in the jet is everywhere zero.

Since the vertical distance between the energy line and the hydraulic grade line at any point is the velocity head at that point, it is evident that

$$\frac{V_5^2}{2g_n} = 30 \qquad \text{and therefore} \qquad V_5 = 24.3 \text{ m/s}$$

Thus

$$Q = 24.3 \times \frac{\pi}{4}(0.125)^2 = 0.3 \, \text{m}^3/\text{s} \blacktriangleleft \qquad \text{(3.11 or 3.12)}$$

From continuity considerations,

$$V_1 = \left(\frac{125}{300}\right)^2 V_5 \qquad \text{(3.12)}$$

and thus

$$\frac{V_1^2}{2g_n} = \left(\frac{125}{300}\right)^4 \frac{V_5^2}{2g_n} = \left(\frac{125}{300}\right)^4 30 = 0.9 \, \text{m}$$

Similarly

$$\frac{V_2^2}{2g_n} = \left(\frac{125}{200}\right)^4 30 = 4.58 \, \text{m}$$

Since the pressure heads are conventionally taken as the vertical distances between the pipe centerline and hydraulic grade line, the pressures in pipe and constriction may be computed as follows[3]:

$$p_1/\gamma = 18 - 0.9 = 17.1 \, \text{m} \qquad p_1 = 167.6 \text{ kPa} \blacktriangleleft$$

$$p_2/\gamma = 3 - 4.58 = -1.58 \, \text{m} \qquad p_2 = 116 \, \text{mm Hg vacuum} \blacktriangleleft$$

$$p_3/\gamma = 12 - 0.9 = 11.1 \, \text{m} \qquad p_3 = 108.8 \text{ kPa} \blacktriangleleft$$

$$p_4/\gamma = 31 - 0.9 = 30.1 \, \text{m} \qquad p_4 = 295.0 \text{ kPa} \blacktriangleleft$$

The velocity of the jet at the top of its trajectory (where there is no vertical component of velocity) is given by

$$V_6 = 24.3 \cos 30° = 21.0 \, \text{m/s}$$

and the elevation here is $90 - (21.0)^2/2g_n = 67.5 \, \text{m}$. $\blacktriangleleft$

With increasing velocity or potential head, the pressure within a flowing fluid drops. However, this pressure does not drop below the absolute zero of pressure as it has been found experimentally that fluids will not sustain the tension implied by pressures less than absolute zero. Thus, a practical physical restriction is placed on the Bernoulli equation. Such a restriction is, in fact, not appropriate for gases because they expand with reduction in pressure, but it frequently assumes great importance in the flow of liquids. Actually, in liquids the absolute pressure can drop only to the vapor pressure of the liquid, whereupon spontaneous vaporization (boiling) takes place. This

---

[3]Note that you are effectively using equation 4.1 here.

vaporization or formation of vapor cavities is called cavitation.[4] The formation, translation with the fluid motion, and subsequent rapid collapse of these cavities produces vibration, destructive pitting, and other deleterious effects on hydraulic machinery, hydrofoils, and ship propellers, for example.

─────────── **Illustrative Problem** ───────────

The barometric pressure is 14.0 psia. What diameter of constriction can be expected to produce incipient cavitation at the throat of the constriction?

Relevant Equations and Given Data

$$Q = AV = \text{Constant} \qquad (3.11 \text{ or } 3.12)$$

$$\frac{p}{\gamma} + \frac{V^2}{2g_n} + z = H = \text{Constant} \qquad (4.1)$$

(apply by use of **E.L.** and **H.G.L.**)

$$p_{\text{barometric}} = 14.0 \text{ psia}$$

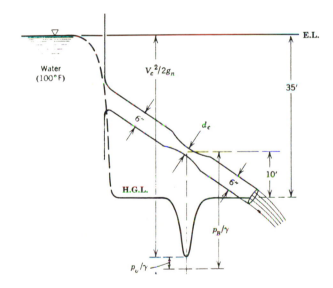

Solution. From Appendix 2, $\gamma = 62.0$ lb/ft³ and $p_v/\gamma = 2.2$ ft.

$$\frac{p_B}{\gamma} = 14.0 \times \frac{144}{62.0} = 32.5 \text{ ft}$$

[4]For a description of cavitation phenomena, see Appendix 7.

Construct **E.L.** and **H.G.L.** as indicated on the sketch. The velocity head at the constriction is given by the distance between the **E.L.** and the **H.G.L.**, that is,

$$\frac{V_c^2}{2g_n} = 35.0 - 10.0 + 32.5 - 2.2 = 55.3 \text{ ft}$$

However,

$$\frac{V_6^2}{2g_n} = 35.0 \text{ ft}$$

Since, from continuity considerations,

$$\left(\frac{d_c}{6}\right)^2 = \frac{V_6}{V_c} \qquad \left(\frac{d_c}{6}\right)^4 = \left(\frac{V_6}{V_c}\right)^2 = \frac{35.0}{55.3} \tag{3.12}$$

yielding $d_c = 5.35$ in. ◀ *Incipient* cavitation must be assumed in such ideal fluid flow problems for losses of head to be negligible; also with small cavitation there is more likelihood of the pipe flowing full at its exit. For a larger cavitation zone the same result will be obtained if losses are neglected and pipes assumed full at exit; however, the low point of the **H.G.L.** will be considerably flattened. See Appendix 7.

The Bernoulli equation is frequently written in terms of pressure rather than head and may be obtained in this form by multiplying equation 4.1 by γ and substituting ρ for γ/g_n; this results in

$$p_1 + \tfrac{1}{2}\rho V_1^2 + \rho g_n z_1 = p_2 + \tfrac{1}{2}\rho V_2^2 + \rho g_n z_2 \tag{4.2}$$

Here the Bernoulli terms p, $\rho V^2/2$, and $\rho g_n z$ are called static pressure, dynamic pressure, and potential pressure, respectively. The *stagnation (or total) pressure*, p_s is defined by

$$p_s = p_o + \tfrac{1}{2}\rho V_o^2 \tag{4.3}$$

and from Fig. 4.5 it can be seen that this is the local pressure at the tip of a pitot tube

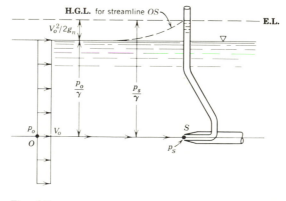

Fig. **4.5**

or, more generally, the pressure at the zero-velocity point on the nose of any solid object in a flow. This point is called appropriately a *stagnation point* because here the flow momentarily stops, or *stagnates*. The variation of pressure and velocity along the central streamline to the nose of a solid object is shown in Fig. 4.5. Note that the pressure rises rather abruptly (but not discontinuously) from p_o to p_s just in front of the object, while the velocity decreases from V_o to zero. With stagnation pressure at s easily measurable, and pressure p_o known or measurable, the velocity V_o of the undisturbed stream can be computed from equation 4.3; this is the essence of the pitot tube principle which finds wide application in many velocity-measuring devices.

Illustrative Problem

This pitot-static tube is carefully aligned with an airstream of density 1.23 kg/m³. If the attached differential manometer shows a reading of 150 mm of water, what is the velocity of the airstream?

Relevant Equation and Given Data

$$p_s = p_o + \tfrac{1}{2}\rho V_o^2 \tag{4.3}$$

$$\rho_a = 1.23 \text{ kg/m}^3 \qquad \gamma_w = 9\ 810 \text{ N/m}^3$$

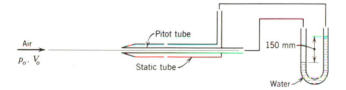

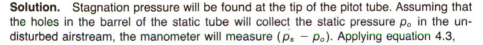

Solution. Stagnation pressure will be found at the tip of the pitot tube. Assuming that the holes in the barrel of the static tube will collect the static pressure p_o in the undisturbed airstream, the manometer will measure $(p_s - p_o)$. Applying equation 4.3,

$$p_s - p_o = \tfrac{1}{2}\rho V_o^2 \tag{4.3}$$

Therefore

$$0.15(9\ 810 - 1.23 \times 9.81) = \tfrac{1}{2}(1.23)V_o^2 \qquad V_o = 48.9 \text{ m/s} \blacktriangleleft$$

A constriction (Fig. 4.6) in a streamtube or pipeline is frequently used as a device for metering fluid flow. Simultaneous application of the continuity and Bernoulli principles to such a constriction allows direct calculation of the flowrate when certain variables are measured.

The equations are

$$Q = A_1 V_1 = A_2 V_2 \tag{3.12}$$

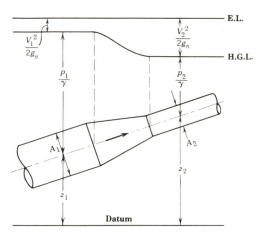

Fig. **4.6**

$$\frac{p_1}{\gamma} + \frac{V_1^2}{2g_n} + z_1 = \frac{p_2}{\gamma} + \frac{V_2^2}{2g_n} + z_2 \qquad (4.1)$$

By simultaneous solution of these equations the flowrate through a constriction can be easily computed if cross-sectional areas of pipe and constriction are known and pressures and heights above datum are measured, either separately or in the combination indicated.

_____ **Illustrative Problem** _____

Calculate the flowrate of gasoline (s.g. = 0.82) through this pipeline, using first the gage readings and then the manometer reading.

Relevant Equations and Given Data

$$Q = AV = \text{Constant} \qquad (3.11 \text{ or } 3.12)$$

$$\frac{p}{\gamma} + \frac{V^2}{2g_n} + z = H = \text{Constant} \qquad (4.1)$$

Solution. With sizes of pipe and constriction given, the problem centers around the computation of the value of the quantity $(p_1/\gamma + z_1 - p_2/\gamma - z_2)$ of the Bernoulli equation.

Taking datum at the lower gage and using the gage readings,

$$\left(\frac{p_1}{\gamma} + z_1 - \frac{p_2}{\gamma} - z_2\right) = \frac{20 \times 144}{0.82 \times 62.4} - \frac{10 \times 144}{0.82 \times 62.4} - 4$$

$$= 24.2 \text{ ft of gasoline}$$

To use the manometer reading, construct the hydraulic grade line levels at 1 and 2. It is evident at once that the difference in these levels is the quantity $(p_1/\gamma + z_1 -$

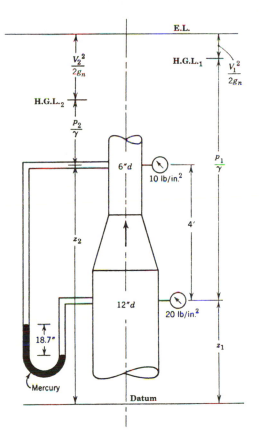

$p_2/\gamma - z_2$) and, visualizing p_1/γ and p_2/γ as liquid columns, it is also apparent that the difference in the hydraulic grade line levels is equivalent to the manometer reading. Therefore

$$\left(\frac{p_1}{\gamma} + z_1 - \frac{p_2}{\gamma} - z_2 \right) = \frac{18.7}{12} \left(\frac{13.57 - 0.82}{0.82} \right) = 24.2 \text{ ft of gasoline}$$

Insertion of 24.2 ft into the Bernoulli equation 4.1 followed by its simultaneous solution with the continuity equation 3.12 yields a flowrate Q of 8.03 cfs. ◀

The Bernoulli principle may be, of course, applied to problems of *open flow* such as the overflow structure of Fig. 4.7. Such problems feature a moving liquid surface in contact with the atmosphere and flow pictures dominated by gravitational action. A short distance upstream from the structure, the streamlines will be straight and parallel and the velocity distribution will be uniform. In this region the quantity $z + p/\gamma$ will be constant, the pressure distribution hydrostatic, and the hydraulic grade line (for all streamtubes) located in the liquid surface; the energy line will be horizontal and located $V_i^2/2g_n$ above the liquid surface. With atmospheric pressure on the liquid surface the

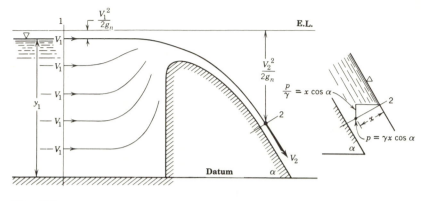

Fig. **4.7**

streamtube in the liquid surface behaves as a free jet, allowing all surface velocities to be computed from the positions of liquid surface and energy line. The prediction of velocities elsewhere in the flowfield where streamtubes are severely convergent or curved is outside the province of one-dimensional flow; suffice it to say that all such velocities are interdependent and the whole flowfield must be established before any single velocity can be computed. At section 2, however (if the streamlines there are assumed straight and parallel), the pressures and velocities may be computed from the one-dimensional assumption.

Illustrative Problem

Refer to Fig. 4.7. At section 2 the water surface is at elevation 30.5, and the 60° spillway surface at elevation 30.0. The velocity in the water surface V_{s2} at section 2 is 6.1 m/s. Calculate the pressure and velocity on the spillway face at section 2. If the bottom of the approach channel is at elevation 29.0, calculate the depth and velocity in the approach channel.

Relevant Equations and Given Data

$$\frac{p}{\gamma} + z = \text{Constant} \qquad \text{(across streamlines: 1–D assumption)} \qquad (2.6)$$

$$q = hV = \text{Constant} \qquad (3.14)$$

$$\frac{p}{\gamma} + \frac{V^2}{2g_n} + z = H = \text{Constant} \qquad (4.1)$$

$$\gamma = 9.8 \times 10^3 \text{ N/m}^3$$

Solution. Thickness of sheet of water at section 2 = $(30.5 - 30.0)/\cos 60° = 1$ m. Apply equation 2.6 to obtain the pressure on spillway face at section 2 = $1 \times 9.8 \times 10^3 \times \cos 60° = 4.9$ kPa. ◀ Elevation of energy line = $30.5 + (6.1)^2/2g_n = 32.4$ m.

$$32.4 = \frac{4.9}{9.8} + \frac{V_{F2}^2}{2g_n} + 30.0 \qquad V_{F2} = 6.1 \text{ m/s} \blacktriangleleft \qquad (4.1)$$

which is to be expected from the one-dimensional assumption. Evidently all velocities through section 2 are 6.1 m/s, so

$$q = 1 \times 6.1 = 6.1 \text{ m}^3/\text{s per metre of spillway length}$$

At section 1, $y_1 + 29.0 = z_1 + p_1/\gamma$; and, applying the Bernoulli equation between sections 1 and 2,

$$y_1 + 29.0 + \frac{V_1^2}{2g_n} = y_1 + 29.0 + \left(\frac{1}{2g_n}\right)\left(\frac{6.1}{y_1}\right)^2 = 32.4 \text{ m} \qquad (4.1)$$

Solving this (cubic) equation by trial yields the roots $y_1 = 3.22$ m, 0.85 m, and -0.69 m. Obviously the second and third roots are invalid here, so the depth in the approach channel will be 3.22 m. ◄ The velocity V_1 may be computed from

$$\frac{V_1^2}{2g_n} = 3.4 - 3.22 \qquad (4.1)$$

or from

$$V_1 = \frac{6.1}{3.22} \qquad (3.14)$$

both of which give $V_1 = 1.9$ m/s. ◄

4.5 The Work-Energy Equation

The previous analysis of a fluid system yielded a useful and practical equation, the Bernoulli equation 4.1, which was applied to establish the relations between elevation, pressure, velocity and total head in a variety of fluid flows. However, many pipelines, for example, contain either pumps or turbines which, respectively, add energy to or extract it from the fluid. There is no satisfactory manner for incorporating these effects in the derivation of the Bernoulli equation, which was based on an analysis of a fluid system. On the other hand, a physically meaningful derivation and additional insight into the physical meanings of the various terms in the Bernoulli equation can be obtained via a control volume analysis.

This control volume analysis utilizes the mechanical work-energy principle and applies it to fluid flow, resulting in a powerful relationship between fluid properties, work done, and energy transported. The Bernoulli equation is then seen to be equivalent to the mechanical work-energy equation for ideal fluid flow.

Consider the streamtube section shown in Fig. 4.8. The flow is steady, and a fluid system occupies zones I and R of the control volume 1221 at time t and zones R and O at time $t + dt$. The streamtube may be of finite or differential size but it is assumed that the flow is one-dimensional at sections 1 and 2. Note that this requirement is to be met only at the entrance and exit of the control volume (Fig. 4.8). There are no restrictions on streamline behavior within the control volume. Custom dictates that p

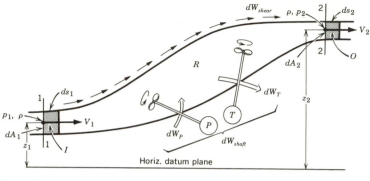

Fig. 4.8

and z are evaluated at the centerline of the flow. The continuity equation 3.12 gives

$$dA_1V_1 = dA_2V_2 \quad \text{or} \quad dA_1ds_1 = dA_2ds_2$$

(see Section 3.5) because ρ is constant here.

From dynamics the mechanical work-energy[5] relation (which is, of course, only an integrated form of Newton's second law) states that the work dW (expressed as a force acting over a distance) done by a system is exactly balanced by an equivalent change in the sum of the kinetic (KE) and potential (PE) energies of the system, that is, in time dt

$$dW + d(KE + PE) = dW + (KE + PE)_{t+dt} - (KE + PE)_t = 0$$

Now,

$$(KE + PE)_t = (KE + PE)_R + (KE + PE)_I$$

$$= (KE + PE)_R + \tfrac{1}{2}(\rho \, dA_1 \, ds_1)V_1^2 + \gamma(dA_1 \, ds_1)z_1$$

$$(KE + PE)_{t+dt} = (KE + PE)_R + (KE + PE)_O$$

$$= (KE + PE)_R + \tfrac{1}{2}(\rho \, dA_2 \, ds_2)V_2^2 + \gamma(dA_2 \, ds_2)z_2$$

because kinetic energy of translation is $\tfrac{1}{2}mv^2$ and potential energy is equivalent to the work of raising the weight of fluid in a zone to a height z above the datum.

The work dW done by the system takes three forms:

1. *Flow work* done via fluid entering or leaving the control volume.
2. *Shaft work* dW_{shaft} done by pumps and turbines.
3. *Shear work* dW_{shear} done by shearing forces acting across the boundary of the system.

---

[5]Heat transfers and internal energy are neglected. If they were included, the result would be the First Law of Thermodynamics (see Chapter 5).

The work done *on* the system by the fluid entering I in time dt is the flow work

$$(p_1 \, dA_1) \, ds_1$$

Hence, the flow work done *by* the system is

$$-(p_1 \, dA_1) \, ds_1$$

The system does work on the fluid in O in time dt, so the flow work by the system is

$$(p_2 \, dA_2) \, ds_2$$

The work done by turbines is $dW_T \geq 0$ because effectively the system does work on the surroundings (energy is extracted from the system), while pump work $dW_P \leq 0$ (work is done on the system; energy is put in). The work transfer is assumed to occur where the machine shaft cuts the boundary and the actual mechanism of, and conditions in, the machine are not of concern.

If the system boundary cuts through a fluid zone so that the system can exert a shear force across the boundary on fluid that is in motion, $dW_{shear} \neq 0$. In all cases studied in this and the next chapter, $dW_{shear} = 0$ because the fluid is inviscid and cannot support shear stress; dW_{shear} is thus neglected. In addition, the flow work done on or by the system is accomplished only at cross sections 11 and 22, as calculated above, because there is no motion perpendicular to the streamtube; therefore, the lateral pressure forces can do no work. Finally, because all internal forces appear in equal and opposite pairs, there is no net work done internally.

In sum, then,

$$dW_{shaft} - (p_1 \, dA_1) \, ds_1 + (p_2 \, dA_2) \, ds_2 + \tfrac{1}{2}(\rho \, dA_2 \, ds_2)V_2^2 + \gamma(dA_2 \, ds_2)z_2$$
$$-\tfrac{1}{2}(\rho \, dA_1 \, ds_1)V_1^2 - \gamma(dA_1 \, ds_1)z_1 = 0$$

Recall that $ds = V \, dt$, define the steady mass flowrate $\dot{m} = \rho \, dA \, V$ through the control volume, and divide by $\dot{m} \, dt = \rho \, dA_1 \, V_1 \, dt = \rho \, dA_1 \, ds_1 = \rho \, dA_2 \, ds_2$ to obtain

$$\frac{1}{\dot{m}} \frac{dW_{shaft}}{dt} - \frac{p_1}{\rho} + \frac{p_2}{\rho} + \tfrac{1}{2}(V_2^2 - V_1^2) + g_n(z_2 - z_1) = 0 \qquad (4.4)$$

Next,

$$\frac{1}{\dot{m}} \frac{dW_{shaft}}{dt} = \frac{1}{\dot{m}} \frac{dW_T}{dt} - \frac{1}{\dot{m}} \frac{dW_P}{dt} = g_n E_T - g_n E_P = \left[\frac{ft \cdot lb}{slug} \right] = \left[\frac{J}{kg} \right]$$

where E_T (ft · lb/lb or N · m/N) and E_P are the energies withdrawn by turbines or added by pumps *per unit weight* of fluid passing through the control volume. Hence, by introducing this nomenclature and dividing by g_n, we obtain

$$\frac{p_1}{\gamma} + \frac{V_1^2}{2g_n} + z_1 + E_P = \frac{p_2}{\gamma} + \frac{V_2^2}{2g_n} + z_2 + E_T \qquad (4.5)$$

which can be interpreted now as a mechanical energy equation. Terms such as p_1/γ, $V_1^2/2g_n$, z, and so forth, have the units of ft · lb/lb or N · m/N = J/N (joules per newton) which represent energy per unit weight of fluid. The energy form of the equation is important in that it lays the basis for later considerations of heat transfer, internal energy, and the like, which do not follow logically from the integrated Euler equation. Of course, if E_P and E_T are zero, equation 4.5 reduces to the Bernoulli equation 4.1, which therefore is the mechanical energy equation in the absence of shaft work.

The addition of mechanical energy (E_P) to a fluid flow by a pump or its extraction (E_T) by a turbine will appear as abrupt rises or falls of the energy line over the respective machines. Usually the engineer requires the total power of such machines, which may be computed from the product of weight flowrate (G) and (E_P or E_T), yielding total power. In terms of horsepower and using English FSS units,

$$\text{Horsepower of machine} = \frac{G(E_P \text{ or } E_T)}{550} = \frac{Q\gamma(E_P \text{ or } E_T)}{550} \tag{4.6}$$

while in SI units,

$$\text{Kilowatts of machine} = \frac{Q\gamma(E_P \text{ or } E_T)}{1\ 000} \tag{4.7}$$

Note 1 hp = 0.746 kW. These equations may also be used as a general equation for converting any unit energy, or head, to the corresponding total power.

_____ **Illustrative Problem** _____

How much power must be supplied for the pump to maintain readings of 250 mm of mercury vacuum and 275 kPa on gages 1 and 2, respectively, while delivering a flowrate of 0.15 m³/s of water?

Relevant Equations and Given Data

$$\frac{p_1}{\gamma} + \frac{V_1^2}{2g_n} + z_1 + E_P = \frac{p_2}{\gamma} + \frac{V_2^2}{2g_n} + z_2 + E_T \tag{4.5}$$

$$\text{Kilowatts of machine} = \frac{Q\gamma(E_P \text{ or } E_T)}{1\ 000} \tag{4.7}$$

$$p_1 = 250 \text{ mm Hg vacuum} \qquad p_2 = 275 \text{ kPa}$$

$$Q = 0.15 \text{ m}^3/\text{s} \qquad \gamma = 9.8 \times 10^3 \text{ N/m}^3$$

Solution. Converting the gage readings to metres of water, $p_1/\gamma = -3.39$ m and $p_2/\gamma = +28.1$ m. Thus the hydraulic grade lines are 3.39 m below and 28.1 m above points 1 and 2, respectively. The velocity heads at these points will be found to be 1.16 m and 3.68 m, respectively, so the energy lines will be these distances above the respective hydraulic grade lines. The rise in the energy line through the pump represents the energy supplied by the pump to each newton of fluid; from the sketch this is

$$E_P = 3.68 + 28.1 + 3 + 3.39 - 1.16 = 37.0 \text{ J/N} \qquad (4.5)$$

$$\text{Pump power} = 0.15 \times 9.8 \times 10^3 \times 37.0/10^3 = 54.4 \text{ kW} \blacktriangleleft \qquad (4.7)$$

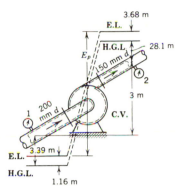

It should be noted that the indicated **H.G.L.**'s are for the pipes only and do not include the pump passages, where the flow is not one-dimensional. The positions of these **H.G.L.**'s give no assurance that the pump will run cavitation-free, since local velocities in the pump passages may be considerably larger than the average velocities in the pipes.

TWO-DIMENSIONAL FLOW

The solution of flowfield problems is much more complex than the solution of one-dimensional flow. Partial differential equations are invariably required for a formal mathematical approach to such problems. In many cases of two-dimensional ideal flow, the theory of complex variables provides exact solutions. In more general cases of real or ideal flow, computer-based numerical solutions of the equations of motion enjoy wide success and engineering applicability. Many approximate techniques also exist for problem solution. The objective of the remainder of this chapter is to present an introduction to certain essentials and practical problems of importance to engineering students. To stress the similarities and differences of one- and two-dimensional flows, the subject is developed in parallel with the preceding treatment of one-dimensional flow. Coverage of advanced mathematical operations or broad generalizations is not attempted at this point;[6] the emphasis is on giving the beginner some appreciation of the intricacies of flowfield problems as compared to the relative simplicity of those of one-dimensional flow.

[6]Some of these are found in Chapter 12

4.6 Euler's Equations

Euler's equations for a vertical two-dimensional flowfield may be derived by applying Newton's second law to a basic differential *system* of fluid of dimensions dx by dz (Fig. 4.9). The forces dF_x and dF_z on such an elemental system have been identified in equations 2.1 and 2.2 of Section 2.1, and with substitution of appropriate pressures they reduce to

$$dF_x = -\frac{\partial p}{\partial x} \, dx \, dz$$

$$dF_z = -\frac{\partial p}{\partial z} \, dx \, dz - \rho g_n \, dx \, dz$$

The accelerations of the system have been derived for *steady flow* (equation 3.4, Section 3.3) as

$$a_x = u\frac{\partial u}{\partial x} + w\frac{\partial u}{\partial z}$$

$$a_z = u\frac{\partial w}{\partial x} + w\frac{\partial w}{\partial z}$$

Applying Newton's second law by equating the differential forces to the products of the mass of the system and respective accelerations gives

$$-\frac{\partial p}{\partial x} \, dx \, dz = \rho \, dx \, dz\left(u\frac{\partial u}{\partial x} + w\frac{\partial u}{\partial z}\right)$$

$$-\frac{\partial p}{\partial z} \, dx \, dz - \rho g_n \, dx \, dz = \rho \, dx \, dz\left(u\frac{\partial w}{\partial x} + w\frac{\partial w}{\partial z}\right)$$

and by cancellation of $dx \, dz$ and slight rearrangement, the Euler equations of two-

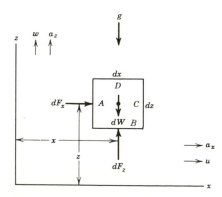

Fig. **4.9**

dimensional flow in a vertical plane are

$$-\frac{1}{\rho}\frac{\partial p}{\partial x} = u\frac{\partial u}{\partial x} + w\frac{\partial u}{\partial z} \tag{4.8}$$

$$-\frac{1}{\rho}\frac{\partial p}{\partial z} = u\frac{\partial w}{\partial x} + w\frac{\partial w}{\partial z} + g_n \tag{4.9}$$

Accompanied by the equation of continuity,

$$\frac{\partial u}{\partial x} + \frac{\partial w}{\partial z} = 0 \tag{3.18}$$

the Euler equations form a set of three simultaneous partial differential equations that are basic to the solution of two-dimensional flowfield problems; complete solution of these equations yields p, u, and w as functions of x and z, allowing prediction of pressure and velocity at any point in the flowfield. It is intriguing that application of a Lagrangian approach, that is, studying the dynamics of a small individual fluid system, has yielded the Eulerian equations of motion.

4.7 Bernoulli's Equation

Bernoulli's equation can be derived by integrating the Euler equations for a uniform density flow, as was done for one-dimensional flow (Section 4.2). By multiplying the first equation by dx and the second by dz and adding them, we find that

$$-\frac{1}{\rho}\left(\frac{\partial p}{\partial x}\,dx + \frac{\partial p}{\partial z}\,dz\right) = u\frac{\partial u}{\partial x}\,dx + w\frac{\partial u}{\partial z}\,dx + u\frac{\partial w}{\partial x}\,dz + w\frac{\partial w}{\partial z}\,dz + g_n\,dz$$

The terms $w(\partial w/\partial x)\,dx$ and $u(\partial u/\partial z)\,dz$ are added to and subtracted from the right-hand side of the equation, and terms are then collected in the following pattern:

$$-\frac{1}{\rho}\left(\frac{\partial p}{\partial x}\,dx + \frac{\partial p}{\partial z}\,dz\right) = \left(u\frac{\partial u}{\partial x}\,dx + w\frac{\partial w}{\partial x}\,dx\right)$$

$$+ \left(u\frac{\partial u}{\partial z}\,dz + w\frac{\partial w}{\partial z}\,dz\right) + (u\,dz - w\,dx)\left(\frac{\partial w}{\partial x} - \frac{\partial u}{\partial z}\right) + g_n\,dz$$

The bracket on the left-hand side of this equation is (see Appendix 5) the total differential dp. The sum of the first two brackets on the right-hand side is easily shown to be $d(u^2 + w^2)/2$, and the third bracket is the vorticity, ξ (section 3.7). Reducing the equation accordingly and dividing it by g_n lead to

$$-\frac{dp}{\gamma} = \frac{d(u^2 + w^2)}{2g_n} + \frac{1}{g_n}(u\,dz - w\,dx)\xi + dz$$

and after integration to

$$\frac{p}{\gamma} + \frac{u^2 + w^2}{2g_n} + z = H - \frac{1}{g_n}\int \xi(u\,dz - w\,dx)$$

in which H is the constant of integration. Because the magnitude of the resultant velocity V at any point in the flowfield is related to its components u and w by $V^2 = u^2 + w^3$, the equation further simplifies to

$$\frac{p}{\gamma} + \frac{V^2}{2g_n} + z = H - \frac{1}{g_n}\int \xi(u\ dz - w\ dx) \qquad (4.10)$$

This equation thus shows that the sum of the Bernoulli terms at *any* point in a steady flowfield is a constant H if the vorticity ξ is zero, that is, *if the flowfield is an irrotational* (or potential) one. Thus *for irrotational flow the same constant applies to all the streamlines of the flowfield* or, in terms of the energy line, all fluid masses in an irrotational flowfield possess the same unit energy. In a rotational flowfield the integral in equation 4.10 must be evaluated. However, along a streamline in any steady flow $dz/dx = w/u$ by definition. Thus $u\ dz = w\ dx$; that is, $u\ dz - w\ dx = 0$ along a streamline. In a rotational flow, the Bernoulli terms are constant along any streamline, but the constant is different for each streamline.

4.8 Application of Bernoulli's Equation

For irrotational flow of an ideal incompressible fluid, the Bernoulli equation may be applied over the flowfield with a single (horizontal) energy line completely describing the energy situation. Figure 4.10 depicts this for two representative points A and B. From the position of the points (above datum) the quantity $(p/\gamma + V^2/2g_n)$ may be determined from the position of the energy line, but the pressures p_A, p_B, and the like, cannot be calculated until the corresponding velocities V_A, V_B, and so forth, are known. However, in a flowfield all the velocities are interdependent and are determined by the streamline definition and by the differential equation (3.18) of continuity; until methods[7] are described for solving these equations, the pressures in the flowfield cannot

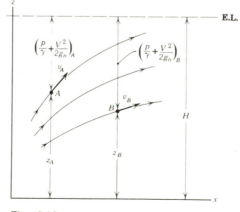

Fig. **4.10**

[7]Some of these methods are cited in Chapter 12.

be accurately predicted. However, the lack of formal mathematical solutions describing the entire velocity field need not deter the engineer from making a semiquantitative approach to such problems; indeed, when no formal solutions exist (as frequently happens) this is the only alternative.

One very useful tool in such approaches to flowfield problems is the effect of flow curvature on the pressure variation across the flow. For any element of streamline (Fig. 4.11) having radius of curvature r, the normal component of acceleration a_r is directed toward the center of curvature and is equal to V^2/r (equation 3.2). Newton's second law, applied to such an elemental fluid system on the streamline, yields a general and useful result in the analysis and interpretation of flowfields. In the radial direction the components of force on the system are

$$(p + dp)\, ds - p\, ds + dW \cos \theta = \frac{(\rho\, dr\, ds)V^2}{r}$$

From the geometry of the system and streamline, $\cos \theta = dz/dr$ and $dW = \gamma\, dr\, ds$. Substituting these values in the equation above, dividing by $\gamma\, ds$, and rearranging produce

$$\frac{d}{dr}\left(\frac{p}{\gamma} + z\right) = \frac{V^2}{g_n r} \tag{4.11}$$

from which it is seen that the gradient of $(p/\gamma + z)$ along r is always positive, or that an increase of $(p/\gamma + z)$ is to be expected along a direction outward from the center of curvature; conversely, a drop of $(p/\gamma + z)$ is expected along a direction toward the center of curvature. Although this equation is formally integrable only for vortex motion (where there is a single center of curvature), it is nevertheless of great value to the engineer in the "reading" of streamline pictures to distinguish regions of high and low $(p/\gamma + z)$. The reader will sense that this development, which predicts the variation of $(p/\gamma + z)$ in the radial direction, will also determine (through the Bernoulli equation) the variation of velocity with radial distance. This variation may be discovered by taking the derivative (with respect to r) of the Bernoulli equation of irrotational flow:

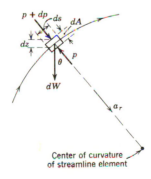

Center of curvature
of streamline element

Fig. **4.11**

$$\frac{d}{dr}\left(\frac{p}{\gamma} + \frac{V^2}{2g_n} + z = H\right) = \frac{d}{dr}\left(\frac{p}{\gamma} + z\right) + \frac{2V}{2g_n}\frac{dV}{dr} = 0$$

or

$$\frac{d}{dr}\left(\frac{p}{\gamma} + z\right) = -\frac{V}{g_n}\frac{dV}{dr}$$

which may be equated to the expression (equation 4.11) obtained from Newton's second law. Thus

$$\frac{V^2}{g_n r} = -\frac{V}{g_n}\frac{dV}{dr} \qquad \text{or} \qquad V\,dr + r\,dV = 0$$

which when integrated (for a flowfield having a single center of curvature) yields

$$Vr = \text{Constant} \tag{4.12}$$

Again it is emphasized that the equation cannot be generally integrated over a whole flowfield because of the numerous centers of curvature of the streamline elements, but it may nevertheless be used to conclude that generally a *decrease*[8] of velocity is to be expected with *increase* of distance from center of curvature in irrotational flowfields. The reader may be familiar with this type of velocity distribution from casual experience with vortices in the atmosphere, tornadoes, or the "bathtub vortex" which frequently develops when a tank is drained through an orifice in the bottom.

Figure 4.12 depicts a curved flow occurring through a passage in a vertical plane. Distortion of the velocity profile from a uniform distribution is shown; pitot tube and piezometer columns also indicate the variation of pressure and velocity throughout the flow. If these facts are supplemented by a streamline picture, streamlines that are uniformly spaced in the straight passages will be crowded together toward the inner wall of the curved passage and widely spaced toward its outer wall; *thus they cannot be concentric circular arcs*.

Another example of the effect of streamline curvature in a flowfield is seen in the convergent-divergent passage of Fig. 4.13. The streamlines *AA* along the walls are most sharply curved, whereas the central streamline possesses no curvature, and the streamlines in the region between walls and centerline are of intermediate curvature. Accordingly, it can be deduced that the velocity profile at section 2 features higher velocity at the walls than on the centerline, and the relative position of the hydraulic grade lines for these streamlines is as shown. In passing, it is of interest to note that, if incipient cavitation occurred at section 2, it would be expected to appear on the *upper wall* of the passage; for both walls at that cross section $(p/\gamma + z)$ is the same but, with z larger for the upper wall, the pressure will be less there. This problem has been treated

---

[8]The beginner, schooled in the mechanics of solid-body rotation where tangential velocity varies linearly with radius, is frequently surprised to discover that this law does not apply to fluid flow. One should not be surprised, however, because the mobility of fluid particles in a curved flow is infinitely greater than that of solid particles which are "locked" together in a rotating object.

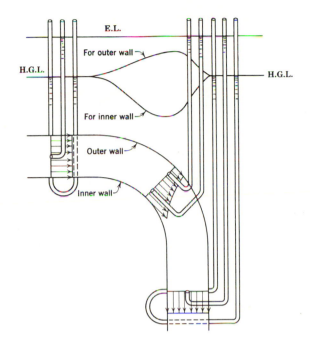

Fig. **4.12**

by one-dimensional methods in Section 4.4; it is clearly a one-dimensional problem at sections 1 and 3, but between these sections there is a flowfield which can be treated as one-dimensional only as an approximation. The application of the continuity equation to such problems is basic and instructive. Clearly

$$Q = A_1V_1 = A_2V_2 = A_3V_3 \tag{3.12}$$

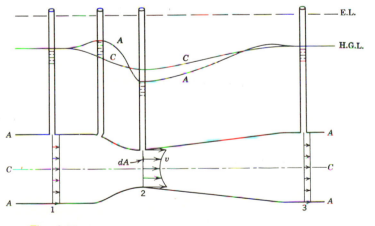

Fig. **4.13**

in which V_1 and V_3 are the respective mean velocities at sections 1 and 3 and also the velocities of individual particles as they pass these sections. At section 2, however, the velocities of fluid particles are very different from the mean velocity, and a relation between these is to be sought. From equation 3.15 the mean velocity V_2 is given by

$$V_2 = \frac{1}{A_2} \int^{A_2} v \, dA$$

It is noted that, although V_2 can be easily calculated from Q, this yields no information on the distribution of velocity, which is determined by the shape and curvature of the passage walls.

Rigorous application of the Bernoulli equation to the flow between sections 1 and 2 is complicated by the nonuniform velocity profile at section 2. Although the same energy line applies to all the streamlines of the flowfield, the engineer is concerned with the separate terms p, z, and v, all of which vary across the flow at section 2. Here the Bernoulli equation must be written in terms of total power rather than unit energy; following the pattern of equations 4.6 and 4.7,

$$\begin{bmatrix} \text{Total power} \\ \text{at section 1} \end{bmatrix} = Q\gamma\left(\frac{p_1}{\gamma} + z_1 + \frac{V_1^2}{2g_n}\right)$$

the quantities $(p/\gamma + z)$ and $V^2/2g_n$ being constant across section 1. At section 2 the total power of the flow must be expressed as an integral, following the same pattern. Using $dQ = v \, dA$,

$$\begin{bmatrix} \text{Total power} \\ \text{at section 2} \end{bmatrix} = \int^{A_2} (v \, dA)\gamma\left(\frac{p}{\gamma} + z + \frac{v^2}{2g_n}\right)$$

Equating the expressions for total power, the "Bernoulli equation" becomes

$$Q\left(\frac{p_1}{\gamma} + z_1 + \frac{V_1^2}{2g_n}\right) = \int^{A_2} v\left(\frac{p}{\gamma} + z\right) dA + \int^{A_2} \frac{v^3}{2g_n} dA \qquad (4.13)$$

for which a method of solution is suggested in the following illustrative problem.

Illustrative Problem

Ideal fluid flows through this symmetrical constriction in the two-dimensional passage. Show how the flowrate may be calculated.

Relevant Equation and Given Data

$$Q\left(\frac{p_1}{\gamma} + z_1 + \frac{V_1^2}{2g_n}\right) = \int^{A_2} v\left(\frac{p}{\gamma} + z\right) dA + \int^{A_2} \frac{v^3}{2g_n} dA \qquad (4.13)$$

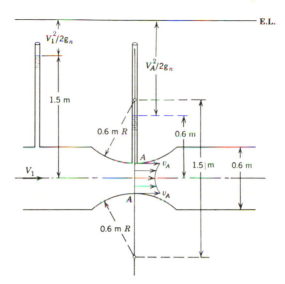

Solution. Draw the energy line from which $V_1^2/2g_n$ and $v_A^2/2g_n$ are identified. Calculate an approximate flowrate by the one-dimensional methods of Section 4.4, assuming v_A to be the mean velocity at section 2. As $v_A > V_2$, the approximate flowrate will be larger than the true flowrate; therefore assume a flowrate smaller than this for the first trial calculation. From the assumed Q, V_1 may be calculated (from the continuity principle), allowing the position of the energy line and the left-hand side of equation 4.13 to be tentatively established. Find a velocity profile of the form indicated that will satisfy the continuity equation for the assumed Q. From this velocity profile calculate the distribution of $(p/\gamma + z)$ across the flow. Finally, carry out the integrations (graphically if necessary) of equation 4.13 to see if this equation is satisfied; on the first trial it will not be, so assume another Q, repeating the procedure until the equation is satisfied. As a guide in the process, $d/dr\,(p/\gamma + z)$ at the walls may be computed from $v_A^2/g_n r$ for assistance in establishing a suitable distribution of $(p/\gamma + z)$.

The identification of stagnation points in flowfields is another useful adjunct to an understanding of the flow process. In ideal flow these points may be expected wherever a streamline is forced to turn a sharp corner; such points will be expected only on solid boundaries. Two stagnation points are shown in the external and internal flowfields of Fig. 4.14. At each stagnation point the velocity is locally zero, so the vertical distance between stagnation point and energy line is a direct measure of the pressure at the stagnation point. With this knowledge, the velocities at other points in the flowfield (if irrotational) may be calculated from static pressure measurements.

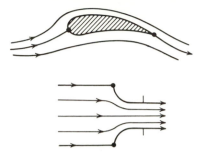

Fig. 4.14

---------------- **Illustrative Problem** ----------------

Ideal fluid of specific weight 50 lb/ft³ flows down this pipe and out into the atmosphere through the "end cap" orifice. Gage B reads 6.0 psi and gage A, 2.0 psi. Calculate the mean velocity in the pipe.

Relevant Equation and Given Data

$$\frac{p}{\gamma} + \frac{V^2}{2g_n} + z = H - \frac{1}{g_n}\int \xi(u\ dz - w\ dx) \tag{4.10}$$

$$p_A = 2.0 \text{ psi} \qquad p_B = 6.0 \text{ psi} \qquad \gamma = 50 \text{ lb/ft}^3$$

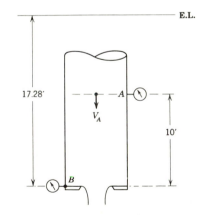

Solution. Identify point B as a stagnation point. The energy line will then be established $6 \times 144/50 = 17.28$ ft above B. Gage A measures the static pressure in the flow; p_A/γ is thus $2.0 \times 144/50 = 5.76$ ft. Therefore, for an irrotational flow where $\xi = 0$,

$$\frac{V_A^2}{2g_n} = 17.28 - 10 - 5.76 = 1.52 \text{ ft} \qquad V_A = 9.90 \text{ fps} \blacktriangleleft \tag{4.10}$$

A typical problem of two-dimensional irrotational open flow is that of the sharp-crested weir, as shown in Fig. 4.15. A short distance upstream from such a structure the streamlines will be essentially straight and parallel, and a one-dimensional situation will therefore exist. Between this section and some point in the falling sheet of liquid where free fall begins, a flowfield occurs about as shown. Because of the velocity (V) of approach to the weir, the energy line may be visualized above the flow picture as indicated. The boundary streamlines BB and AA (downstream from the weir crest) are called *free streamlines*, their precise position in space being unknown, but the pressure on them is everywhere constant, in this case zero (gage); once their position is estab-lished (by analysis or experiment), the velocity at any point on them may be calculated because the vertical distance between any point and the energy line is the velocity head ($v^2/2g_n$) at the point. The pressure distribution in the flow at section 1 is hydrostatic, the bottom pressure at A' being simply γy. The only stagnation point in the flow is noted at A'', at which the pressure will be $\gamma(y + V^2/2g_n)$. At any other point (C) in the flowfield, ($v_c^2/2g_n + p_c/\gamma$) can be computed from the positions of point and energy line; however, the pressure there is not calculable without the velocity v_c, which is interrelated with all other velocities and not generally predictable until complete details of the whole flowfield are known. The pressure distribution in the plane of the weir plate is qualitatively predictable because the pressures at both boundaries (free streamlines) are zero; with the streamlines of sharpest curvature nearest to the weir crest, $d/dr\,(p/\gamma + z)$ is largest here, producing a positive pressure in the flow as shown. Downstream from this cross section the pressure within the falling sheet diminishes, becoming essentially zero as the streamlines become straighter and more parallel.

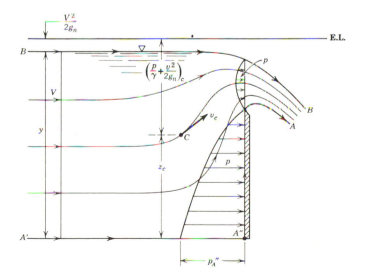

Fig. **4.15**

--- **Illustrative Problem** ---

Calculate the flowrate through this two-dimensional nozzle discharging to the atmosphere and designate any stagnation points in the flow. Assume irrotational flow.

Relevant Equations and Given Data

$$q = hV = \text{Constant} \tag{3.14}$$

$$V = \frac{1}{A}\int^A v \, dA \tag{3.15}$$

$$\frac{p_1}{\gamma} + \frac{V_1^2}{2g_n} + z_1 = \frac{p_2}{\gamma} + \frac{V_2^2}{2g_n} + z_2 \qquad \text{when } \xi = 0 \tag{4.10}$$

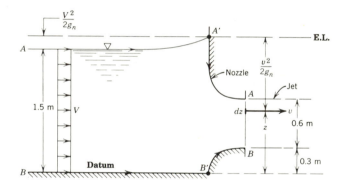

Solution. From the energy line drawn a distance $V^2/2g_n$ above the horizontal liquid surface, and with the pressure assumed zero throughout the free jet at the nozzle exit,

$$1.5 + \frac{V^2}{2g_n} = z + \frac{v^2}{2g_n} \qquad v = \sqrt{2g_n\left(1.5 + \frac{V^2}{2g_n} - z\right)} \tag{4.10}$$

The flowrate q may then be expressed

$$q = 1.5V = \int_{0.3}^{0.9} v \, dz = \int_{0.3}^{0.9}\sqrt{2g_n\left(1.5 + \frac{V^2}{2g_n} - z\right)} \, dz$$

$$(3.14 \text{ and } 3.15)$$

which may be solved by trial to yield $q = 2.81 \text{ m}^3/\text{s} \cdot \text{m}$. ◀

Stagnation points on the boundary streamlines AA and BB are to be expected. The one at B' needs no comment. At some point on the plane $A'B'$ there must be a stagnation point on the top boundary streamline; assume that this is somewhere below A'. Such a point could not be a stagnation point since its distance below the energy line would indicate a velocity head and thus a velocity there. Accordingly it is concluded that the stagnation point must be at A' and the liquid surface must rise to this point. ◀

Frequently it is possible to obtain the complete kinematics of a flowfield by mathematical methods that yield specific equations for the streamlines and velocities, from which accelerations and pressure variations may be predicted. A classic and useful example of this is the (irrotational) flowfield about a cylinder of radius R in a rectilinear flow of velocity U, which is shown in Fig. 4.16. The radial and tangential components of velocity anywhere in the flowfield may be shown[9] to be

$$v_r = U\left(1 - \frac{R^2}{r^2}\right)\cos\theta \qquad \text{and} \qquad v_t = -U\left(1 + \frac{R^2}{r^2}\right)\sin\theta$$

In such problems the velocity along the surface of the body is of greatest interest; here $r = R$, so $v_r = 0$ and $v_t = -2U\sin\theta$. For $\theta = 0$ and $\theta = \pi$, v_t will be zero, thus confirming the expected stagnation points at the head and tail of the body. Applying the Bernoulli equation between the undisturbed flow and any point on the body contour,

$$\frac{p_o}{\gamma} + \frac{U^2}{2g_n} + z_o = \left(\frac{p}{\gamma} + z\right) + \frac{(-2U\sin\theta)^2}{2g_n}$$

it is seen that $(p/\gamma + z)$ at any point on the cylinder may be predicted from the properties $(p_o, z_o, \text{and } U)$ of the undisturbed flow; from the cylinder size, z is determined and thus computation of the pressures on the cylindrical surface may be accomplished. By similar methods any pressure in the flowfield may be computed. Unfortunately, this ideal fluid motion is quite different from the observed real fluid motion (see Chapter 13), and the results of the present analysis are approximately valid for real flow only on a portion of the front face of the cylinder.

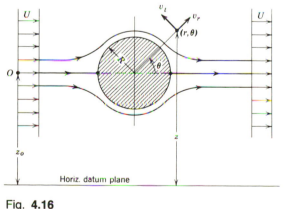

Fig. **4.16**

---

[9]See equations 12.18 and 12.19, Section 12.4.

---------------------- **Illustrative Problems** ----------------------

A cylinder of 6 in. diameter extends horizontally across the test section of large open-throat wind tunnel through which the air flows at a velocity of 100 fps, pressure 0 psi, and specific weight 0.08 lb/ft³. Calculate theoretical values of the velocity and pressure in the flowfield at the point (θ = 120°, r = 6 in.).

Relevant Equations and Given Data

$$\frac{p}{\gamma} + \frac{V^2}{2g_n} + z = H \qquad \text{(when } \xi = 0\text{)} \tag{4.10}$$

$$v_r = U\left(1 - \frac{R^2}{r^2}\right) \cos \theta \tag{12.18}$$

$$v_t = -U\left(1 + \frac{R^2}{r^2}\right) \sin \theta \tag{12.19}$$

U = 100 ft/s R = 3 in. r = 6 in.

θ = 120° p_o = 0 γ = 0.08 lb/ft³

Solution.

$$v_r = -100[1 - (\tfrac{3}{6})^2]0.5 = -37.5 \text{ fps} \tag{12.18}$$

$$v_t = -100[1 + (\tfrac{3}{6})^2]0.866 = -108.1 \text{ fps} \tag{12.19}$$

$$V = \sqrt{(37.5)^2 + (108.1)^2} = 114.4 \text{ fps} \blacktriangleleft$$

Applying the Bernoulli equation, taking the horizontal datum plane through the center of the cylinder,

$$0 + \frac{(100)^2}{2g_n} + 0 = \frac{p}{0.08} + \frac{(114.4)^2}{2g_n} + 0.5 \times 0.866 \tag{4.10}$$

$$\frac{p}{0.08} = -47.9 - 0.43 \qquad p = -3.87 \text{ psf} \blacktriangleleft$$

In gas flow problems it is customary to neglect the z terms in the Bernoulli equation; had this been done here, the 0.43 would not have appeared and the pressure would have been −3.83 psf, less than 1% from that calculated. In problems with larger velocities, the z terms are of even less importance.

Find a general relation for the pressure difference between an undisturbed airflow and any point on the cylinder in Fig. 4.16, and plot a representative distribution.

Relevant Equations

See list for previous problem.

Solution. From the analysis above

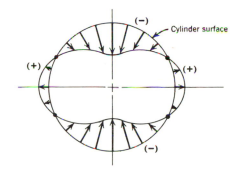

$$p_o + \tfrac{1}{2}\rho U^2 + \rho g_n z_o = p + \rho g_n z + \tfrac{1}{2}\rho(-2U \sin \theta)^2$$

By assuming $\rho g_n z_o$ and $\rho g_n z$ are negligible in the airflow, we find that

$$p - p_o = \tfrac{1}{2}\rho[U^2 - (2U \sin \theta)^2]$$

$$= \tfrac{1}{2}\rho U^2[1 - 4 \sin^2\theta]$$

A *nondimensional* pressure coefficient is conveniently defined as

$$C_p = \frac{p - p_o}{\tfrac{1}{2}\rho U^2} = 1 - 4 \sin^2\theta \quad \blacktriangleleft$$

This result is easily portrayed on a polar diagram, on which C_p is plotted relative to the surface of the cylinder.

The flow through a sharp-edged opening (Fig. 4.17) produces a flowfield which has many engineering applications. Jet contraction and flow curvature are produced by the (approximately) radial approach of fluid to the orifice, the streamlines becoming essentially straight and parallel at a section (termed the vena contracta)[10] a short distance downstream from the opening. Here and at other sections downstream the pressure through the jet is essentially zero, as explained in Section 4.4. Elsewhere the pressure is zero only on the free streamlines which bound the jet, but, with the centers of curvature of streamlines A, B, and C in the vicinity of O and the pressure increasing away from the center of curvature, it is apparent that, in the plane of the opening, pressures increase and velocities decrease toward the centerline; thus within the curved portion of the jet the pressures are expected to be larger than zero. In engineering practice this problem is usually treated by one-dimensional methods, which are entirely adequate at the vena contracta; applied to the flow cross section in the plane of the opening, they are quite meaningless and lead only to contradictions.

---

[10]Using advanced analytical methods, Kirchhoff showed the width of the vena contracta to be $(\pi/\pi + 2) \times$ (width of opening) for discharge at high velocity.

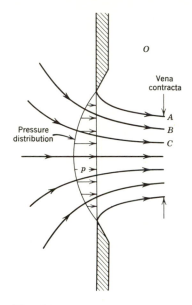

Fig. **4.17**

Illustrative Problem

A two-dimensional flow of liquid discharges from a large reservoir through the sharp-edged opening; a pitot tube at the center of the vena contracta produces the reading indicated. Calculate the velocities at points A, B, C, and D, and the flowrate.

Relevant Equations and Given Data

$$q = hV \qquad (3.14)$$

$$V = \frac{1}{A}\int_{A}^{A} v \, dA \qquad (3.15)$$

$$\frac{p}{\gamma} + \frac{V^2}{2g_n} + z = H \qquad \text{(when } \xi = 0) \qquad (4.10)$$

Solution. The pitot tube reading determines the position of the energy line (and also that of the free surface in the reservoir). Since the pressures at points A, B, C, and D are all zero, the respective velocity heads are determined by the vertical distances between the points and the energy line. The velocities at A, B, C, and D are 4.66 m/s, 5.12 m/s, 4.75 m/s, and 5.03 m/s, respectively. ◀

The flowrate may be computed by integrating the product $v \, dA$ over the flow cross section CD, in which $v = \sqrt{2g_n h}$:

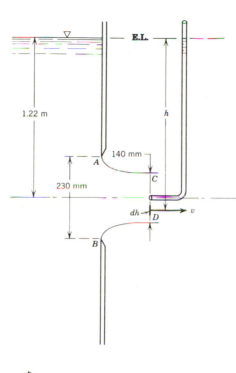

$$q = \int_C^D v\,dA = \int_{h_C}^{h_D} \sqrt{2g_n h}\,\,dh = \sqrt{2g_n}\tfrac{2}{3}h^{3/2}\Big|_{1.15}^{1.29} = 0.685\ \text{m}^3/\text{s} \cdot \text{m} \blacktriangleleft$$

(3.14 & 3.15)

The limits of integration are then the vertical distances (in metres) from energy line to points D and C, respectively. Assuming that the mean velocity at section CD is at the center and thus measured by the pitot tube, $V = \sqrt{2g_n} \times 1.22 = 4.89$ m/s; the (approximate) flowrate, q, may be computed from $q = 4.89 \times 0.140 = 0.685$ m³/s·m, giving the same result as the more refined calculation. Actually the center velocity is greater than the mean, but for large ratios of head to orifice opening there is a negligible difference between them. For small ratio of head to orifice opening the flowfield is greatly distorted by drooping of the jet, producing sharply curved streamlines and ill-defined vena contracta; for such conditions the foregoing methods of calculation may be used only as crude approximations.

Films

Rouse, H. Mechanics of fluids: fluid motion in a gravitational field. Film No. U45961, Media Library, Audiovisual Center, Univ. of Iowa.

Shapiro, A. H. Pressure fields and fluid acceleration. NCFMF/EDC Film No. 21609, Encyclopaedia Britannica Educ. Corp.

Problems

4.1. Integrate the one-dimensional Euler equation for a perfect gas at constant temperature.

4.2. Integrate the one-dimensional Euler equation for an isentropic, gas process.

4.3. On an overflow structure of 50° slope the water depth (measured normal to the surface of the structure) is 1.2 m (4 ft) and the streamlines are essentially straight and parallel. Calculate the pressure on the surface of the structure.

4.4. Water flows in a 1 m (3.3 ft) diameter horizontal supply pipe. The flow is turbulent with a maximum centerline velocity of 2 m/s (6.6 ft/s). What is the difference between the pressures at the top and the bottom of the pipe?

4.5. Water flows in a pipeline. At a point in the line where the diameter is 7 in., the velocity is 12 fps and the pressure is 50 psi. At a point 40 ft away the diameter reduces to 3 in. Calculate the pressure here when the pipe is (*a*) horizontal, and (*b*) vertical with flow downward.

4.6. In a pipe 0.3 m in diameter, 0.3 m^3/s of water are pumped up a hill. On the hilltop (elevation 48) the line reduces to 0.2 m diameter. If the pump maintains a pressure of 690 kPa at elevation 21, calculate the pressure in the pipe on the hilltop.

4.7. In a 3 in. horizontal pipeline containing a pressure of 8 psi, 100 gal/min of liquid hydrogen flow. If the pipeline reduces to 1 in. diameter, calculate the pressure in the 1 in. section.

4.8. If crude oil flows through this pipeline and its velocity at *A* is 2.4 m/s, where is the oil level in the open tube *C*?

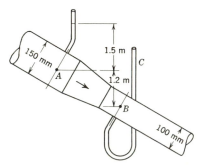

Problem **4.8**

4.9. Air of specific weight 12.6 N/m^3 flows through a 100 mm constriction in a 200 mm pipeline. When the flowrate is 11.1 N/s, the pressure in the constriction is 100 mm of mercury vacuum. Calculate the pressure in the 200 mm section, neglecting compressibility.

4.10. A pump draws water from a reservoir through a 12 in. pipe. When 12 cfs are being pumped, what is the pressure in the pipe at a point 8 ft above the reservoir surface, in psi? In feet of water?

4.11. If the pressure in the 0.3 m pipe of problem 3.21 is 70 kPa, what pressures exist in the branches, assuming all pipes are in the same horizontal plane? Water is flowing.

4.12. Water is flowing. Calculate *H* (m) and *p* (kPa).

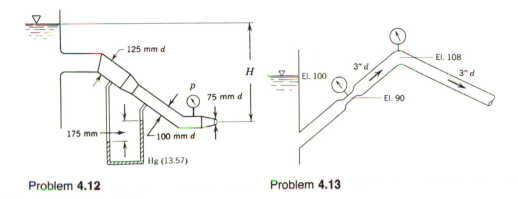

Problem **4.12**

Problem **4.13**

4.13. If each gage shows the same reading for a flowrate of 1.00 cfs, what is the diameter of the constriction?

4.14. Derive a relation between A_1 and A_2 so that for a flowrate of 0.28 m³/s the static pressure will be the same at sections 1 and 2. Also calculate the manometer reading for this condition.

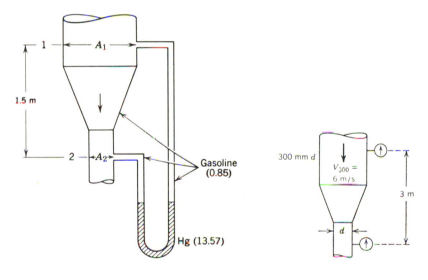

Problem **4.14**

Problem **4.15**

4.15. Water is flowing. Calculate the required pipe diameter, d, for the two gages to read the same.

4.16. If the pipe of problem 4.15 is part of a water supply system in a moon station, find d. Recall $g_m \approx g_n/6$.

4.17. For a flowrate of 75 cfs of air ($\gamma = 0.076\ 3$ lb/ft³) what is the largest A_2 that will cause water to be drawn up to the piezometer opening? Neglect compressibility effects.

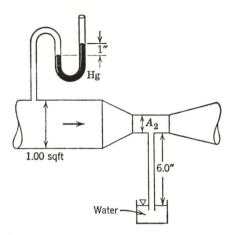

$$\text{Problem } \mathbf{4.17}$$

4.18. A smooth nozzle of 50 mm (2 in.) diameter is connected to a water tank. Connected to the tank at the same elevation is an open U-tube manometer containing mercury and showing a reading of 625 mm (25 in.). The lower mercury surface is 500 mm (20 in.) below the tank connection. What flowrate will be obtained from the nozzle? Water fills the tube between tank and mercury.

4.19. Water discharges from a tank through a nozzle of 50 mm (2 in.) diameter into an air tank where the pressure is 35 kPa (5 psi). If the nozzle is 4.5 m (14.8 ft) below the water surface, calculate the flowrate.

4.20. Water discharges through a nozzle of 25 mm (1 in.) diameter under a 6 m (20 ft) head into a tank of air in which a vacuum of 250 mm (10 in.) of mercury is maintained. Calculate the rate of flow.

4.21. Air is pumped through the tank as shown. Neglecting effects of compressibility, compute the velocity of air in the 100 mm pipe. Atmospheric pressure is 91 kPa and the specific weight of air 11 N/m³.

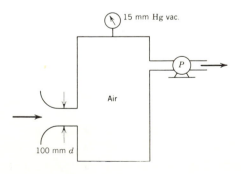

$$\text{Problem } \mathbf{4.21}$$

4.22. A closed tank contains water with air above it. The air is maintained at a pressure of 103 kPa (14.9 psi) and 3 m (9.8 ft) below the water surface a nozzle discharges into the atmosphere. At what velocity will water emerge from the nozzle?

4.23. Water is flowing. The flow picture is axisymmetric. Calculate the flowrate and manometer reading.

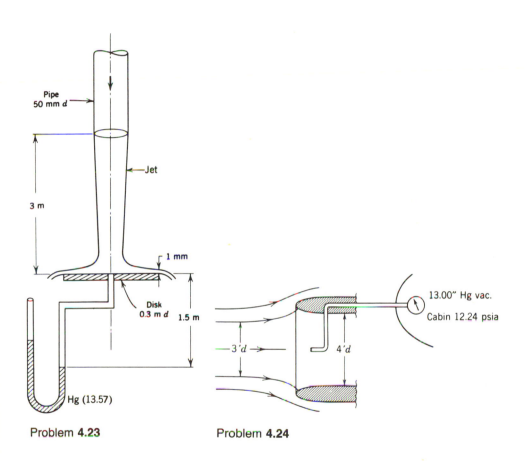

Problem **4.23**   Problem **4.24**

4.24. An airplane flies at altitude 30 000 ft where the air density is 0.000 9 slug/ft³ and the air pressure 4.37 psia. A pitot tube inside the entrance of its jet engine is connected to a conventional bourdon pressure gage inside the airplane. This gage reads 13.00 in. of mercury vacuum when the cabin is pressurized to 12.24 psia. If the streamline configuration is as shown, how much air (cfs) is passing through the engine? Assume the air incompressible.

4.25. A section of a freely falling water jet is observed to taper from 1 300 mm² to 975 mm² in a vertical distance of 1.2 m. Calculate the flowrate in the jet.

4.26.   Calculate the rate of flow through this pipeline and the pressures at $A,B,C,$ and D.

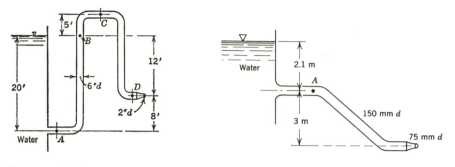

Problem **4.26**   Problem **4.27**

4.27.   Calculate the pressure in the flow at A: (a) for the system shown, and (b) for the pipe without the nozzle.

4.28.   A siphon consisting of a 25 mm (1 in.) hose is used to drain water from a tank. The outlet end of the hose is 2.4 m (7.9 ft) below the water surface, and the bend in the hose is 0.9 m (3 ft) above the water surface. Calculate the pressure in the bend and flowrate.

4.29.   A 75 mm horizontal pipe is connected to a tank of water 1.5 m below the water surface. The pipe is gradually enlarged to 88 mm diameter and discharges freely into the atmosphere. Calculate the flowrate and the pressure in the 75 mm pipe.

4.30.   Calculate the pressure head in the 20 mm section when $h = 0.16$ m. Calculate the largest h at which the divergent tube can be expected to flow full.

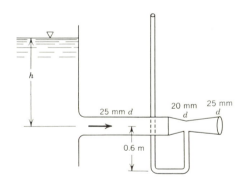

Problem **4.30**

4.31.   The head of water on a 50 mm diameter smooth nozzle is 3 m. If the nozzle is directed upward at angles of (a) 30°, (b) 45°, (c) 60°, and (d) 90°, how high above the nozzle will the jet rise, and how far from the nozzle will the jet pass through the horizontal plane in which the nozzle lies? What is the diameter of the jet at the top of the trajectory?

4.32. A fire hose nozzle which discharges water at 46 m/s (150 ft/s) is to throw a stream through a window 30 m (100 ft) above and 30 m (100 ft) horizontally from the nozzle. What is the minimum angle of elevation of the nozzle that will accomplish this?

4.33. The jet passes through a point A. Calculate the flowrate.

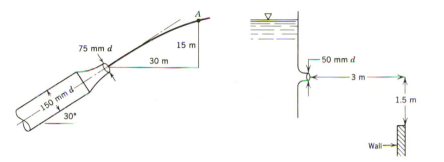

Problem 4.33   **Problem 4.34**

4.34. Calculate the minimum flowrate that will pass over the wall.

4.35. From Bernoulli's equation derive a relation between the velocities at any two points on the trajectory of a free jet in terms of the vertical distance between the points.

4.36. A jet of water falling vertically has a velocity of 24 m/s (79 ft/s) and a diameter of 50 mm (2 in.) at elevation 12 m (40 ft); calculate its velocity at elevation 4.5 m (15 ft).

4.37. Applying free trajectory theory to the centerline of a free jet discharging from a horizontal nozzle under a head h, show that the radius of curvature of the jet at the tip of the nozzle is $2h$.

4.38. For the two orifices discharging as shown, prove that $h_1 y_1 = h_2 y_2$.

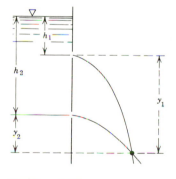

Problem 4.38

4.39. Water flows from one reservoir in a 200 mm (8 in.) pipe, while water flows from a second reservoir in a 150 mm (6 in.) pipe. The two pipes meet in a "tee" junction with a 300 mm (12 in.) pipe that discharges to the atmosphere at an elevation of 20 m (66 ft). If the water surface in the reservoirs is at 30 m (98 ft) elevation, what is the total flowrate?

4.40. What size must one set for the pipe from the second reservoir of problem 4.39 so that the flowrate from the second reservoir is twice that from the first? What is the total flowrate in this case?

4.41. A constriction of 150 mm diameter occurs in a horizontal 250 mm water line. It is stated that the pressure in the 250 mm pipe is 125 mm of mercury vacuum when the flowrate is 170 l/s. Is this possible? Why or why not?

4.42. Water flows through a 1 in. constriction in a horizontal 3 in. pipeline. If the water temperature is 150°F and the pressure in the line is maintained at 40 psi, what is the maximum flowrate that may occur? Barometric pressure is 14.7 psia.

4.43. The liquid has a specific gravity of 1.60 and negligible vapor pressure. Calculate the flowrate for incipient cavitation in the 75 mm section, assuming that the tube flows full. Barometric pressure is 100 kPa.

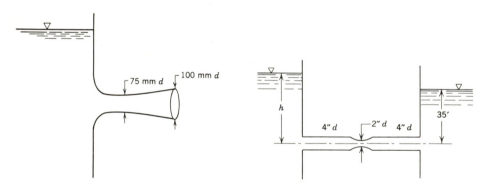

Problem **4.43**   Problem **4.44**

4.44. Water flows between two open reservoirs. Barometric pressure is 13.4 psia. Vapor pressure is 6.4 psia. Above what value of h will cavitation be expected in the 2 in. constriction?

4.45. Barometric pressure is 101.3 kPa. For $h > 0.6$ m, cavitation is observed at the 50 mm section. If the pipe is horizontal and flows full throughout, what is the vapor pressure of the water?

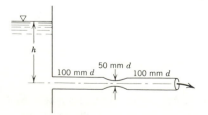

Problem **4.45**

4.46. If cavitation is observed in the 50 mm section, what is the flowrate? Barometric pressure is 100 kPa.

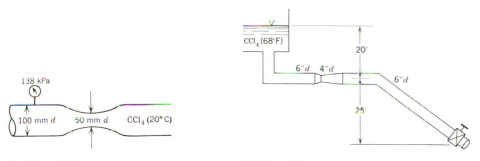

Problem **4.46**

Problem **4.47**

4.47. Barometric pressure is 14.0 psia. What is the maximum flowrate that can be obtained by opening the valve?

4.48. A variety of combinations of d and h will allow maximum possible flowrate to occur through this system. Derive a relationship between d and h that will always produce this. Barometric pressure is 101 kPa.

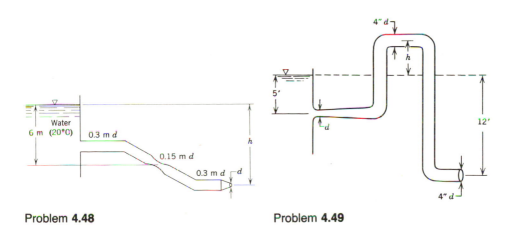

Problem **4.48**

Problem **4.49**

4.49. Calculate the maximum h and the minimum d that will permit cavitation-free flow through this frictionless pipe system. Atmospheric pressure = 14.5 psia; vapor pressure = 1.5 psia. Water is flowing.

4.50. For a flowrate of 0.28 m³/s derive a relation between h and d that will indicate incipient cavitation in the top of this siphon pipe. Over what range of h is this relation applicable?

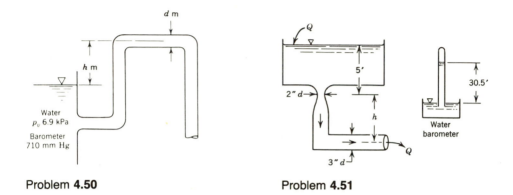

Problem **4.50**   Problem **4.51**

4.51. At what h will cavitation be incipient in the 2 in. section? Assume the same water in system and barometer.

4.52. The pressure in this closed tank is gradually increased by pumping. Calculate the gage reading at which cavitation will appear in the 25 mm constriction. The barometric pressure is 94 kPa.

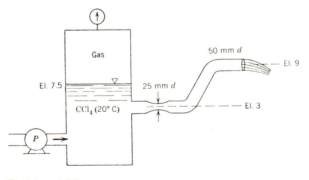

Problem **4.52**

4.53. If cavitation anywhere in this pipe system is to be avoided, what is the diameter of the largest nozzle which may be used? Atmospheric pressure = 100 kPa; vapor pressure = 10.3 kPa.

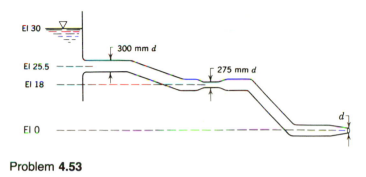

Problem **4.53**

4.54. What is the smallest nozzle diameter (d) that will produce the maximum possible flowrate through this frictionless pipe system? Atmospheric pressure is 14.3 psia; vapor pressure of the water is 1.3 psia.

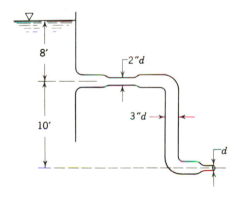

Problem **4.54**

4.55. A convergent-divergent tube is connected to a large tank as in problem 4.43. It is to be proportioned so that the negative pressure head at the throat is half the height of the water surface above the centerline of the tube. For no cavitation and assuming that the tube flows full, what ratio of exit diameter to throat diameter is required?

4.56. Cavitation occurs in this convergent-divergent tube as shown. The right-hand side of the manometer is connected to the cavitation zone. The water in the right-hand tube has all evaporated leaving only vapor. Assuming an ideal fluid at 40°C, calculate the gage reading if the local atmospheric pressure is 750 mm of mercury.

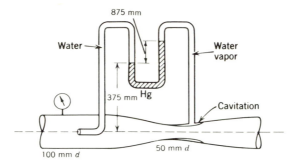

Problem 4.56

4.57.   Water at 70°C (158°F) is being siphoned from a tank through a hose in which the velocity is 4.5 m/s (15 ft/s). What is the maximum theoretical height of the high point ("crown") of the siphon above the water surface that will allow this flow to occur? Assume standard barometric pressure.

4.58.   Barometric pressure is 101.3 kPa and vapor pressure of the water is 32.4 kPa. Calculate the elevation of end B for maximum flowrate through this pipeline.

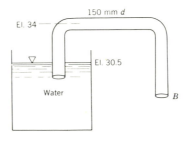

Problem 4.58

4.59.   The pressure in the test section of a wind tunnel is −27 mm (−1.1 in.) of water when the velocity is 97 km/h (60 mph). Calculate the pressure on the nose of an object when placed in the test section. Assume the specific weight for air 12.0 N/m³ (0.076 3 lb/ft³).

4.60.   The pressure in a 100 mm pipeline carrying 4 500 l/min of fluid weighing 11 kN/m³ is 138 kPa. Calculate the pressure on the upstream end of a small object placed in this pipeline.

4.61.   A submarine moves at 46 km/h (25 knots) through saltwater (s.g. = 1.025) at a depth of 15 m (50 ft). Calculate the pressure on the nose of the submarine.

4.62. Calculate the flowrate of water in this pipeline.

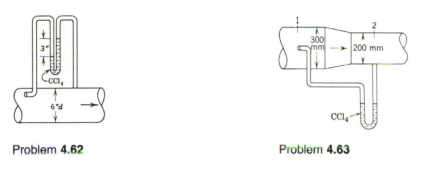

Problem **4.62**   Problem **4.63**

4.63. Three tenths of a cubic metre per second of water are flowing. Calculate the manometer reading (*a*) using the sketch as shown, and (*b*) when the pitot tube is at section 2 and the static pressure connection is at section 1.

4.64. Calculate the flowrate through this pipeline.

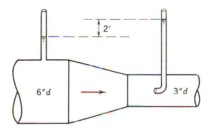

Problem **4.64**

4.65. Water is flowing. Calculate the flowrate.

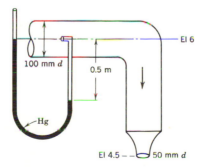

Problem **4.65**

4.66.   Gasoline (s.g. 0.85) is flowing. Calculate gage readings and flowrate.

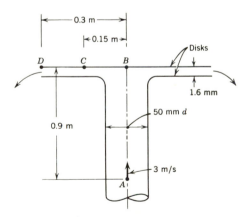

Problem **4.66**

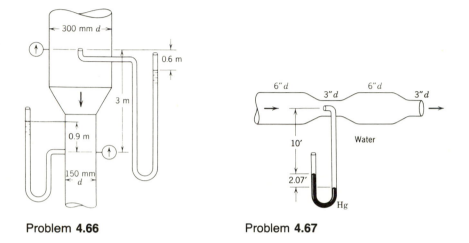

Problem **4.67**

4.67.   Calculate the velocity head in the 3 in. constriction.

4.68.   Water is flowing. Assume the flow between the disks to be radial and calculate the pressures at A, B, C, and D. The flow discharges to the atmosphere.

Problem **4.68**

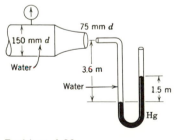

Problem **4.69**

4.69.   Calculate the gage reading.

4.70. Calculate the flowrate of water through this nozzle.

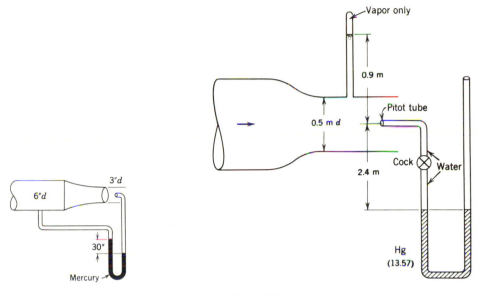

Problem **4.70**   Problem **4.71**

4.71. A flowrate of 1.54 m³/s of water (specific weight 9.81 kN/m³; vapor presure 6.9 kPa) passes through this water tunnel. The cock is now closed. Calculate magnitude and direction of the manometer reading after the cock is opened. Atmospheric pressure is 100kPa.

4.72. Water is flowing. Calculate the flowrate and gage reading.

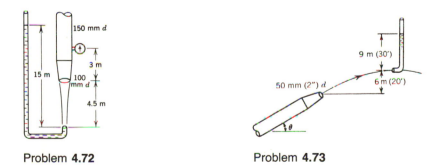

Problem **4.72**   Problem **4.73**

4.73. The tip of the pitot tube is at the top of the jet. Calculate the flowrate and the angle θ.

4.74. Air discharges from a duct of 300 mm diameter through a 100 mm nozzle into the atmosphere. The pressure in the duct is found by a draft gage to be 25 mm of water. Assuming the specific weight of air to be constant at 11.8 N/m³, what is the flowrate?

4.75. If in problem 4.8 the vertical distance between the liquid surfaces of the piezometer tubes is 0.6 m, what is the flowrate?

4.76. Water flows through a 75 mm constriction in a horizontal 150 mm pipe. If the pressure in the 150 mm section is 345 kPa and that in the constriction 207 kPa, calculate the velocity in the constriction and the flowrate.

4.77. Ten cubic feet per second of liquid of specific weight 50 lb/ft³ are flowing. Calculate the manometer reading.

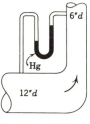

Problem **4.77**

4.78. In the preceding problem calculate the flowrate when the manometer reading is 10 in.

4.79. Carbon tetrachloride flows downward through a 50 mm constriction in a 75 mm vertical pipeline. If a differential manometer containing mercury is connected to the constriction and to a point in the pipe 1.2 m above the constriction and this manometer reads 350 mm, calculate the flowrate. (Carbon tetrachloride fills manometer tubes to the mercury surfaces.)

4.80. Calculate the flowrate.

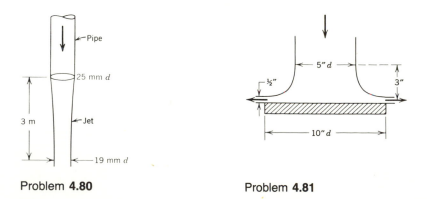

Problem **4.80**   Problem **4.81**

4.81. If a free jet of fluid strikes a circular disk and produces the flow picture shown, what is the flowrate?

4.82. Through a transition structure between two rectangular open channels the width narrows from 2.4 m to 2.1 m, the depth decreases from 1.5 m to 1.05 m, and the bottom rises 0.3 m. Calculate the flowrate.

4.83. This "Venturi flume" is installed in a horizontal frictionless open channel of 10 ft width and water depth 10 ft. In the "throat" of the flume where the width has been narrowed to 8 ft, the water depth is observed to be 8 ft. Calculate the flowrate in the channel.

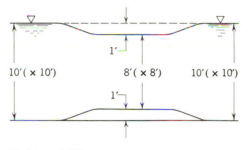

10′(× 10′)   8′(× 8′)   10′(× 10′)

Problem **4.83**

4.84. Through a transition structure between two rectangular open channels the width increases from 3 m to 3.6 m while the water surface remains horizontal. If the depth in the upstream channel is 1.5 m, what is the depth in the downstream channel?

4.85. If the two-dimensional flowrate over this sharp crested weir is 10 cfs/ft, what is the thickness of the sheet of falling water at a point 3 ft below the weir crest?

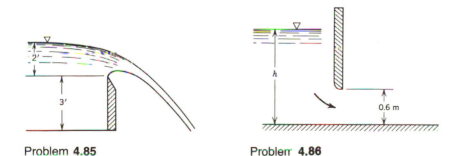

Problem **4.85**   Problem **4.86**

4.86. Calculate the two-dimensional flowrate through this frictionless sluice gate when the depth h is 1.5 m. Also calculate the depth h for a flowrate of 3.25 m³/s · m.

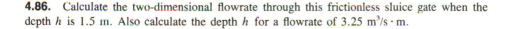

4.87. The flow in an open drainage channel passes into a pipe that carries it through a highway embankment. If the channel flows full, what is the flowrate?

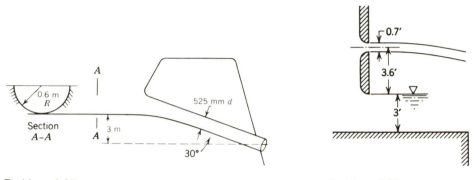

Problem **4.87**   Problem **4.88**

4.88. This two-dimensional gate contains a two-dimensional nozzle (slot) as shown. For a flowrate of 80 cfs/ft in the channel, predict the flowrates through the slot and under the gate.

4.89. Channel and gate are 1 m wide (normal to the plane of the paper). Calculate q_1, q_2, and Q_3.

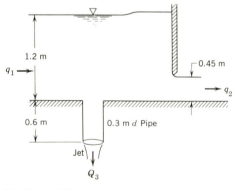

Problem **4.89**

4.90. A pump having a 100 mm suction pipe and 75 mm discharge pipe pumps 32 l/s of water. At a point on the suction pipe a vacuum gage reads 150 mm of mercury; on the discharge pipe 3.6 m above this point a pressure gage reads 331 kPa. Calculate the power supplied by the pump.

4.91. A pump of what power is theoretically required to raise 900 l/min (240 gpm) of water from a reservoir of surface elevation 30 m (98 ft) to one of surface elevation 75 m (250 ft)?

4.92. If 340 l/s of water are pumped over a hill through a 450 mm pipeline, and the hilltop is 60 m above the surface of the reservoir from which the water is being taken, calculate the pump power required to maintain a pressure of 175 kPa on the hilltop.

4.93. Water is pumped from a large lake into an irrigation canal of rectangular cross section 3 m wide, producing the flow situation shown. Calculate the required pump power assuming ideal flow.

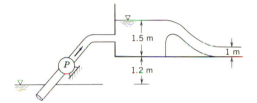

Problem **4.93**

4.94. A pump has 8 in. inlet and 6 in. outlet. If the hydraulic grade line rises 60 ft across the pump when the flowrate is 5.0 cfs of sodium at 1 000°F, how much horsepower is the pump delivering to the fluid? For the same rise in hydraulic grade line, what flowrate will be maintained by an expenditure of 60 hp?

4.95. Water is flowing. Calculate the pump power for a flowrate of 28 l/s.

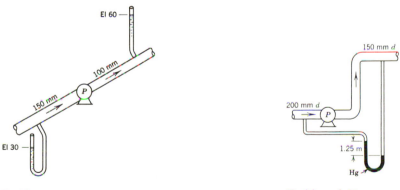

Problem **4.95**   Problem **4.96**

4.96. One hundred and twenty litres per second of jet fuel (JP-4) are flowing. Calculate the pump power.

4.97. If pitot tubes replace the static pressure connections of the preceding problem and give the same manometer reading, what is the flowrate if the pump is supplying 7.5 kW to the fluid?

4.98. A pump takes water from a tank and discharges it into the atmosphere through a horizontal 50 mm nozzle. The nozzle is 4.5 m above the water surface and is connected to the pump's 100 mm discharge pipe. What power must the pump have to maintain a pressure of 276 kPa just upstream from the nozzle?

4.99.   Calculate the pump power, assuming that the diverging tube flows full.

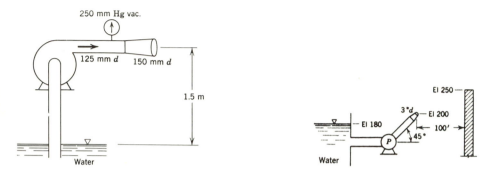

Problem **4.99**   Problem **4.100**

4.100.   Calculate the minimum pump horsepower that will send the jet over the wall.

4.101.   Calculate the pump power.

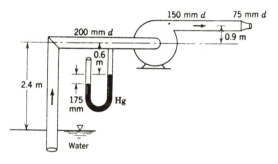

Problem **4.101**

4.102.   Compute the pump horsepower required to maintain a flowrate of 4.0 cfs if the barometric pressure is 14.3 psia and the vapor pressure is 1.0 psia. Calculate the maximum possible x for reliable operation.

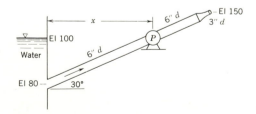

Problem **4.102**

4.103. Referring to the sketch of problem 4.102 above, what flowrate will be expected when 50 horsepower are supplied to the pump?

4.104. Two cubic feet per second of ethyl alcohol are flowing in a moon station supply system. Calculate the pump horsepower.

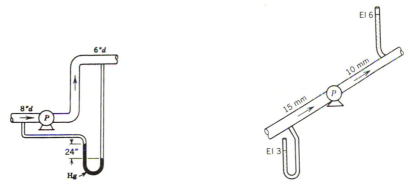

Problem **4.104**   Problem **4.105**

4.105. Water is flowing in this moon station supply pipeline. Calculate the pump power for a flowrate of 3 l/s.

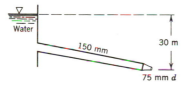

Problem **4.106**

4.106. A pump is to be installed to increase the flowrate through this pipe and nozzle by 20%. Calculate the required power.

4.107. Through a 4 in. pipe, 1.0 cfs of water enters a small hydraulic motor and discharges through a 6 in. pipe. The inlet pipe is lower than the discharge pipe, and at a point on the inlet pipe a pressure gage reads 70 psi; 14 ft above this on the discharge pipe a pressure gage reads 40 psi. What horsepower is developed by the motor? For the same gage reading, what flowrate would be required to develop 10 hp?

4.108. A 7.5 kW pump is used to draw air from the atmosphere (pressure = 101.3 kPa) through a smooth bell-mouth nozzle with a 250 mm diameter throat and into a very large tank where the absolute pressure is 1.5 bar. If the air density is essentially constant at 1.25 kg/m³, what is the velocity in the nozzle throat?

4.109. A hydraulic turbine in a power plant takes 3 m³/s (106 ft³/s) of water from a reservoir of surface elevation 70 m (230 ft) and discharges it into a river of surface elevation 20 m (66 ft). What theoretical power is available in this flow?

4.110. Calculate the h that will produce a flowrate of 85 l/s and a turbine output of 15 kW.

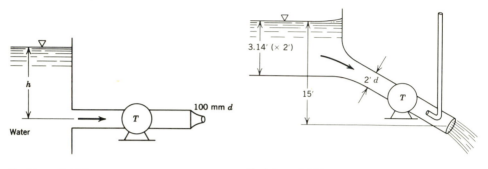

Problem **4.110**   Problem **4.111**

4.111. Water flows from the rectangular channel into a 2 ft diameter pipe where the turbine extracts 10 ft · lb of energy from every pound of fluid. How far above the pipe outlet will the water surface be in the column connected to the pitot tube?

4.112. The turbine extracts from the flowing water half as much energy as remains in the jet at the nozzle exit. Calculate the power of the turbine.

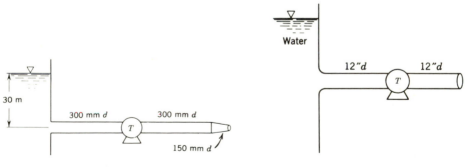

Problem **4.112**   Problem **4.113**

4.113. This turbine develops 100 horsepower when the flowrate is 20 cfs. What flowrate may be expected if the turbine is removed?

4.114. Calculate the power output of this turbine.

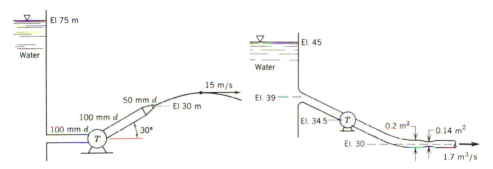

Problem 4.114   **Problem 4.115**

4.115. What is the maximum power the turbine can extract from the flow before cavitation will occur at some point in the system? Barometric pressure is 102 kPa, and vapor pressure of the water is 3.5 kPa.

4.116. A flowrate of 7 m³/s of water passes through this hydraulic turbine. The static pressure at the *top* of the inlet pipe is 345 kPa and across a 1.5 m diameter of the outlet pipe ("draft tube") the stagnation pressure is 250 mm of mercury vacuum. How much power may be expected from the machine?

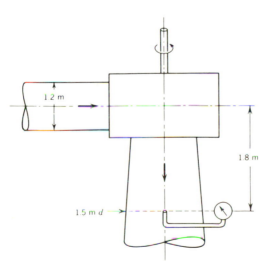

Problem 4.116

4.117. Assuming a very large tank and density of air constant at 0.002 5 slug/ft³, calculate (*a*) pressure just upstream from the turbine, (*b*) the turbine horsepower, and (*c*) the jet velocity when the turbine is removed.

Problem **4.117**   Problem **4.118**

4.118. Calculate the power available in the jet of water issuing from this nozzle.

4.119. Water flows in a 0.1 m (0.33 ft) wide gap between two very large flat plates; assume the flow is two-dimensional. The flow is laminar and so the velocity profile is parabolic with a centerline velocity of 0.2 m/s (0.67 ft/s). Under the assumption that the streamlines of the flow are straight and parallel, find the change in the sum of the Bernoulli terms in equation 4.10 with distance from the flow centerline. Use two different methods, one of which uses the vorticity. Show that the methods are equivalent.

4.120. If the irrotational flow of problem 3.3*f* is in a horizontal plane and the pressure head at the origin of coordinates is 3 m (10 ft), what is the pressure head at point (2,2)? What is the pressure head when the flowfield is in a vertical plane?

4.121. If the irrotational flowfields of problem 3.3*l* and 3.3*m* are in a horizontal plane and the velocity and pressure head at a radius of 0.6 m are 4.5 m/s and 3 m, respectively, what pressure head exists at a radius of 0.9 m? When these flowfields are in vertical planes and the pressure head at ($\theta = 90°$, $r = 0.6$) is 3 m, what is the pressure head at ($\theta = 0°$, $r = 0.9$)?

4.122. If the irrotational flowfield of a free vortex is described by the equations $v_t = c/r$ and $v_r = 0$, derive an equation for the profile of the liquid surface.

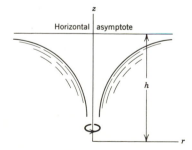

Problem **4.122**

4.123. A 180° circular bend occurs in a horizontal plane in a passage 1 m (3 ft) wide. The radius of the inner wall is 0.3 m (1 ft) and that of the outer wall 1.3 m (4 ft). If the velocity (of the water) halfway around the bend on the outer wall is 3 m/s (10 ft/s) and the pressure there is 35 kPa (5 psi), what will be the velocity and pressure at the corresponding point on the inner wall (using the relation of equation 4.12)? Note that this equation is an approximation because the centers of curvature of all streamlines are not coincident with the center of curvature of the bend. Also calculate the approximate flowrate in the passage.

4.124. If the bend of the preceding problem is the top of a siphon, and thus in a vertical plane, what are the velocity and pressure? What is the approximate flowrate?

4.125. Estimate the peripheral velocity 30 m from the center of a tornado if the peripheral velocity 120 m from the center is 80 km/h. If the barometer reads 710 mm of mercury at the latter point, what reading can be expected at the former? Assume air density 1.23 kg/m³.

4.126. Predict the flowrate through this two-dimensional outflow structure for a water depth of 1.5 m in the canal. Also calculate the water depth on the upstream face of the structure and the pressure at A.

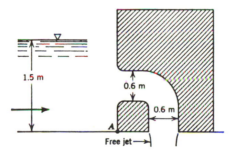

Problem **4.126**

4.127. Complete the solution of the first Illustrative Problem of Section 4.8.

4.128. Assuming an ideal fluid, this flowfield would occur when deep open flow passes over a submerged semicylindrical weir. If the local velocities through section 2 may be expressed by

$$v = V_1 \left[1 + \left(\frac{R}{y} \right)^n \right]$$

determine the two-dimensional flowrate and the pressure at the top of the weir.

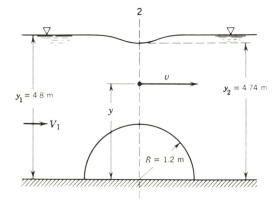

Problem **4.128**

4.129. The two-dimensional flowrate over this submerged semicylinder is 11.3 m³/s · m. The ideal fluid flowing has the density of water. The velocity over the surface of the semicylinder may be closely approximated by $v = 6 \sin \theta$. Derive an expression for the pressure on the semicylinder. Integrate this pressure distribution in such ways that it will yield the vertical and horizontal components of force exerted by the water on the cylinder. Assume the flowfield perfectly symmetrical about the line AA.

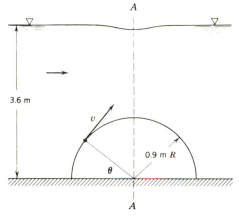

Problem **4.129**

4.130. If a gently curved overflow structure placed in a high velocity flow produces the flowfield shown, and if the indicated velocity distribution is assumed at section 2, what flowrate is indicated?

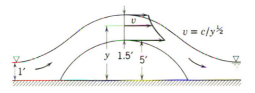

Problem **4.130**

4.131. The flowrate is 0.75 m³/s. The gage pressure in the corner at *A* is 83 kPa. Calculate the gage pressure at *B* and the power of the turbine.

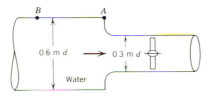

Problem **4.131**

4.132. A small circular cylinder spans the test section of a large water tunnel. A water and mercury differential manometer is connected to two small openings in the cylinder, one facing directly into the approaching flow the other at 90° to this point. For a manometer reading of 10 in. (250 mm) calculate the velocity approaching the cylinder.

4.133. For the flowfield described by equations 12.18 and 12.19, derive the equation of a line at all points of which the pressure is p_o. See Fig 4.16.

4.134. If the depth at the vena contracta caused by this sluice gate is 0.6 m, calculate the two-dimensional flowrate and the water depth on the upstream face of the gate.

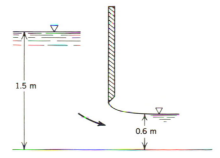

Problem **4.134**

4.135. Water is flowing. The pressure at A is 1.75 psi. Calculate the flowrate, diameter of vena contracta, and pressures at points B, C, and D.

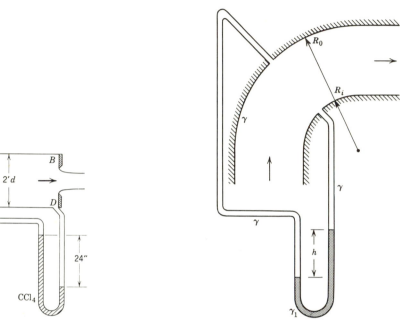

Problem **4.135**   Problem **4.136**

4.136. Liquid of specific weight γ flows through a *two-dimensional* "elbow-meter" similar to that of Fig 11.32. A differential manometer is connected between the two pressure taps and contains liquid of specific weight γ_1. Assuming the velocity distribution at the section containing the pressure taps is given by equation 4.12, derive an expression for flowrate in terms of specific weights, R_i, R_o, and manometer reading, h.

4.137. In the throat of a passage with curved walls the distribution of velocity will be axisymmetric and somewhat as shown. If this velocity distribution is characterized mathematically by $v = 3 + 13.3r^2$, what is the flowrate? If the flow is irrotational, what pressure difference is to be expected between the top and the center of the passage?

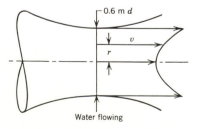

Water flowing

Problem **4.137**

4.138. Referring to the sketch of Fig 4.15, show that the force exerted by the water on the weir can be calculated (approximately) by

$$\gamma P\left[\frac{P}{2} + H + \frac{q^2}{2g_n(P + H)^2}\right]$$

in which P is the weir height and h the head on the weir. Will this approximate force be expected to be more or less than the exact force on the weir?

4.139. A circular disk of diameter d is placed (normal to the flow) in a free stream of pressure p_o, velocity V_o, and density ρ. If the pressure is assumed to fall parabolically from stagnation pressure at the center of the disk to p_2 at the edges and the pressure p_2 is assumed to be distributed uniformly over the downstream side of the disk, derive an expression for the drag force on the disk including only p_o, p_2, and d. Note that the drag force on such a disk may be predicted more reliably using equation 13.1 and $C_D = 1.12$ from Fig 13.13.

4.140. The study of the flowfield around the elliptic cylinder shows that the local velocity at the midsection is $3V_o/2$ and that this is the maximum velocity in the whole flowfield. This cylinder is placed (with its axis vertical) in a large water tunnel as shown on the sketch. For a gage reading of 70 kPa (10.0 psi), calculate the V_o at which cavitation will be incipient. Atmospheric pressure = 96 kPa (14.0 psi). Vapor pressure = 6.9 kPa (1.0 psi).

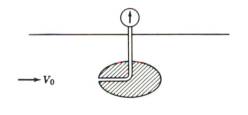

Problem **4.140**

5

FLOW OF A COMPRESSIBLE IDEAL FLUID

The intent of the following treatment is to give the beginner some feeling for the subject, an understanding of some of its difficulties, and an appreciation of the striking differences between compressible and incompressible flow. A complete understanding of the problems of compressible fluid flow cannot be acquired from fluid mechanics alone but depends on successful synthesis of fluid mechanics with thermodynamics. The authors assume that the reader has not yet had formal training in thermodynamics. Accordingly, some basic thermodynamics is included here with explanations which, although incomplete for comprehensive understanding, are adequate for pursuit of the subject at hand. All the thermodynamics required for the remainder of the text is presented just below and in Section 5.1. Ideal fluid motion is treated in the remainder of this chapter, but shock waves and compressible flows in pipes are discussed in Chapters 6 and 9, respectively. Compressibility effects on measurements are considered in Section 11.6 and on drag and lift in Chapter 13.

To begin, it is necessary to define or redefine certain terms in their thermodynamic context. A fluid *system* remains some fixed, identifiable, quantity of matter. A system *property* is an observable or measurable characteristic of the system, for example, temperature, density, pressure, and so forth. The *state* of a system is established by examination of the properties. If at two different times the properties of the system are all the same, the system is said to be at the same state at both times. When a system's state changes the system has undergone a *process* which usually involves transfers across the system's boundaries and work done by or on the system. If a process leads to a final state that is the same as the initial state, this process is a *cycle*.

The two basic laws of thermodynamics are stated for processes. The First Law is a conservation law for processes and leads to definition of a new property, called the *internal energy of the fluid*. The Second Law prescribes the permitted direction in which a process may proceed and leads to the new property, called *entropy*. Implicit in these laws is also the concept of the reversible process. A process is *reversible* if and only if, upon completion of the process, the system and all of its surroundings can be returned to their initial states, that is, the process can be completely undone. All real fluid flow processes are *irreversible*, but some can be approximated by reversible processes, when the causes of irreversibility, that is, viscous action and heat conduction across finite temperature differences, do not dominate the motion.

In this chapter, the assumption of an ideal (inviscid) fluid restricts the discussion to *frictionless* flow processes, as in Chapter 4 for the imcompressible fluid. A further

176

restriction often made is the assumption of no heat transfer to or from the fluid, which is the definition of an *adiabatic* process. In thermodynamics a *frictionless adiabatic* process is called an *isentropic* one, because as seen below it is accompanied by no change of entropy. Since this process involves no heat transfer or friction, it is, therefore, a reversible process. Such processes are closely approximated in practice if they occur with small friction and with such rapidity that there is little opportunity for heat transfer, for example, high-velocity gas flow (*gas dynamics*) over short distances (i.e., nozzles rather than pipes) where in real problems friction and heat transfer are actually small. Pipes are treated in Chapter 9.

High-velocity gas flow is associated in general with large changes of pressure, temperature, and density, but changes of these variables are of course much smaller in low-velocity flow; therefore many of the latter problems may be treated approximately yet satisfactorily by the methods of Chapter 4. However, there is no precise boundary between high-velocity and low-velocity motion, and whether a gas may be treated as an incompressible[1] fluid depends on the accuracy of the results required. Only experience with the equations of Chapters 4, 5, 6, and 9 will provide an answer to this question.

5.1 The Laws of Thermodynamics

Thermodynamics is concerned with the study of the interactions of work, heat, and system properties. The First Law of Thermodynamics is an empirical law that expresses the conservation of energy in any process (barring nuclear mass-energy conversions and electromagnetic effects). From Section 4.5, it is known that a fluid system possesses both kinetic and potential energy and can do work on its surroundings. From thermodynamics, it is known that energy transfers occur across system boundaries when there is a temperature difference there; this energy in transition as a result of a temperature difference is called *heat*. Furthermore, a fluid system possesses energy as a result of the kinetic energy of its molecules and the forces between them; this property of a system is known as *internal energy* and manifests itself in temperature, high or low temperature implying high or low internal energy, respectively. The First Law states that in a process for a system

$$dQ = dW + dE$$

where dQ is the heat transferred to the system, dW is the work done by the system, and dE is the change in the *total* energy of the system.

Just as the work-energy principle and a control volume analysis were used in Section 4.5 to derive the work-energy equation 4.5, the First Law of Thermodynamics and a control volume analysis are now used to derive an energy equation (which, in fact, reduces to equation 4.5 in the absence of heat transfer and of density variations).

---

[1] For example, in the aerodynamics of low-speed pleasure aircraft it is a sufficiently accurate approximation to consider the air incompressible; for current commercial jet aircraft, missiles, jet engines, etc., such an assumption would obviously be unsatisfactory.

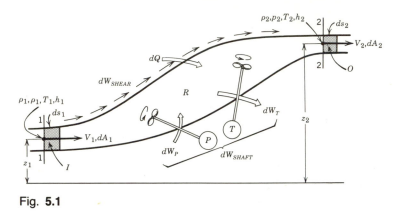

Fig. 5.1

Consider the streamtube section shown in Fig. 5.1. The flow is steady and the fluid system occupies zones I and R of the control volume 1221 at time t and zones R and O at time $t + dt$. The streamtube may be of finite or differential size, but it is assumed the flow is one-dimensional at sections 1 and 2. In the period dt an amount of heat dQ is transferred to the fluid system and an amount of work dW is done by the system. The work dW takes the three forms described in Section 4.5: (a) flow work, (b) shaft work dW_{shaft}, and (c) shear work dW_{shear}. Recall from Section 4.5 that in all cases studied in Chapters 4 and 5, $dW_{\text{shear}} = 0$ because the fluid is inviscid and cannot support shear stress. Thus, if the internal energy of the fluid is IE, the First Law yields

$$dQ = dW_{\text{shaft}} - (p_1 \, dA_1) \, ds_1 + (p_2 \, dA_2) \, ds_2 + d(KE + PE + IE) \quad (5.1)$$

where the second and third terms on the right are the flow work terms (see Section 4.5).

Now, for this steady flow (refer to Section 4.5)

$$(KE + PE + IE)_{t+dt} = (KE + PE + IE)_R + \tfrac{1}{2}(\rho_2 \, dA_2 \, ds_2)V_2^2$$
$$+ \; \gamma_2(dA_2 \, ds_2)z_2 + (\rho_2 \, dA_2 \, ds_2)ie_2$$

$$(KE + PE + IE)_t = (KE + PE + IE)_R + \tfrac{1}{2}(\rho_1 \, dA_1 \, ds_1)V_1^2$$
$$+ \; \gamma_1(dA_1 \, ds_1)z_1 + (\rho_1 \, dA_1 \, ds_1)ie_1$$

where ie is the internal energy per unit mass of fluid.[2] As $d(KE + PE + IE) = (KE + PE + IE)_{t+dt} - (KE + PE + IE)_t$, equation 5.1 yields

$$dQ = dW_{\text{shaft}} - p_1 \, dA_1 \, ds_1 + p_2 \, dA_2 \, ds_2 + \tfrac{1}{2}\rho_2 \, dA_2 \, ds_2 V_2^2 + \gamma_2 \, dA_2 \, ds_2 z_2$$
$$+ \; \rho_2 \, dA_2 \, ds_2 ie_2 - \tfrac{1}{2}\rho_1 \, dA_1 \, ds_1 V_1^2 - \gamma_1 \, dA_1 \, ds_1 z_1 - \rho_1 \, dA_1 \, ds_1 ie_1 \quad (5.2)$$

According to the continuity equation 3.8

$$\rho_1 \, dA_1 V_1 = \rho_2 \, dA_2 V_2 \qquad \text{or} \qquad \rho_1 \, dA_1 \, ds_1 = \rho_2 \, dA_2 \, ds_2$$

---

[2] For perfect gases, thermodynamics shows ie to be a function of temperature only and change of internal energy related to change of temperature by $ie_1 - ie_2 = c_v(T_1 - T_2)$.

because $ds = V\,dt$. By defining the steady mass flow rate $\dot{m} = \rho\,dA\,V$ through the control volume and dividing equation 5.2 by $\dot{m}\,dt$, we obtain

$$\frac{1}{\dot{m}}\frac{dQ}{dt} = \frac{1}{\dot{m}}\frac{dW_{\text{shaft}}}{dt} - \frac{p_1}{\rho_1} + \frac{p_2}{\rho_2} + \tfrac{1}{2}(V_2^2 - V_1^2) + g_n(z_2 - z_1) + ie_2 - ie_1 \quad (5.3)$$

In thermodynamics, it is common to rename certain terms in equation 5.3. First,

$$q_H = \frac{1}{\dot{m}}\frac{dQ}{dt} = \left[\frac{\text{N}\cdot\text{m}}{\text{kg}}\right] = \left[\frac{\text{joule}}{\text{kg}}\right] = \left[\frac{\text{ft}\cdot\text{lb}}{\text{slug}}\right]$$

is the heat added to the fluid in the control volume per unit of mass passing through the control volume 1221. Second, from Section 4.5,

$$\frac{1}{\dot{m}}\frac{dW_{\text{shaft}}}{dt} = \frac{1}{\dot{m}}\frac{dW_T}{dt} - \frac{1}{\dot{m}}\frac{dW_P}{dt} = g_n E_T - g_n E_P = \left[\frac{\text{J}}{\text{kg}}\right] = \left[\frac{\text{ft}\cdot\text{lb}}{\text{slug}}\right]$$

where E_T and E_P are energies withdrawn by turbines or added by pumps *per unit weight* of fluid passing through the control volume. By using the common combination of terms $p/\rho + ie$, called the *specific enthalpy h*, equation 5.3 can be converted to the steady flow energy equation

$$q_H = g_n(E_T - E_P) + h_2 - h_1 + \tfrac{1}{2}(V_2^2 - V_1^2) + g_n(z_2 - z_1) \quad (5.4)$$

or on division by g_n and with rearrangement to

$$\hat{h}_1 + z_1 + \frac{V_1^2}{2g_n} + E_P + \frac{q_H}{g_n} = \hat{h}_2 + z_2 + \frac{V_2^2}{2g_n} + E_T \quad (5.5)$$

where $\hat{h} = h/g_n$ (compare to equations 4.1 and 4.5). The enthalpy depends only on the system properties including p, ρ, and temperature T, and so enthalpy is also a property. The units of h are J/kg or ft$\cdot$lb/slug.

──────────────────── **Illustrative Problem** ────────────────────

Air flows as a perfect gas in a pipe without friction between the points indicated. If heat $q_H = -1 \times 10^5$ J/kg is lost between these points from each unit mass of fluid, find V_2.

Relevant Equations and Given Data

$$p/\rho = RT \quad (1.3)$$

$$c_p - c_v = R$$

$$ie = c_v T \qquad h = \frac{p}{\rho} + ie$$

$$q_H = g_n(E_T - E_P) + h_2 - h_1 + \tfrac{1}{2}(V_2^2 - V_1^2) + g_n(z_2 - z_1) \quad (5.4)$$

$$T_1 = 200°C \qquad T_2 = 100°C \qquad V_1 = 100 \text{ m/s}$$

$$q_H = -1 \times 10^5 \text{ J/kg} \qquad c_p = 1\,003 \text{ J/kg}\cdot\text{K}$$

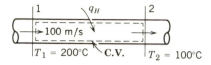

Solution. From equation 1.3 and the other definitions

$$h_2 - h_1 = (R + c_v)(T_2 - T_1) = c_p(T_2 - T_1)$$

Here, $E_T = E_P = 0$ and $z_2 - z_1 = 0$. Thus,

$$-1 \times 10^5 = 1\,003(100 - 200) + \tfrac{1}{2}(V_2^2 - 100^2) \qquad (5.4)$$

$$V_2 = 103 \text{ m/s} \blacktriangleleft$$

The Second Law of Thermodynamics is interpreted in terms of a quantity called the *entropy S* defined as

$$dS = dQ_{\text{reversible}}/T$$

where T is the absolute temperature at the point of transfer and $dQ_{\text{reversible}}$ is the heat transferred to a system during a reversible process (the temperature differences across the boundary must be vanishingly small). For a *nonflow process* in which a fluid system of uniform and invariant chemical composition undergoes only a simple reversible change in volume by working against the boundaries and receiving heat in time dt, the First Law yields [only the pressure acting during the volume change $d(m/\rho)$ does work]

$$dQ = pd\left(\frac{m}{\rho}\right) + d(mie)$$

or

$$T\frac{dS}{m} = pd\left(\frac{1}{\rho}\right) + d(ie) \qquad (5.6)$$

where m is the system mass, ρ its density, and ie its internal energy per unit mass. If the *specific entropy* $s = S/m$ and the specific enthalpy h are introduced

$$ds = \frac{1}{T}\left(dh - \frac{1}{\rho}dp\right) \qquad (5.7)$$

As all the variables on the right in the above equations 5.6 and 5.7 are properties of the fluid, it follows that *entropy is* also a *fluid property*. In consequence the equations must hold for *any process* even though entropy is defined only in terms of heat transfer in a reversible process. By using equation 5.7 and an arbitrarily prescribed datum for s (only differences are important), the state of fluids can be described very usefully on entropy versus enthalpy (*Mollier*) diagrams as will be seen in the subsequent sections.

The Second Law can be stated in many useful forms. Here only an extension of the law is needed, namely, that in any process

$$dS \geq \frac{dQ}{T} \tag{5.8}$$

Equality holds for a reversible process, and hence, the *entropy cannot decrease in an adiabatic ($dQ = 0$) process.* In all natural or real adiabatic processes

$$(dS)_{\text{adiabatic}} > 0 \tag{5.9}$$

A reversible, adiabatic process is, thus, properly called *isentropic ($dS = 0$).*

The entropy relations for a perfect gas ($p/\rho = RT$ and c_p and c_v are constant) are easy to derive and instructive. First, by definition the specific heats are

$$c_v = \left(\frac{\partial ie}{\partial T}\right)_v$$

$$c_p = \left(\frac{\partial h}{\partial T}\right)_p = \frac{\partial}{\partial T}\left(ie + \frac{p}{\rho}\right)_p = c_v + \frac{\partial}{\partial T}(RT)_p = c_v + R$$

so, as discovered above,

$$h_2 - h_1 = c_p(T_2 - T_1)$$
$$ie_2 - ie_1 = c_v(T_2 - T_1)$$

Then, from equation 5.7,

$$ds = \frac{1}{T}\left(c_p \, dT - \frac{1}{\rho}dp\right) = c_p\frac{dT}{T} - \frac{R}{p}dp$$

and

$$s_2 - s_1 = c_p \ln\left(\frac{T_2}{T_1}\right) - R \ln\left(\frac{p_2}{p_1}\right)$$

$$= c_v\left[\ln\left(\frac{T_2}{T_1}\right) + \frac{R}{c_v}\ln\left(\frac{T_2}{T_1}\right) - \frac{R}{c_v}\ln\left(\frac{p_2}{p_1}\right)\right]$$

$$= c_v \ln\left[\left(\frac{T_2}{T_1}\right)^{1+(R/c_v)}\left(\frac{p_2}{p_1}\right)^{R/c_v}\right]$$

Now, as $k = c_p/c_v = 1 + (R/c_v)$,

$$s_2 - s_1 = c_v \ln\left[\left(\frac{T_2}{T_1}\right)^k\left(\frac{p_2}{p_1}\right)^{1-k}\right] \tag{5.10}$$

and the entropy change which occurs when a system undergoes a process can be computed. Because entropy is a property, the entropy change undergone by fluid systems as they pass through the control volume of Fig. 5.1, for example, can be obtained via equation 5.10 by use of the property values known (the state of the fluid) at sections 1 and 2.

--------------------- **Illustrative Problem** ---------------------

Calculate the entropy change between sections 1 and 2 in the previous Illustrative Problem if $\rho_1 = 0.74$ kg/m³ and $\rho_2 = 0.47$ kg/m³.

Relevant Equations and Given Data

See diagram for previous problem.

$$p/\rho = RT \tag{1.3}$$

$$k = c_p/c_v$$

$$s_2 - s_1 = c_v \ln\left[\left(\frac{T_2}{T_1}\right)^k \left(\frac{\rho_2}{\rho_1}\right)^{1-k}\right] \tag{5.10}$$

$T_1 = 200°C \qquad T_2 = 100°C \qquad \rho_1 = 0.74$ kg/m³

$\rho_2 = 0.47$ kg/m³ $\qquad c_p = 1\,003$ J/kg·K

$R = 286.8$ J/kg·K $\qquad k = 1.40 \qquad$ (see Table 2)

Solution. From the data, $c_v = 1\,003/1.40 = 716.4$ J/kg·K. From equation 1.3, the absolute pressures

$$p_1 = 0.74 \times 286.8(200 + 273.2) = 100.4 \text{ kPa} \tag{1.3}$$

$$p_2 = 0.47 \times 286.8(100 + 273.2) = 50.3 \text{ kPa} \tag{1.3}$$

Thus,

$$s_2 - s_1 = 716.4 \ln\left[\left(\frac{373.2}{473.2}\right)^{1.4}\left(\frac{50.3}{100.4}\right)^{-0.4}\right] = -40 \text{ J/kg·K} \blacktriangleleft \tag{5.10}$$

ONE-DIMENSIONAL FLOW

5.2 Euler's Equation and the Energy Equation

In the previous section a control volume analysis led to the general energy equation 5.4 for a streamtube. On the other hand, a development of an Euler equation for one-dimensional flow of a compressible ideal fluid can be carried out by analysis of a system just as was done in Section 4.1 for an incompressible fluid. This leads to an energy equation for one-dimensional flow in the absence of heat transfer and shear or shaft work.

Consider a streamline and a small cylindrical fluid system for analysis, as shown in Fig. 5.2. Although fluid density will vary along the streamline, the mean density ρ of the differential fluid system shown differs negligibly from that at its ends and thus

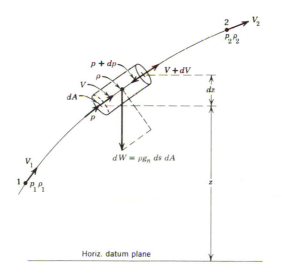

Fig. **5.2**

may be taken as constant throughout the system.[3]

Euler's equation is therefore (as before)

$$\frac{dp}{\rho} + V\,dV + g_n\,dz = 0$$

For compressible flow, however, the term $g_n\,dz$ is usually dropped and the Euler equation written

$$\frac{dp}{\rho} + V\,dV = 0 \qquad \text{or} \qquad \frac{dp}{\gamma} + \frac{V\,dV}{g_n} = 0 \qquad (5.11)$$

This simplification is justified by the fact that compressible-flow problems are usually concerned with gases of light weight and with flows of small vertical extent in which changes of pressure and velocity are predominant and changes of elevation negligible[4] by comparison.

When there is no heat transfer and no work done by pumps and turbines in the flow of an ideal fluid, the motion is isentropic and the steady flow energy equation 5.4 for the streamline in Fig. 5.2 becomes

$$h_1 + \tfrac{1}{2}V_1^2 = h_2 + \tfrac{1}{2}V_2^2 \qquad \text{or} \qquad ie_1 + \frac{p_1}{\rho_1} + \tfrac{1}{2}V_1^2 = ie_2 + \frac{p_2}{\rho_2} + \tfrac{1}{2}V_2^2 \qquad (5.12)$$

when $z_2 - z_1$ is neglected. In differential form, the second of these energy equations can be written as

---

[3] A similar situation is discussed in Section 2.1.
[4] See the fourth Illustrative Problem of Section 4.8.

$$d(ie) + pd\left(\frac{1}{\rho}\right) + \frac{1}{\rho}dp + V\,dV = 0$$

and, because of equation 5.6 with $dS = 0$, the energy equation reduces to

$$\frac{dp}{\rho} + V\,dV = 0$$

which is recognized as the Euler equation. *Thus, the energy and Euler equations are identical for isentropic flow.*

For application to perfect gases, equations 5.12 may be written in other useful terms. Recall that $ie = c_v T$, $p/\rho = RT$, $h = (c_v + R)T$, and $c_v + R = c_p$. Thus, substituting these relations into the first of equations 5.12 gives

$$c_p T_1 + \frac{V_1^2}{2} = c_p T_2 + \frac{V_2^2}{2} \tag{5.13}$$

The foregoing equations actually apply equally well to frictional and frictionless flow (the derivation of the energy equation 5.4 did not exclude friction effects), but only for the adiabatic case in which no heat is added to or extracted from the fluid; also they provide only for the situation where no mechanical energy is added to or extracted from the fluid by pump or turbine. For the remainder of this chapter their application will be confined to frictionless adiabatic (*isentropic*) flow.

5.3   Integration of the Euler Equation

When the Euler equation is integrated along the streamline (or small streamtube) for isentropic flow of perfect gases, it becomes

$$\frac{V_2^2 - V_1^2}{2} = \int_{p_2}^{p_1} \frac{dp}{\rho} = \frac{p_1}{\rho_1}\frac{k}{k-1}\left[1 - \left(\frac{p_2}{p_1}\right)^{(k-1)/k}\right]$$

$$\text{or} \quad \frac{p_2}{\rho_2}\frac{k}{k-1}\left[\left(\frac{p_1}{p_2}\right)^{(k-1)/k} - 1\right] \tag{5.14}$$

in which the integration has been performed by substitution of the isentropic relation $p_1/\rho_1^k = p_2/\rho_2^k$. The same result can also be obtained directly from the isentropic energy equation (5.13) by rearranging it and substituting[5] $kp/\rho(k-1)$ for $c_p T$ and $(p_2/p_1)^{1/k}$ for ρ_2/ρ_1. From equations 5.13 and 5.14, the variation of absolute pressure and temperature (and thus density) with velocity may be predicted along the streamline of Fig. 5.2 or the streamtube of Fig. 5.1.

Equation 5.13, which shows $(c_p T + V^2/2)$ to be constant along any streamline in adiabatic flow, is frequently written

$$c_p T_1 + \frac{V_1^2}{2} = c_p T_2 + \frac{V_2^2}{2} = c_p T_s \tag{5.13}$$

[5]See Section 1.4.

At a stagnation point where V is zero, T is the *stagnation temperature* T_s, which is seen to be constant[6] for all points on the streamline.

For the isentropic flow of vapors other than perfect gases, equation 5.12 may be used directly; the specific enthalpy h is a funciton of pressure and temperature and may be obtained from appropriate tables and diagrams; for isentropic flow, values of h_1 and h_2 for the same entropy must be used.

Illustrative Problems

At one point on a streamline in an airflow the velocity, absolute pressure, and temperature are 30 m/s, 35 kPa, and 150°C, respectively. At a second point on the same streamline the velocity is 150 m/s. If the process along the streamline is assumed isentropic, calculate the pressure and temperature at the second point.

Relevant Equations and Given Data

$$c_p T_1 + \frac{V_1^2}{2} = c_p T_2 + \frac{V_2^2}{2} = c_p T_s \tag{5.13}$$

$$\frac{V_2^2 - V_1^2}{2} = \int_{p_2}^{p_1} \frac{dp}{\rho} = \frac{p_1}{\rho_1} \frac{k}{k-1} \left[1 - \left(\frac{p_2}{p_1} \right)^{(k-1)/k} \right] \tag{5.14}$$

$$V_1 = 30 \text{ m/s} \qquad p_1 = 35 \text{ kPa}$$

$$T_1 = 150°C \qquad V_2 = 150 \text{ m/s}$$

Solution. From Table 2, $c_p = 1\ 003$ J/kg · K

$$\frac{(150)^2 - (30)^2}{2} = 1\ 003(T_1 - T_2) \tag{5.13}$$

$$T_1 - T_2 = 10.8°C \qquad T_2 = 139.2°C \blacktriangleleft$$

Using equation 5.14 (with RT_1 substituted for p_1/ρ_1),

$$\frac{(150)^2 - (30)^2}{2} = 288(150 + 273.2)\frac{1.4}{0.4}\left[1 - \left(\frac{p_2}{p_1}\right)^{0.286}\right] \tag{5.14}$$

$$\left(\frac{p_2}{p_1}\right)^{0.286} = 0.974\ 7 \qquad \frac{p_2}{p_1} = 0.914\ 1$$

$$p_2(\text{absolute}) = 32 \text{ kPa} \blacktriangleleft$$

Steam in a large tank is at an absolute pressure and temperature of 2 400 kPa and 500°C, respectively. The steam flows from the tank through a smooth nozzle and into a passage, and at a point in the passage the absolute pressure is observed to be 1 500 kPa. Determine the temperature and velocity at this point assuming an isentropic process.

---

[6]Thus stagnation temperature in adiabatic flow is analogous to the total head of ideal incompressible flow.

Relevant Equation and Given Data

$$h_1 + \tfrac{1}{2}V_1^2 = h_2 + \tfrac{1}{2}V_2^2 \tag{5.12}$$

$$p_1 = 2\,400 \text{ kPa} \qquad T_1 = 500°C$$

$$p_2 = 1\,500 \text{ kPa}$$

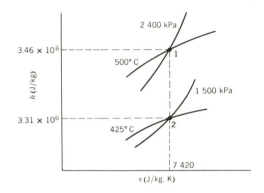

Solution. Refer to the *Mollier diagram* contained in any text on thermodynamics; this is a plot of specific enthalpy h against specific entropy s for various pressures and temperatures. At the intersection of the 2 400 kPa and 500°C lines on the diagram, h is found to be 3.46 MJ/kg. Now drop vertically down the chart (at constant entropy) to the 1 500 kPa line. From the temperature line passing through point 2 the temperature may be read; it is 425°C. For this point the specific enthalpy h is found to be 3.31 MJ/kg. Substituting h_1 and h_2 into equation 5.12, noting that the velocity V_1 in the large tank will be zero,

$$\frac{V_2^2}{2} = 3.46 \times 10^6 - 3.31 \times 10^6 \qquad V_2 = 548 \text{ m/s} \blacktriangleleft \tag{5.12}$$

5.4 The Stagnation Point

In gas dynamics, equation 5.14 is usually expressed in terms of Mach number, **M**. Using the first form of the equation, this may be easily accomplished by recalling (equation 1.10) that the acoustic (sonic) velocity a is given by $\sqrt{kp/\rho}$ and thus $a_1^2 = kp_1/\rho_1$. By substituting this into equation 5.14 and rearranging, we find that

$$\frac{V_2^2}{a_1^2} = \mathbf{M}_1^2 + \frac{2}{k-1}\left[1 - \left(\frac{p_2}{p_1}\right)^{(k-1)/k}\right] \tag{5.15}$$

Now consider the application of this equation to a stagnation point (S) in a compressible flow (Fig. 5.3). With the fluid compressible, the rise of pressure at the stagnation point causes compression of the fluid, producing a higher density (ρ_s) and

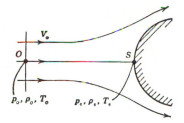

Fig. **5.3**

temperature (T_s) there. Evidently, the extent of these compression effects depends primarily on the magnitude of the stream velocity V_o; they are large at high velocities and small (often negligible) at low velocities. At the stagnation point, $V_2 = 0$ and $p_2 = p_s$; substituting these values and $p_1 = p_o$ and $\mathbf{M}_1 = \mathbf{M}_o$ into equation 5.15, the stagnation pressure p_s is given by

$$\frac{p_s}{p_o} = \left[1 + \mathbf{M}_o^2 \frac{k-1}{2}\right]^{k/(k-1)} \qquad (5.16)^7$$

If the right-hand side of this equation is expanded by the binomial theorem there results (retaining the first three terms)

$$p_s = p_o + \tfrac{1}{2}\rho_o V_o^2[1 + \tfrac{1}{4}\mathbf{M}_o^2 + \ldots] \qquad (5.17)^7$$

Comparison of this equation with equation 4.3 of Section 4.4 shows that the effects of compressibility have been isolated in the bracketed quantity and that these effects depend only on the Mach number. The bracketed quantity may thus be considered a "compressibility correction factor" and the effect of compressibility on $(p_s - p_o)$ calculated with fair precision.[8]

From equation 5.16 it can be observed that measurements of p_s and p_o allow calculation of the "free stream" Mach number $\mathbf{M}_o$ of the undisturbed flow. However, to obtain the velocity V_o a temperature measurement is also required, and in practice the temperature T_s is measured at the stagnation point. From T_s, p_s, and p_o, the velocity V_o may be handily calculated by using the second form of equation 5.14 between points O and S; this equation becomes (RT_s having been substituted for p_s/ρ_s)

$$\frac{V_o^2}{2} = c_p T_s\left[1 - \left(\frac{p_o}{p_s}\right)^{(k-1)/k}\right] \qquad (5.18)^7$$

from which V_o may be calculated directly.

---

[7]The use of equations 5.16, 5.17, and 5.18 is restricted to $\mathbf{M}_o < 1$. For $\mathbf{M}_o > 1$ a shock wave exists across the streamline between points O and S and the flow is no longer isentropic. See Sections 6.7 and 11.6.
[8]The precision is excellent for small $\mathbf{M}_o$ but decreases with increasing $\mathbf{M}_o$ as the neglected terms of the binomial expansion become significant.

——————————— **Illustrative Problem** ———————————

An airplane flies at 400 mph (586 ft/s) through still air at 13.0 psia and 0°F. Calculate pressure, temperature, and air density at the stagnation points (on nose of fuselage and leading edges of wings).

Relevant Equations and Given Data

$$\rho = p/RT \tag{1.3}$$

$$a = \sqrt{kRT} \tag{1.11}$$

$$c_p T_1 + \frac{V_1^2}{2} = c_p T_2 + \frac{V_2^2}{2} \tag{5.13}$$

$$\frac{p_s}{p_o} = \left[1 + M_o^2 \frac{k-1}{2}\right]^{k/(k-1)} \tag{5.16}$$

$$p_s = p_o + \tfrac{1}{2}\rho_o V_o^2 [1 + \tfrac{1}{4}M_o^2 + \ldots] \tag{5.17}$$

$$p_o = 13.0 \text{ psia} \qquad V_o = 586 \text{ ft/s} \qquad T_o = 0°F$$

$$c_p = 6\,000 \text{ ft} \cdot \text{lb/slug} \cdot °R$$

$$R = 1\,715 \text{ ft} \cdot \text{lb/slug} \cdot °R \qquad k = 1.40$$

Solution.

$$\rho_o = \frac{13.0 \times 144}{1\,715 \times 460} = 0.002\,37 \text{ slug/ft}^3 \tag{1.3}$$

$$a_o = \sqrt{1.4 \times 1\,715 \times 460} = 1\,052 \text{ ft/s} \tag{1.11}$$

$$M_o = \frac{586}{1\,052} = 0.557$$

$$\text{(approx.) } p_s = 13.0 + \frac{1}{2}\frac{0.002\,37 \times (586)^2}{144}[1 + \tfrac{1}{4}(0.557)^2]$$

$$= 16.05 \text{ psia} \blacktriangleleft \tag{5.17}$$

$$\text{(exact) } p_s = 13.0\left[1 + (0.557)^2\frac{1.4-1}{2}\right]^{3.5} = 16.18 \text{ psia} \blacktriangleleft \tag{5.16}$$

$$\frac{(586)^2}{2} = 6\,000(T_s - 460) \qquad T_s = 488.5°R \qquad T_s = 28.5°F \blacktriangleleft \tag{5.13}$$

$$\rho_s = \frac{16.18 \times 144}{488.5 \times 1\,715} = 0.002\,78 \text{ slug/ft}^3 \blacktriangleleft \tag{1.3}$$

5.5   The One-Dimensional Assumption

The foregoing equations of this chapter may be applied successfully to passages of finite cross section when the streamlines are essentially straight and parallel. Consistent with neglecting the difference in the z terms in the Euler and energy equations, the pressure is taken to be constant[9] over the flow cross section; the absence of friction permits no variation of velocity. With pressure and velocity constant throughout the flow cross section, constancy of temperature and density will follow from equations 5.13 and 5.14.

When streamlines are not essentially straight and parallel, variations of pressure, velocity, temperature, and density are to be expected. As with incompressible flow,[10] increase of pressure and decrease of velocity will be found with increasing distance from center of curvature; in compressible flow, such variations of pressure and velocity will produce variations of temperature and density as well. Although mean values of the variables may be visualized and computed, their use in the equations of one-dimensional flow is not recommended except for approximate calculations.

5.6 Subsonic and Supersonic Velocities

Combination of the continuity and Euler equations yields information on the superficial shapes of passages required to produce changes of flow velocity when such velocities are subsonic or supersonic. These equations are

$$\frac{dA}{A} + \frac{d\rho}{\rho} + \frac{dV}{V} = 0 \tag{3.9}$$

$$\frac{dp}{\rho} + VdV = 0 \tag{5.11}$$

Multiplying the first term of equation 5.11 by $d\rho/d\rho$, a^2 is recognized (equation 1.9) as $dp/d\rho$, and $d\rho/\rho$ is obtained as $-V\,dV/a^2$ or $-V^2dV/a^2V$. Substituting this in equation 3.9, identifying V/a as the Mach number **M,** and rearranging, we obtain

$$\frac{dA}{A} = \frac{dV}{V}(\mathbf{M}^2 - 1) \tag{5.19}$$

From this equation we can deduce some far-reaching and somewhat surprising conclusions. Analysis of the equation for $dV/V > 0$ shows that: for $\mathbf{M} < 1$, $dA/A < 0$; for $\mathbf{M} = 1$, $dA/A = 0$; for $\mathbf{M} > 1$, $dA/A > 0$. This means that, for subsonic flow ($\mathbf{M} < 1$), a reduction of cross-sectional area (Fig. 5.4) is required for an increase of velocity; however, for supersonic flow ($\mathbf{M} > 1$), *an increase of area is required to produce an increase of velocity*. For flow at sonic speed ($\mathbf{M} = 1$) the rate of change of area must be zero; that is, this might occur at a maximum or minimum cross section of the streamtube; use of the preceding conclusions will show this to be restricted to a minimum section (*a throat*) only. However, equation 5.19 does *not* allow the conclusion that a throat will always produce a flow of sonic velocity because **M** is *not necessarily one* when $dA/A = 0$. If, at a throat ($dA/A = 0$) $\mathbf{M} \gtrless 1$, it follows that

---

[9]For the incompressible fluid ($p/\gamma + z$) is constant over the flow cross section; see Section 4.3.
[10]See Section 4.8.

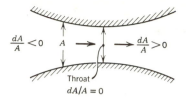

Fig. **5.4**

$dV/V = 0$, implying a maximum or minimum velocity there. Upstream from the throat $dA/A < 0$, so here $dV/V < 0$ for $\mathbf{M} > 1$ and $dV/V > 0$ for $\mathbf{M} < 1$. Accordingly it can be concluded that if the throat velocity is not sonic it will be a maximum in subsonic flow and a minimum in supersonic flow.

5.7 The Convergent Nozzle

Consider now the frictionless flow of a gas from a large tank ($A_1 \sim \infty$, $V_1 \sim 0$) through a convergent nozzle (Fig. 5.5) into a region of pressure, p_2'. Using the second form of equation 5.14 and substituting a_2^2 for kp_2/ρ_2 and $\mathbf{M}_2$ for V_2/a_2, we obtain

$$\mathbf{M}_2^2 = \frac{2}{k-1}\left[\left(\frac{p_1}{p_2}\right)^{(k-1)/k} - 1\right] \tag{5.20}$$

In view of the preceding developmemt, supersonic velocities are not expected in this problem, because the fluid starts from rest and there are no divergent passages; accordingly $\mathbf{M}_2 \lesseqgtr 1$. If the pressure difference ($p_1 - p_2'$) is large enough to produce sonic velocity, this velocity must exist at the throat of the nozzle and $\mathbf{M}_2 = 1$; placing this value in equation 5.20 and solving for p_2/p_1 gives the critical pressure ratio (p_2/p_1)$_c$:

$$\left(\frac{p_2}{p_1}\right)_c = \left(\frac{2}{k+1}\right)^{k/(k-1)} \tag{5.21}$$

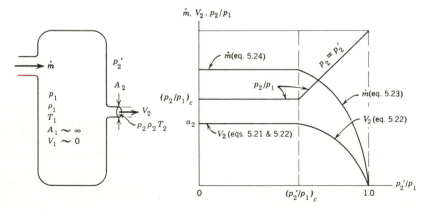

Fig. **5.5**

Thus, if the sonic velocity is attained by the fluid, the absolute pressure, p_2, in the minimum section is in fixed ratio to the absolute pressure, p_1, in the tank and *therefore independent of the pressure p_2'.* It follows that in a free jet moving at sonic velocity *the pressure within the jet at the nozzle exit is never less, and is usually more, than the pressure which surrounds it;* the outside pressure, p_2', tends to penetrate the jet at the sonic speed but cannot distribute itself over the whole jet cross section at the nozzle exit because the fluid there is moving at this same high velocity.

Of course, the sonic velocity is not attained unless the pressure drop between the inside and outside of the tank is large enough. For small pressure drops (i.e., for pressure ratios, p_2/p_1, above the critical) the pressures p_2' and p_2 are the same, and the velocity at the nozzle exit may be computed (from equation 5.14) by

$$\frac{V_2^2}{2} = \frac{p_1}{\rho_1}\frac{k}{k-1}\left[1 - \left(\frac{p_2}{p_1}\right)^{(k-1)/k}\right]$$
(5.22)

A graphic summary of these facts is given in Fig. 5.5.

In many problems (especially those of fluid metering) the flowrate is the most important quantity to be computed in flow through nozzles. The computation can be easily made by combining some of the foregoing equations.

If the pressure ratio p_2'/p_1 is more than the critical, the mass flowrate can be calculated from $\dot{m} = A_2\rho_2 V_2$, using V_2 from equation 5.13 and the isentropic relation $p_2/p_1 = (\rho_2/\rho_1)^k$. The result is

$$\dot{m} = A_2 \sqrt{\frac{2k}{k-1}p_1\rho_1\left[\left(\frac{p_2}{p_1}\right)^{2/k} - \left(\frac{p_2}{p_1}\right)^{(k+1)/k}\right]}$$
(5.23)

If the pressure ratio p_2'/p_1 is less than the critical, $p_2/p_1 = [2/(k+1)]^{k/(k-1)}$, and substituting this in equation 5.23, along with p_1/RT_1 for ρ_1, yields

$$\dot{m} = \frac{A_2 p_1}{\sqrt{T_1}} \sqrt{\frac{k}{R}\left(\frac{2}{k+1}\right)^{(k+1)/(k-1)}}$$
(5.24)

in which the large square root is obviously a characteristic constant of the gas, dependent on R and k; this allows a simple calculation for flowrate and is good reason for selecting metering nozzles of such proportions that sonic velocities are produced.

───────────── **Illustrative Problem** ─────────────

Air discharges from a large tank, in which the pressure is 700 kPa and temperature 40°C, through a convergent nozzle of 25 mm tip diameter. Calculate the flowrates when the pressure outside the jet is (a) 200 kPa, and (b) 550 kPa, and the barometric pressure is 101.3 kPa. Also calculate the pressure, temperature, velocity, and sonic velocity at the nozzle tip for these flowrates.

Relevant Equations and Given Data

$$\rho = p/RT$$
(1.3)

$$a = \sqrt{kRT} \qquad (1.11)$$

$$c_p T_1 + \frac{V_1^2}{2} = c_p T_2 + \frac{V_2^2}{2} \qquad (5.13)$$

$$\frac{V_2^2 - V_1^2}{2} = \frac{p_1}{\rho_1} \frac{k}{k-1} \left[1 - \left(\frac{p_2}{p_1}\right)^{(k-1)/k} \right] \qquad (5.14)$$

$$\left(\frac{p_2}{p_1}\right)_c = \left(\frac{2}{k+1}\right)^{k/(k-1)} \qquad (5.21)$$

$$\dot{m} = A_2 \sqrt{\frac{2k}{k-1} p_1 \rho_1 \left[\left(\frac{p_2}{p_1}\right)^{2/k} - \left(\frac{p_2}{p_1}\right)^{(k+1)/k} \right]} \qquad (5.23)$$

$$\dot{m} = \frac{A_2 p_1}{\sqrt{T_1}} \sqrt{\frac{k}{R} \left(\frac{2}{k+1}\right)^{(k+1)/(k-1)}} \qquad (5.24)$$

$p_1 = 700$ kPa (gage) $= 801.3$ kPa (abs) $\qquad T_1 = 40°C = 313$ K $\qquad d_2 = 25$ mm

Solution. First calculate the density of the air in the tank and the critical pressure ratio.

$$\rho_1 = \frac{801.3 \times 10^3}{286.8 \times 313} = 8.92 \text{ kg/m}^3 \qquad (1.3)$$

$$\left(\frac{p_2}{p_1}\right)_c = \left(\frac{2}{2.4}\right)^{3.5} = 0.528 \qquad (5.21)$$

(a) $\quad p_2' = 200$ kPa (gage): $\qquad \dfrac{p_2'}{p_1} = \dfrac{301.3}{801.3} = 0.376 < 0.528$

so

$$p_2 \text{ (abs.)} = 0.528 \times 801.3 = 423.1 \text{ kPa} \blacktriangleleft$$

and substituting $k = 1.40$, $R = 286.8$, $A_2 = 4.91 \times 10^{-4}$, $p_1 = 801.3 \times 10^3$, and $T_1 = 313$ into equation 5.24: $\dot{m} = 0.9$ kg/s. $\blacktriangleleft$ Using the first form of equation 5.14, with $V_1 = 0$ and $p_2/p_1 = 0.528$: $V_2 = 323.9$ m/s $\blacktriangleleft$,which is also the sonic velocity. Using equation 5.13 with $V_1 = 0$, $V_2 = 323.9$ m/s, and $T_1 = 313$ K: $T_2 = 261$ K. $\blacktriangleleft$

(b) $\quad p_2' = 550$ kPa (gage): $\qquad \dfrac{p_2'}{p_1} = \dfrac{651.3}{801.3} = 0.813 > 0.528$

so

$$p_2 \text{(abs.)} = 651.3 \text{ kPa} \blacktriangleleft$$

and, substituting $A_2 = 4.91 \times 10^{-4}$, $k = 1.40$, $p_1 = 801.3 \times 10^3$, $\rho_1 = 8.92$, and $p_2/p_1 = 0.813$ into equation 5.23: $\dot{m} = 0.72$ kg/s. $\blacktriangleleft$ Using the first form of equation 5.14 with $V_1 = 0$ and $p_2/p_1 = 0.813$: $V_2 = 190$ m/s. $\blacktriangleleft$ Using equation 5.13 with $V_1 = 0$, $V_2 = 190$ m/s, and $T_1 = 313$ K: $T_2 = 295$ K. $\blacktriangleleft$ Calculating the sonic velocity from equation 1.11, $a_2 = 344.2$ m/s, giving $\mathbf{M_2 = 0.552}$.

5.8 Constriction in Streamtube

When a compressible fluid flows through a constriction in a steamtube or pipeline (Fig. 5.6), the equations of Section 5.7 are not applicable unless the constriction is very small (compared to the pipe). When the constriction is larger, the velocity V_1 is no longer negligible compared to V_2 (see equation 5.14), and adjustments must be made in the foregoing equations to account for this. However, when sonic velocities are attained in the constriction, it has usually been chosen small enough that V_1 is negligible and thus adjustment of equation 5.24 is not usually necessary and will not be discussed here.

For flow of gases well above the critical pressure ratio, however, V_1 usually cannot be neglected. For this case an equation for flowrate may be derived by simultaneous solution of

$$\frac{V_2^2 - V_1^2}{2} = \frac{p_1}{\rho_1}\frac{k}{k-1}\left[1 - \left(\frac{p_2}{p_1}\right)^{(k-1)/k}\right] \tag{5.14}$$

$$\dot{m} = A_1\rho_1 V_1 = A_2\rho_2 V_2 \tag{3.10}$$

$$p_2/p_1 = (\rho_2/\rho_1)^k \tag{1.7}$$

This yields

$$\dot{m} = \frac{A_2}{\sqrt{1 - \left(\frac{p_2}{p_1}\right)^{2/k}\left(\frac{A_2}{A_1}\right)^2}}\sqrt{\frac{2k}{k-1}p_1\rho_1\left[\left(\frac{p_2}{p_1}\right)^{2/k} - \left(\frac{p_2}{p_1}\right)^{(k+1)/k}\right]} \tag{5.25}$$

Comparison of this equation with equation 5.23 indicates that the effect of including V_1 is concentrated in the first square root, which can be seen to approach 1.00 rapidly as the area ratio, A_2/A_1, decreases.

Equation 5.25, because of its unwieldy form, is sometimes solved by the use of an *expansion factor*, Y, applied to the simpler and analogous solution for incompressible flow,[11] which is

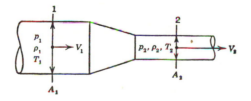

Fig. **5.6**

[11] See Section 4.4

$$Q = \frac{A_2}{\sqrt{1 - \left(\frac{A_2}{A_1}\right)^2}} \sqrt{2g_n \left(\frac{p_1 - p_2}{\gamma}\right)} \tag{5.26}$$

Y is defined as the factor that, when multiplied into the product of ρ_1 and equation 5.26, will yield equation 5.25. Thus

$$\dot{m} = \frac{YA_2\rho_1}{\sqrt{1 - \left(\frac{A_2}{A_1}\right)^2}} \sqrt{2g_n \left(\frac{p_1 - p_2}{\gamma_1}\right)} \tag{5.27}$$

An expression for Y can be derived by equating equations 5.25 and 5.27; it is found to be a function of the three variables p_2/p_1, A_2/A_1, and k, and thus can be computed once for all and presented in tables or plots. The equation for Y and its tabulated values are found in Appendix 8.

——————————— **Illustrative Problem** ———————————

Air flows through a 25 mm constriction in a 37.5 mm pipeline. In the pipe the pressure and temperature of the air are 689.5 kPa and 38°C, respectively. Calculate the flowrate if the pressure in the constriction is 551.6 kPa. Barometric pressure is 101.3 kPa.

Relevant Equations and Given Data

$$\rho = p/RT \tag{1.3}$$

$$\dot{m} = \frac{A_2}{\sqrt{1 - \left(\frac{p_2}{p_1}\right)^{2/k}\left(\frac{A_2}{A_1}\right)^2}} \sqrt{\frac{2k}{k-1}p_1\rho_1\left[\left(\frac{p_2}{p_1}\right)^{2/k} - \left(\frac{p_2}{p_1}\right)^{(k+1)/k}\right]} \tag{5.25}$$

$$\dot{m} = \frac{YA_2\rho_1}{\sqrt{1 - \left(\frac{A_2}{A_1}\right)^2}} \sqrt{2g_n \left(\frac{p_1 - p_2}{\gamma_1}\right)} \tag{5.27}$$

$$p_{bar} = 101.3 \text{ kPa} \qquad d_1 = 37.5 \text{ mm} \qquad d_2 = 25 \text{ mm}$$

$$p_1 = 689.5 \text{ kPa} \qquad T_1 = 38°C \qquad p_2 = 551.6 \text{ kPa}$$

Solution. $p_2/p_1 = 0.825$, $p_1 = 790.8$ kPa, $\rho_1 = 8.86$ kg/m³. Substituting these quantities in equation 5.25 gives a flowrate of 0.776 kg/s. ◀

The flowrate may also be calculated by interpolating a value of Y (0.874) from Appendix 8 and using it in equation 5.27; then

$$\dot{m} = \frac{0.874 \times \frac{\pi}{4}(0.025)^2 \times 8.86}{\sqrt{1 - (2/3)^4}} \sqrt{2g_n \frac{137.9 \times 1\,000}{86.9}} = 0.776 \text{ kg/s} \blacktriangleleft$$

$$\tag{5.27}$$

5.9 The Convergent-Divergent Nozzle

From the study of the flow of compressible fluid through a convergent-divergent passage (De Laval nozzle), much may be learned of basic phenomena and engineering application. For simplicity consider the discharge from a large reservoir through such a passage with pressure p_1 and temperature T_1 in the reservoir (Fig. 5.7). If sonic velocity exists at the throat, the flowrate $\dot{m}$ is determined by equation 5.24, the throat area A_2 being known. Assumption of the pressure p_3 in the jet at the nozzle exit allows computation of the exit area A_3 from equation 5.23, using A_3 for A_2 and p_3 for p_2; however, it will be found that *two very different p_3's will yield the same area A_3 and flowrate $\dot{m}$*. The higher of these pressures (p_3'') will cause subsonic velocity in the diverging part of the passage, the lower one (p_3''') will produce supersonic velocity there. The variations of pressure along the passage for these two conditions are shown in Fig. 5.7.

It is of fundamental importance to examine such a nozzle at the same reservoir conditions (p_1 and T_1) but with "back pressure" p_3' other than those used for the determination of A_3. Reduction of p_3' below p_3''' cannot affect conditions at the throat, so no change of flowrate is produced and no change of nozzle performance is to be expected (except that the compressed gas passing the nozzle exit will expand rapidly on emergence into a region of lower pressure). Raising the back pressure above p_3'', as shown on Fig. 5.8, causes a reduction of flowrate, velocities throughout the nozzle become subsonic, and the upper pressure distribution exists in the nozzle.

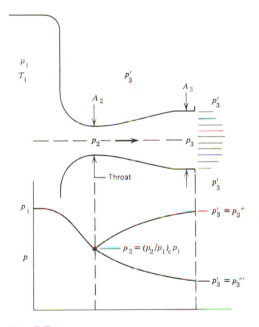

Fig. **5.7**

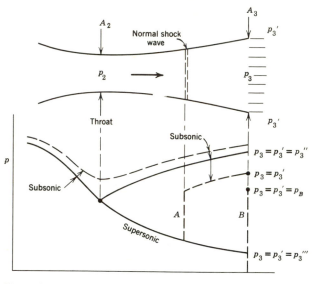

Fig. **5.8**

Between p_3'' and p_3''' there are an infinite number of "back pressures," none of which can satisfy the equations of isentropic flow; this is due to the formation of a *normal shock wave* [12] (with considerable internal friction and increase of entropy) in the divergent passage. Through such a wave the velocity drops abruptly from supersonic to subsonic and the pressure jumps abruptly about as shown on curve A. For the curve A and for others similar to it, the flow will emerge from the nozzle at subsonic velocity and the pressure p_3 will equal p_3'. At some lower value of p_3' (curve B) the shock wave will occur at the nozzle exit, which is the limiting case to be treated by one-dimensional methods; for pressures p_3' between p_B and p_3''' the jet will emerge from the nozzle with $p_3 = p_3'''$ and the shock wave will be in the flowfield downstream from the nozzle exit. Since such flowfields are either two-or three-dimensional, they cannot be described by the foregoing one-dimensional equations.

————————————— **Illustrative Problem** —————————————

Air discharges through a convergent-divergent passage (of 25 mm throat diameter) into the atmosphere. The pressure and temperature in the reservoir are 700 kPa and 40°C, respectively; the barometric pressure is 101.3 kPa. Calculate the nozzle tip diameter required for $p_3 = 101.3$ kPa. Calculate the flow velocity, sonic velocity, and Mach number at the nozzle exit. Determine the pressure p_3'' which will yield the same flowrate, and the pressure p_B which will produce a normal shock wave at the nozzle exit.

[12] See Section 6.7.

Relevant Equations and Given Data

$$\rho = p/RT \tag{1.3}$$

$$\dot{m} = \rho A V \tag{3.10}$$

$$c_p T_1 + \frac{V_1^2}{2} = c_p T_2 + \frac{V_2^2}{2} \tag{5.13}$$

$$\frac{V_2^2}{2} = \frac{p_1}{\rho_1} \frac{k}{k-1} \left[1 - \left(\frac{p_2}{p_1} \right)^{(k-1)/k} \right] \tag{5.22}$$

$$\dot{m} = A_2 \sqrt{\frac{2k}{k-1} p_1 \rho_1 \left[\left(\frac{p_2}{p_1} \right)^{2/k} - \left(\frac{p_2}{p_1} \right)^{(k+1)/k} \right]} \tag{5.23}$$

$$\frac{p_2 - p_1}{p_1} = \frac{2k}{k+1} (M_1^2 - 1) \tag{6.18}$$

$p_1 = 700$ kPa (gage) $p_{bar} = 101.3$ kPa $d_2 = 25$ mm

$T_1 = 40°C$ $p_3 = 101.3$ kPa (abs.)

Since supersonic velocities will occur in the divergent portion of the nozzle, sonic velocity will be expected at the throat. As these are the conditions of the Illustrative Problem of Section 5.7, the flowrate $\dot{m}$ is 0.9 kg/s as before.

Solution.

$$\frac{V_3^2}{2} = \frac{1.4}{0.4} \times \frac{801.3 \times 10^3}{8.92} \left[1 - \left(\frac{101.3}{801.3} \right)^{0.280} \right] \qquad V_3 = 525.8 \text{ m/s} \blacktriangleleft \tag{5.22}$$

$$(525.8)^2/2 = 1\ 003(T_1 - T_3) \qquad T_1 - T_3 = 137.8 \text{ K} \qquad T_3 = 175.2 \text{ K} \tag{5.13}$$

$$a_3 = \sqrt{1.4 \times 286.8 \times 175.2} = 265 \text{ m/s} \blacktriangleleft \qquad M_3 = \frac{525.8}{265} = 1.98 \blacktriangleleft$$

$$\rho_3 = \frac{101.3 \times 10^3}{286.8 \times 175.2} = 2.02 \text{ kg/m}^3 \tag{1.3}$$

$$A_3 = \frac{0.9}{2.02 \times 525.8} = 8.5 \times 10^{-4} \text{m}^2 \qquad d_3 = 33 \text{ mm} \blacktriangleleft \tag{3.10}$$

$$0.9 = 8.5 \times 10^{-4} \times \sqrt{\frac{2 \times 1.4}{0.4} 801.3 \times 10^3 \times 8.92 \left[\left(\frac{p_3''}{p_1} \right)^{1.43} - \left(\frac{p_3''}{p_1} \right)^{1.715} \right]} \tag{5.23}$$

Solving by trial,

$$\frac{p_3''}{p_1} = 0.91 \qquad p_3''(\text{abs.}) = 0.91 \times 801.3 \times 10^3 = 729.2 \text{ kPa} \blacktriangleleft$$

$$p_B (\text{abs.}) = 101.3 \times 10^3 \left[1 + \frac{2 \times 1.4}{1.4 + 1} (1.98^2 - 1) \right] = 446.4 \text{ kPa} \blacktriangleleft \tag{6.18}$$

TWO-DIMENSIONAL FLOW

The study of two-dimensional fields of compressible flow presents the same difficulties as those of incompressible flow (Sections 4.6 to 4.8) and a further complication: variation of density over the flowfield—which means another variable in the equations and increased difficulty of solution. Although formal mathematical solution of such problems is not possible, special techniques (particularly numerical solution of problems on digital computers) have been invented for the solution of certain problems of engineering interest; however, review of these methods is outside the scope of an elementary text. The intent of the following treatment is merely to develop the basic equations, to describe certain flowfields, and to discuss the applicability and limitations of the equations.

5.10 Euler's Equations and Their Integration

Euler's equations for the two-dimensional flow of an ideal compressible fluid are the same as those for the incompressible fluid except for the neglect of g_n, which was justified in Section 5.2. This neglection allows the flowfield to be considered to be in a horizontal $(x - y)$ plane and the Euler equations to be written as

$$-\frac{1}{\rho}\frac{\partial p}{\partial x} = u\frac{\partial u}{\partial x} + v\frac{\partial u}{\partial y} \tag{4.8}$$

$$-\frac{1}{\rho}\frac{\partial p}{\partial y} = u\frac{\partial v}{\partial x} + v\frac{\partial v}{\partial y} \tag{4.9}$$

They may be rearranged and combined in the same pattern as that of Section 4.7 to give

$$-\frac{dp}{\rho} = d\left(\frac{u^2 + v^2}{2}\right) + (u\,dy - v\,dx)\left(\frac{\partial v}{\partial x} - \frac{\partial u}{\partial y}\right)$$

in which $(\partial v/\partial x - \partial u/\partial y)$ is recognized as the vorticity ξ, and integrated to yield

$$\int\frac{dp}{\rho} + \frac{V^2}{2} = C + \int\xi(u\,dy - v\,dx)$$

which reduces, *for irrotational flow* $(\xi = 0)$, to

$$\int\frac{dp}{\rho} + \frac{V^2}{2} = C$$

in which C is a constant for all points in the flowfield (Fig. 5.9). For an isentropic flowfield this equation may be integrated by using the isentropic relation between p and ρ of equation 1.7; the result is

$$\frac{V_2^2 - V_1^2}{2} = \frac{k}{k-1}\frac{p_1}{\rho_1}\left[1 - \left(\frac{p_2}{p_1}\right)^{(k-1)/k}\right] \tag{5.28}$$

Although equations 5.14 and 5.28 are identical, application of the former is restricted

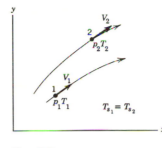

Fig. **5.9**

to points on the same streamline, whereas in equation 5.28 points 1 and 2 may be any points in the flowfield. Use of the isentropic relation between pressure and temperature in equation 5.28, along with $p_1/\rho_1 = RT_1$, yields

$$c_p T_1 + \frac{V_1^2}{2} = c_p T_2 + \frac{V_2^2}{2} = c_p T_s \qquad (5.29)$$

which is identical to equation 5.13 but shows that stagnation temperature is constant not only along single streamlines but also at all points in an isentropic flowfield. The limitations of equations 5.28 and 5.29 deserve emphasis: they have been derived for frictionless and adiabatic (i.e., isentropic) irrotational flow and hence cannot be expected to apply to real flowfields with heat transfer, boundary friction, and shock waves. They do, however, provide a useful method of approach to many compressible flow problems.

Consider now (for comparison with the foregoing) a particular compressible flowfield which is *nonisentropic* but throughout which the stagnation temperature is constant. With no change of stagnation temperature along the streamlines, equation 5.13 may be applied to conclude that there is no exchange of heat energy between adjacent streamtubes; however, frictional processes may vary from streamtube to streamtube with accompanying variability of entropy between them. This situation is closely approximated downstream from a curved shock wave. *Crocco's theorem,* [13] a classic synthesis of thermodynamic and fluid mechanics principles, shows that such a nonisentropic flowfield cannot be irrotational.

5.11   Application of the Equations

Application of equations 5.28 and 5.29 to a flowfield is straightforward enough if all the velocities are known, along with the temperature and pressure at one point in the flowfield; from these, all pressures and temperatures throughout the field may be computed, and from pressure and temperature the fluid density at any point may be predicted.

---

[13] Development and discussion of Crocco's theorem will be found in the references at the end of this chapter.

Illustrative Problem

Air approaches this streamlined object at the speed, pressure, and temperature shown. Calculate the pressures, temperatures, and Mach numbers at points A and B, where the velocities are 800 and 900 ft/s, respectively.

Relevant Equations and Given Data

$$p = \rho RT \tag{1.3}$$

$$\frac{p}{\rho^k} = \text{Constant} \tag{1.7}$$

$$a = \sqrt{kRT} \tag{1.11}$$

$$c_p T_1 + \frac{V_1^2}{2} = c_p T_2 + \frac{V_2^2}{2} = c_p T_s \tag{5.29}$$

$$R = 1\ 715 \text{ ft} \cdot \text{lb/slug} \cdot {}^\circ R \qquad c_p = 6\ 000 \text{ ft} \cdot \text{lb/slug} \cdot {}^\circ R$$

$$k = 1.4 \qquad V_A = 800 \text{ ft/s} \qquad V_B = 900 \text{ ft/s}$$

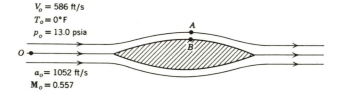

$V_o = 586$ ft/s
$T_o = 0°$F
$p_o = 13.0$ psia
$a_o = 1052$ ft/s
$M_o = 0.557$

Solution. Refer to the Illustrative Problem of Section 5.4 for some preliminary calculations and for pressure and temperature at the stagnation point. Then

$$\frac{(800)^2}{2} = 6\ 000(488.5 - T_A) \qquad T_A = 435.3°R \blacktriangleleft \tag{5.29}$$

$$\frac{(900)^2}{2} = 6\ 000(488.5 - T_B) \qquad T_B = 421.0°R \blacktriangleleft \tag{5.29}$$

From the isentropic relationship between pressure and temperature (equations 1.3 and 1.7),

$$\frac{p_A}{16.18} = \left(\frac{435.3}{488.5}\right)^{3.5} \qquad \frac{p_B}{16.18} = \left(\frac{421.0}{488.5}\right)^{3.5}$$

$$p_A = 10.85 \text{ psia} \qquad p_B = 9.65 \text{ psia} \blacktriangleleft$$

$$a_A = \sqrt{1.4 \times 1\ 715 \times 435.3} = 1\ 021 \text{ ft/s} \tag{1.11}$$

$$a_B = \sqrt{1.4 \times 1\ 715 \times 421.0} = 983 \text{ ft/s} \tag{1.11}$$

$$\mathbf{M_A} = \frac{800}{1\ 021} = 0.783 \qquad \mathbf{M_B} = \frac{900}{983} = 0.915 \blacktriangleleft$$

Formal mathematical solution of the inverse of the foregoing problem, when boundary conditions are specified and the flowfield through a passage or about an object is to be determined (i.e., velocities, pressures, etc., predicted), is impossible; answers to such problems are obtained by approximations, linearizations, numerical integration on a computer of the equations of the motion, and so on, most of which are beyond the scope of this elementary book. However, it is useful to consider such a problem and some of the methods and limitations of its solution.

The flowfield around the simple streamlined object of Fig. 5.10 is to be predicted. The shape and orientation of the object are known, and the *boundary condition* of velocity, pressure, and density of the fluid approaching the object is also known. The available independent equations are

$$\frac{\partial}{\partial x}(\rho u) + \frac{\partial}{\partial y}(\rho v) = 0 \tag{3.17}$$

$$-\frac{1}{\rho}\frac{\partial p}{\partial x} = u\frac{\partial u}{\partial x} + v\frac{\partial u}{\partial y} \tag{4.8}$$

$$-\frac{1}{\rho}\frac{\partial p}{\partial y} = u\frac{\partial v}{\partial x} + v\frac{\partial v}{\partial y} \tag{4.9}$$

$$\frac{p}{\rho^k} = \text{Constant} \tag{1.7}$$

The objective in solving these equations is to obtain u and v as functions of x and y.

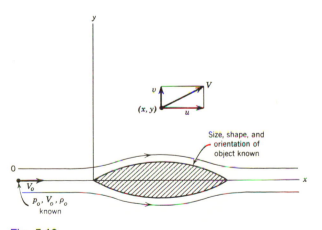

Fig. **5.10**

If such a solution can be obtained, pressures, temperatures, and fluid densities anywhere in the flowfield may also be predicted. The form and orientation of the object,
of course, affects the flow picture; this enters the solution of the problem as a *boundary
condition* expressing the fact that the velocity V along the surface of the body is
everywhere tangent to the body. Mathematically, this means

$$\left[\frac{dy}{dx} = \frac{v}{u}\right]_{\text{at surface of body}}$$

The unknowns in the problem are seen to be u, v, x, y, ρ, and p, and with four equations
and two boundary conditions the unknowns are seen to be determinable; however,
analytic integration of such equations is not possible, so a formal mathematical solution
of the problem cannot be obtained. However, it is possible to attack such problems by
successive approximations in which the flow solution is improved systematically in a
sequence of trials or by direct computer numerical solution of the so-called finite
difference representations of the governing differential equations. The numerical methods are very powerful and have yielded a significant number of useful results and
insights to the details of the motion.

The isentropic treatment of such compressible flowfields around solid objects is
limited to the case where velocities are everywhere subsonic. If the approaching and
leaving velocities (Fig. 5.10) are both subsonic and at a point on the body (near its
midsection) the velocity becomes supersonic, a shock wave is to be expected downstream from this point where the velocity becomes subsonic again. Analysis of such
problems is exceedingly complex because of the unknown position and extent of the
shock wave, and its nonisentropic and discontinuous nature. Upstream from and outboard of the shock wave the flow of the ideal compressible fluid is (whether subsonic
or supersonic) irrotational and isentropic, but downstream from such waves (in general)
the flow is rotational and nonisentropic;[14] however, if such shock waves are straight (or
essentially so), the flowfield downstream from them may be shown to be irrotational
and isentropic; this simplifies the problem somewhat.

Flowfields within passages present the same difficulties as the external flowfields
described above, so further examples need not be cited here. In general, such internal
flowfields are less amenable to the use of the ideal fluid because of the pervasiveness
of wall frictional effects which render the isentropic (irrotational) assumption invalid.
However, such methods may be effective for short passages, duct inlets, and in other
cases where wall friction is of small importance.

References

Chapman, A. J., and Walker, W. F. 1971. *Introductory gas dynamics*. New York: Holt,
   Rinehart and Winston.
Hatsopoulos, G. N., and Keenan, J. H. 1965. *Principles of general thermodynamics*. New York:
   Wiley.

---

[14] See Section 5.10.

Liepmann, H. W., and Roshko, A. 1957. *Elements of gasdynamics*. New York: Wiley.

Mises, R. von. 1958. *Mathematical theory of compressible fluid flow*. New York: Academic Press.

Oswatitsch, K. 1956. *Gas dynamics*. New York: Academic Press.

Reynolds, W. C., and Perkins, H. C. 1977. *Engineering thermodynamics*. 2nd ed. New York: McGraw-Hill.

Shapiro, A. H. 1953. *The dynamics and thermodynamics of compressible fluid flow*. Vol. I. New York: Ronald Press.

Spalding, D. B., and Cole, E. H. 1973. *Engineering thermodynamics*. 3rd ed. London: Edward Arnold.

Thompson, P. A. 1972 *Compressible-fluid dynamics*. New York: McGraw-Hill.

Films

Coles, D. Channel flow of a compressible fluid. NCFMF/EDC Film No. 21616, Encyclopaedia Britannica Educ. Corp.

Rouse, H. Mechanics of fluids: effects of fluid compressibility. Film No. 36960, Media Library, Audiovisual Center, Univ. of Iowa.

Shell Oil Co. Approaching the speed of sound; Beyond the speed of sound; Transonic flight. (3 films) Shell Film Library, 1433 Sadlier Circle, West Drive, Indianapolis, Ind. 46239.

Problems

5.1. Describe the processes that occur in the internal combustion engine. Are any close to being reversible? Can you find a cycle? Is the thermodynamic cycle the same as the cycle in a "four-stroke cycle" engine?

5.2. Calculate the energy delivered to the turbine per unit mass of airflow when the transfer in the heat exchanger is zero. Then, how does the energy delivered depend on q_H through the exchanger if all other conditions remain the same? Assume air is a perfect gas.

Problem **5.2**   Problem **5.3**

5.3. If hydrogen flows as a perfect gas without friction between stations 1 and 2 while $q_H = 7.5 \times 10^5$ J/kg, find V_2.

5.4. Air flows in a 2 in. pipe without friction. If the velocity and temperature at a section are 100 ft/s and 200°F, construct a relationship which shows the various amounts of heat or pump energy that must be added to produce a velocity and temperature of 90 ft/s and 205°F farther downstream.

5.5. The velocity and temperature are measured at two points on the same streamline in a flow of carbon dioxide and found to be 60 m/s, 40°C and 120 m/s, 35°C. This flow is *not* adiabatic. How much heat has been added to or extracted from the fluid between the two points?

5.6. Considering the following gases as perfect gases, calculate the entropy change in each process:

$$
\begin{array}{llll}
(a)\ \text{Nitrogen} & T_1 = 250\ \text{K} & p_1 = 150\ \text{bar} \\
& T_2 = 330\ \text{K} & p_2 = 10\ \text{bar} \\
(b)\ \text{Air} & T_1 = 200\ \text{K} & p_1 = 1\ \text{bar} \\
& T_2 = 233\ \text{K} & p_2 = 200\ \text{bar}
\end{array}
$$

5.7. Derive the perfect gas equation for enthalpy change in an isentropic process.

5.8. Calculate the entropy change in problem 5.2

5.9. Carbon dioxide flows at a speed of 10 m/s (30 ft/s) in a pipe and then through a nozzle where the velocity is 50 m/s (150 ft/s). What is the change in gas temperature between pipe and nozzle? Assume this is an adiabatic flow of a perfect gas.

5.10. Methane gas is flowing in an insulated pipe where the temperature and velocity are 30°C (86°F) and 22.5 m/s (74 ft/s), respectively. Using the perfect gas laws and assuming ideal flow, construct a graph of velocity versus temperature for other points in the line. Are there any limits on this graph?

5.11. The velocity and temperature at a point in an isentropic flow of helium are 90 m/s and 90°C, respectively. Predict the temperature on the same streamline where the velocity is 180 m/s. What is the ratio between the pressure at the two points?

5.12. At a point in an adiabatic flow of nitrogen the velocity is 200 m/s (650 ft/s). Between this point and another one on the same streamline the rise of temperature is 10°C (18°F). Calculate the velocity at the second point.

5.13. Derive equation 5.17 from equation 5.16.

5.14. Revise equation 5.17 by including the fourth term in the binomial expansion.

5.15. Carbon dioxide flows in a duct at a velocity of 90 m/s, absolute pressure 140 kPa, and temperature 90°C. Calculate pressure and temperature on the nose of a small object placed in this flow.

5.16. If nitrogen at 15°C is flowing and the stagnation temperature on the nose of a small object in the flow is measured as 38°C, what is the velocity in the pipe?

5.17. Oxygen flows in a passage at a pressure of 25 psia. The pressure and temperature on the nose of a small object in the flow are 28 psia and 150°F, respectively. What is the velocity in the passage?

5.18. Calculate the stagnation pressure in an airstream of absolute pressure, temperature, and velocity, 101.3 kPa, 15°C, and 320 m/s, respectively, using (a) equation 5.17 and (b) equation 5.16. Compare results.

5.19. What is the pressure on the nose of a bullet moving through standard sea level air at 300 m/s (985 ft/s) according to (a) equation 5.17 and (b) equation 5.16? Compare results.

5.20. With equation 5.16 invalid for $M_o > 1$, derive an expression for the minimum allowable value of p_o/p_s which may be used for velocity computation in equation 5.18.

5.21. Assume air flows in a nozzle at 500 m/s, $\rho = 2$ kg/m³ and $T = 300°C$. Is an increase or decrease in area required to further increase the flow velocity?

5.22. In a given duct flow $M = 2.0$; the velocity undergoes a 20% decrease. What percent change in area was needed to accomplish this? What would be the answer if $M = 0.5$?

5.23. Derive equations (a) 5.23 and (b) 5.24.

5.24. Nitrogen flows from a large tank, through a convergent nozzle of 2 in. tip diameter, into the atmosphere. The temperature in the tank is 200°F. Calculate pressure, velocity, temperature, and sonic velocity in the jet; and calculate the flowrate when the tank pressure is (a) 30 psia and (b) 25 psia. Barometric pressure is 15.0 psia. What is the lowest tank pressure that will produce sonic velocity in the jet? What is this velocity and what is the flowrate?

5.25. Air flows from the atmosphere into an evacuated tank through a convergent nozzle of 38 mm tip diameter. If atmospheric pressure and temperature are 101.3 kPa and 15°C, respectively, what vacuum must be maintained in the tank to produce sonic velocity in the jet? What is the flowrate? What is the flowrate when the vacuum is 254 mm of mercury?

5.26. Oxygen discharges from a tank through a convergent nozzle. The temperature and velocity in the jet are $-20°C$ (0°F) and 270 m/s (900 ft/s), respectively. What is the temperature in the tank? What is the temperature on the nose of a small object in the jet?

5.27. Carbon dioxide discharges from a tank through a convergent nozzle into the atmosphere. If the tank temperature and pressure are 38°C and 140 kPa, respectively, what jet temperature, pressure, and velocity can be expected? Barometric pressure is 101.3 kPa.

5.28. In the Illustrative Problem of Section 5.7, calculate the pressure, temperature, velocity, and sonic velocity at a point in the nozzle where the diameter is 50 mm.

5.29. Air (at 100°F and 100 psia) in a large tank flows into a 6 in. pipe, whence it discharges to the atmosphere (15.0 psia) through a convergent nozzle of 4 in. tip diameter. Calculate pressure, temperature, and velocity in the pipe.

5.30. Carbon dioxide flows through a convergent nozzle in a wall between two large tanks in which the absolute pressures are 450 kPa and 210 kPa. Calculate the jet velocity if the jet temperature is 10°C. What is the temperature in the upstream tank?

5.31. Calculate the required diameter of a convergent nozzle to discharge 5.0 lb/s of air from a large tank (in which the temperature is 100°F) to the atmosphere (14.7 psia) if the pressure in the tank is: (a) 25.0 psia, and (b) 30.0 psia.

5.32. Air flows from a 150 mm pipe, through a 25 mm convergent nozzle, into the atmosphere. Calculate the absolute pressure in the pipe when the absolute pressure in the jet is 138 kPa.

5.33. Derive equation 5.25.

5.34. Carbon dioxide flows through a 4 in. constriction in a 6 in. pipe. The pressures in pipe and constriction are 40 and 35 psia, respectively. The temperature in the pipe is 100°F. Calculate (a) flowrate, (b) temperature in the constriction, (c) sonic velocity in the constriction, and (d) velocities in pipe and constriction.

5.35. Nitrogen discharges from a 100 mm pipe through a 50 mm nozzle into the atmosphere. Barometric pressure is 101.3 kPa, and the pressure in the pipe is 0.3 m of water. The temperature in the pipe is 15.5°C. Calculate the flowrate, (a) neglecting expansion of the gas and (b) including expansion. What percent error is induced by neglecting expansion?

5.36. Five pounds of air per second discharge from a tank through a convergent-divergent nozzle into another tank where a vacuum of 10 in. of mercury is maintained. If the pressure and temperature in the upstream tank are 100 in. of mercury absolute and 100°F, respectively, what nozzle-exit diameter must be provided for full expansion? What throat diameter is required? Calculate pressure, temperature, velocity, and sonic velocity in throat and nozzle exits. Barometric pressure is 30 in. of mercury.

5.37. Carbon dioxide flows from a tank through a convergent-divergent nozzle of 25 mm throat and 50 mm exit diameter. The absolute pressure and temperature in the tank are 241.5 kPa and 37.8°C, respectively. Calculate the mass flowrate when the absolute pressure p'_3 is (a) 172.5 kPa and (b) 221 kPa.

5.38. If the nozzle of the Illustrative Problem of Section 5.9 is a convergent one of 33 mm tip diameter, calculate velocity and sonic velocity in the jet, and the mass flowrate. Compare results, and note the effect of changing the shape of a nozzle without changing its exit diameter.

5.39. If the nozzle in the Illustrative Problem of Section 5.9 has a divergent passage 100 mm long, calculate the required passage diameter, at three equally spaced points between throat and nozzle tip, for a linear pressure drop through this part of the passage.

5.40. A convergent-divergent nozzle of 50 mm tip diameter discharges to the atmosphere (103.2 kPa) from a tank in which air is maintained at an absolute pressure and temperature of 690 kPa and 37.8°C, respectively. What is the maximum mass flowrate that can occur through this nozzle? What throat diameter must be provided to produce this mass flowrate?

5.41. Atmospheric air (at 98.5 kPa and 20°C) is drawn into a vacuum tank through a convergent-divergent nozzle of 50 mm throat diameter and 75 mm exit diameter. Calculate the largest mass flowrate that can be drawn through this nozzle under these conditions.

5.42. The exit section of a convergent-divergent nozzle is to be used for the test section of a supersonic wind tunnel. If the absolute pressure in the test section is to be 140 kPa (20 psia), what pressure is required in the reservoir to produce a Mach number of 5 in the test section? For the air temperature to be −20°C (0°F) in the test section, what temperature is required in the reservoir? What ratio of throat area to test section area is required to meet these conditions?

5.43. In the nozzle of problem 5.36, what range of pressures in the tank will cause a normal shock wave in the nozzle?

5.44. A perfect gas ($k = 2.0$, and $R = 3\ 758$ ft · lb/slug °R) discharges from a tank through a convergent-divergent nozzle into the atmosphere (barometric pressure 28 in. of mercury). The gage pressure in the tank is 50 psi, and the temperature is 100°F. The throat and exit diameters of the nozzle are 1.00 in. and 1.05 in., respectively. Show calculations to prove whether a normal shock wave is to be expected in this nozzle.

5.45. Air discharges through a convergent-divergent nozzle which is attached to a large reservoir. At a point in the nozzle a normal shock wave is detected across which the absolute pressure jumps from 69 to 207 kPa (10 to 30 psia). Calculate the pressures in the throat of the nozzle and in the reservoir.

THE IMPULSE-MOMENTUM PRINCIPLE

The Euler equations were derived in previous chapters by application of scalar forms of Newton's second law. However, Newton's law is a vector relation and it can be written for an elemental fluid system or particle in terms of the impulse provided by external forces (shears, pressures, gravity, etc.) and the resulting change of the linear momentum of the system as

$$\left(\sum \mathbf{F}\right) dt = d(m\mathbf{V}_c) \qquad \text{or} \qquad \sum \mathbf{F} = \frac{d}{dt}(m\mathbf{V}_c) \qquad (6.1)$$

where $\mathbf{V}_c$ is the velocity of the center of mass of the system of mass m, $m\mathbf{V}_c$ is the *linear momentum,* and $(\sum \mathbf{F})\, dt$ is the impulse in time dt from the sum of all external forces acting on the system. For any system, its mass

$$m = \int_{\text{SYS}} dm$$

while

$$\mathbf{V}_c = \frac{1}{m} \int_{\text{SYS}} \mathbf{V}\, dm$$

and the integrals are taken over all elements of mass in the system. If the forces acting on a system produce a net *torque* $\mathbf{T}$, the torque impulse causes a change in the *angular momentum* or the *moment of momentum* according to Newton's second law in the form

$$\mathbf{T} = \sum (\mathbf{r} \times \mathbf{F}) = \frac{d}{dt}(\mathbf{r} \times m\mathbf{V}_c) \qquad (6.2)$$

where $\mathbf{r}$ is the radius vector from the origin O of a fixed coordinate system to the point of application of a force or to the mass center.

For plane flows in which, for example, the motion occurs and forces are applied in the x-z plane (Fig. 6.1), the vector cross product[1] in equation 6.2 is easily visualized. Let $\mathbf{e}_r$ and $\mathbf{e}_t$ be unit vectors in the radial and tangential directions (and lying in the x-z plane as shown in Fig. 6.1), while $\mathbf{e}_y$ is a vector of unit magnitude directed along the y-axis. Then, the torque vector $\mathbf{T}_y$, can be written as

$$\mathbf{T} = \mathbf{r} \times \mathbf{F} = r\mathbf{e}_r \times (F_r\mathbf{e}_r + F_t\mathbf{e}_t) = rF_t\mathbf{e}_y = \overset{\frown}{\mathbf{T}_y}$$

because $\mathbf{e}_r \times \mathbf{e}_r = 0$, but $\mathbf{e}_r \times \mathbf{e}_t = \mathbf{e}_y$. It follows that $\mathbf{T}_y$ has the direction of the third

[1] See Appendix 5.

207

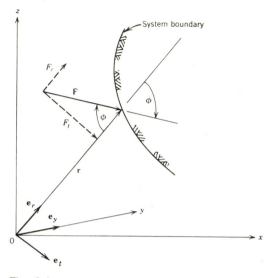

Fig. **6.1**

coordinate y of a right-handed system if the torque is clockwise. If the angle ϕ is positive in the clockwise direction (as shown) when measured from $\mathbf{r}$ to $\mathbf{F}$, the value $F_t = |\mathbf{F}| \sin \phi$ is the component of $\mathbf{F}$ perpendicular to $\mathbf{r}$; the radial component $F_r = |\mathbf{F}| \cos \phi$ has no moment or torque about the origin as expected. Finally, if $\mathbf{T}_y = T_y \mathbf{e}_y$, then

$$T_y = rF_t = |\mathbf{r}|\,|\mathbf{F}| \sin \phi$$

for this force and motion in the x-z plane.

Equations 6.1 and 6.2 comprise the statement of the impulse-momentum principle for systems; their extension to fluid flow problems with control volumes is described below. The impulse-momentum principle, along with the continuity and energy principles, provides a third basic tool for the solution of fluid flow problems. Sometimes its application leads to the solution of problems that cannot be solved by the energy principle alone; more often it is used in conjunction with the energy principle to obtain more comprehensive solutions of engineering problems.

The following treatment of the subject is limited to steady flow; its objective is to develop the principle and demonstrate its application to a variety of practical problems. Treatments of more complex problems will be found in the References which are listed at the end of chapter.

6.1 Development of the Principle for Control Volumes

The conversion of the impulse-momentum principle to control volume form is made by considering a general control volume and repeating the type of analysis used to derive the two-dimensional conservation of mass (continuity) equation in Section 3.6. Before proceeding, however, the reader should review that derivation and compare it to the one-dimensional type of control volume analyses for the continuity (Section 3.5), work-energy (Section 4.5), and energy (Section 5.1) equations. The analysis is carried through here in two dimensions for simplicity and pictorial clarity only. The vector results are explicitly applicable to three-dimensional flows.

Consider the steady flow past some solid object and the control volume in Fig. 6.2. The object can be any fixed device that deflects or turns the flow, for example, a guide vane in a hydraulic turbine (Section 6.12) or in the stator of an impulse steam turbine (Section 6.11). In contrast to the case in Chapters 4 and 5, here the fluid may be real (viscous). It may be compressible also. For this flow and system, equation 6.1 is

$$\sum \mathbf{F} = \frac{d}{dt}(m\mathbf{V}_c) = \frac{d}{dt}\left(\int_{SYS} \mathbf{V}\,dm\right)$$

By looking at Fig. 6.2 and using the definition of a derivative, we obtain

$$\frac{d}{dt}\left(\int_{SYS} \mathbf{V}\,dm\right) = \lim_{\Delta t \to 0}\left\{\frac{\left(\int_{SYS} \mathbf{V}\,dm\right)_{t+\Delta t} - \left(\int_{SYS} \mathbf{V}\,dm\right)_{t}}{\Delta t}\right\}$$

and, because the momentum in R never changes in the steady flow,

$$\left(\int_{SYS} \mathbf{V}\,dm\right)_{t+\Delta t} = \int_{R} \mathbf{V}\,dm + \int_{O} \mathbf{V}\,dm$$

$$\left(\int_{SYS} \mathbf{V}\,dm\right)_{t} = \int_{R} \mathbf{V}\,dm + \int_{I} \mathbf{V}\,dm$$

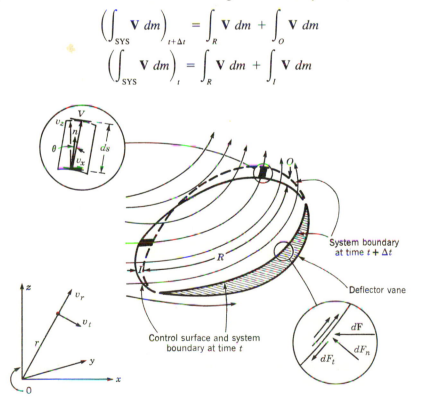

System boundary
at time $t + \Delta t$

Deflector vane

Control surface and system
boundary at time t

Fig. **6.2**

From the continuity derivation of Section 3.6, for the zone O in Fig. 6.2 (where fluid flows out of the control volume through the control surface),

$$dm = \rho \, dA \, \Delta s \cos \theta = \rho \, dA \, |\mathbf{V}| \cos \theta \, \Delta t$$

$$= \rho \mathbf{V} \cdot \mathbf{n} \, dA \, \Delta t = \rho \mathbf{V} \cdot d\mathbf{A} \, \Delta t$$

Thus, integrating over the portion of the control surface through which fluid is flowing out

$$\int_O \mathbf{V} \, dm = \Delta t \int_{\text{C.S.}_{\text{OUT}}} \mathbf{V}(\rho \mathbf{V} \cdot d\mathbf{A})$$

Similarly, where the fluid flows in through the control surface

$$\int_I \mathbf{V} \, dm = \Delta t \left[-\int_{\text{C.S.}_{\text{IN}}} \mathbf{V}(\rho \mathbf{V} \cdot d\mathbf{A}) \right]$$

and

$$\frac{d}{dt}\left(\int_{\text{SYS}} \mathbf{V} \, dm \right) = \lim_{\Delta t \to 0} \left\{ \left(\int_{\text{C.S.}_{\text{OUT}}} \mathbf{V}(\rho \mathbf{V} \cdot d\mathbf{A}) + \int_{\text{C.S.}_{\text{IN}}}^{\prime} \mathbf{V}(\rho \mathbf{V} \cdot d\mathbf{A}) \right) \frac{\Delta t}{\Delta t} \right\}$$

$$= \oint_{\text{C.S.}} \mathbf{V}(\rho \mathbf{V} \cdot d\mathbf{A})$$

In the limit as $\Delta t \to 0$ the external forces acting on the system are equivalent to those acting on the fluid in the control volume. Therefore, for a control volume in steady flow

$$\sum \mathbf{F} = \oint_{\text{C.S.}} \mathbf{V}(\rho \mathbf{V} \cdot d\mathbf{A}) = \oint_{\text{C.S.}} \mathbf{V}(\rho \mathbf{V} \cdot \mathbf{n} \, dA)$$

$$= \oint_{\text{C.S.}} \mathbf{V} \, d\dot{m} \tag{6.3}$$

where $d\dot{m} = \rho \mathbf{V} \cdot d\mathbf{A} = \rho V_n \, dA = \rho \, dQ$ is the mass flowrate (kg/s or slug/s) *out* of the control volume through a particular area dA. The quantity $\mathbf{V} \, d\dot{m} = [\text{kg} \cdot \text{m/s}^2 = \text{N}$ or slug $\cdot$ ft/s^2 = lb] is the *flux of momentum* through the elemental area dA. If a net external force acts on the fluid in the control volume, there must be a net momentum flux from the control volume according to equation 6.3.

The forces acting on the fluid in a control volume are either (a) *body forces,* such as the gravitational attractive force, that act on *each element of mass* (solid or fluid) *in the control volume* or (b) *surface forces* that act at the control surface. (All internal surface forces cancel.) For a real fluid flow there is no slip along the solid body in Fig. 6.2, but the body does exert a shear, as well as a normal, force on the fluid in the control volume. Shafts that penetrate the control surface can contribute both normal and shear force to the fluid by virtue of their action in the interior, but these forces may be accounted for at the point where the shaft intersects the control surface. It is counted an advantage of the impulse-momentum principle that the details of the motion inside the control surface need not be known (the actual configuration of solid and fluid regions must be known if body forces are to be computed, however). The work-energy and the energy equations (Sections 4.5 and 5.1) share the same advantage.

The basic vector relation, equation 6.3, is easily decomposed to scalar form, that is, from Appendix 5,

$$\sum \mathbf{F} = \sum F_x \mathbf{e}_x + \sum F_z \mathbf{e}_z = \oint_{\text{c.s.}} \{V_x \mathbf{e}_x + V_z \mathbf{e}_z\} \, d\dot{m}$$

where (F_x, V_x) and (F_z, V_z) are the x- and z-components of external forces and the velocity. Then, because the $\mathbf{e}_x$ components on each side of the equation must balance,

$$\sum F_x = \oint_{\text{c.s.}} V_x \, d\dot{m} = \oint_{\text{c.s.}} V_x(\rho \mathbf{V} \cdot d\mathbf{A}) \qquad (6.4)$$

and

$$\sum F_z = \oint_{\text{c.s.}} V_z \, d\dot{m} = \oint_{\text{c.s.}} V_z(\rho \mathbf{V} \cdot d\mathbf{A}) \qquad (6.5)$$

For the moment of momentum, equation 6.2 gives for Fig. 6.2,

$$\mathbf{T} = \sum (\mathbf{r} \times \mathbf{F}) = \frac{d}{dt} \left(\int_{\text{SYS}} \mathbf{r} \times \mathbf{V} \, dm \right)$$

It follows, then, without repeating the details from above, that for a control volume

$$\mathbf{T} = \sum (\mathbf{r} \times \mathbf{F}) = \oint_{\text{c.s.}} (\mathbf{r} \times \mathbf{V})(\rho \mathbf{V} \cdot d\mathbf{A}) = \oint_{\text{c.s.}} (\mathbf{r} \times \mathbf{V}) \, d\dot{m} \qquad (6.6)$$

where $d\dot{m} > 0$ for flow *out* of the control volume. In a plane flow with force and velocity components defined in the radial (along r) and tangential (perpendicular to r) directions (see Fig. 6.1), equation 6.6 becomes

$$T = \sum \overset{+\curvearrowright}{T_y} = \sum r F_t = \oint_{\text{c.s.}} (r v_t) \, d\dot{m} \qquad (6.7a)$$

or, for example,

$$T = \sum (z F_x - x F_z) = \oint_{\text{c.s.}} (z V_x - x V_z) \, d\dot{m} \qquad (6.7b)$$

when $|\mathbf{r}| = r = (x^2 + z^2)^{1/2}$, $r v_t = z V_x - x V_z$, and so forth.

In the remaining sections of this chapter the momentum, work-energy (energy), and continuity principles are used to solve a number of representative problems. The value of having three tools available and the means of choosing the best for each problem are demonstrated. Although only nonviscous flows are discussed in this chapter, a particularly useful and powerful application of equation 6.4 is given in Section 13.3 where boundary layer flow is analyzed.

ELEMENTARY APPLICATIONS

In the next sections, we discuss a number of applications of the impulse-momentum principle to situations where the flows across the control surface are one-dimensional, that is, the streamlines are essentially straight and parallel. Recall that the inherent advantage of the impulse-momentum principle is that only flow conditions at inlets and exits of the control volume are needed for successful application; detailed (and often complex) flow processes within the control volume need not be known to apply the principle. It should be emphasized that efficient application of the principle depends greatly on the judicious selection of a convenient control volume, with streamlines essentially straight and parallel at inlet and exit.

6.2 Pipe Bends, Enlargements, and Contractions

The force exerted by a flowing fluid on a bend, enlargement, or contraction in a pipeline may be readily computed by application of the impulse-momentum principle without detailed information on shape of passage or pressure distribution therein.

  The reducing pipe bend of Fig 6.3 is typical of many of the problems encountered in fluid mechanics. The object is to calculate the force exerted by the fluid on the bend between sections 1 and 2. The control volume *ABCD* encloses the fluid within the bend and the flow at sections 1 and 2 is assumed one-dimensional. **F** is the force exerted *by the bend on the fluid* (see the free-body diagram in Fig 6.3). The flowrate, pressures and velocities at sections 1 and 2, and the superficial flow geometry are known; **F** is to be found.

  For streamlines essentially straight and parallel at sections 1 and 2, the forces F_1 and F_2 result from hydrostatic pressure distributions in the flowing fluid. If the mean pressures p_1 and p_2 are large and the pipe areas are relatively small, $F_1 = p_1A_1$ and $F_2 = p_2A_2$ can be assumed with little error to act at and along the centerline of the pipe.[2] The summation of body forces acting on the fluid in *ABCD* will, of course, yield the total weight of fluid (W). Thus, W is dependent only on the specific weight of the fluid and the control volume geometry.[3] The force **F** exerted by the bend on the fluid is the resultant of the pressure distribution over the entire interior of the bend between sections 1 and 2; although this distribution is usually unknown in detail, its resultant can be predicted by use of equations 6.4 and 6.5. The force exerted *by the fluid on the bend* is then the equal and opposite of this resultant.

  When equations 6.4 and 6.5 are applied to the control volume *ABCD*, the results are

$$\sum F_x = p_1A_1 - p_2A_2 \cos \alpha - F_x$$

and

$$\oint_{\text{C.S.}} V_x \, d\dot{m} = \oint_{\text{C.S.}} V_x(\rho\mathbf{V}\cdot d\mathbf{A}) = V_{x\text{OUT}}\rho Q - V_{x\text{IN}}\rho Q = (V_2 \cos \alpha)\rho Q - V_1\rho Q$$

so, from equation 6.4,

---

[2] In Fig 6.3, the points (x_1', z_1') and (x_2', z_2') identify the *centers of pressure* (not the centroids) of sections 1 and 2. However, these will be coincident for uniform pressure distributions. See Section 2.4

[3] Frequently in engineering problems this force is small compared to the forces on the surfaces of the control volume, but the decision to neglect it requires either preliminary calculations or considerable experience. It obviously plays no part if flow is in a horizontal plane, since its direction is perpendicular to this plane.

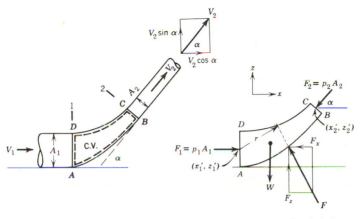

Fig. **6.3**

$$p_1 A_1 - p_2 A_2 \cos \alpha - F_x = (V_2 \cos \alpha - V_1)\rho Q$$

while

$$\sum F_z = -W - p_2 A_2 \sin \alpha + F_z$$

and

$$\oint_{\text{c.s.}} V_z \, d\dot{m} = \oint_{\text{c.s.}} V_z (\rho \mathbf{V} \cdot d\mathbf{A}) = V_{z\,\text{OUT}}\rho Q - V_{z\,\text{IN}}\rho Q = (V_2 \sin \alpha)\rho Q - 0$$

so, from equation 6.5,

$$-W - p_2 A_2 \sin \alpha + F_z = (V_2 \sin \alpha)\rho Q$$

Note that $d\dot{m} = \rho V_n \, dA = 0$ on all surfaces except at cross sections 1 and 2 where[4] $d\dot{m}_1 = -\rho V_1 \, dA_1 = -\rho \, dQ$, $d\dot{m}_2 = \rho \, dQ$ and V_{x_1} and V_{x_2} are assumed constant over the cross sections. Only F_x and F_z are unknown in the above and so can be obtained. The magnitude and direction of $\mathbf{F}$ are then readily established. The position of $\mathbf{F}$ depends on the location of W acting at the center of gravity of the fluid within the bend; this in turn will depend on the geometry of the bend. Use of equation 6.7 will allow prediction of the location of $\mathbf{F}$. Thus, taking moments about the center of section 1 (AD) and taking r as the effective lever arm of $\mathbf{F}$ about the center of section AD (Fig 6.3) and $F = |\mathbf{F}|$:

$$T = \sum (zF_x - xF_z)$$

$$= -rF + (x_w - x_1')W + (z_2' - z_1')(-F_2) \cos \alpha - (x_2' - x_1')(-F_2) \sin \alpha$$

[4] Even if the flow is compressible, continuity requires that $\rho Q = $ constant in steady flow in a (finite) streamtube (see Section 3.5).

and

$$\oint_{\text{C.S.}} (zV_x - xV_z) \, d\dot{m} = (zV_x - xV_z)_{\text{OUT}}\rho Q - (zV_x - xV_z)_{\text{IN}}\rho Q$$

$$= [(z_2' - z_1')V_2 \cos \alpha - (x_2' - x_1')V_2 \sin \alpha]\rho Q - 0$$

so, by using equation 6.7, we find that

$$-rF + (x_w - x_1')W - (z_2' - z_1')F_2 \cos \alpha + (x_2' - x_1')F_2 \sin \alpha$$

$$= [(z_2' - z_1')V_2 \cos \alpha - (x_2' - x_1')V_2 \sin \alpha]\rho Q$$

Note that contributions to moments by F_1 and V_1 are zero because the coordinate origin for moments was taken in the center of section 1 (section AD).

───────────────── **Illustrative Problem** ─────────────────

When 300 l/s of water flow through this vertical 300 mm by 200 mm pipe bend, the pressure at the entrance is 70 kPa. Calculate the force by the fluid on the bend if the volume of the bend is 0.085 m³

Relevant Equations and Given Data

$$Q = AV \tag{3.11}$$

$$\frac{p}{\gamma} + \frac{V^2}{2g_n} + z = H = \text{Constant} \tag{4.1}$$

$$\sum F_x = \oint_{\text{C.S.}} V_x(\rho \mathbf{V} \cdot d\mathbf{A}) = \oint_{\text{C.S.}} V_x \, d\dot{m} \tag{6.4}$$

$$\sum F_z = \oint_{\text{C.S.}} V_z(\rho \mathbf{V} \cdot d\mathbf{A}) = \oint_{\text{C.S.}} V_z \, d\dot{m} \tag{6.5}$$

$$T = \sum (zF_x - xF_z) = \oint_{\text{C.S.}} (zV_x - xV_z) \, d\dot{m} \tag{6.7b}$$

$Q = 0.3$ m³/s $d_1 = 300$ mm $d_2 = 200$ mm

$p_1 = 70$ kPa   Vol. bend $= 0.085$ m³ $\gamma = 9.79$ kN/m³

$\rho = 998$ kg/m³

Solution.   From the continuity equation 3.11: $V_1 = 4.24$ m/s and $V_2 = 9.55$ m/s, and from the Bernoulli equation 4.1: $p_2 = 33.4$ kPa.

Now, for the free-body diagram, the pressure forces F_1 and F_2 can be computed:

$$F_1 = \frac{\pi}{4}(0.3)^2 70 \times 10^3 = 4\ 948 \text{ N} \qquad F_2 = \frac{\pi}{4}(0.2)^2 33.4 \times 10^3 = 1\ 049 \text{ N}$$

The fluid weight in the bend $W = \gamma$ (vol. bend) $= 833$ N. With this and the velocity

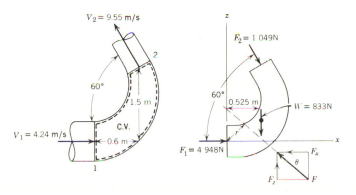

diagram, equations 6.4 and 6.5 may be applied:

$$4\ 948 - F_x + (1\ 049)0.5 = (-9.55 \times 0.5 - 4.24)(998 \times 0.3) \qquad (6.4)$$

$$F_z - (1\ 049)0.866 - 833 = (9.55 \times 0.866 - 0)(998 \times 0.3) \qquad (6.5)$$

$$F_x = +8\ 172\ N \qquad F_z = +4\ 218\ N \qquad F = 9\ 196\ N \qquad \theta = 27.3°$$

The plus signs confirm the direction assumptions for F_x and F_z. Therefore the force *on the bend* is 9.2 kN downward to the right at 27.3° with the horizontal. ◀

Now, assuming that the bend is of such shape that the center of gravity of the fluid therein is 0.525 m to the right of section 1 and that F_1 and F_2 act at the centroids of sections 1 and 2, respectively, equation 6.7b may be used to find the location of F. Taking moments about the center of section 1,

$$-r(9\ 196) + 0.525(833) + 1.5 \times (1\ 049 \times 0.5) - 0.6 \times (-1\ 049 \times 0.866)$$

$$= 1.5 \times (-9.55 \times 0.5) \times (998 \times 0.3)$$

$$- 0.6 \times (9.55 \times 0.866) \times (998 \times 0.3) \qquad (6.7b)$$

$$r = 0.59\ m\ ◀$$

6.3 Structures in Open Flow

The forces exerted by flowing water on overflow or underflow structures (weirs or gates) may frequently be estimated by application of the impulse-momentum principle. However, the location of such forces depends on the flowfield produced and cannot be obtained from one-dimensional methods because the velocity distributions in inlet and outlet sections and the actual shapes of the free surfaces affect the result. The sluice gate of Fig 6.4 is an example of this. From the depths y_1 and y_2 simultaneous application of the Bernoulli and continuity principles yields the flowrate and the mean velocities V_1 and V_2. Selecting a control volume bounded by sections 1 and 2 and applying equation 6.4,[5]

---

[5] Note that in open channel flows, the water depth is denoted by y.

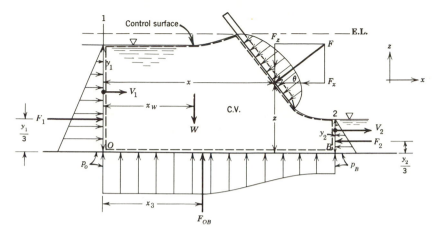

Fig. **6.4**

$$\sum F_x = F_1 - F_2 - F_x = \frac{\gamma y_1^2}{2} - \frac{\gamma y_2^2}{2} - F_x = (V_2 - V_1)\rho q$$

in which F_1 and F_2 have been computed from hydrostatics principles and unit width (normal to the plane of the paper) has been assumed so $q = y_1 V_1 = y_2 V_2$. For a plane gate surface and an ideal fluid, F is normal to the gate and thus may be computed directly from $F = F_x / \cos \theta$; similarly, $F_z = F_x \tan \theta$. Application of equation 6.5 gives

$$\sum F_z = F_{OB} - W - F_z = 0$$

in which W is the weight of fluid between sections 1 and 2, and F_{OB} the force exerted by the surface OB on the fluid. With the magnitude and location of W dependent on the detailed form of the control volume and the location of F dependent on its (unknown) pressure distribution, it is evident that computations can be carried no further until the complete geometry of the flowfield is known. However, equation 6.7b may still be applied to demonstrate the method. Taking moments about O,

$$-x_3(F_{OB}) + z(-F_x) - x(-F_z) + \frac{y_2}{3}(-F_2) + \frac{y_1}{3}(F_1) - x_w(-W)$$

$$= \frac{y_2}{2}V_2\rho q + \frac{y_1}{2}V_1(-\rho q)$$

Note that F_1 and F_2 act at the centers of pressure, not at the centroids of sections 1 and 2 (see Section 2.4).

--- **Illustrative Problem** ---

This two-dimensional overflow structure (shape and size unknown) produces the flowfield shown. Calculate the magnitude and direction of the horizontal component of force on the structure.

Relevant Equations and Given Data

$$q = y_1V_1 = y_2V_2 \tag{3.14}$$

$$\frac{p}{\gamma} + \frac{v^2}{2g_n} + z = H = \text{Constant} \tag{4.1}$$

$$\sum F_x = \oint_{\text{C.S.}} V_x \, d\dot{m} \tag{6.4}$$

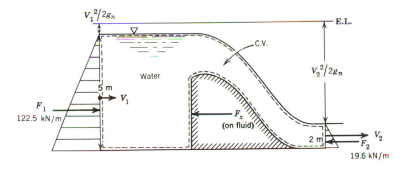

Solution. Assuming ideal fluid, construct a horizontal energy line above the liquid surface, from which (equation 4.1)

$$\frac{V_1^2}{2g_n} + 5 = \frac{V_2^2}{2g_n} + 2 \tag{4.1}$$

and from continuity (equation 3.14)

$$q = 5V_1 = 2V_2 \tag{3.14}$$

By solving these simultaneous equations, we obtain

$$V_1 = 3.33 \text{ m/s} \qquad V_2 = 8.33 \text{ m/s} \qquad q = 16.65 \text{ m}^3/\text{s} \cdot \text{m}$$

Next construct a control volume of fluid between sections of 5 m and 2 m depth, through which streamlines are straight and parallel. If $\gamma = 9.8$ kN/m^3,

$$F_1 = \frac{9.8(5)^2}{2} = 122.5 \text{ kN/m} \qquad F_2 = \frac{9.8(2)^2}{2} = 19.6 \text{ kN/m}$$

Now, by applying equation 6.4, we find that (for $\rho = 1\,000$ kg/m^3)

$$\sum F_x = 122\,500 - F_x - 19\,600 = (8.33 - 3.33)\,(1\,000 \times 16.65) \qquad (6.4)$$

whence F_x, the force component desired (and exerted by the structure on the fluid), is 19.65 kN/m. ◀ The direction of this force component *on the structure* is therefore in a *downstream* direction. Because this force is the integral of the horizontal components of an *unknown* pressure distribution over a structure of *undefined* form, the effectiveness of the impulse-momentum principle is again demonstrated.

6.4   Abrupt Enlargement in a Closed Passage

The impulse-momentum principle can be employed to predict the fall of the energy line (that is, the energy loss due to a rise in the internal energy of the fluid caused by viscous dissipation) at an abrupt axisymmetric enlargement in a passage (Fig. 6.5). Consider the control surface *ABCD,* drawn to enclose the zone of momentum change. Assume a one-dimensional flow. The flowrate Q through the control volume is

$$Q = A_1 V_1 = A_2 V_2 \qquad (3.12)$$

The rate of change of momentum is

$$\oint_{ABCD} V_x \, d\dot{m} = (V_x \rho Q)_{\text{OUT}} - (V_x \rho Q)_{\text{IN}} = (V_2 - V_1)\frac{Q\gamma}{g_n} = (V_2 - V_1)\frac{A_2\gamma V_2}{g_n}$$

The forces producing this change of momentum act on the surfaces *AB* and *CD*; if these are assumed to result from hydrostatic[6] pressure distributions over the areas, they may be calculated as

$$\sum F_x = p_1 A_2 - p_2 A_2 = (p_1 - p_2) A_2$$

By applying equation 6.4, we obtain

$$\frac{p_1 - p_2}{\gamma} = \frac{V_2(V_2 - V_1)}{g_n}$$

However, from the energy line,

$$\frac{p_1 - p_2}{\gamma} = \frac{V_2^2}{2g_n} - \frac{V_1^2}{2g_n} + \Delta\,(\text{E.L.})$$

Equating these expressions for $(p_1 - p_2)/\gamma$ and solving for $\Delta\,(\text{E.L.})$,

$$\Delta\,(\text{E.L.}) = \frac{(V_1 - V_2)^2}{2g_n} \qquad (6.8)$$

---

[6] This is a good assumption for area *CD* because the conditions of one-dimensional flow are satisfied. For area *AB* it is an approximation because of the dynamics of the eddies in the "dead water" zone. Accordingly the result of the analysis will be approximate and will require experimental verification. See Section 9.9.

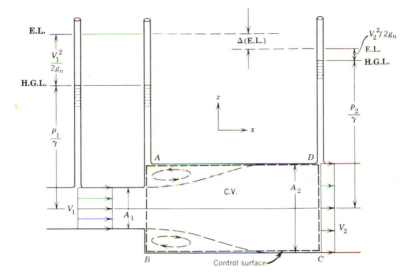

Fig. **6.5**

This analysis is a classic one of early analytical hydraulics; $\Delta(\mathbf{E.L.})$ is frequently termed Borda-Carnot *head loss*, after those who contributed to its original development.

6.5 The Hydraulic Jump

Frequently in open channel flow, when liquid at high velocity discharges into a zone of lower velocity, a rather abrupt rise (a standing wave) occurs in the liquid surface and is accompanied by violent turbulence, eddying, air entrainment, and surface undulations; such a wave is known as a *hydraulic jump*.[7] In spite of the foregoing complications and the consequential large head loss, application of the impulse-momentum principle gives results very close to those observed in field and laboratory; the engineering problem is to find the relation between the depths for a given flowrate. Assume a plane flow situation with flowrate q and construct a control volume (between two sections 1 and 2 where the streamlines are straight and parallel; Fig. 6.6) enclosing the jump. The only horizontal forces are seen to be (neglecting friction) the hydrostatic ones of equation 2.12. Applying equation 6.4 gives

$$\sum F_x = F_1 - F_2 = \frac{\gamma y_1^2}{2} - \frac{\gamma y_2^2}{2} = (V_2 - V_1)\frac{\gamma q}{g_n}$$

Substituting the continuity relations $V_2 = q/y_2$ and $V_1 = q/y_1$ and rearranging give the desired relationship among y_1, y_2, and q:

[7] Further discussion of the hydraulic jump is found in Section 10.10.

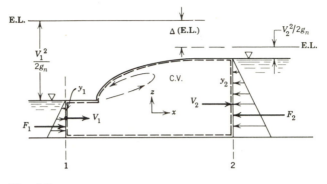

Fig. 6.6

$$\frac{q^2}{g_n y_1} + \frac{y_1^2}{2} = \frac{q^2}{g_n y_2} + \frac{y_2^2}{2}$$

and solving for y_2/y_1 yields

$$\frac{y_2}{y_1} = \frac{1}{2}\left[-1 + \sqrt{1 + \frac{8q^2}{g_n y_1^3}}\right] = \frac{1}{2}\left[-1 + \sqrt{1 + \frac{8V_1^2}{g_n y_1}}\right] \qquad (6.9)$$

From this equation it will be seen that: (a) for $V_1^2/g_n y_1 = 1$, $y_2/y_1 = 1$; (b) for $V_1^2/g_n y_1 > 1$, $y_2/y_1 > 1$; and (c) $V_1^2/g_n y_1 < 1$, $y_2/y_1 < 1$.

Condition (c), that of a *fall of the liquid surface,* although satisfying the impulse-momentum and continuity equations, will be found to produce a rise of the energy line through the jump, and is thus physically impossible. Accordingly it may be concluded that, for a hydraulic jump to occur, the upstream conditions must be such that $V_1^2/g_n y_1 > 1$. Later (Section 8.1) $V_1^2/g_n y_1$ will be shown to be the square of the *Froude number,* the ratio of flow velocity to surface wave velocity, and thus analogous to the Mach number of compressible flow.

———————————— **Illustrative Problem** ————————————

Water flows in a horizontal open channel at a depth of 0.6 m; the flowrate is 3.7 m³/s · m of width. If a hydraulic jump is possible, calculate the depth just downstream from the jump and the power dissipated in it.

Relevant Equations and Given Data

$$q = y_1 V_1 = y_2 V_2 \qquad (3.14)$$

$$\frac{p_1}{\gamma} + \frac{V_1^2}{2g_n} + y_1 = \frac{p_2}{\gamma} + \frac{V_2^2}{2g_n} + y_2 + \Delta(\text{E.L.}) \qquad (4.5)$$

$$P = q\gamma E_{\text{LOSS}} \qquad (4.7)$$

$$\frac{y_2}{y_1} = \frac{1}{2}\left[-1 + \sqrt{1 + \frac{8q^2}{g_ny_1^3}}\right] = \frac{1}{2}\left[-1 + \sqrt{1 + \frac{8V_1^2}{g_ny_1}}\right] \tag{6.9}$$

$q = 3.7 \text{ m}^3/\text{s·m} \qquad y_1 = 0.6 \text{ m} \qquad p_1 = p_2 = 0 \text{ at surfaces}$

Solution.

$$V_1 = \frac{3.7}{0.6} = 6.17 \text{ m/s} \tag{3.14}$$

$$\frac{V_1^2}{g_ny_1} = \frac{38.05}{9.81 \times 0.6} = 6.46$$

which is greater than 1, so a jump is possible.

$$y_2 = \frac{0.6}{2}[-1 + \sqrt{1 + 8 \times 6.46}] = 1.88 \text{ m} \blacktriangleleft \tag{6.9}$$

$$V_2 = \frac{3.7}{1.88} = 1.97 \text{ m/s}$$

$$\Delta(\mathbf{E.L.}) = 0.6 + \frac{(6.17)^2}{2g_n} - 1.88 - \frac{(1.97)^2}{2g_n} = 0.46 \text{ m} \tag{4.5}$$

Power dissipated (per metre of channel width):

$$P = 3.7 \times 9.8 \times 10^3 \times 0.46 = 16.7 \text{ kW} \blacktriangleleft \tag{4.7}$$

which shows the hydraulic jump to be an excellent energy dissipator; frequently it is used for this purpose in engineering designs.

6.6   The Oblique Standing Wave

Under certain circumstances in open flow with $V_1 > \sqrt{g_ny_1}$, a diagonal standing wave will form whose properties can be analyzed by the impulse-momentum principle. A wedge-shaped bridge pier (in a river) producing such a wave is shown in Fig. 6.7. Upstream from the wave front the liquid depth is y_1, and downstream it is y_2. By isolating a control volume $ABCD$ of length 1 m parallel to the wave front, application of the impulse-momentum and continuity principles is readily made.

Apply the continuity principle by noting that, because (from symmetry) the flowrate across AB is equal to that across CD, the flowrates across AD and BC are also equal. Accordingly,

$$q = y_1V_1 \sin \beta = y_2V_2 \sin (\beta - \theta) \tag{6.10}$$

The impulse-momentum principle applied along the tangential t-direction shows (from symmetry and the absence of body forces in the horizontal plane) that the forces on the surfaces AB and CD are equal and thus the t-component of velocity upstream and downstream from the wave must also be equal; therefore

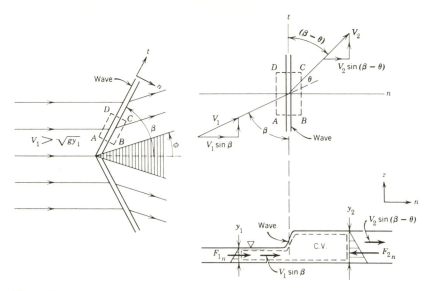

Fig. **6.7**

$$V_1 \cos \beta = V_2 \cos (\beta - \theta)$$

Apply the impulse-momentum principle along the normal direction n; the difference in the hydrostatic forces must account for the rate of change of momentum per unit time. This is written (see equation 6.4)

$$\sum F_n = (F_1 - F_2)_n = \frac{\gamma y_1^2}{2} - \frac{\gamma y_2^2}{2}$$

$$= [V_2 \sin (\beta - \theta) - V_1 \sin \beta]\frac{\gamma q}{g_n} \tag{6.11}$$

and, because $F_{2n} > F_{1n}$, $V_2 \sin (\beta - \theta) < V_1 \sin \beta$, and the indicated direction of deflection of the streamlines through the wave is therefore justified.

Combination of these equations yields

$$\frac{\tan^2(\beta - \theta)}{\tan^2\beta} = \frac{1 + 2(V_2^2/g_n y_2) \sin^2(\beta - \theta)}{1 + 2(V_1^2/g_n y_1) \sin^2\beta} \tag{6.12}$$

which shows that (for $V_1^2/g_n y_1 > 1$): for small values of β, $V_2^2/g_n y_2 > 1$ and, for large values of β, $V_2^2/g_n y_2 < 1$. Thus the oblique standing wave resembles the hydraulic jump of Section 6.5 for large values of β but is quite unlike it for small values of β.

A relationship[8] analogous to that of equation 6.9 for the hydraulic jump may be set down directly by comparing the development leading to equation 6.9 with that leading to equations 6.10 and 6.11. These are identical except for the terms $\sin \beta$ and

---

[8]For other relationships, see Sec. 15-17 of V. T. Chow, *Open-Channel Hydraulics*, McGraw-Hill, 1959.

$\sin(\beta - \theta)$ which accompany the velocities V_1 and V_2, respectively. Hence, for the oblique standing wave, equation 6.9 may be adjusted to read

$$\frac{y_2}{y_1} = \frac{1}{2}\left[-1 + \sqrt{1 + \frac{8V_1^2 \sin^2 \beta}{g_n y_1}}\right] \qquad (6.13)$$

────────────────── **Illustrative Problem** ──────────────────

Water flows in an open channel at a depth of 0.6 m and velocity 6 m/s. A wedge-shaped pier is to be placed in the flow to produce a standing wave of angle (β) 40°. Calculate the required wedge angle (2θ) and the depth just downstream from the wave.

Relevant Equations and Given Data

$$q = y_1 V_1 \sin \beta = y_2 V_2 \sin(\beta - \theta) \qquad (6.10)$$

$$\frac{\tan^2(\beta - \theta)}{\tan^2 \beta} = \frac{1 + 2(V_2^2/g_n y_2)\sin^2(\beta - \theta)}{1 + 2(V_1^2/g_n y_1)\sin^2 \beta} \qquad (6.12)$$

$$\frac{y_2}{y_1} = \frac{1}{2}\left[-1 + \sqrt{1 + \frac{8V_1^2 \sin^2 \beta}{g_n y_1}}\right] \qquad (6.13)$$

$$y_1 = 0.6\,\text{m} \qquad V_1 = 6\,\text{m/s} \qquad \beta = 40°$$

Solution.

$$\frac{V_1^2}{g_n y_1} = \frac{(6)^2}{9.81 \times 0.6} = 6.12$$

$$y_2 = \frac{0.6}{2}[-1 + \sqrt{1 + 8(6.12)(0.642)^2}] = 1.08\,\text{m} \blacktriangleleft \qquad (6.13)$$

$$V_2 \sin(\beta - \theta) = \left(\frac{0.6}{1.08}\right) \times 6 \times 0.642 = 2.14\,\text{m/s} \qquad (6.10)$$

$$\tan^2(\beta - \theta) = 0.704\left[\frac{1 + 2(2.14)^2/9.8 \times 1.08}{1 + 2(6.12)\text{x}(0.642)^2}\right] \qquad (6.12)$$

$$\tan(\beta - \theta) = 0.466 \qquad (\beta - \theta) = 25° \qquad \theta = 15° \qquad 2\theta = 30° \blacktriangleleft$$

A more difficult, but reasonable, question to answer is: given y_1, V_1, and θ; find V_2, y_2, and β.

──

Shock Waves

In Section 5.9 it was found that, for given entry conditions and configuration, there existed a range of exit pressures for which isentropic flow through a converging-diverging nozzle is not possible. Experiment shows that under these conditions the flow in the nozzle undergoes, at some point in the diverging section, an abrupt change from supersonic to subsonic velocity. This change is accompanied by large and abrupt rises in pressure, density, and temperature. The zone in which these changes take place is so thin that, for computations outside the zone, it may be considered to be a single line, that is, a discontinuity in the flow. This discontinuity is called a *normal shock wave* (that is, a shock wave perpendicular to the flow direction). The actual thickness of a normal shock wave is estimated to be of the order of the mean free path of the fluid molecules, that is, a micrometre (or between 10^{-4} and 10^{-5} inches). In this wave the gradients of velocity and density are so steep that viscous action, heat conduction, and mass diffusion are all appreciable with the result that the flow also undergoes a large entropy increase as it passes through the wave.

The normal shock wave is actually only a special case of the broader class of flow discontinuities called *oblique shock waves* that are found in most supersonic flows, both internal (in ducts, pipes, jet engine intakes, and compressors) and external (over the surfaces of wings, spacecraft, and so forth). These compression shock waves are analogous to the hydraulic jump and oblique standing waves, that were discussed in Sections 6.5 and 6.6, and, comparable to the nozzle flow, occur because, for a given flowrate and downstream and upstream water depths, there exist no ideal flow (that is, isentropic) solutions to the governing equations. On the basis of this similarity, many meaningful model studies of supersonic gas flows have been made in water channels by use of the *so-called* hydraulic analogy.

6.7   The Normal Shock Wave

By applying the continuity principle to the normal shock wave of Fig. 6.8, we find that

$$\dot{m} = A_1\rho_1V_1 = A_2\rho_2V_2 \tag{6.14}$$

The impulse-momentum principle, applied as in the hydraulic jump, gives

$$\sum F_x = p_1A_1 - p_2A_2 = (V_2 - V_1)\dot{m}$$

After eliminating $\dot{m}$ and noting that $A_1 = A_2$, combination of these equations yields

$$p_2 - p_1 = (\rho_1V_1^2 - \rho_2V_2^2) \tag{6.15}$$

which allows the pressure jump $(p_2 - p_1)$ across the wave to be computed. However, this requires the use of another equation to obtain V_2 from p_1, γ_1, and V_1. The adiabatic energy equation

$$\frac{V_2^2}{2} - \frac{V_1^2}{2} = h_1 - h_2 \tag{5.12}$$

may be used, because, although internal friction (with increase of entropy) is to be expected in the shock wave, there is no flow of heat to or from a fluid control volume enclosing the shock. For perfect gases, (see equation 5.13) $h_1 - h_2 = c_p(T_1 - T_2)$. With $c_p = Rk/(k - 1)$ and $RT = p/\rho$, equation 5.12 can be written

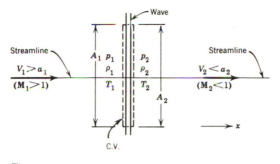

Fig. **6.8**

$$\frac{V_2^2}{2} - \frac{V_1^2}{2} = \frac{k}{k-1}\left(\frac{p_1}{\rho_1} - \frac{p_2}{\rho_2}\right) \tag{6.16}$$

and solved simultaneously with equation 6.15 to yield, after some algebraic manipulation, a relationship between the Mach numbers M_1 and M_2. This is

$$M_2^2 = \frac{1 + \dfrac{k-1}{2}M_1^2}{kM_1^2 - \dfrac{k-1}{2}} \tag{6.17}$$

from which M_2 is found, given M_1. As this equation is satisfied when $M_1 = M_2 = 1$, all other solutions must have the property: for $M_1 > 1$, $M_2 < 1$ and for $M_1 < 1$, $M_2 > 1$; however, it will be shown that only the first of these is physically possible—the second solution implies a *loss* of entropy through the wave and thus violates the second law of thermodynamics (see equation 5.9). Accordingly it may be concluded that through a normal shock wave the velocity must fall from supersonic to subsonic.

The pressure jump through the shock wave may also be computed in terms of the *shock strength* $(p_2 - p_1)/p_1$ to yield

$$\frac{p_2 - p_1}{p_1} = \frac{2k}{k+1}(M_1^2 - 1) \tag{6.18}$$

and the velocity ratio, V_2/V_1,

$$\frac{V_2}{V_1} = \frac{(k-1)M_1^2 + 2}{(k+1)M_1^2} \tag{6.19}$$

and, from p_2, V_2, and M_2, temperature, density, and sonic velocity downstream from the wave may be readily obtained.

A firm understanding of why a normal shock must always lead to $M_2 < 1$ from $M_1 > 1$ is simply obtained by constructing an enthalpy-entropy plot of the process. This exercise also leads to concepts that enjoy wide use for other compressible flow problems.[9] Such plots can be constructed for any gas; however, only perfect gases are considered here.

First, construct a line on an h-s plot (Fig. 6.9) for the adiabatic shock process of the locus of all states that satisfy only the continuity equation 6.14 and the energy equation 5.12. This is called a *Fanno* line. For given initial conditions p_1, V_1, ρ_1, T_1, and h_1, equation 5.12 gives

$$h_0 = h_1 + \frac{V_1^2}{2} = h + \frac{V^2}{2} \tag{5.12}$$

where h_0 is the stagnation (maximum) enthalpy. For a perfect gas $T = h/c_p$ and $p = \rho RT$. As the continuity equation 6.14 gives $\rho_1 V_1 = \rho V$ because the areas are the same on both sides of the shock, the entropy relation 5.10 for a perfect gas is

$$s - s_1 = c_v \ln\left[\left(\frac{T}{T_1}\right)^k \left(\frac{p}{p_1}\right)^{1-k}\right]$$

$$= c_v \ln\left[\left(\frac{h}{h_1}\right)\left(\frac{V_1}{V}\right)^{1-k}\right] \tag{6.20}$$

Now, equation 5.12 gives V as a function of h and the initial (or equivalently, the stagnation) conditions; therefore, there is an explicit relation between s and h that can be plotted by varying h. A typical resulting Fanno line is sketched on Fig. 6.9.

Next, construct a line on Fig. 6.9 of the locus of all states that satisfy only the

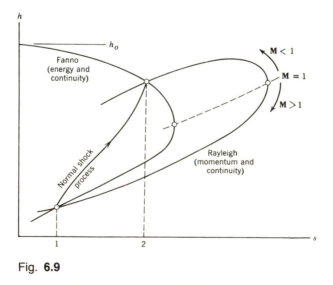

Fig. 6.9

[9]See A. H. Shapiro, *The dynamics and thermodynamics of compressible fluid flow*, vol. I., The Ronald Press, 1953.

continuity equation 6.14 and the impulse-momentum equation 6.15. This is called a *Rayleigh* line. From equation 6.20

$$s - s_1 = c_v \ln \left[\left(\frac{h}{h_1}\right)^k \left(\frac{p}{p_1}\right)^{1-k} \right]$$

(6.21)

As $\rho_1 V_1 = \rho V$ from continuity, equation 6.15 yields

$$\frac{p}{p_1} = 1 + \frac{\rho_1 V_1}{p_1}(V_1 - V)$$

Given an initial state (ρ_1, p_1, V_1, T_1 and $h_1 = c_p T_1$), then $\rho = \rho(V)$, $p/p_1 = f(V)$, and from the perfect gas relation $p/\rho = RT$ so $T = T(V)$ and $h = h(V)$. Accordingly, there is an explicit relationship between s and h that can be obtained by varying the parameter V. A typical resulting Rayleigh line is sketched on Fig. 6.9.

The continuity, impulse-momentum, and energy equations are all satisfied across a shock wave. Therefore, the normal shock process can occur only between the two states where the Rayleigh and Fanno lines intersect. Further study of the above equations reveals that on each line the point where s is a maximum (where also $ds/dh = 0$) corresponds to $V = a$; that is, $\mathbf{M} = 1$. As the maximum value of h is h_0 when $V = 0$, the upper branches of the lines must correspond to $\mathbf{M} < 1$ and the lower branches to $\mathbf{M} > 1$. The Second Law of Thermodynamics (Section 5.1) requires that the entropy cannot decrease in an adiabatic process (equation 5.9); therefore, the normal shock process must go from $\mathbf{M} > 1$ (supersonic) to $\mathbf{M} < 1$ (subsonic) in every case.

Illustrative Problem

Upstream from a normal shock wave in an airflow the pressure, velocity, and sonic velocity are 14.7 psia, 1 732 ft/s, and 862 ft/s, respectively. Calculate these quantities just downstream from the wave, and the rise in temperature through the wave.

Relevant Equations and Given Data

$$a = \sqrt{kRT}$$

(1.11)

$$\mathbf{M}_2^2 = \frac{1 + \dfrac{k-1}{2}\mathbf{M}_1^2}{k\mathbf{M}_1^2 - \dfrac{k-1}{2}}$$

(6.17)

$$\frac{p_2 - p_1}{p_1} = \frac{2k}{k+1}(\mathbf{M}_1^2 - 1)$$

(6.18)

$$\frac{V_2}{V_1} = \frac{(k-1)\mathbf{M}_1^2 + 2}{(k+1)\mathbf{M}_1^2}$$

(6.19)

$$p_1 = 14.7 \text{ psia} \qquad V_1 = 1\,732 \text{ ft/s} \qquad a_1 = 862 \text{ ft/s}$$

$$k = 1.4 \qquad R = 1\,715 \text{ ft} \cdot \text{lb/slug} \cdot {}^\circ\text{R}$$

Solution.

$$M_1 = \frac{1\,732}{862} = 2.01$$

$$M_2^2 = \frac{1 + 0.4(2.01)^2/2}{1.4(2.01)^2 - 0.4/2} = 0.331 \qquad M_2 = 0.58 \tag{6.17}$$

$$p_2 = 14.7\left[1 + \frac{2(1.4)}{2.4}(2.01^2 - 1)\right] = 66.8 \text{ psia} \blacktriangleleft \tag{6.18}$$

$$V_2 = 1\,732\left[\frac{0.4(2.01)^2 + 2}{2.4(2.01)^2}\right] = 646 \text{ ft/s} \blacktriangleleft \tag{6.19}$$

$$a_2 = \frac{646}{0.58} = 1\,144 \text{ ft/s} \blacktriangleleft$$

$$1\,144 = \sqrt{1.4 \times 1\,715 \times T_2} \qquad T_2 = 517°R \tag{1.11}$$

$$862 = \sqrt{1.4 \times 1\,715 \times T_1} \qquad T_1 = 310°R \tag{1.11}$$

The rise of temperature through the wave is $T_2 - T_1 = 207°F$. ◀ The conditions which will produce this particular shock wave in a nozzle are shown in the Illustrative Problem of Section 5.9.

6.8 The Oblique Shock Wave

In an age of high-speed flight the reader has had casual acquaintance with the shock waves produced by objects traveling at supersonic speeds.[10] The geometry of such two-dimensional waves may be studied by application of the continuity, impulse-momentum, and energy principles in the following manner.

Construct a control surface $ABCD$ as indicated in Fig. 6.10, taking $BC = AD = 1$

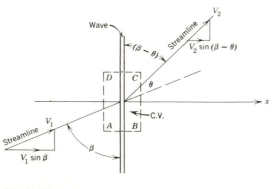

Fig. **6.10**

[10]See Section 13.10 for photograph and analytical justification for such waves.

unit, and also consider the control volume to be 1 unit deep perpendicular to the plane of the paper. From symmetry it is seen that the flowrates across AB and CD are equal; from this, it follows that the mass flowrates, $\dot{m}$, across BC and AD are also equal. These are

$$\dot{m} = \rho_1 V_1 \sin \beta = \rho_2 V_2 \sin (\beta - \theta) \qquad (6.22)$$

Because the forces caused by the pressures on surfaces AB and CD are equal and opposite, the impulse-momentum equation applied in the y-direction discloses that the components of velocity parallel to the wave front are equal, giving

$$V_1 \cos \beta = V_2 \cos (\beta - \theta) \qquad (6.23)$$

The impulse-momentum principle applied in the x-direction gives

$$p_1 - p_2 = [V_2 \sin (\beta - \theta) - V_1 \sin \beta] \dot{m} \qquad (6.24)$$

The energy equation (without addition or extraction of heat) applied across the wave is (as in Section 6.7)

$$\frac{V_2^2}{2} - \frac{V_1^2}{2} = \frac{k}{k-1} \left(\frac{p_1}{\rho_1} - \frac{p_2}{\rho_2} \right) \qquad (6.16)$$

These equations may be combined and manipulated to yield a variety of useful relationships, one of which is

$$M_2^2 \sin^2(\beta - \theta) = \frac{1 + \dfrac{k-1}{2} M_1^2 \sin^2\beta}{k M_1^2 \sin^2\beta - \dfrac{k-1}{2}} \qquad (6.25)^{[11]}$$

This equation reveals that (for $M_1 > 1$): $M_2 < 1$ for large β and $M_2 > 1$ for small β. The so-called *strong shock* (featured by large β) is similar to the normal shock of Section 6.7 in that the flow changes from supersonic to subsonic in passing through the wave, whereas for small β (*weak shock*) the flow is supersonic on both sides of the wave. Through all such waves, however, there is internal friction with increase of entropy; the flow is rotational (see Section 5.10), and $M_2 < M_1$.

————————————— **Illustrative Problem** —————————————

Upstream from an oblique shock wave in an airflow, the pressure, velocity, and sonic velocity are 14.7 psia, 1 732 ft/s, and 862 ft/s, respectively. The wave angle (β) is 40°. Calculate angle (θ) required to produce this wave, the pressure, velocity, and sonic velocity just downstream from the wave, and the rise of temperature through the wave.

———

[11]This result and others for V_2/V_1 and $(p_2 - p_1)/p_1$ may be derived directly from equations 6.17, 6.18, and 6.19 in the following manner. Compare simultaneous equations 6.14, 6.15 and 6.16 with equations 6.22, 6.24, and 6.16. These are identical except for $\sin \beta$ and $\sin (\beta - \theta)$. However, these quantities are multiplied into V_1 and V_2, respectively, and hence will appear in derived results in this pattern and also multiplied into M_1 and M_2, respectively.

Relevant Equations and Given Data

$$a = \sqrt{kRT} \tag{1.11}$$

$$c_p T_1 + \frac{V_1^2}{2} = c_p T_2 + \frac{V_2^2}{2} \tag{5.13}$$

$$\frac{p_2 - p_1}{p_1} = \frac{2k}{k+1}(M_1^2 - 1) \tag{6.18}$$

$$\frac{V_2}{V_1} = \frac{(k-1)M_1^2 + 2}{(k+1)M_1^2} \tag{6.19}$$

$$V_1 \cos \beta = V_2 \cos (\beta - \theta) \tag{6.23}$$

$$M_2^2 \sin^2(\beta - \theta) = \frac{1 + \dfrac{k-1}{2} M_1^2 \sin^2\beta}{k M_1^2 \sin^2\beta - \dfrac{k-1}{2}} \tag{6.25}$$

$p_1 = 14.7 \text{ psia}$ $\quad V_1 = 1\,732 \text{ ft/s}$ $\quad a_1 = 862 \text{ ft/s}$ $\quad \beta = 40°$

$R = 1\,715 \text{ ft} \cdot \text{lb/slug} \cdot °R$ $\quad c_p = 6\,000 \text{ ft} \cdot \text{lb/slug} \cdot °R$ $\quad k = 1.4$

Solution.

$$M_1 = \frac{1\,732}{862} = 2.01$$

$$862 = \sqrt{1.4 \times 1\,715 \times T_1} \qquad T_1 = 310°R \tag{1.11}$$

Applying equation 6.18, appropriately adjusted with $\sin \beta = 0.642$,

$$p_2 = 14.7 \left[1 + \frac{2(1.4)}{2.4}(2.01^2 \times 0.642^2 - 1) \right] = 26.0 \text{ psia} \blacktriangleleft \tag{6.18}$$

Applying equation 6.19, appropriately adjusted with $\sin \beta$ and $\sin (\beta - \theta)$,

$$\frac{V_2 \sin (\beta - \theta)}{V_1 \sin \beta} = \frac{(k-1)M_1^2 \sin^2\beta + 2}{(k+1)M_1^2 \sin^2\beta} \tag{6.19}$$

and, using $V_2/V_1 = \cos \beta/\cos (\beta - \theta)$ from equation 6.23,

$$\frac{\tan (\beta - \theta)}{\tan 40°} = \frac{0.4(2.01 \times 0.642)^2 + 2}{2.4(2.01 \times 0.642)^2} \qquad \beta - \theta = 29.3° \qquad \theta = 10.7° \blacktriangleleft$$

$$M_2^2 = \frac{1}{(0.489)^2} \times \frac{1 + (0.4/2)(2.01 \times 0.642)^2}{1.4(2.01 \times 0.642)^2 - 0.4/2} = 2.635 \tag{6.25}$$

$$M_2 = 1.62$$

$$V_2 = \frac{1\,732 \cos 40°}{\cos 29.3°} = 1\,523 \text{ ft/s} \blacktriangleleft \tag{6.23}$$

$$a_2 = \frac{1\,523}{1.62} = 940 \text{ ft/s} \blacktriangleleft$$

$$940 = \sqrt{1.4 \times 1\ 715 \times T_2} \qquad T_2 = 367.5°R \tag{1.11}$$

or

$$\frac{(1\ 732)^2 - (1\ 523)^2}{2} = 6\ 000(T_2 - T_1) \tag{5.13}$$

$$T_2 - T_1 = 57.5° \qquad T_2 = 367.5°R \blacktriangleleft$$

Flow Machines

Many machines are characterized by the transfer of energy, forces, or torques between a moving stream of fluid and the solid elements of the machine. These are *flow machines*. They include a large and important selection of common units, such as jet engines, rocket motors, airplane and marine propellers, steam turbines for electricity generation and ship propulsion, garden insecticide sprays (jet pumps), industrial centrifugal pumps, and the massive turbines of hydroelectric plants.

The energy and force transfers in flow machines can be either to or from the fluid and can be accomplished in a number of configurations. Thus, it is useful to further categorize flow machines. In *turbines* energy is taken from the fluid stream; in *pumps* (*compressors, blowers,* or *fans*) energy is added to the fluid stream. If the transfers occur between the fluid stream and a machine element that rotates about a fixed axis (a compressor rotor, turbine runner, or pump impeller), the machine is classed as a *turbomachine*. In turbomachines, the flow may be essentially *axial* (parallel to the axis of rotation of the rotating element), *radial*, or *mixed* (a combination of radial and axial).

In the following five sections a number of typical flow machines are briefly described and analyzed in an elementary manner by application of the impulse-momentum and energy principles. In Chapter 8 the operating characteristics of a few turbomachines are considered by application of similitude and dimensional analysis. No attempt has been made in this elementary treatment to present a design manual for flow machines; the reader must refer to the references for detailed studies of specific classes of machine.

6.9   Jet Propulsion

In a modern aircraft jet-propulsion system (Fig. 6.11), air is drawn in at the upstream end and its pressure, density, and temperature are raised by a compressor (usually axial flow). Just downstream from the compressor, fuel is injected and burned, adding energy to the fluid. The mixture then expands and passes through a gas turbine (which drives the compressor) on its way to the nozzle; from here it emerges into the atmosphere at high velocity and at a pressure equal to or greater than that surrounding the nozzle. A cross section of a modern jet engine is shown in Fig. 6.12.

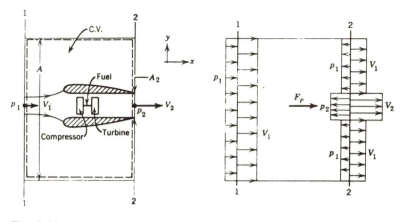

Fig. **6.11**

An impulse-momentum analysis of such a unit allows its propulsive force to be computed. Here a control volume of indefinite horizontal and vertical extent that is fixed to the jet-propulsion unit[12] can be used, as shown in Fig. 6.11, section 1 being located well upstream of the unit and section 2 in the plane of the nozzle exit. The speed of the jet unit relative to still air is V_1. Thus, in a coordinate frame that moves with the jet unit, the incoming air velocity far from the unit is V_1. If it is assumed that air which does not pass through the unit experiences no net change of velocity or pressure, the pressures and velocities at all points of sections 1 and 2 will be p_1 and V_1, respectively, except in the nozzle exit, where they are p_2 and V_2. Taking the summation of forces on the fluid within the control volume gives

$$\sum F_x = p_1 A - p_1(A - A_2) - p_2 A_2 + F_P = (p_1 - p_2)A_2 + F_P$$

in which F_P is the propulsive force exerted *by the unit on the fluid*. Now the fuel is added through the control surface and so must be considered in the fluid momentum analysis. Between the point of fuel injection and nozzle exit the velocity of the mass flowrate ($\dot{m}_f$) of fuel increases from zero to V_2; between sections 1 and 2 the velocity of the mass flowrate ($\dot{m}_a$) of air increases from V_1 to V_2. Accordingly

$$\oint_{C.S.} V_x \, d\dot{m} = V_2(\dot{m}_f) + (V_2 - V_1)(\dot{m}_a)$$

---

[12] In this and several other applications, it is desirable to fix the control volume to the object being analyzed rather than to hold the control volume fixed in space as has been assumed to this point. From particle dynamics it is recalled that Newton's second law is applicable in an *inertial reference frame*. A reference frame that is fixed in space is an inertial frame; but it can be shown that a reference frame which is moving at a *constant velocity* relative to a fixed frame is also an inertial frame. An inspection of the derivations of all the equations based on Newton's second law, which have been carried out to this point, reveals that, if all velocities and accelerations, and so on, are referenced to an inertial frame, then all the results remain valid and the equations are, indeed, unchanged. Accordingly, when it is useful to do so, control volumes will be fixed on the object being analyzed, but only if it is moving with a constant velocity.

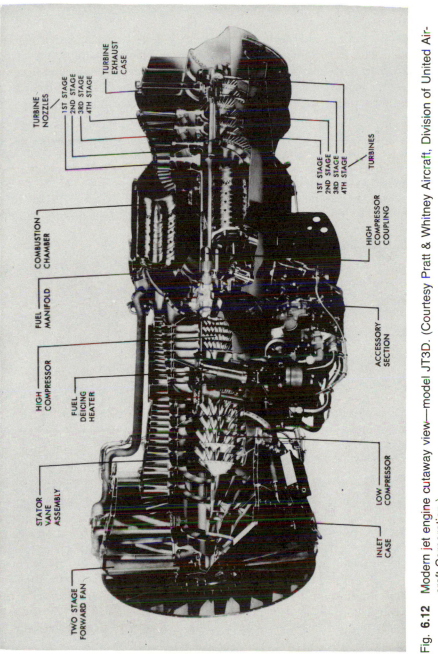

Fig. **6.12** Modern jet engine cutaway view—model JT3D. (Courtesy Pratt & Whitney Aircraft, Division of United Aircraft Corporation.)

Using equation 6.4 yields the propulsive force, F_P, obtainable from the unit; it is

$$F_P = (\dot{m}_a)(V_2 - V_1) + (\dot{m}_f)V_2 + (p_2 - p_1)A_2 \tag{6.26}$$

Although this force is fundamental to preliminary design calculations, it yields no information about the optimum design of passages, compressor, turbine, and so forth, for which other principles must be used. It does, however, demonstrate again the strength of the impulse-momentum principle in the bypassing of internal complexities and obtaining useful results from upstream and downstream flow conditions alone.

———————— Illustrative Problem ————————

A jet-propulsion unit is to develop a propulsive force of 200 kN when moving through the U.S. Standard Atmosphere (Appendix 4) at 10 km altitude at a speed of 250 m/s. The velocity, absolute pressure, and area at the nozzle exit are 1.2 km/s, 35 kPa, and 1.4 m², respectively. How much air must be drawn through the unit? Fuel may be neglected.

Relevant Equation and Given Data

$$F_P = (\dot{m}_a)(V_2 - V_1) + (\dot{m}_f)V_2 + (p_2 - p_1)A_2 \tag{6.26}$$

$$F_P = 200 \text{ kN} \qquad V_1 = 250 \text{ m/s} \qquad V_2 = 1.2 \text{ km/s}$$

$$p_2 = 35 \text{ kPa} \qquad A_2 = 1.4 \text{ m}^2 \qquad \text{altitude} = 10 \text{ km: } p_1 = 26.5 \text{ kPa}$$

Solution.

$$200 \times 10^3 = (\dot{m}_a)(1\ 200 - 250) + (35.0 - 26.5)10^3 \times 1.4 \tag{6.26}$$

$$\dot{m}_a = 198 \text{ kg/s} \blacktriangleleft$$

A suitable compressor must then be designed to provide this flowrate.

6.10   Propellers and Windmills

Although the screw propellers of ships or aircraft cannot be analyzed with the impulse-momentum and energy principles alone, application of these principles to the problem will lead to some of the laws which characterize their design.

A pair of controllable pitch ship propellers are shown in Fig. 6.13. An idealized screw propeller with its slipstream is shown in Fig. 6.14. For such a propeller operating in an unconfined fluid, the pressures p_1 and p_4 at some distance ahead of and behind the propeller, and the pressures over the slipstream boundary, are the same. However, from the shape of the slipstream (using the continuity and Bernoulli principles) the mean pressure p_2 just upstream from the propeller is smaller than p_1, and the pressure p_3 just downstream from the propeller is larger than p_4.[13] When the fluid in the slipstream

---

[13] Applying this, and the fact of the preceding sentence, at the propeller tip apparently results in two different pressures there. This is due to the assumption of a uniform pressure distribution; in reality the pressure is not uniform over the propeller disk because at the propeller tip it must equal that surrounding the slipstream.

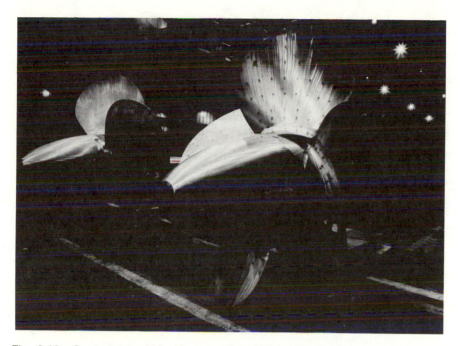

Fig. **6.13** Controllable pitch ship propellers (4.2 m d.). (Courtesy Lips N. V. Propeller Works, Drunen, Holland.)

between sections 1 and 4 is isolated, it is observed that the only force acting is that exerted by the propeller on the fluid.[14] This may be computed either from the pressure difference $(p_3 - p_2)$ or from the gain in momentum flux between sections 1 and 4. Therefore

$$(p_3 - p_2)A = F = (V_4 - V_1)\rho Q = (V_4 - V_1)A\rho V \qquad (6.27)$$

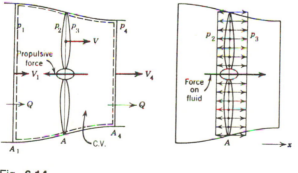

Fig. **6.14**

[14] The same pressure distributed over sections 1, 4, and the slipstream boundary can produce no net force.

in which V is the mean velocity through the propeller disk; cancelling A's in this equation gives

$$p_3 - p_2 = (V_4 - V_1)\rho V \qquad (6.28)$$

Now, by applying the Bernoulli principle between sections 1 and 2,

$$p_1 + \tfrac{1}{2}\rho V_1^2 = p_2 + \tfrac{1}{2}\rho V_2^2 \qquad (4.2)$$

and between sections 3 and 4,

$$p_3 + \tfrac{1}{2}\rho V_3^2 = p_4 + \tfrac{1}{2}\rho V_4^2 \qquad (4.2)$$

and by using $p_1 = p_4$, another expression may be derived for $(p_3 - p_2)$; it is

$$p_3 - p_2 = \tfrac{1}{2}\rho(V_4^2 - V_1^2) \qquad (6.29)$$

By equating equations 6.28 and 6.29,

$$V = (V_1 + V_4)/2 \qquad (6.30)$$

which shows that the velocity through the propeller disk is the numerical average of the velocities at some distance ahead of and behind the propeller; in other words, there is the same increase of velocity ahead of the propeller as behind it. This result, modified slightly to allow for friction, rotational effects, and so forth, is one of the basic assumptions of propeller design.

The useful power output, P_o, derived from a propeller, is the thrust, F, multiplied by the velocity, V_1, at which the propeller is moving forward.

$$P_o = FV_1 = (V_4 - V_1)\rho QV_1 \qquad (6.31)$$

The power input, P_i, is that required to maintain continual increase of velocity of the slipstream from V_1 to V_4. From equation 4.6 or 4.7,

$$P_i = \frac{\rho Q}{2}(V_4^2 - V_1^2) = \rho Q(V_4 - V_1)\left(\frac{V_4 + V_1}{2}\right) = \rho Q(V_4 - V_1)V \qquad (6.32)$$

The ideal efficiency, η, of the propeller is then

$$\eta = \frac{P_o}{P_i} = \frac{V_1}{V} \qquad (6.33)$$

and, because V is always greater than V_1, it may be concluded that the efficiency of a propeller even in an ideal fluid can never be 100%.[15]

There are many similarities between propeller and windmill, but their purposes are quite different. The propeller is designed primarily to create a propulsive force or thrust and acts as a pump. The windmill is designed to extract energy from the wind and, hence, is a turbine. Because of the different objectives of windmill and propeller, their

---

[15] When $V = V_1$, a propeller of 100% efficiency results, but such a propeller produces no propulsive force! (See equation 6.27.) Practical ship and airplane propellers may have efficiencies of 80%.

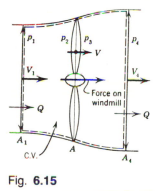

Fig. **6.15**

efficiencies are calculated differently. However, comparison of Figs. 6.14 and 6.15 shows the windmill to be, as far as the flow picture is concerned, the inverse of the propeller. In the windmill the "slipstream" widens as it passes the machine and the pressure p_2 is greater than pressure p_3. However, by applying the Bernoulli and impulse-momentum principles as before, the velocity through the windmill disk, as through the propeller, may be shown to be the numerical average of V_1 and V_4.

In a frictionless machine, the power delivered to the windmill must be exactly that extracted from the air, which in turn is represented by the decrease of the kinetic energy of the slipstream between sections 1 and 4. This is the *output* of the machine and is given by

$$P_o = \frac{\rho Q}{2}(V_1^2 - V_4^2) \tag{6.34}$$

It is customary to define windmill efficiencies as the ratio of this power output to the total power *available* in a streamtube of cross-sectional area A and wind velocity V_1. Thus the efficiency of an ideal windmill is

$$\eta = \frac{P_o}{P_a} = \frac{(V_1^2 - V_4^2)AV\rho/2}{AV_1\rho V_1^2/2} = \frac{(V_1 + V_4)(V_1^2 - V_4^2)}{2V_1^3} \tag{6.35}$$

$(V_1 + V_4)/2$ having been substituted for V. The maximum efficiency is found by differentiating η with respect to V_4/V_1 and setting the result equal to zero. This gives a value of $V_4/V_1 = \frac{1}{3}$, which when substituted in equation 6.35 produces a maximum efficiency of $\frac{16}{27}$, or 59.3%. Because of friction and other losses this efficiency is, of course, not realized in practice; the highest possible efficiency for a real windmill appears to be around 50%. While the traditional "Dutch windmill" with large sail-like blades operates at an efficiency around 15%, the world fuel crisis has revived interest in wind energy. Since the mid-1970s major research and development efforts have focused on developing windmills, such as those with airfoil-shaped (propellerlike) blades, with acceptable efficiencies (up to 47%).

_____ **Illustrative Problem** _____

The engine of an airplane flying through still air (specific weight 12.0 N/m³) at 320 km/h (88.9 m/s) delivers 1 120 kW to an ideal propeller 3 m in diameter. Calculate slipstream velocity, velocity through the propeller disk, and the diameter of the slipstream ahead of and behind the propeller. Also calculate thrust and efficiency.

Relevant Equations and Given Data

$$(p_3 - p_2)A = F = (V_4 - V_1)\rho Q = (V_4 - V_1)A\rho V \tag{6.27}$$

$$V = (V_1 + V_4)/2 \tag{6.30}$$

$$P_i = \frac{\rho Q}{2}(V_4^2 - V_1^2) = \rho Q(V_4 - V_1)\left(\frac{V_4 + V_1}{2}\right) = \rho Q(V_4 - V_1)V \tag{6.32}$$

$$\eta = \frac{P_o}{P_i} = \frac{V_1}{V} \tag{6.33}$$

$$P_i = 1.120 \text{ MW} \qquad V_1 = 88.9 \text{ m/s} \qquad d = 3 \text{ m} \qquad \gamma = 12.0 \text{ N/m}^3$$

Solution.

$$P_i = 1\,120 \times 10^3 = \frac{\pi}{4}(3)^2\left(\frac{V_4 + 88.9}{2}\right)\frac{12.0}{9.81}\left(\frac{V_4^2 - 88.9^2}{2}\right) \tag{6.32}$$

$$V_4 = 103 \text{ m/s} \blacktriangleleft$$

$$V = \frac{(103 + 88.9)}{2} = 95.95 \text{ m/s} \blacktriangleleft \qquad Q = 678 \text{ m}^3/\text{s} \tag{6.30}$$

$$A_1 = \frac{678}{88.9} = 7.63 \text{ m}^2 \qquad d_1 = 3.12 \text{ m} \blacktriangleleft$$

$$A_4 = \frac{678}{103} = 6.58 \text{ m}^2 \qquad d_4 = 2.9 \text{ m} \blacktriangleleft$$

$$F = (103 - 88.9)\,678\left(\frac{12.0}{9.81}\right) = 11.7 \text{ kN} \blacktriangleleft \tag{6.27}$$

$$\eta = \frac{11.7 \times 88.9}{1\,120} = 92.8\% \qquad \text{or} \qquad \eta = \frac{88.9}{95.95} = 92.8\% \blacktriangleleft \tag{6.33}$$

6.11 Deflectors and Blades—The Impulse Turbine

When a free jet is deflected by a blade surface, a change of momentum occurs and a force is exerted on the blade. If the blade is allowed to move, this force will act through a distance, and power may be derived from the moving blade; this is the basic principle of the impulse turbine.

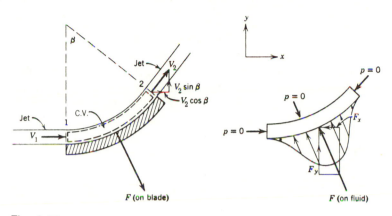

F (on blade)   F (on fluid)

Fig. **6.16**

The jet of Fig. 6.16 is deflected by a fixed blade and may be assumed to be in a horizontal plane.[16] With the control surface drawn around the region of momentum change, it is seen at once that the force exerted by the blade is the only force acting on the fluid. Therefore, from equations 6.4 and 6.5,

$$\sum F_x = -F_x = (V_2 \cos \beta - V_1)\rho Q \tag{6.36}$$

$$\sum F_y = F_y = (V_2 \sin \beta - 0)\rho Q \tag{6.37}$$

If this blade now moves (Fig. 6.17) with a constant velocity u in the same direction

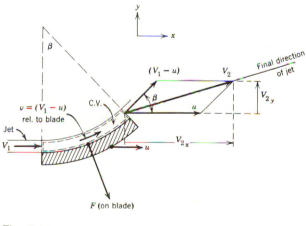

F (on blade)

Fig. **6.17**

[16] The difference in elevation between beginning and end of blade is usually negligible in practical blade problems.

as the original jet, the jet is no longer deflected through an angle β because the leaving velocity, V_2, is now the resultant of the blade velocity and the velocity of the fluid over the blade. The velocity, v, of fluid relative to the blade is $(V_1 - u)$, and if friction is neglected, this relative velocity is the same at the entrance and exit of the blade system. Therefore, as the jet leaves the blade, it has an absolute velocity, V_2, equal to the vector sum of u and $(V_1 - u)$. Then, from Fig. 6.17 and equations 6.4 and 6.5, with the control volume fixed to the blade,

$$\sum F_x = -F_x = (V_{2x} - V_{1x})\rho Q = -(V_1 - u)(1 - \cos \beta)\rho Q \qquad (6.38)$$

$$\sum F_y = F_y = (V_{2y} - V_{1y})\rho Q = (V_1 - u)\sin \beta \rho Q \qquad (6.39)$$

Engineers frequently prefer to treat such problems with the relative velocities v_1 and v_2, both of which are equal to $(V_1 - u)$, thus reducing the problem to that of the stopped blade of Fig. 6.16; the validity of this may be seen from the velocity triangles and by noting that the substitution of $(V_1 - u)$ for V_1 and V_2 in equations 6.36 and 6.37 will yield equations 6.38 and 6.39, respectively.

For a single moving blade the flowrate, Q, in these equations is not the flowrate in the jet, since less[17] fluid is deflected per second by a single blade than is flowing in the jet. However, in the practical application of this theory to a turbine, where a series of blades is used, the fluid deflected *by the blade system* per second is the same as that flowing.

Problems of this type should be recognized as unsteady-flow problems for which the equations have been written for time-average conditions. As a blade moves through the jet a time-varying force will act on it, rising from zero and falling back to zero. From this it is clear that the foregoing equations can give no information about the maximum force exerted on the blade or the force on the blade at any instant of time, or the optimum spacing of blades to secure a required deflection of fluid; nevertheless, the equations may be used effectively in preliminary design and performance calculations.

In an impulse turbine (Fig. 6.18) a series of blades of the type above is mounted on the periphery of a wheel. Consider the idealized blade in Fig. 6.19; while a blade is in the jet it is moving in a direction approximately parallel to that of the jet; thus the equations above may be applied directly to find the power characteristics of the machine.[18] The force component F_y does no work, since it acts through no distance; when the component F_x is multiplied by u, the power transferred from jet to turbine is obtained. In a frictionless machine this is the output and is given by

$$P = Q\rho (V_1 - u)(1 - \cos \beta)u \qquad (6.40)$$

---

[17] For a blade moving in the same direction as the jet.

[18] Notice that the actual Pelton wheel blades or buckets split the jet and deflect it symmetrically to the sides. It is easy to see that the simple analysis for a single blade (Fig. 6.19) is valid for each Pelton bucket. Indeed, the control volume shown there surrounds all the moving blades and is fixed in space. As long as the jet path shown is the correct time-averaged picture, one can apply equation 6.4 directly to obtain F_x. The result is precisely that from equation 6.38, which came from a moving control volume analysis.

Fig. **6.18**   Runner of the six-nozzle vertical Pelton-type turbine
"Castaic" in U.S.A. (Courtesy of Escher Wyss Limited,
Zurich, Switzerland.)

From this it is evident (1) that no power is obtained from the machine when $u = 0$
(machine stopped) and when $u = V_1$ (runaway speed) and (2) that, with Q, ρ, β, and
V_1 constant, the relationship between P and u is a parabolic one. Taking $dP/du = 0$
to obtain the properties of maximum output shows that this will occur when $u = V_1/2$
and is given by

$$P_{max} = Q\rho \frac{V_1^2}{4}(1 - \cos \beta) \qquad (6.41)$$

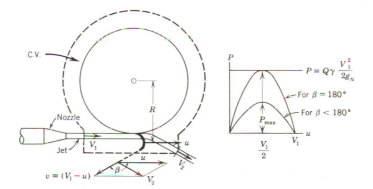

Fig. **6.19**

For a blade angle of 180°, this becomes

$$P_{max} = \frac{Q\rho V_1^2}{2} = \frac{Q\gamma V_1^2}{2g_n} \tag{6.42}$$

As this is exactly the power of the free jet (see equation 4.7), it may be concluded that all the jet power may be theoretically extracted and transferred to the machine (at 100% efficiency), if (1) the peripheral speed is one-half the jet speed and if (2) the blade angle is 180°. In practice, blade angles will be found to be around 165°, operating peripheral speeds to be about 48% of jet velocity, with resulting peak efficiencies near 90% (see Section 8.3).

Another and more general type of free-jet turbine (Fig. 6.20) may be successfully analyzed by using the energy and impulse-momentum principles. In Fig. 6.20a a jet plays upon a series of blades mounted on the periphery of a rotor wheel. The sketch represents, for example, one stage of an impulse steam turbine (Fig. 6.20b). The power P transmitted from jet to blade system, called a *cascade,* is (neglecting friction) exactly that lost from the jet (see equation 4.7):

$$P = \frac{Q\rho(V_1^2 - V_2^2)}{2} \tag{6.43}$$

which shows that $V_2 < V_1$. This power is calculable from $P = F_y u,$ in which F_y is the working component of force exerted by the jet on the moving blade system. From the impulse-momentum principle, F_y (on the fluid) is given by

$$\sum F_y = -F_y = (V_2 \sin \alpha_2 - V_1 \sin \alpha_1)\rho Q \tag{6.44}$$

showing that $V_2 \sin \alpha_2 < V_1 \sin \alpha_1$. These requirements (without regard to the details of the blade geometry) justify the relative magnitudes and positions of V_1 and V_2 shown in Fig. 6.20a.

Suitable blades to produce these changes of velocity may now be designed by observing that the absolute velocity (V) of the fluid is the vector sum of relative velocity (v) and blade velocity (u), the relative velocity being everywhere tangent to the blade (that is, the fluid does not separate from the blade) for good design. Accordingly, the velocity triangles are as shown and the inlet and exit blade angles, β_1 and β_2, determined. A further requirement of a frictionless system is that the relative velocities v_1 and v_2 must be equal, as in Fig. 6.17.

The foregoing equations and requirements allow the preliminary design of free-jet turbines to produce a given power at a given flowrate and required speed; this is illustrated in the following problems.

––––––––––––––––   **Illustrative Problems**   ––––––––––––––––

An impulse turbine of 6 ft diameter is driven by a water jet of 2 in. diameter moving at 200 ft/s. Calculate the force on the blades and the horsepower developed at 250 rpm. The blade angles are 150°.

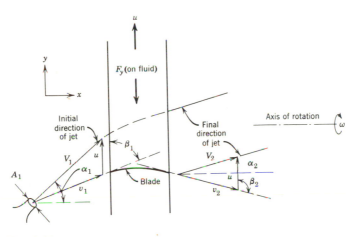

Fig. **6.20a**

Fig. **6.20b** Steam turbine rotor and blade cascades. (Courtesy of DELAVAL Turbine, Inc.

Relevant Equations and Given Data

$$Q = AV \tag{3.12}$$

$$\sum F_x = -F_x = (V_{2x} - V_{1x})\rho Q = -(V_1 - u)(1 - \cos \beta)\rho Q \tag{6.38}$$

$$P = \frac{Q\rho(V_1^2 - V_2^2)}{2} \tag{6.43}$$

$R = 3 \text{ ft} \qquad \omega = 250 \text{ rev/min.} \qquad \beta = 150°$

$d = 2 \text{ in.}$

Solution.

$$u = \left(\frac{250}{60}\right) 2\pi \times 3 = 78.6 \text{ ft/s}$$

$$v_1 = v_2 = V_1 - u = 200 - 78.6 = 121.4 \text{ ft/s}$$

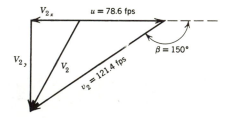

From the velocity diagram,

$$V_{2x} = -(121.4)0.866 + 78.6 = -26.6 \text{ ft/s}$$

$$V_{2y} = (121.4)0.50 = 60.7 \text{ ft/s}$$

$$V_2 = 66.5 \text{ ft/s}$$

The flowrate is

$$Q = \left(\frac{\pi}{4}\right)\left(\frac{2}{12}\right)^2 200 = 4.36 \text{ ft}^3/\text{s} \tag{3.12}$$

The working component of force on the fluid is

$$-F_x = [-26.6 - 200]1.936 \times 4.36 = -1\,915 \text{ lb} \blacktriangleleft \tag{6.38}$$

The power developed may be computed in two ways:

From mechanics,

$$P = \frac{1\,915 \times 78.6}{550} = 274 \text{ hp}$$

From equation 6.43,

$$P = \frac{4.36 \times 1.936(200^2 - 66.5^2)}{2 \times 550} = 274 \text{ hp} \blacktriangleleft \qquad (6.43)$$

A free-jet impulse turbine is to produce 74.6 kW at a blade speed of 23 m/s. A water jet having V_1 = 46 m/s, diameter = 51 mm, and α_1 = 60° is used to drive the machine. Calculate the required blade angles β_1 and β_2.

Relevant Equations and Given Data

$$Q = AV \qquad (3.12)$$

$$P = \frac{Q\rho(V_1^2 - V_2^2)}{2} \qquad (6.43)$$

$$\sum F_y = -F_y = (V_2 \sin \alpha_2 - V_1 \sin \alpha_1)\rho Q \qquad (6.44)$$

P = 74.6 kW d = 51 mm V_1 = 46 m/s u = 23 m/s α_1 = 60°

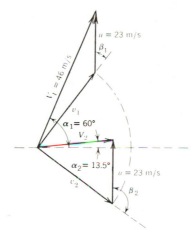

Solution.   See Fig. 6.20*a*.

$$Q = \left(\frac{\pi}{4}\right)(0.051)^2 46 = 0.094 \text{ m}^3/\text{s} \qquad (3.12)$$

$$F_y = 74.6 \times \frac{1\ 000}{23} = 3.24 \text{ kN}$$

$$74.6 \times 1\ 000 = 0.094 \times \frac{1\ 000(46^2 - V_2^2)}{2} \qquad V_2 = 23 \text{ m/s} \qquad (6.43)$$

$$3.24 \times 1\ 000 = 0.094 \times 1\ 000(46 \times 0.866 - 23 \sin \alpha_2) \qquad \alpha_2 = 13.5° \quad (6.44)$$

From the geometry of the velocity triangles,

$$\beta_1 = 54° \qquad \beta_2 = 128.3° \blacktriangleleft$$

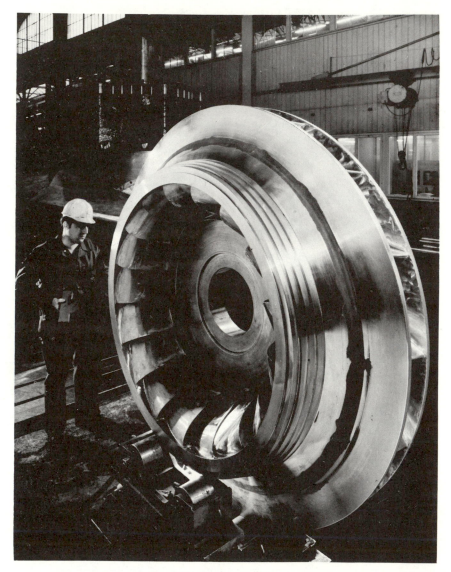

Fig. **6.21**   Reaction turbine runner—58 MW high pressure Francis turbine for the Ros-shag power station of the Oesteneichische Tauemkroftwerke AG (operates under 672 m head). (Courtesy of Escher Wyss Limited, Zurich, Switzerland.)

6.12   Reaction Turbine and Centrifugal Pump

One of the most important engineering applications of the impulse-momentum principle is the design of turbines, pumps, and other turbomachines such as fluid drives and torque converters, all of which involve very complex three-dimensional flowfields beyond the scope of this book. However, the principle may be applied to an assumed two-dimensional flowfield in a reaction turbine and centrifugal pump (Figs 6.21 and 6.22) to demonstrate the method and to gain some basic understanding of the design and operation of such machines.

Figure 6.23 is a definition sketch for idealizations of both of these machines, which feature a moving blade system symmetrical about the axis of rotation. This blade system is the essential part of the rotating element of the machine, called the *runner* for the turbine and the *impeller* for the pump. As fluid flows through these blade systems, the tangential component of the absolute velocity changes, decreasing through the runner of the turbine, increasing through the impeller of the pump. In constructing control surfaces (concentric circles) around the blade systems, section 1 becomes the inlet and section 2 the outlet of the control volume for both machines. The impulse-momentum principle may now be applied. But, while the other analyses above dealt

Open impeller   Volute case

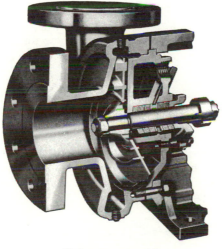

Cutaway assembly

Fig. **6.22** Centrifugal pumps. (Courtesy of Peerless Pump Division, FMC Corporation.)

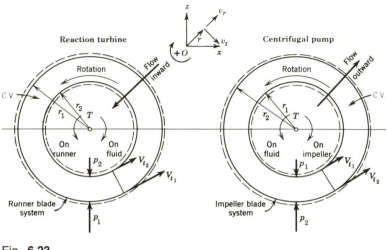

Fig. **6.23**

with linear momentum, now the moment of momentum is the key to the analysis. By resolving the velocities (**V**) into tangential and radial components, $\oint_{c.s.}(rv_t)\ d\dot{m}$ may be easily computed because the radial components of velocity pass through the center of moments; the result (for both machines) is

$$\oint_{C.S.} (rv_t)\ d\dot{m} = (-V_{t_2}r_2 + V_{t_1}r_1)Q\rho \tag{6.45}$$

The forces exerted on the fluid within the control volume are produced by the pressures on sections 1 and 2, and by the blades of the moving blade system; as the former are wholly radial, they can have no moment about O. The radial components of the latter also cancel for the same reason, leaving only the tangential components exerted by blade system on fluid to be considered. Although these forces are not identified as to magnitude and location, their total effect is a *torque, T, by the blade system on the fluid*. Applying equation 6.7a,

$$T = -[V_{t_2}r_2 - V_{t_1}r_1]Q\rho \tag{6.46}$$

From the relative magnitude of $V_{t_1}r_1$ and $V_{t_2}r_2$ it is evident that T (*on the fluid*) is clockwise ($T > 0$) for the turbine and counterclockwise ($T < 0$) for the pump; the torques *on runner and impeller* will necessarily be in the opposite directions.

The change of unit energy, E, associated with either of these machines may be calculated from equation 6.46 by recalling from mechanics that $P = T\omega$ and from equation 4.7 that $P = Q\gamma E$. The result is

$$E = \frac{\omega}{g_n}[V_{t_1}r_1 - V_{t_2}r_2] \tag{6.47}$$

which gives some insight into the relation between the internal dynamics of the machine

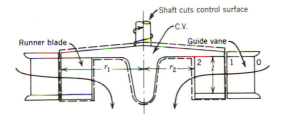

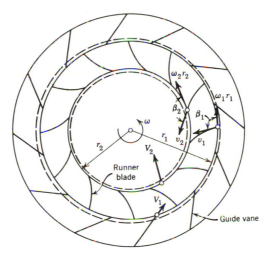

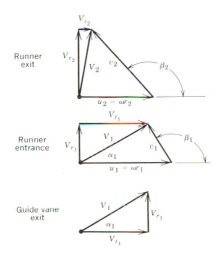

Fig. **6.24**   Francis-type reaction turbine schematic

and the head, E_P or E_T, which was emphasized in Section 4.5. For a pump, where $V_{t_2}r_2 > V_{t_1}r_1$, equation 6.47 is written

$$E_P = \frac{\omega}{g_n}[V_{t_2}r_2 - V_{t_1}r_1] \tag{6.48}$$

but for the turbine, where $V_{t_2}r_2 < V_{t_1}r_1$,

$$E_T = \frac{\omega}{g_n}[V_{t_1}r_1 - V_{t_2}r_2] \tag{6.49}$$

The alternative to this adjustment in the equation is to set up a special sign convention; this is convenient in more advanced analysis but hardly necessary here.

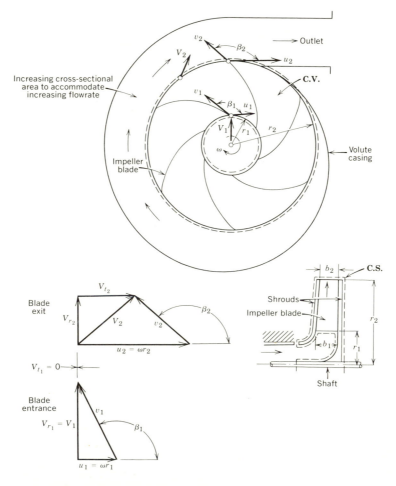

Fig. **6.25** Centrifugal pump schematic

An idealized two-dimensional reaction turbine is shown in Fig. 6.24 with accompanying velocity diagrams. Fixed guide vanes exert a torque on the fluid which gives it a tangential velocity component; because these guide vanes do not move, this torque does no work. The fluid then passes through the moving runner, through which the tangential component of velocity decreases, producing a torque according to equation 6.46. As in Section 6.11, the blades of the runner must be designed for smooth flow to accomplish this. Blade angles are determined from the velocity diagrams as before, the size of such diagrams being dictated by the equation of continuity ($q = 2\pi r_1 V_{r_1} = 2\pi r_2 V_{r_2}$), the angular speed of the runner through $u = \omega r$, and the guide vane angle, α. Important features of the velocity triangles are: (1) the increase of V_r through the runner and (2) the same V_1 at guide vane exit and runner entrance. The latter indicates no abrupt change in magnitude and direction of the velocity as it passes from guide vanes to moving runner; because this is imperative for good design, runner blades are shaped to accomplish it.

A section through an idealized two-dimensional centrifugal pump is shown in Fig. 6.25. With no guide vanes upstream from the impeller, the tangential component of absolute velocity is zero at the inlet and it increases through the impeller as torque is exerted on the fluid. This power is transferred from impeller to fluid and the energy of the fluid increases; both pressure and kinetic energies of the fluid increase through the impeller. As in the turbine, blade angles required to produce specified heads and flowrates for specified rotational speeds may be deduced from the velocity triangles. After the fluid leaves the impeller, it enters the pump (volute) casing where the flow is decelerated, converting the high kinetic energy of the fluid into pressure head, and the fluid is channeled (with as few losses as possible) to the outlet pipe.

Illustrative Problems

A two-dimensional reaction turbine has $r_1 = 5$ ft, $r_2 = 3.5$ ft, $\beta_1 = 60°$, $\beta_2 = 150°$, and thickness of 1 ft parallel to the axis of rotation. With a guide vane angle of 15° and a flowrate of 333 cfs of water, calculate the required speed of the runner for smooth flow at the inlet. For this condition calculate also the torque exerted on the runner, the power developed, the energy extracted from each pound of fluid, and the pressure drop through the runner.

Relevant Equations and Given Data

$$Q = AV = 2\pi r l V_r \tag{3.12}$$

$$\frac{p_1}{\gamma} + \frac{V_1^2}{2g_n} + z_1 = \frac{p_2}{\gamma} + \frac{V_2^2}{2g_n} + z_2 + E_T \tag{4.5}$$

$$P = \frac{Q\gamma E_T}{550} \tag{4.6}$$

$$T = -[V_{t_2}r_2 - V_{t_1}r_1]Q\rho \tag{6.46}$$

$$r_1 = 5 \text{ ft} \qquad r_2 = 3.5 \text{ ft} \qquad \beta_1 = 60° \qquad \beta_2 = 150°$$

$$l = 1 \text{ ft} \qquad \alpha_1 = 15° \qquad Q = 333 \text{ ft}^3/s \qquad \gamma = 62.4 \text{ lb/ft}^3$$

$$\rho = 1.936 \text{ slug/ft}^3$$

Solution.

$$V_{r_1} = \frac{333}{2\pi(5)} = 10.6 \text{ ft/s} \tag{3.12}$$

$$V_{r_2} = \frac{333}{2\pi(3.5)} = 15.15 \text{ ft/s} \tag{3.12}$$

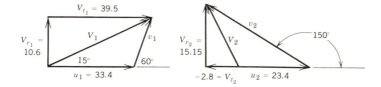

From the velocity triangles,

$$V_{t_1} = 10.6 \cot 15° = 39.5 \text{ ft/s}$$

$$u_1 = \omega r_1 = 39.5 - 10.6 \tan 30° = 33.4 \text{ ft/s}$$

$$\omega = \frac{33.4}{5} = 6.68 \text{ rad/s} = 63.8 \text{ r/min}$$

$$u_2 = \omega r_2 = 6.68 \times 3.5 = 23.4 \text{ ft/s}$$

$$V_{t_2} = 23.4 - 15.15 \cot 30° = -2.8 \text{ ft/s}$$

$$T = -[(-2.8)3.5 - 39.5 \times 5]333 \times 1.936 = 133\,300 \text{ ft} \cdot \text{lb} \blacktriangleleft \tag{6.46}$$

$$\text{Horsepower developed} = \frac{133\,300 \times 6.68}{550} = 1\,620 \text{ hp} \blacktriangleleft$$

$$\frac{333 \times 62.4 \times E_T}{550} = 1\,620 \qquad E_T = 42.9 \text{ ft} \cdot \text{lb/lb} \blacktriangleleft \tag{4.6}$$

From the diagrams, $V_1 = 40.8$ ft/s, $V_2 = 15.4$ ft/s, and, applying the Bernoulli equation between sections 1 and 2,

$$\frac{p_1}{\gamma} + \frac{(40.8)^2}{2g_n} = \frac{p_2}{\gamma} + \frac{(15.4)^2}{2g_n} + 42.9 \qquad p_1 - p_2 = 9.0 \text{ psi} \blacktriangleleft \tag{4.5}$$

A two-dimensional centrifugal pump impeller has $r_1 = 0.3$ m, $r_2 = 1$ m, $\beta_1 = 120°$, $\beta_2 = 135°$, and thickness of 0.1 m parallel to the axis of rotation. If it delivers 2 m³/s with no tangential velocity component at the entrance, what is its rotational speed? For this condition, calculate torque and power of the machine, the energy given to each newton of water, and the pressure rise through the impeller.

Relevant Equations and Given Data

See the above problem for equations 3.12, 4.5, and 6.46.

$$P = Q\gamma E_P \qquad (4.7)$$

$$E_P = \frac{\omega}{g_n}[V_{t_2}r_2 - V_{t_1}r_1] \qquad (6.48)$$

$r_1 = 0.3$ m $\qquad r_2 = 1$ m $\quad \beta_1 = 120°$ $\qquad \beta_2 = 135°$ $\qquad b_1 = b_2 = 0.1$ m

$Q = 2$ m³/s $\quad \gamma = 9.8$ kN/m³ $\quad \rho = 1\,000$ kg/m³

Solution.

$$V_1 = V_{r_1} = \frac{2}{(2\pi \times 0.3 \times 0.1)} = 10.6 \text{ m/s} \qquad (3.12)$$

$$V_{r_2} = 3.18 \text{ m/s}$$

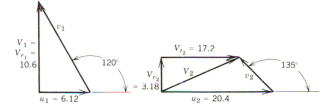

From the velocity triangles,

$$u_1 = \omega r_1 = 10.6 \tan 30° = 6.12 \text{ m/s}$$

$$\omega = \frac{6.12}{0.3} = 20.4 \text{ rad/s} = 195 \text{ r/min}$$

$$u_2 = \omega r_2 = 20.4 \times 1 = 20.4 \text{ m/s}$$

$$V_{t_2} = 20.4 - \frac{3.18}{\tan 45°} = 17.2 \text{ m/s}$$

$$T = -(17.2 \times 1 - 0) \times 2 \times 1\,000 = -34.4 \text{ kN·m} \blacktriangleleft \qquad (6.46)$$

$$E_P = \frac{\omega}{g_n}[V_{t_2}r_2 - V_{t_1}r_1] = \frac{-T\omega}{Q\gamma} = \frac{34.4 \times 10^3 \times 20.4}{2 \times 9.8 \times 10^3} = 35.8 \text{ J/N} = 3.58 \text{ m}$$
$$(6.48)$$

$$P = Q\gamma E_P = 2 \times 9.8 \times 10^3 \times 35.8 = 700 \text{ kN·m/s} = 700 \text{ kW} \blacktriangleleft \qquad (4.7)$$

$$\frac{p_1}{\gamma} + \frac{(10.6)^2}{2g_n} + 35.8 = \frac{p_2}{\gamma} + \frac{(17.5)^2}{2g_n} \qquad p_2 - p_1 = 254 \text{ kPa} \blacktriangleleft \qquad (4.5)$$

6.13 Rocket Propulsion

In the flow machines described so far, the energy transfers occurred to or from a stream of fluid moving through the machine. Thus, in the jet engine, a small amount of fuel, added to the stream and ignited, produces a large thrust. A rocket does not depend on the use of an external fluid stream, but carries a mass of fuel (or *propellant*) that is burned and exhausted at high speed to create thrust.

Normally, a rocket starts from rest or a uniform motion state and, while the engine is on and burning fuel, accelerates continuously. As the fuel mass is exhausted from the engine, the mass of the rocket decreases. The process and motion are unsteady and not amenable to treatment by the steady-state, impulse-momentum equations derived earlier. However, a very fundamental and useful analysis can be made by returning to the system concept.

Consider the rocket shown at time t in Fig. 6.26. It has a mass m and velocity v. For simplicity let the rocket be in the vacuum of deep space where air resistance and gravity attraction forces are negligible. There are, therefore, no external forces on the rocket. Suppose that the rocket engine is running in the interval t to $t + dt$ and exhausts a mass $(-dm)$ with an effective velocity[19] v_{ex} relative to the main rocket (as $m > 0$ and decreases with time, $dm < 0$).

Because there are no external forces acting on the system composed of the rocket frame, payload, and fuel (burned and unburned), conservation of momentum must hold for the system. Thus,

$$\sum (mv)_t = \sum (mv)_{t+dt} \tag{6.50}$$

$$mv = (m + dm)(v + dv) + (-dm)(v - v_{ex})$$

$$= mv + m\, dv + dm\, v_{ex} + dm\, dv$$

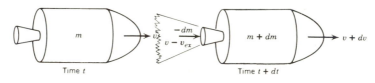

Fig. **6.26**

---

[19] Generally, to obtain the greatest exhaust velocity, rockets employ a convergent-divergent (De Laval) nozzle (Section 5.9). Thus, the exhaust pressure at the nozzle end is not always equal to the ambient or back pressure $p_a(p_a = 0$ in space). This results in a net pressure force (due to the pressure differential) that acts on the rocket over the nozzle outlet area and, accordingly, the exhaust jet must either expand or contract to reach equilibrium with the surroundings. Commonly, v_{ex} is defined, to account for this, as the effective (equilibrium) exhaust velocity

$$v_{ex} = v_o + \frac{(p_o - p_a)A_o}{\dot{m}}$$

where v_o, p_o, A_o are, respectively, the nozzle outlet velocity, pressure, and area. The term $(p_o - p_a)A_o/\dot{m}$ accounts for the effect of the pressure force on the nozzle.

Canceling mv and neglecting the higher order term $dm\, dv$ yield

$$v_{ex}\, dm = -m\, dv \tag{6.51}$$

or

$$a = \frac{dv}{dt} = -v_{ex}\frac{1}{m}\frac{dm}{dt} \tag{6.52}$$

This relation can be integrated from an initial state $(v_i,\, m_i)$ to obtain

$$\Delta v = v - v_i = v_{ex}\,\ln\frac{m_i}{m}$$

The greatest Δv obtainable is

$$\Delta v = v_{ex}\,\ln\frac{m_i}{m_{emp}} \tag{6.53}$$

where m_i is the initial mass of the rocket frame, fuel, and payload and m_{emp} is the mass of the empty frame and payload alone.

Because the main rocket has a mass m at time t, the thrust Th exerted by the engine is easily deduced to be

$$Th = ma = -v_{ex}\frac{dm}{dt} = v_{ex}\dot{m} \tag{6.54}$$

Here $\dot{m} = -dm/dt$ is the mass flowrate from the engine. Notice that the thrust is dependent, not on the rocket speed, but only on the effective *relative* speed of the exhaust and on the mass flowrate.

References

Rocket and Jet Propulsion

Hill, P. G., and Peterson, C. R. 1965. *Mechanics and thermodynamics of propulsion.* New York: Addison-Wesley.

Lancaster, O. E., Ed. 1959. *Jet propulsion engines.* Vol. XII in *High speed aerodynamics and jet propulsion.* Princeton: Princeton Univ. Press.

Loh, W. H. T. 1968. *Jet, rocket, nuclear, ion and electric propulsion: theory and design.* New York: Springer-Verlag.

Smith, G. G. 1950. *Gas turbines and jet propulsion for aircraft.* 5th ed. London: Illiffe.

Sutton, G. P., and Ross, D. M. 1976. *Rocket propulsion elements.* 4th ed. New York: Wiley

Zucrow, M. J. 1958. *Aircraft and missile propulsion.* New York: Wiley

Propellers and Windmills

Eldridge, F. R. 1980. *Wind machines.* 2nd ed. New York: Van Nostrand Reinhold Co.

Glauert, H. 1947. *Elements of airfoil and airscrew theory.* 2nd ed. New York: Macmillan.

Meacock, F. T. 1947. *Elements of aircraft propeller design*. London: E. and F. N. Spon.
Simmons, D. M. 1975. *Wind power*. New Jersey: Noyes Data Corp.
Taylor, D. W. 1933. *The speed and power of ships*. Rev. ed. Washington, D.C.: Ramsdell.
Theodorsen, T. 1948. *Theory of propellers*. New York: McGraw-Hill.
Torrey, V. 1976. *Wind-catchers*. Brattleboro, Vt.: S. Greene Press.

Turbines and Pumps

Benaroya, A. 1978. *Centrifugal pumps*. Tulsa, Okla.: Petroleum Pub. Co.
Dixon, S. L. 1978. *Fluid mechanics, thermodynamics of turbomachinery*. 3rd ed. New York: Pergamon Press.
Hawthorne, W. R., Ed. 1964. *Aerodynamics of turbines and compressors*. Vol. X of *High speed aerodynamics and jet propulsion*. Princeton: Princeton Univ. Press.
Hicks, T. G., and Edwards, T. W. 1971. *Pump application engineering*. New York: McGraw-Hill.
Hydraulic Institute. 1965. *Standards of the hydraulic institute*. 11th ed. New York.
Karassik, I. J., Krutzsch, W. C., Fraser, W. H., and Messina, J. P., Eds. 1976. *Pump Handbook*. New York: McGraw-Hill.
Kovalev, N. N. 1965. *Hydroturbines*. Washington, D.C.: Office of Technical Services, U.S. Dept. of Commerce.
Mosonyi, E. 1960, 1963. *Water power development*. Vols. I and II. Budapest: Akadémiai Kiado.
Shepherd, D. G. 1956. *Principles of turbomachinery*. New York: Macmillian.

Problems

6.1.   Derive equation 6.6.

6.2.   Derive equations 6.7 from first principles.

6.3.   A horizontal 150 mm pipe, in which 62 l/s of water are flowing, contracts to a 75 mm diameter. If the pressure in the 150 mm pipe is 275 kPa, calculate the magnitude and direction of the horizontal force exerted on the contraction.

6.4.   A horizontal 50 mm pipe, in which 1 820 l/min of water are flowing, enlarges to a 100 mm diameter. If the pressure in the smaller pipe is 138 kPa, caclulate magnitude and direction of the horizontal force on the enlargement.

6.5.   A conical enlargement in a vertical pipeline is 5 ft long and enlarges the pipe from 12 in. to 24 in. diameter. Calculate the magnitude and direction of the vertical force on this enlargement when 10 cfs of water flow upward through the line and the pressure at the smaller end of the enlargement is 30 psi.

6.6.   A conical diverging tube is horizontal, 0.3 m long, has 75 mm throat diameter, 100 mm exit diameter, and discharges 28.3 l/s of water into the atmosphere. Calculate the magnitude and direction of the force components exerted by the water on the tube.

6.7.   A 100 mm nozzle is bolted (with 6 bolts) to the flange of a 300 mm horizontal pipeline and discharges water into the atmosphere. Calculate the tension load on each bolt when the pressure in the pipe is 600 kPa. Neglect vertical forces.

6.8.   For the nozzle of problem 6.7, what flowrate will produce a tension force of 7 kN in each bolt?

6.9. Calculate the force exerted by the water on this orifice plate. Assume that water in the jet between orifice plate and vena contracta weighs 4.0 lb.

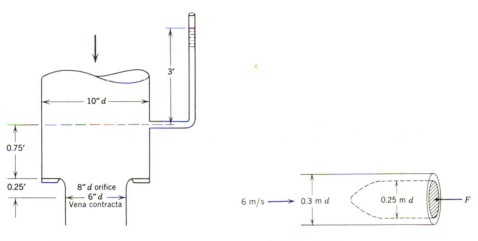

Problem **6.9**   Problem **6.10**

6.10. The projectile partially fills the end of the 0.3 m pipe. Calculate the force required to hold the projectile in position when the mean velocity in the pipe is 6 m/s.

6.11. This "needle nozzle" discharges a free jet of water at a velocity of 30 m/s. The tension force in the stem is measured experimentally and found to be 4 448 N. Predict the horizontal force on the bolts.

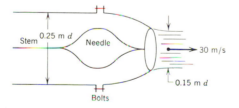

Problem **6.11**

6.12. A 90° bend occurs in a 0.3 m horizontal pipe in which the pressure is 276 kPa. Calculate the magnitude and direction of the horizontal force on the bend when 0.28 m³/s of water flow therein.

6.13. A 6 in. horizontal pipeline bends through 90° and while bending changes its diameter to 3 in. The pressure in the 6 in. pipe is 30 psi. Calculate the magnitude and direction of the horizontal force on the bend when 2.0 cfs of water flow therein. Both pipes are in the same horizontal plane.

6.14. A 100 mm by 50 mm 180° pipe bend lies in a horizontal plane. Find the horizontal force of the water on the bend when the pressures in the 100 mm and 50 mm pipes are 105 kPa and 35 kPa, respectively.

6.15. Calculate the force on the bolts. Water is flowing.

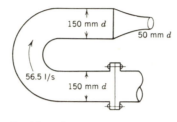

150 mm d

50 mm d

56.5 l/s

150 mm d

Problem **6.15**

6.16. For the configuration of problem 6.15, calculate the torque about the pipe's centerline in the plane of the bolted flange that is caused by the flow through the nozzle. The nozzle centerline is 0.3 m above the flange centerline. What is the effect of this torque on the force on the bolts calculated in problem 6.15? Neglect the effects of the weights of the pipe and the fluid in the pipe.

6.17. The axes of the pipes are in a vertical plane. The flowrate is 2.83 m³/s of water. Calculate the magnitude, direction, and location of the resultant force of the water on the pipe bend.

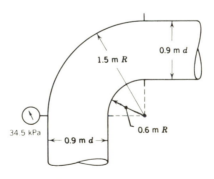

0.9 m d

1.5 m R

34.5 kPa

0.9 m d

0.6 m R

Problem **6.17**

6.18. Water flows through a tee in a horizontal pipe system. The velocity in the stem of the tee is 15 ft/s, and the diameter is 12 in. Each branch is of 6 in. diameter. If the pressure in the stem is 20 psi, calculate magnitude and direction of the force of the water on the tee if the flow-rates in the branches are the same.

6.19. Two types of gasoline are blended by passing them through a horizontal "wye" as shown. Calculate the magnitude and direction of the force exerted on the "wye" by the gasoline. The pressure $p_3 = 145$ kPa.

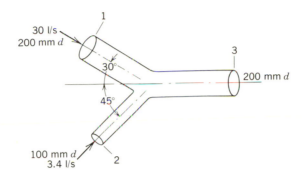

30 l/s
200 mm d

1

30°

45°

3

200 mm d

100 mm d
3.4 l/s

2

Problem **6.19**

6.20. If the two pipes from the reservoirs of problem 4.39 join through the (unequally-sized) branches of a horizontal tee and the discharge pipe leaves along the stem of the tee, find the magnitude and direction of the force of the water on the tee. The tee is at an elevation of 25 m (82 ft).

6.21. A fire truck is equipped with a 20 m (66 ft) long extension ladder which is attached at a pivot and raised to an angle of 45°. A 100 mm (4 in.) diameter fire hose is laid up the ladder and a 50 mm (2in.) diameter nozzle is attached to the top of the ladder so that the nozzle directs the stream horizontally into the window of a burning building. If the flowrate is 30 l/s (1 ft³/s), compute the torque exerted about the ladder pivot point. The ladder, hose, and the water in the hose weigh about 150 N/m (10 lb/ft).

6.22. Calculate the torque exerted on the flange joint by the fluid flow as a function of the pump flowrate. Neglect the weight of the 100 mm diameter pipe and the fluid in the pipe.

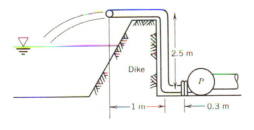

2.5 m

Dike

P

1 m

0.3 m

Problem **6.22**

6.23. A nozzle of 50 mm tip diameter discharges 0.018 7 m³/s of water vertically upward. Calculate the volume of water in the jet between the nozzle tip and a section 3.6 m above this point.

6.24. The block weighs 1 lb and is held up by the water jet issuing from the nozzle. Calculate H for a flowrate of 0.054 54 cfs. Ignore the small quantity of water above plane AA.

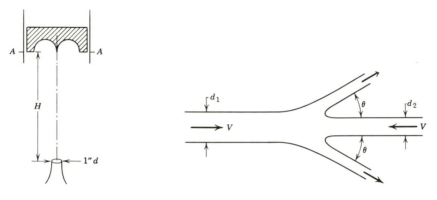

Problem **6.24**   Problem **6.25**

6.25. When round jets of the same velocity meet head on this flow picture results. Derive (ignoring gravity effects) an expression for θ in terms of d_1 and d_2.

6.26. The lower tank weighs 224 N, and the water in it weighs 897 N. If this tank is on a platform scale, what weight will register on the scale beam?

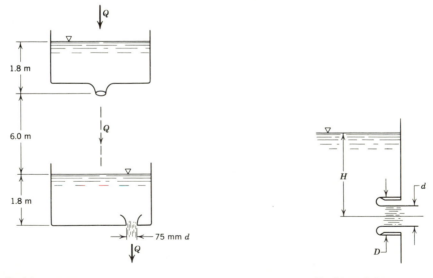

Problem **6.26**   Problem **6.27**

6.27. A free jet issues form this "Borda orifice" into the atmosphere. Calculate d/D.

6.28. The pressure difference results from head loss caused by eddies downstream from the orifice plate. Wall friction is negligible. Calculate the force exerted by the water on the orifice plate. The flowrate is 7.86 cfs.

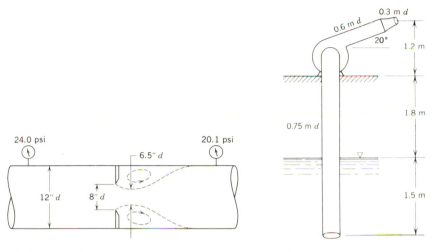

Problem **6.28**   Problem **6.29**

6.29. The pump, suction pipe, discharge pipe, and nozzle are all welded together as a single unit. Calculate the horizontal component of force (magnitude and direction) exerted by the water on the unit when the pump is developing a head of 22.5 m.

6.30. When the pump is started, strain gages at A and B indicate longitudinal tension forces in the pipes of 23 and 100 lb, respectively. Assuming a frictionless system, calculate flowrate and pump horsepower.

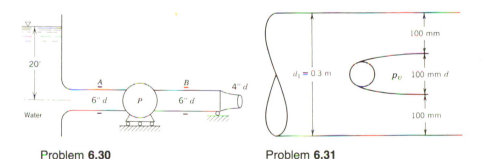

Problem **6.30**   Problem **6.31**

6.31. A sphere of 50 mm diameter placed on the centerline of a water (40°C) tunnel produces a vapor cavity as shown. The cavity occurs in such high speed flows because the pressure behind the body is reduced to the vapor pressure. Calculate the force exerted by water and vapor on the sphere if the velocity at section 1 is 30 m/s.

6.32. Calculate the drag on the streamlined axisymmetric body. The velocity defect downstream of the body marks the *wake* caused by friction effects on the body.

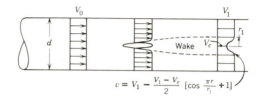

$$v = V_1 - \frac{V_1 - V_c}{2} \left[\cos \frac{\pi r}{r_1} + 1 \right]$$

Problem **6.32**

6.33. Referring to the sketch of problem 4.85, prove that the thickness (measured *vertically*) of the falling sheet of water is constant for all locations far from the weir crest.

6.34. The flowrate passing over this sharp-crested weir in a channel of 1 ft width is 3.5 cfs. Calculate the magnitude and direction of the force exerted by the water on the weir plate.

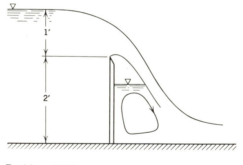

Problem **6.34**

6.35. The passage is 1.2 m wide normal to the paper. What will be the horizontal component of force exerted by the water on the structure?

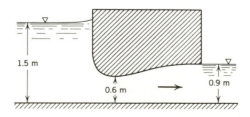

Problem **6.35**

6.36. If the two-dimensional flowrate through this sluice gate is 50 cfs/ft, calculate the horizontal and vertical components of force on the gate, neglecting wall friction.

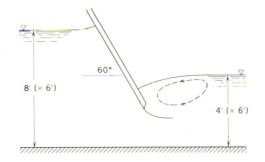

Problem **6.36**

6.37. Calculate the magnitude and direction of the horizontal component of force exerted by the flowing water on this (hatched) outflow structure. Assume velocity distribution uniform where streamlines are straight and parallel.

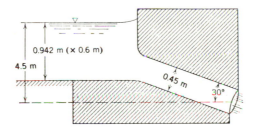

Problem **6.37**

6.38. Calculate the horizontal component of force exerted by the water on this "submerged sluice gate." The pressure distribution at section 2 *may be assumed hydrostatic.* Between sections 2 and 3 head losses are large because of diffusion and roller. All wall and bottom friction may be neglected. Consider the flow field two-dimensional.

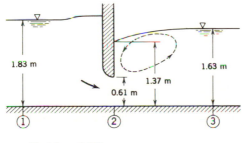

Problem **6.38**

6.39. Flow from the end of a two-dimensional open channel is deflected vertically downward by the gate *AB*. Calculate the force exerted by the water on the gate. At (and downstream from) *B* the flow may be considered a free jet.

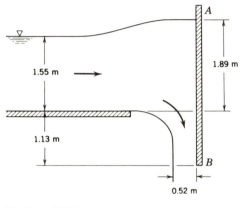

Problem **6.39**

6.40. Calculate the magnitude and direction of the horizontal force exerted by the water on the frictionless "drop structure" *AB*. Assume the structure to be 1 ft wide normal to the paper.

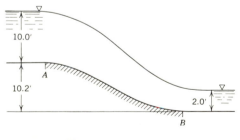

Problem **6.40**

6.41. Calculate the magnitude and direction of the horizontal and vertical components of resultant force exerted by the flowing water on the "flip bucket" *AB*. Assume that the water between sections *A* and *B* weighs 2.69 kN and that downstream from *B* the moving fluid may be considered to be a free jet.

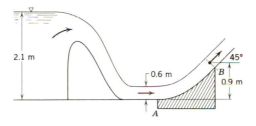

2.1 m

0.6 m

45°

B

0.9 m

A

Problem **6.41**

6.42. Calculate F_x exerted by the water on the block which has been placed at the end of this horizontal open channel. The channel and block are 4 ft wide normal to the paper.

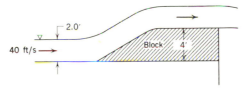

2.0′

40 ft/s

Block   4′

Problem **6.42**

6.43. This two-dimensional overflow structure (hatched) at the end of an open channel produces a free jet as shown. The water depth in the channel is 1.5 m and the thickness of the jet at the top of its trajectory is 0.6 m. Predict the horizontal component of force by water on structure if the structure is 3 m wide normal to the plane of the paper.

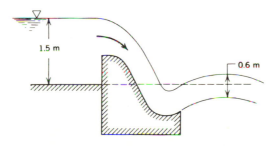

1.5 m

0.6 m

Problem **6.43**

6.44. In sketch (a) the "cylinder gate" is closed. In sketch (b) the gate has been raised 2 ft and the upstream water depth increased 2 ft. Calculate the magnitudes of the horizontal force components by water on gate. Compare the magnitudes of the vertical components. Which of these latter will be the larger? Why? Will both resultant forces pass through the center of the gate? Why or why not?

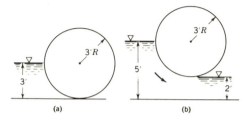

(a)   (b)

Problem **6.44**

6.45. Calculate the force, F, required to drive the scoop (hatched) at such velocity that the top of the jet centerline is 3 m above the channel bottom. Assume the scoop and channel 1.5 m wide normal to the paper and that the scoop extracts all of the water from the channel.

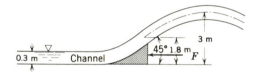

Problem **6.45**

6.46. Upstream from an axisymmetric abrupt enlargement in a horizontal passage the mean pressure is 140 kPa (20 psi) and the diameter is 150 mm (6 in.). Downstream from the enlargement the mean pressure is 210 kPa (30 psi) and the diameter is 300 mm (12 in.). Estimate the flowrate through the passage and the force on the enlargement.

6.47. Find the pressure change in and force on the abrupt contraction. $Q = 0.2$ m³/s, $A_1 = 0.1$ m², $A_2/A_1 = 0.4$, $p_1 = 200$ kPa, $\Delta(\mathbf{E.L.}) = 0.3\ V_2^2/2g_n$ (see Section 9.9), and water (5°C) is flowing.

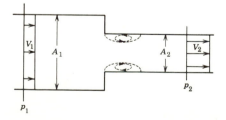

Problem **6.47**

6.48. Derive equations for the pressure change and energy loss in the abrupt enlargement. Compare the results with those of Section 6.4. The velocity profiles are parabolas, representing a laminar flow.

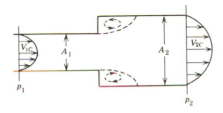

Problem **6.48**

6.49. At a point in a rectangular channel 6 m (20 ft) wide just downstream from a hydraulic jump, the depth is observed to be 3.6 m (12 ft) when the flowrate is 60 m³/s (2 000 ft³/s). What were the depth and velocity just upstream from the jump?

6.50. A hydraulic jump is observed in an open channel 3 m (10 ft) wide. The approximate depths (such depths are very difficult to measure accurately) upstream and downstream from the jump are 0.6 m (2 ft) and 1.5 m (5 ft), respectively. Estimate the flowrate in the channel.

6.51. This corrugated ramp is used as an energy dissipator in a two-dimensional open channel flow. For a flowrate of 5.4 m³/s·m calculate the head lost, the power dissipated, and the horizontal component of force exerted by the water on the ramp.

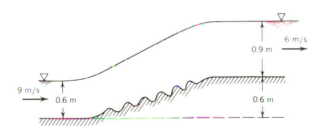

Problem **6.51**

6.52. In an open channel 1.5 m (5 ft) wide, the depth of water is 0.3 m (1 ft). Small standing waves (caused by imperfections in the sidewalls) are observed on the water surface at an angle of 30° with the walls. Estimate the flowrate in the channel.

6.53. Calculate the drop in the energy line (the head loss) produced by the oblique wave described in the Illustrative Problem of Section 6.6.

6.54. In an open rectangular channel 3 m (10 ft) wide the water depth is 0.3 m (1 ft). A standing wave 0.15 m (0.5 ft) high is produced by narrowing the channel with a straight diagonal wall. The angle between the wave and the original channel wall is observed to be 60°. Estimate the flowrate in the channel. What must be the angle of the diagonal wall to produce such a wave?

6.55. Derive equations 6.17, 6.18, and 6.19.

6.56. A normal shock wave exists in an airflow. The absolute pressure, velocity, and temperature just upstream from the wave are 207 kPa, 610 m/s, and $-17.8°C$, respectively. Calculate the pressure, velocity, temperature, and sonic velocity just downstream from the shock wave.

6.57. If through a normal shock wave (in air) the absolute pressure rises from 275 to 410 kPa (40 to 60 psia) and the velocity diminishes from 460 to 346 m/s (1 500 to 1 125 ft/s), what temperatures are to be expected upstream and downstream from the wave?

6.58. The stagnation temperature in an airflow is 149°C upstream and downstream from a normal shock wave. The absolute stagnation pressure downstream from the shock wave is 229.5 kPa. Through the wave the absolute pressure rises from 103.4 to 138 kPa. Determine the velocities upstream and downstream from the wave.

6.59. Assuming that air is a perfect gas, calculate the entropy change across the normal shock wave in the Illustrative Problem in Section 6.7.

6.60. Carry out the derivations suggested by footnote 11 (Section 6.8).

6.61. A wedge-shaped object in a supersonic wind tunnel is observed to produce an oblique shock wave of angle (β) 60°. The absolute pressure through the wave rises from 210 to 280 kPa (30 to 40 psia) and the temperature of the air upstream from the object is 40°C (100°F). Calculate the velocity of the air at this point.

6.62. The velocities and absolute pressures upstream and downstream from an oblique shock wave in an airflow are found to be 610 m/s, 138 kPa, and 457 m/s, 207 kPa, respectively. Calculate the Mach numbers upstream and downstream from the wave if the wave angle (β) is 30°.

6.63. An airplane flies at twice the speed of sound in the U.S. Standard Atmosphere (Appendix 4) at an altitude of 10 km (30 000 ft). If the leading edge of the wing may be approximated by a wedge of angle (2θ) 20°, what pressure rise through the oblique shock wave is to be expected?

6.64. Calculate the thrust and power of the jet engine of the Illustrative Problem, Section 6.9, as functions of air speed for the given nozzle conditions. Plot the results and find the point of maximum power delivery.

6.65. At high speeds the compressor and turbine of the jet engine may be eliminated entirely. The result is called a *ram-jet* (a subsonic configuration is shown). Here the incoming air is slowed and the pressure increases; the air is heated in the widest part by the burning of injected fuel. The heated air exhausts at high velocity from the converging nozzle. What nozzle area A_2 is needed to deliver a 90 kN thrust at an air speed of 270 m/s if the exhaust velocity is the sonic velocity for the heated air, which is at 1 000 K. Assume that the jet operates at an altitude of 12 km and neglect the fuel mass and pressure differentials.

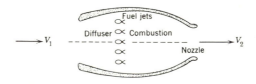

Problem **6.65**

6.66. A horizontal convergent-divergent nozzle is bolted to a water tank from which it discharges under a 3.6 m head. The throat and tip diameters of the nozzle are 125 mm and 150 mm, respectively. Determine the net horizontal force exerted on the tank-and-nozzle combination by the flowing fluid.

6.67. Calculate the force exerted on the tank and nozzle by the airflow in the Illustrative Problem of Section 5.7 if discharge is to the atmosphere.

6.68. Calculate the force exerted on the tank and nozzle in the Illustrative Problem of Section 5.9 for flow without the shock wave.

6.69. The pump maintains a pressure of 10 psi at the gage. Calculate the tension force in the cable.

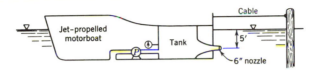

Problem **6.69**

6.70. A motorboat moves up a river at a speed of 9 m/s (30 ft/s) (relative to the land). The river flows at a velocity of 1.5 m/s (5 ft/s). The boat is powered by a jet-propulsion unit which takes in water at the bow and discharges it (beneath the surface) at the stern. Measurements in the jet show its velocity (relative to the boat) to be 18 m/s (60 ft/s). For a flowrate through the unit of 0.15 m³/s (5.3 ft³/s), calculate the propulsive force produced.

6.71. A head of 1.8 m is maintained on the nozzle. What is the propulsive force: (*a*) on the car, (*b*) on the blade, (*c*) on the tank and nozzle?

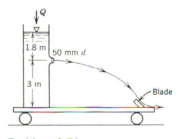

Problem **6.71**

6.72. A ship moves up a river at 32 km/h (20 mph) relative to the shore. The river current has a velocity of 8 km/h (5 mph). The velocity of the water a short distance behind the propellers is 64 km/h (40 mph) relative to the ship. If the velocity of 2.8 m³/s (100 cfs) of water is changed by the propeller, calculate its thrust.

6.73. An airplane flies at 200 km/h through still air of specific weight 12 N/m³. The propeller is 2.4 m in diameter, and its slipstream has a velocity of 290 km/h relative to the fuselage. Calculate: (*a*) the propeller efficiency, (*b*) the velocity through the plane of the propeller, (*c*) the power input, (*d*) the power output, (*e*) the thrust of the propeller, and (*f*) the pressure difference across the propeller disk.

6.74. A propeller must produce a thrust of 9 kN to drive an airplane at 280 km/h. An ideal propeller of what size must be provided if it is to operate at 90% efficiency? Assume air density 1.23 kg/m³.

6.75. A "flying platform" is supported by four downward-directed jets, one at each corner of the platform for stability. The platform is to support a gross load of 10 kN (2 250 lb). What flowrate of standard sea-level air must be delivered by each fan? How much power must be supplied by each motor? Consider the air to be frictionless and incompressible. Assume the platform to be stationary and well off the ground.

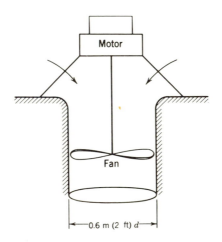

Problem **6.75**

6.76. This ducted propeller unit drives a ship through still water at a speed of 4.5 m/s. Within the duct the mean velocity of the water (relative to the unit) is 15 m/s. Calculate the propulsive force produced by the unit. Calculate the force exerted on the fluid by the propeller. Account for the difference between these forces.

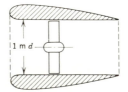

Problem **6.76**

6.77. Show that this ducted propeller system when moving forward at velocity V_1 will have an efficiency given by $2V_1/(V_4 + V_1)$. If for a specific design and point of operation, $V_2/V_1 = 9/4$ and $V_4/V_2 = 5/4$, what fraction of the propulsive force will be contributed: (*a*) by the propeller, and (*b*) by the duct?

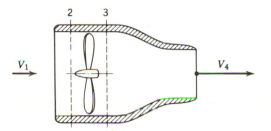

Problem **6.77**

6.78. This ducted propeller unit (now operating as a turbine) is towed through still water at a speed of 7.5 m/s. Calculate the maximum power that the propeller can develop. Neglect all friction effects.

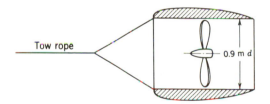

Problem **6.78**

6.79. A propeller type of turbine wheel is installed near the end of a horizontal pipeline discharging water to the atmosphere. Derive an expression for the power to be expected from the turbine in terms of the gage pressure upstream from the turbine disk, and the flowrate.

6.80. What is the maximum power that can be expected from a windmill 30 m (100 ft) in diameter in a wind of 50 km/h (30 mph)? Assume air density 1.225 kg/m³ (0.002 38 slug/ft³).

6.81. If an ideal windmill is operating at best efficiency in a wind of 48 km/h, what is the velocity through the disk and at some distance behind the windmill? What is the thrust on this windmill, assuming a diameter of 60 m and an air density of 1.23 kg/m³? What are the mean pressures just ahead of and directly behind the windmill disk?

6.82. The flowrate through a 3 m (10 ft) diameter windmill is 170 m³/s (6 000 ft³/s). The mean pressures just upstream and downstream from the windmill plane are 240 Pa (5 lb/ft²) and -190 Pa (-3.9 lb/ft²), respectively. Assuming the air density 1.3 kg/m³ (0.002 5 slug/ft³), calculate the wind velocity, the axial force on the windmill, and the power of the machine.

6.83. Transverse thrusters are used to make large ships fully maneuverable at low speeds without tugboat assistance. A transverse thruster consists of a propeller mounted in a duct; the unit is then mounted below the waterline in the bow or stern of the ship. The duct runs completely across the ship. Calculate the thrust developed by a 1 865 kW unit (supplied to the propeller) if the duct is 2.8 m in diameter and the ship is stationary.

6.84. A 16 l/s horizontal jet of water of 25 mm diameter strikes a stationary blade which deflects it 60° from its original direction. Calculate the vertical and horizontal components of force exerted by the liquid on the blade.

6.85. The jet shown strikes the semicylindrical vane, which is in the vertical plane. Calculate the horizontal component of force on the vane. Neglect all friction.

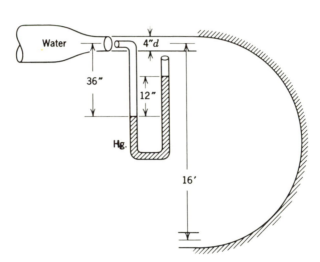

Problem **6.85**

6.86. Calculate the resultant force of the water on this plate. The flowrate is 42.5 l/s.

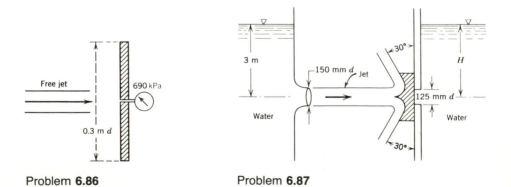

Problem **6.86**   Problem **6.87**

6.87. The plate covers the 125 mm diameter hole. What is the maximum H that can be maintained without leaking?

6.88. Calculate the magnitude and direction of the vertical and horizontal components and the total force exerted on this stationary blade by a 50 mm jet of water moving at 15 m/s.

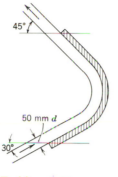

45°

50 mm d

30°

Problem **6.88**

6.89. A smooth nozzle of 50 mm (2 in.) tip diameter is connected to the bottom of a large water tank and discharges vertically downward. The water surface in the tank is 1.5 m (5 ft) above the tip of the nozzle. Three metres (10 ft) below the nozzle tip there is a horizontal disk of 100 mm (4 in.) diameter surface. Calculate the force exerted by the water jet on the disk.

6.90. A jet falls from a reservoir into a weightless dish that floats on the surface of a second reservoir. If the submerged volume of the dish is 0.12 m³, calculate the weight of the water within the dish (i.e., below the line AB).

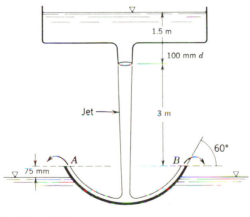

1.5 m

100 mm d

Jet →

3 m

60°

A

B

75 mm

Problem **6.90**

6.91. This water jet of 50 mm diameter moving at 30 m/s is divided in half by a "splitter" on the stationary flat plate. Calculate the magnitude and direction of the force on the plate. Assume that flow is in a horizontal plane.

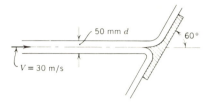

Problem **6.91**

6.92. If the splitter is removed from the plate of the preceding problem and sidewalls are provided on the plate to keep the flow two-dimensional, how will the jet divide after striking the plate?

6.93. Calculate the gage reading when the plate is pushed horizontally to the left at a constant speed of 25 ft/s. Also calculate the force and power required to push the plate at this speed.

Problem **6.93**   Problem **6.94**

6.94. The blade is one of a series. Calculate the force exerted by the jet on the blade system.

6.95. This blade is one of a series. What force is required to move the series horizontally against the direction of the jet at a velocity of 15 m/s (or 50 ft/s)? What power is required to accomplish this motion?

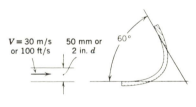

Problem **6.95**

6.96. From a water jet of 2 in. (50 mm) diameter, moving at 200 ft/s (60 m/s), 180 hp (135 kW) are to be transferred to a blade system (Fig. 6.17) which is moving in the direction of the jet at 50 ft/s (15 m/s). Calculate the required blade angle.

6.97. A series of blades (Fig. 6.17), moving in the same direction as a water jet of 1 in. (25 mm) diameter and of velocity 150 ft/s (46 m/s), deflects the jet 75° from its original direction.

What relation between blade velocity and blade angle must exist to satisfy this condition? What is the force on the blade system?

6.98. If a system of blades (Fig. 6.17) is free to move in a direction parallel to that of a jet, prove that the direction of the force on the blade system is the same for all blade velocities.

6.99. A crude impulse turbine has flat radial blades and is in effect a "paddle wheel." If the 25 mm (1 in.) diameter water jet has a velocity of 30 m/s (100 ft/s) and is tangent to the rim of the wheel, calculate the approximate power of the machine when the blade velocity is 12 m/s (40 ft/s).

6.100. For a flowrate of 12 l/s and turbine speed of 65 r/min, estimate the power transferred from jet to turbine wheel.

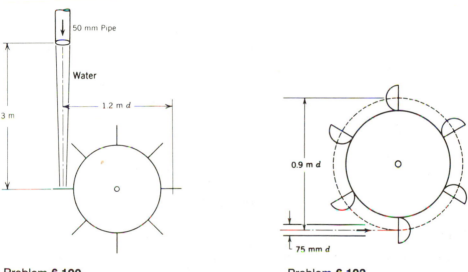

Problem **6.100**　　　　　　　　　　Problem **6.102**

6.101. A 150 mm pipeline equipped with a 50 mm nozzle supplies water to an impulse turbine 1.8 m in diameter having blade angles of 165°. Plot a curve of theoretical power versus r/min when the pressure behind the nozzle is 690 kPa. What is the force on the blades when the maximum power is being developed? Plot a curve of force on the blades versus r/min.

6.102. The velocity of the water jet driving this impulse turbine is 45 m/s. The jet has a 75 mm diameter. After leaving the buckets the (absolute) velocity of the water is observed to be 15 m/s in a direction 60° to that of the original jet. Calculate the mean tangential force exerted by jet on turbine wheel and the speed (r/min) of the wheel.

6.103. When an air jet of 1 in. diameter strikes a series of blades on a turbine rotor, the (absolute) velocities are as shown. If the air is assumed to have a constant specific weight of 0.08 lb/ft³, what is the force on the turbine rotor? How much horsepower is transferred to the rotor? What must be the velocity of the blade system?

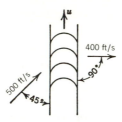

Problem **6.103**

6.104. This system of blades develops 149 kW under the influence of the jet shown. Calculate the blade velocity.

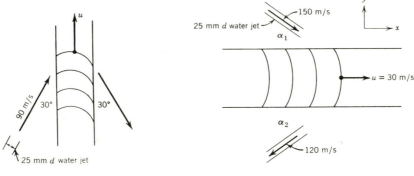

Problem **6.104**   Problem **6.105**

6.105. If $\alpha_1 = \alpha_2$, calculate: α, F_x on the blade system, and the power developed.

6.106. In passing through this blade system, the absolute jet velocity decreases from 136 to 73.8 ft/s (41.5 to 22.5 m/s). If the flowrate is 2.0 ft³/s (57 l/s) of water, calculate the power transferred to the blade system and the vertical force component exerted on the blade system.

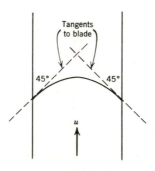

Problem **6.106**

6.107. If $u = v_1 = v_2 = 100$ ft/s (30 m/s) for the blade system of the preceding problem, calculate the absolute velocities of the jet entering and leaving the blade system. If the flowrate is 5 ft³/s (140 l/s), how much power may be expected from the machine?

6.108. The flowrate is 57 l/s (2.0 ft³/s) of water, $u = 9$ m/s (30 ft/s), and $v_1 = v_2 = 12$ m/s (40 ft/s). Calculate: the absolute velocity of the water entering and leaving the system, F_x and F_y (magnitude and direction) on the blade system, and the power transferred from jet to blade system.

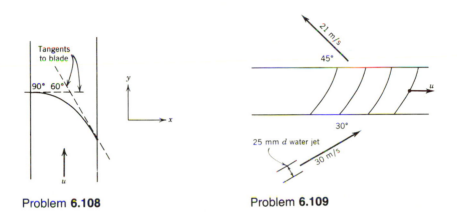

Problem **6.108**   Problem **6.109**

6.109. The (absolute) velocities and directions of the jets entering and leaving the blade system are as shown. Calculate the power transferred from jet to blade system and the blade angles required.

6.110. This stationary blade is pivoted at point O. Calculate the torque (about O) exerted thereon when a 2 in. water jet moving at 100 ft/s passes over it as shown.

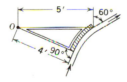

Problem **6.110**

6.111. Eight thousand cubic feet (225 m³) per second of water flow into a hydraulic turbine whose guide vanes, set at an angle of 15°, have an exit circle of 8 ft (2.4 m) radius and are 6 ft (1.8 m) high. In the draft tube the flow is observed to have no tangential component. What is the torque on the runner? If the runner is rotating at 150 r/min, how much power is being delivered to the turbine and how much energy is being extracted from each pound (newton) of water?

6.112. A simple reaction turbine has $r_1 = 0.9$ m, $r_2 = 0.6$ m, and its flow cross section is 0.3 m high. The guide vanes are set so that $\alpha_1 = 30°$. When 2.83 m³/s of water flow through this turbine the angle α_2 is found to be 60°. Calculate the torque exerted on the turbine runner. If the angle β_2 is 150°, calculate the speed of rotation of the turbine runner and the angle β_1 necessary for smooth flow into the runner. Calculate the power developed by the turbine at the above speed.

6.113. The nozzles are all of 25 mm diameter and each nozzle discharges 7 l/s of water. If the turbine rotates at 100 r/min, calculate the power developed.

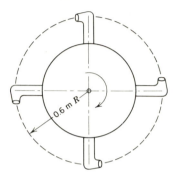

Problem **6.113**

6.114. A centrifugal pump impeller having r_1 = 50 mm, r_2 = 150 mm, and width b = 37.5 mm is to pump 225 l/s of water and supply 12.2 J of energy to each newton of fluid. The impeller rotates at 1 000 r/min. What blade angles are required? What power is required to drive this pump? Assume smooth flow at the inlet of the impeller.

6.115. A centrifugal pump impeller having dimensions and angles as shown rotates at 500 r/min. Assuming a radial direction of velocity at the blade entrance, calculate the flowrate, the pressure difference between inlet and outlet of blades, and the torque and power required to meet these conditions.

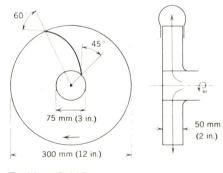

Problem **6.115**

6.116. If the impeller of the preceding problem rotates between horizontal planes of infinite extent and the flowrate is 25 l/s (0.86 ft³/s), what rise of pressure may be expected between one point having r = 150 mm (6 in.) and another having r = 225 mm (9 in.)?

6.117. At the outlet of a pump impeller of diameter 0.6 m and width 150 mm, the (absolute) velocity is observed to be 30 m/s at an angle of 60° with a radial line. Calculate the torque exerted on the impeller.

6.118. Derive an expression for the thrust against the supports of a stationary rocket if fuel is supplied at a mass rate $\dot{m}_f$ and the exit velocity is v_{ex}.

6.119. A vehicle in deep space has a mass of 1 000 kg (69 slug); if v_{ex} is 1 km/s (0.62 mi/s), how much fuel must be burned to increase the speed by 500 m/s (1 600 ft/s)?

6.120. A space vehicle is designed so that the fuel to payload ratio is 25 to 1. What is the ratio of the maximum vehicle velocity to the engine exhaust velocity if the vehicle is initially at rest in deep space (no external or body forces act on it)?

7

FLOW OF A REAL FLUID

One of the major objectives of Chapter 1 was to characterize the fluid state, and an important observed fluid property mentioned there was its viscosity which permits the development of shear stresses and resistance to flow. When the fluid is static (no flow), only the normal stress or pressure exists, and Chapter 2 focused on these cases of fluid statics. Of course, viscosity played no role. In Chapter 3 kinematics was studied, and the equation of continuity (the principle of conservation of mass) was derived. Although not excluded, viscous effects again played no role because kinematics is concerned with motions without regard for the forces which cause them.

The flow of an ideal fluid was considered in Chapters 4 (incompressible) and 5 (compressible). An ideal fluid was defined to be inviscid, that is, devoid of viscosity. Thus, there were no frictional effects between moving fluid layers or between the fluid and bounding walls. The observed *no-slip* condition for a real fluid at a solid wall did not apply, and the ideal fluid slipped freely along bounding walls. The Euler, Bernoulli, and work-energy equations as they were derived in Chapter 4 are strictly applicable only to ideal fluid flows. However, in Chapter 5 (Section 5.1), the analysis of the First Law of Thermodynamics led to an energy equation that included heat transfer and the internal energy of the fluid in addition to the terms in the work-energy equation. It will be seen later in this chapter that for many flows the energy equations 5.4 or 5.5 can be applied directly to the real fluid flows because, although shear work was neglected in their derivation, the frictional forces at fixed walls (e.g., in bounded flows, such as those in pipes, etc.) do no work. On the other hand, forces acting on a fluid do cause momentum changes even though no work is done. Therefore, a crucial set of forces, namely, shear forces at solid boundaries, was not included when the impulse-momentum principle was applied in Chapter 6 to derive equations 6.4, 6.5, and 6.7 for ideal fluid flows. This neglect is trivial to fix, however; for real fluids one merely includes the appropriate frictional forces on the left-hand side of these equations when the forces acting on the control volume are being summed in any problem solution.

In a real fluid then, viscosity introduces resistance to motion by causing shear or friction forces between fluid particles and between these and boundary walls. For flow to take place, work must be done against these resistance forces, and in the process energy is converted into heat. The inclusion of viscosity also allows the existence of two physically distinct flow regimes (as described in Section 1.5) and, in addition (by causing separation and secondary flows), it frequently produces flow situations entirely different from those of the ideal fluid. The effects of viscosity on the velocity profile also render invalid the assumption of a uniform velocity distribution. The derivation of the Euler equations (e.g., see Section 4.6 for a two-dimensional analysis) can be altered to include the shear stresses in a real fluid in addition to the normal stress or pressure

280

already included there. The result is a set of nonlinear, second-order partial differential equations, called the Navier-Stokes equations. Unfortunately, few useful analytic solutions to these equations have been found. Therefore, the engineer must resort to experimental results, semiempirical methods, and numerical simulations to solve problems. This requires a good basic understanding of a variety of physical phenomena, which are described in this chapter and which are the basic manifestations of fluid friction.

This chapter is divided into several functional sections. First, the concepts of laminar and turbulent flow and the conditions under which they occur are examined. Next, the influence of solid boundaries on the flow is outlined (more detail is provided in Chapter 9). Finally, with these concepts as a base, a range of physical situations which can occur are surveyed. For the survey the flows are collected into two useful classes: *internal* (fully bounded) and *external* flows. This chapter provides the key for what follows in Chapters 8, 9, 10, and 13.

7.1 Laminar and Turbulent Flow

In laminar flow, agitation of fluid particles is of a molecular nature only (and, hence, at a length scale of the order of the mean free path of the molecules). On the usual macroscopic scale of observation, these particles appear then to be constrained to motion in essentially parallel paths by the action of viscosity. The shearing stress between adjacent moving layers is determined in laminar flow by the viscosity and is completely defined by the differential equation (Section 1.5)

$$\tau = \mu \frac{dv}{dy} \qquad (1.12)$$

the stress being the product of viscosity and velocity gradient (Fig. 7.1). If the laminar flow is disturbed by wall roughness or some other obstacle, the disturbances are rapidly damped by viscous action, and downstream the flow is smooth again. A laminar flow is stable against such disturbances, but a turbulent flow is not. In turbulent flow, fluid particles do not remain in layers, but move in a heterogeneous fashion through the flow, sliding past other particles and colliding with some in an entirely haphazard manner that results in a rapid and continuous macroscopic mixing of the flowing fluid, with length scales which are very much greater than the molecular scales in laminar flow.

The random motion and the observed eddies in a turbulent flow suggest that both the inertia forces, associated with the accelerations during the motion, and the viscous

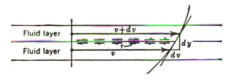

Fig. **7.1**

forces, induced by the action of the viscosity, may be important. When the viscous forces are dominant, the flow might be expected to be laminar. When the inertia forces are dominant, the flow might well be turbulent. These characteristics were demonstrated by Reynolds,[1] with an apparatus similar to that of Fig. 7.2. Water flows from a tank through a bell-mouthed glass pipe, the flow being controlled by the valve A. A thin tube, B, leading from a reservoir of dye, C, has its opening within the entrance of the glass pipe. With low velocities in the glass pipe, a thin filament of dye issuing from the tube did not diffuse but formed a thin (essentially) straight line parallel to the axis of the pipe. As the valve was opened and greater velocities were attained, the dye filament wavered and broke, eventually diffusing through the water flowing in the pipe. Reynolds found that the mean velocity at which the filament of dye began to break up (termed *critical velocity*) was dependent on the degree of quiescence of the water in the tank, higher critical velocities being obtainable with increased quiescence. He also discovered that if the dye filament had once diffused it became necessary to decrease the velocity in order to restore it, but that the restoration always occurred at approximately the same mean velocity in the pipe.

Since rapid mixing of fluid particles during flow would cause diffusion of the dye filament, Reynolds deduced that at low velocities this mixing was absent and that the fluid particles moved in parallel layers, or laminae, sliding past adjacent laminae but not mixing with them;[2] this is the regime of *laminar flow*. Since at higher velocities the dye filament diffused through the pipe, it was apparent that rapid intermingling of fluid particles was occurring and the flow was *turbulent*. Laminar flow broke down into turbulent flow at some critical velocity above that at which turbulent flow was restored to the laminar condition, the former velocity being an *upper critical velocity*, and the latter, a *lower critical velocity*.

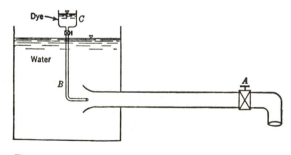

Fig. **7.2**

[1] O. Reynolds, "An Experimental Investigation of the Circumstances Which Determine Whether the Motion of Water Shall Be Direct or Sinuous and of the Law of Resistance in Parallel Channels," *Phil. Trans. Roy. Soc.* vol. 174, part III, p. 935, 1883. See also *Osborne Reynolds and Engineering Science Today,* Manchester University Press (Barnes & Noble, Inc.), 1970.
[2] Actually, molecular diffusion does occur, but it is too slow to have any noticeable effect in this experiment. See Section 1.5.

Other evidence of the existence of two flow regimes can be obtained from the simple experiment illustrated in Fig. 7.3. Here the fall of pressure between two points in a section of a long straight pipe is measured by the manometer reading h and correlated with the mean velocity V. For the small values of V a plot of h against V yields a straight line ($h \propto V$), but at higher values of V a nearly parabolic curve ($h \tilde{\propto} V^2$) results. Evidently the flow is laminar in the first case, but turbulent in the second. Between the two flow regimes lies an interesting *transition region;* as V is increased, data follows the line OABCD but with diminishing V follows $DCAO$. From these results and Reynolds' observations it may be deduced that points A and B define the lower and upper critical velocities, respectively.

Reynolds was able to generalize his conclusions from his dye stream experiments by the introduction of a dimensionless term **R**, later called the Reynolds number, which was defined by

$$\mathbf{R} = \frac{Vd\rho}{\mu} \quad \text{or} \quad \frac{Vd}{\nu} \tag{7.1}$$

in which V is the mean velocity in the pipe, d the pipe diameter, and ρ and μ the density and viscosity of the flowing fluid. Reynolds found that certain critical values of the Reynolds number, $\mathbf{R}_c$, defined the upper and lower critical velocities for all fluids flowing in all sizes of pipes, and thus deduced that single numbers define the limits of laminar and turbulent pipe flow *for all fluids.*

The upper limit of laminar flow is indefinite, being dependent on several incidental conditions such as: (1) initial quiescence of the fluid, (2) shape of pipe entrance, and (3) roughness of pipe, and these values are of little practical interest. The lower limit of turbulent flow, defined by the lower critical Reynolds number, is of greater engineering importance; it defines a condition below which all turbulence entering the flow from any source will eventually be damped out by viscosity. This lower critical Reynolds number thus sets a limit below which laminar flow will always occur; many experiments have indicated the lower critical Reynolds number to have a value of approximately 2 100.

The concept of a critical Reynolds number delineating the regimes of laminar and turbulent flow is indeed a useful one in promoting concise generalization of certain flow

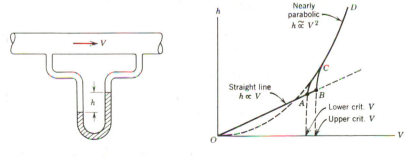

Fig. **7.3**

phenomena. Applying this concept to the flow of *any fluid in cylindrical pipes,* the engineer can predict that the flow will be laminar if $\mathbf{R} < 2\ 100$ and turbulent if $\mathbf{R} \gg 2\ 100$.[3] However, *critical Reynolds number is very much a function of boundary geometry.* For flow between parallel walls (using mean velocity V, and spacing d) $\mathbf{R}_c \cong 1\ 000$; for flow in a wide open channel (using mean velocity V and water depth d) $\mathbf{R}_c \cong 500$; for flow about a sphere (using approach velocity V and sphere diameter d) $\mathbf{R}_c \cong 1$. Such critical Reynolds numbers must be determined experimentally; because of the complex origins of turbulence, analytical methods for predicting critical Reynolds numbers have yet to be developed.

It is shown in Chapter 8 that $\mathbf{R}$, in fact, is the ratio of the inertia forces to the viscous forces in the flow. When $\mathbf{R}$ is small, viscous forces dominate. When $\mathbf{R}$ is large, inertia forces dominate. More important, from the form of the expression for $\mathbf{R}$ and the above discussion, it is clear that the laminar or turbulent regimes are not determined just by flow velocity. Laminar flows (small $\mathbf{R}$) are characterized by low velocities, small length scales (e.g., small diameter pipes), and fluids with high kinematic viscosity. Turbulent flows (large $\mathbf{R}$) are characterized by high velocities, large length scales, and fluids of low kinematic viscosity.

—————————————— **Illustrative Problem** ——————————————

Water of kinematic viscosity $1.15 \times 10^{-6}\,\mathrm{m^2/s}$ flows in a cylindrical pipe of 30 mm diameter. Calculate the largest flowrate for which laminar flow can be expected. What is the equivalent flowrate for air?

Relevant Equations and Given Data

$$Q = AV \tag{3.11}$$

$$\mathbf{R} = \frac{Vd}{\nu} \tag{7.1}$$

$$\nu_{water} = 1.15 \times 10^{-6}\ \mathrm{m^2/s} \qquad d = 0.03\ \mathrm{m}$$

Solution. Taking $\mathbf{R}_c = 2\ 100$ as the conservative upper limit for laminar flow,

$$2\ 100 = \frac{V(0.03)}{0.000\ 001\ 15} \qquad V = 0.080\ 5\ \mathrm{m/s} \tag{7.1}$$

$$Q_{water} = 0.080\ 5 \times \left(\frac{\pi}{4}\right)(0.03)^2 = 5.69 \times 10^{-5}\ \mathrm{m^3/s} \blacktriangleleft \tag{3.11}$$

$$\nu_{air} = 1.37 \times 10^{-5}\ \mathrm{m^2/s}$$

$$Q_{air} = 6.78 \times 10^{-4}\ \mathrm{m^3/s} \cong 12\ Q_{water} \blacktriangleleft \tag{7.1)(3.11}$$

———

[3] See Chapter 9. When $2\ 100 < \mathbf{R} < 4\ 000$, a transition occurs and the flow can be either laminar or turbulent.

7.2 Turbulent Flow and Eddy Viscosity

If the Reynolds number is high enough, virtually every type of flow will be turbulent. Turbulence is found in the atmosphere, in the ocean, in the flow about aircraft and missles, in most pipe flows, in estuaries and rivers, and in the wakes of moving vehicles. This turbulence is generated primarily by friction effects at solid boundaries or by the interaction of fluid streams that are moving past each other with different velocities. Despite its ubiquitous appearance, turbulence is difficult to define. Tennekes and Lumley[4] suggest that the best one can do is to list characteristics of turbulent flow. They list the following:

1. Irregularity or randomness in time and space.
2. Diffusivity or rapid mixing.
3. High Reynolds number.
4. Three-dimensional vorticity fluctuations (see Section 3.7).
5. Dissipation of the kinetic energy of the turbulence by viscous shear stresses.
6. Turbulence is a continuum phenomenon even at the smallest scales.
7. Turbulence is a feature of fluid flows, not a property of the fluids themselves.

The first three characteristics were mentioned earlier in the chapter, the rapid mixing being significant because it implies high rates (compared to laminar flow) of momentum and heat transfer through the flow. The vorticity fluctuations symbolize the essential three-dimensional nature of turbulence. The characteristic of dissipation makes it clear that there must be a continuous energy supply, for example, from the mean flow to the turbulence, or it will decay. In contrast, the essentially irrotational motion in the long water waves on the ocean involves little dissipation, explaining why great storm waves can travel over whole oceans without losing their identity.

To analyze turbulence it is useful to focus on the fluid particles. These particles are observed to travel in *randomly* moving fluid masses of varying sizes called *eddies*; these cause at any point in the flow, a rapid and irregular pulsation of velocity about a well-defined mean value. This may be visualized as in Fig. 7.4, where v is the time mean velocity and v' is the instantaneous velocity, necessarily a function of time. The instantaneous velocity v' may be considered to be composed of the vector sum of the mean velocity and the components of the pulsations v_x and v_y, both functions of time; by defining and isolating v_x and v_y in this manner, certain essentials of turbulence can be fruitfully studied. Measurements of v_x and v_y by hot-wire anemometer[5] (Section 11.8) yield a record similar to that of Fig. 7.4, which because of the random nature of turbulence discloses no regular period or amplitude; nevertheless such records allow the definition of certain turbulence characteristics. The root-mean-square (rms) values $(\overline{v_x^2})^{1/2}$ or $(\overline{v_y^2})^{1/2}$ are a measure of the violence of turbulent fluctuations, that is, the magnitude of departure of v' from v; the value $(\overline{v_x^2})^{1/2}/v$ is the *relative intensity* of turbulence. The mean time interval[6] between reversals in the sign of v_x (or v_y) is a

---

[4] H. Tennekes and J. L. Lumley, *A First Course in Turbulence*, The MIT Press, 1972.

[5] The time mean velocity $v = \overline{v'}$, where the overbar denotes a time average value, can be measured directly with a small pitot tube.

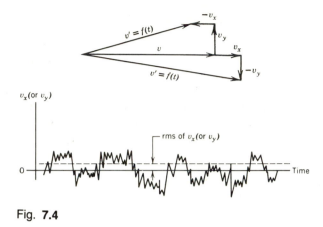

Fig. **7.4**

measure of the *scale* of the turbulence because it is a measure of the size of the turbulent eddies passing the point. In general, the intensity of turbulence increases with velocity, and scale increases with boundary dimensions. The former is easily imagined from the more rapid diffusion of the dye filament in a Reynolds apparatus with increased velocity; the latter can be visualized from the expectation that turbulent eddies will be larger in a large canal than in a small pipe for the same mean velocity. Indeed, as a rule of thumb, the largest eddy size expected is equal to a characteristic length of the flow, that is, the pipe radius, a channel width or depth, a boundary layer thickness, and so forth.

Because turbulence is an entirely chaotic motion of small fluid masses through short distances in every direction as flow takes place, the motion of individual fluid particles is impossible to trace and characterize mathematically. However, mathematical relationships may be obtained by considering the average motion of aggregations of fluid particles or by statistical methods.

Shearing stress in turbulent flow may be visualized by considering two adjacent points in a flow cross section (Fig. 7.5); at one of these points the mean velocity is v, at the other, $v + \Delta v$. If the small transverse distance between the points is l, a velocity gradient dv/dy (of the mean velocity) is implied. If v and $v + \Delta v$ are now taken to

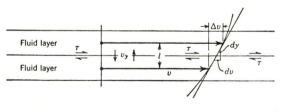

Fig. **7.5**

---

[6]This would correspond to the half-period of a simple harmonic vibration.

be the mean velocities of (fictitious) fluid layers, the turbulence velocity v_y represents the observed transverse motion of small fluid masses between layers, such mass being transferred in one direction or in the other. However, before transfer these fluid masses have velocities (v and $v + \Delta v$) which after transfer become $v + \Delta v$ and v, respectively; this means that their *momentum is changed* during the transfer process—tending to speed up the slower layer and slow down the faster one—just as if there were a shearing stress between them. Thus the existence of shearing stress in turbulent flow is deducible from momentum considerations.

For a century, engineers have grappled with the problem of developing useful and accurate expressions (in terms of mean velocity gradients and other flow properties) for turbulent shear stress. Some progress has been made, but perfection has not been attained (indeed, it does not appear attainable at this time). The following is only a brief discussion of a few essentials to provide perspective.[7]

The first attempt to express turbulent shear stress in mathematical form was made by Boussinesq,[8] who followed the pattern of the laminar flow equation and wrote

$$\tau = \varepsilon \frac{dv}{dy} \tag{7.2}$$

where the *eddy viscosity*, ε, was a property of the flow (not of the fluid alone) which depended primarily on the structure of the turbulence. From its definition the eddy viscosity can be seen to have the disadvantageous feature of varying from point to point throughout the flow. Nevertheless this first expression for turbulent shear is used frequently today because of the comparison between μ and ε and because it has been possible to develop useful, if not theoretically satisfying, expressions for ε in many engineering problems. Now the equation is usually written

$$\tau = (\mu + \varepsilon) \frac{dv}{dy}$$

to cover the combined situation where both viscous action and turbulent action are present in a flow. For the limiting conditions where the flow is entirely laminar or entirely turbulent, ε or μ is taken to be zero, respectively, and the foregoing equation reverts to equation 1.12 or 7.2.

Taking the fluctuating velocities of fluid particles due to turbulence as v_y and v_x, respectively, normal to and along the direction of general mean motion, Reynolds[9] confirmed that these turbulence velocities caused an effective mean shearing stress in turbulent flow and he showed that the stress could be written as

$$\tau = -\rho \, \overline{v_x v_y}$$

[7] H. Tennekes and J. L. Lumley, *A First Course in Turbulence,* The MIT Press, 1972, provides a complete guide for the novice.

[8] J. Boussinesq, "Essay on the Theory of Flowing Water," French Academy of Sciences, 1877.

[9] O. Reynolds, "On the Dynamical Theory of Incompressible Viscous Fluids and the Determination of the Criterion," *Phil. Trans. Roy. Soc.,* A1, vol. 186, p. 123, 1895.

in which $\overline{v_x v_y}$ is the mean value of the product of $v_x v_y$. Terms of the form $-\rho \overline{v_x v_y}$ are now called *Reynolds stresses*. Prandtl[10] succeeded in relating the velocities of turbulence to the general flow characteristics by proposing that small aggregations of fluid particles are transported by turbulence a certain mean distance,[11] l, from regions of one velocity to regions of another and in so doing suffer changes in their general velocities of motion (see Fig. 7.5). Prandtl termed the distance l the *mixing length* and suggested that the change in velocity, Δv, incurred by a fluid particle moving through the distance l was proportional to v_x and to v_y, that is, $l \, dv/dy \propto v_x$ and $l \, dv/dy \propto v_y$.

From this he suggested

$$\tau = -\rho \overline{v_x v_y} = \rho l^2 \left(\frac{dv}{dy}\right)^2 \tag{7.3}$$

as a valid equation for shearing stress in turbulent flow. Then, from equation 7.2, the eddy viscosity in a turbulent flow is

$$\varepsilon = \rho l^2 (dv/dy) \tag{7.4}$$

These expressions, although satisfactory in many respects, have the disadvantage that l is an unknown function of y that, in case of flows near a boundary or wall, presumably becomes smaller as the boundary or wall is approached. Of course, l must be chosen so that the right-hand side of equation 7.3 agrees with the actual stress.

In the case of flow near a bounding wall, the turbulence is strongly influenced by the wall, and v_x and v_y must be zero at the wall. It is intuitive then to let the mixing length l vary directly with the distance from the wall y. This gives

$$l = \kappa y \tag{7.5}$$

The constant κ is the so-called von Kármán constant which has been determined by use of experimental data. The nominal value of the Kármán constant is 0.4. Incorporating assumption 7.5 into equation 7.3 yields

$$\tau = \rho \kappa^2 y^2 \left(\frac{dv}{dy}\right)^2 \tag{7.6}$$

The linear variation of l near a wall is found to give a mean velocity profile that agrees closely with experiments.[12]

----------- **Illustrative Problems** -----------

Show that, if the velocity profile in laminar flow is parabolic, the shear stress profile must be a straight line.

----

[10] L. Prandtl, "Ueber die ausgebildete Turbulenz," *Proc. 2nd Intern. Congr., Appl. Mech.*, Zurich, p. 62, 1926.

[11] This concept is analogous to that of the mean free path in molecular theory.

[12] See Chapter 9.

Relevant Equation

$$\tau = \mu \frac{dv}{dy} \qquad (1.12)$$

Solution. For a parabolic relation between v and y, $v = c_1 y^2 + c_2$; therefore $dv/dy = 2c_1 y$. Since $\tau = \mu\, dv/dy$, $\tau \propto y$. ◄

A turbulent flow of water occurs in a pipe of 2 m diameter. The velocity profile is measured experimentally and found to be closely approximated by the equation $v = 10 + 0.8 \ln y$, in which v is in metres per second and y (the distance from the pipe wall) is in metres. The shearing stress in the fluid at a point $\frac{1}{3}$ m from the wall is calculated analytically from measurements of pressure drop (see Section 7.10) to be 103 Pa. Calculate the eddy viscosity, mixing length, and turbulence constant at this point.

Relevant Equations and Given Data

$$\tau = \epsilon \frac{dv}{dy} \qquad (7.2)$$

$$\tau = -\rho \overline{v_x v_y} = \rho l^2 \left(\frac{dv}{dy}\right)^2 \qquad (7.3)$$

$$\tau = \rho \kappa^2 y^2 \left(\frac{dv}{dy}\right)^2 \qquad (7.6)$$

$$v = 10 + 0.8 \ln y \qquad d = 2\text{ m} \qquad y = 1/3\text{ m}$$

$$\tau(y = 1/3) = 103\text{ Pa}$$

Solution.

$$\frac{dv}{dy} = \frac{d}{dy}(10 + 0.8 \ln y) = \frac{0.8}{y} = \frac{0.8}{(0.33)} = 2.4$$

$$103 = \epsilon(2.4) \qquad \epsilon = 42.9\text{ Pa·s} \blacktriangleleft \qquad (7.2)$$

$$103 = 1\,000 l^2 (2.4)^2 \qquad l = 0.134\text{ m} \blacktriangleleft \qquad (7.3)$$

$$103 = 1\,000 \kappa^2 \left(\frac{1}{3}\right)^2 (2.4)^2 \qquad \kappa = 0.401 \blacktriangleleft \qquad (7.6)$$

The magnitude of the eddy viscosity ϵ when compared with the viscosity μ (approximately 0.001 Pa·s) is of special interest in that it provides a direct comparison between the (large) turbulent shear and (small) laminar shear for the same velocity gradient. The mixing length, l, when compared with the pipe radius is found to be about 10% of the latter dimension; this is a nominal value of correct order of magnitude, as is the turbulence constant, κ.

7.3   Fluid Flow Past Solid Boundaries

A knowledge of flow phenomena near a solid boundary is of great value in engineering problems because, in practice, flow is always affected to some extent by the solid boundaries over which it passes; for example, the classic aeronautical problem is the external flow of fluid over the surfaces of an object such as a wing or fuselage, and in many other branches of engineering the problem of internal flow *between* solid boundaries, as in pipes and channels, is of paramount importance.

For a real fluid, experimental evidence shows that the velocity of the layer adjacent to the surface is zero (relative to the surface). This means that a velocity profile must show a velocity of zero at the boundary. In visualizing the flow over a boundary surface it is well to imagine a very thin layer of fluid, possibly having the thickness of but a few molecules, adhering to the surface with a continuous increase of velocity of the fluid as one moves farther away from the surface, the magnitude of the velocity being dependent on the shear in the fluid. For rough surfaces this simple picture is somewhat compromised because small eddies tend to form between the roughness projections, causing local unsteadiness of the flow.

Laminar flow occurring over smooth[13] or rough boundaries (Fig. 7.6) possesses essentially the same properties, the velocity being zero at the boundary surface and the shear stress throughout the flow being given by equation 1.12. Thus, in laminar flow, *surface roughness has no effect* on the flow picture as long as the roughness projections are small relative to the flow cross section size.

In turbulent flow, however, the roughness of the boundary surface will affect the physical properties of the fluid motion. When turbulent flow occurs over *smooth* solid boundaries, it is always separated from the boundary by a *sublayer* of viscosity-

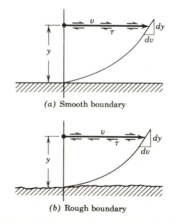

(a) Smooth boundary

(b) Rough boundary

Fig. **7.6**   Laminar flow over boundaries that *act* smooth.

---

[13] In the dynamics of solids, a "smooth" surface is often assumed to be a frictionless one; this concept is irrelevant here.

dominated flow (Fig. 7.7). This sublayer has been observed experimentally, and its existence may be justified theoretically by the following simple reasoning: The presence of a boundary in a turbulent flow will curtail the freedom of the turbulent mixing process by reducing the available mixing length, and, in a region very close to the boundary, the available mixing length is reduced to zero (i.e., the turbulence is completely extinguished) and a film of viscous flow over the boundary results.

In the viscous sublayer the shear stress, τ, is given by the viscous or laminar flow equation 1.12, and at a distance from the boundary, where turbulence is fully developed, by equations 7.3 or 7.6. Between the latter region and the viscous sublayer lies a transition zone in which shear stress results from a complex combination of both turbulent and viscous action, turbulent mixing being inhibited by the viscous effects due to the proximity of the wall. The fact that there is a transition from fully developed turbulence to no turbulence at the boundary surface shows that the viscous sublayer, although given (for convenience) an arbitrary thickness,[14] δ_v (Fig. 7.7), does *not* imply a sharp line of demarcation between the laminar and turbulent regions.

As stated above, the roughness of boundary surfaces will affect the physical properties of turbulent flow, and the effect of this roughness is dependent on the relative size of roughness and viscous sublayer. A boundary surface is said to be *smooth* if its projections or protuberances are so completely submerged in the viscous sublayer (Fig. 7.7b) that they have no effect on the structure of the turbulence. However, experiments

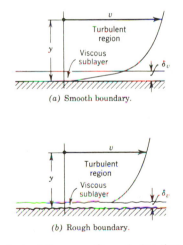

(a) Smooth boundary.

(b) Rough boundary.

Fig. **7.7** Turbulent flow over boundaries that *act* smooth.

[14] Intensive research on the viscous sublayer has shown that its thickness varies with time, showing the sublayer flow to be unsteady, this unsteadiness being associated with eddy formation adjacent to the surface. As these eddies or spots of turbulence are carried away from the wall, their presence accounts for the increased turbulence observed in the transition zone. See J. T. Davies, *Turbulence Phenomena*, Academic Press, 1972.

have shown that roughness heights larger than about one-third of sublayer thickness will augment the turbulence and have some effect on the flow. Thus the thickness of the viscous sublayer is the criterion of effective roughness, and, because the thickness of this sublayer also depends on certain properties of the flow, it is quite possible for the same boundary surface to behave as a smooth one or a rough one, depending on the size of the Reynolds number and of the viscous sublayer which tends to form over it.[15]

EXTERNAL FLOWS

Each of us has a mental image of internal and external fluid flows. Beyond the obvious physical differences, there are marked differences in the results sought in analysis of these flows. In *external flows,* one seeks the flow pattern around an object immersed in the fluid (over a wing or a flat plate, etc.), the lift and drag (resistance to motion) on the object, and perhaps the patterns of viscous action in the fluid as it passes around a body. For *internal flows,* the focus is often not on lift or drag, but on energy or head losses, pressure drops, and cavitation where energy is dissipated, because in internal flow, energy or work is used to move fluid through passages. In external flow cases, energy or work is used typically to move the object through the fluid.

7.4 Boundary Layers

Analyses made on the basis of ideal (inviscid) flow over streamlined bodies produce two results that are contrary to observation. First, the calculated drag on the body is negligible, while the observed drag is not. Second, the ideal fluid slips smoothly by the body; real fluid does not. For fluids with small viscosity (e.g., air and water), Prandtl[16] first suggested the boundary layer concept to explain the resistance of streamlined bodies, flat plates parallel to the flow, and so forth. The essential point is that the *frictional aspects of the flow are confined to the boundary layer* and perhaps a wake behind a body (where the flow is *rotational*), but outside the boundary layer the viscosity of the fluid is essentially inoperative, that is, the flow is effectively frictionless and *irrotational*. The boundary layer idea has found wide application in numerous problems of fluid dynamics and has provided a powerful tool for the analysis of problems of fluid resistance; it has probably contributed more to progress in modern fluid mechanics than any other single idea.

The mechanism of boundary layer growth can be described as follows. As the fluid flows past a body, fluid particles at the body walls remain at rest, and a high velocity gradient (dv/dy) is developed in the vicinity of the boundary. These high velocity gradients are associated with large frictional stresses in the boundary layer which "eat

---

[15] It will be shown later that sublayer thickness decreases with increasing Reynolds number. Usually, in practice, the change of a smooth surface to a rough one results from an increase of Reynolds number brought about by an increase of velocity.

[16] *Proc. Third Intern. Math. Congress,* Heidelberg, 1904.

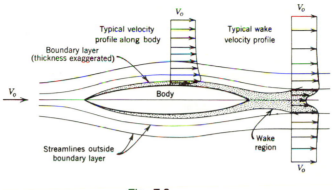

Fig. **7.8**

their way" into the flow downstream by slowing down successive fluid elements. Thus the boundary layers steadily thicken downstream along the body. A typical flow pattern on a streamlined body is shown schematically in Fig. 7.8.

The flow in a boundary layer can be either laminar or turbulent. Figure 7.9, which shows a boundary layer developing on a smooth flat plate, is used to visualize the situation. A laminar boundary layer is to be expected at the upstream end where the boundary layer is thin, because of the inhibiting effect of the smooth wall on the development of turbulence, and the dominance of viscous action. One of two Reynolds numbers is widely used to define the character of flow in a boundary layer; they are, in terms of the undisturbed velocity V_o, boundary layer thickness δ and distance from the edge of the plate x,

$$\mathbf{R}_x = \frac{Vx}{\nu} \qquad \text{and} \qquad \mathbf{R}_\delta = \frac{V\delta}{\nu} \tag{7.7}$$

and experiments on flat plates (Fig. 7.9) have shown that their nominal critical values are approximately 500 000 and 4 000 respectively. Below these values, laminar boundary layers are to be expected, whereas Reynolds numbers above these values define turbulent boundary layers. Comparison of flat plate boundary layers with those of the streamlined body (Fig. 7.8) reveals certain differences between them. The superficial similarity between Figs. 7.8 and 7.9 is noted immediately. However, for the stream-

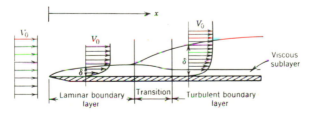

Fig. **7.9**

lined body: (a) its surface has a curvature that may affect the boundary layer development either due to inertial effects or induced separation (see Section 7.5 below) if the body is particularly blunt, and (b) the velocity in the irrotational flow just outside the boundary layer changes continuously along the body because of the disturbance to the overall flow offered by the body of finite width (this is verified by solution of the problem of ideal flow past a body).

In Sections 13.3 and 13.5 boundary layers are discussed further in the context of external flows or flows about immersed objects.

7.5 Separation

Separation of moving fluid from boundary surfaces is another important difference between the flow of ideal and real fluids. The mathematical theory of the ideal fluid yields no information about the expectation of separation even in simple cases where intuition alone would predict separation with complete certainty. Examples are shown in Fig. 7.10 for a sharp projection on a wall and a flat plate normal to a rectilinear flow. For the ideal fluid the flowfields will be found to be symmetrical upstream and downstream from such obstructions, the fluid rapidly accelerating toward the obstruction and decelerating in the same pattern downstream from it. However, the engineer would reason that the inertia of the moving fluid would prevent its following the sharp corners of such obstructions and that consequently separation of fluid from boundary surface is to be expected there, resulting in asymmetric flowfields featured by eddies and wakes downstream from the obstructions. Motion pictures of such eddies disclose that they are basically unsteady—forming, being swept away, and re-forming—thus absorbing energy from the flow and dissipating it in heat as they decay in an extensive zone downstream from the obstruction; thus the sketches of Fig. 7.10 are to be taken as time-

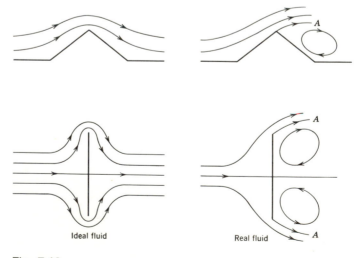

Ideal fluid   Real fluid

Fig. **7.10**

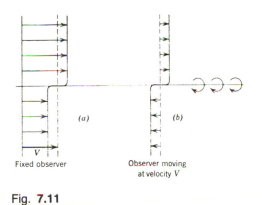

(a)   (b)

Fixed observer   Observer moving
at velocity V

Fig. **7.11**

average flow pictures which are intended to convey the essentials but not the complete details of flow separation.

Surfaces of discontinuity (indicated by A on Fig. 7.10) divide the live stream from the adjacent and more sluggishly moving eddies. Across such surfaces there will be high velocity gradients and accompanying high shear stress, but no discontinuity of pressure. The tendency for surfaces of discontinuity to break up into smaller eddies may be seen from the simplified velocity profile of Fig. 7.11a. An observer moving at velocity V would see the relative velocity profile of Fig. 7.11b from which the tendency for eddy formation is immediately evident.

Under special circumstances the streamlines of a surface of discontinuity become *free streamlines*, [17] which are streamlines along which the pressure is constant. It is apparent that the surfaces of discontinuity of Fig. 7.10 do not quite satisfy this requirement, and thus their streamlines may be considered free streamlines only as a crude approximation. Another example of interest is the cavitation zone behind a disk (Fig. 7.12) or other sharp-cornered object in a liquid flowfield; here the constant pressure imposed on the free streamline is the vapor pressure of the liquid. Bernoulli's equation applied along any free streamline shows that, if pressure is constant, $z + v^2/2g_n$ is also constant; if variations in z are negligible or zero, free streamlines are also lines of constant velocity. Free streamline theory is concerned with the prediction of form and

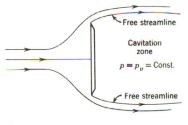

Free streamline

Cavitation
zone

$p = p_v = $ Const.

Free streamline

Fig. **7.12**

---

[17] Some examples of free streamlines were cited in Section 4.8.

positions of such lines from their unique properties and the hydrodynamical equations of flowfield theory.

Although the prediction of separation may be quite simple for sharp-cornered obstructions, this is a considerably more complex matter for gently curved (streamlined) objects or surfaces. Suppose that separation of flow from the streamlined strut of Fig. 7.13 does not occur. This is a very reasonable assumption for small ratios of thickness to length but much less reasonable for high values of this ratio. For no separation, the flowfield is virtually identical with that of the ideal fluid except for the growth of thin boundary layers between the strut and the remainder of the flowfield (cf., Fig. 7.8). The boundary layers coalesce at the trailing edge of the strut, producing a narrow wake of fine-grained eddies. Proceeding along the surface of the strut from (the stagnation point) A to B, the pressure falls because the flow is accelerating, producing a favorable pressure gradient which "strengthens" the boundary layer. From B to C, however, the pressure rises as the flow decelerates because the body is thinning and its effect on the flow diminishes, producing an adverse (unfavorable) pressure gradient which may "weaken" the boundary layer sufficiently to cause separation. The likelihood of separation is enhanced with increase of thickness-to-length ratio of the strut, which from the increased divergence of the streamlines adjacent to BC will produce a larger unfavorable pressure gradient. This adverse pressure gradient penetrates the boundary layer and serves to produce a force opposing the motion of its fluid; if the gradient is large enough, the slowly moving fluid near the wall will be brought to rest and begin to accumulate, diverting the live flow outward from the surface and producing an eddy accompanied by separation of the flow from the body surface. After separation has occurred, the flowfield near the separation point will appear about as shown in Fig. 7.14. Obviously the analytical prediction of separation point location is an exceedingly difficult problem requiring accurate quantitative information on the phenomena cited above; for this reason the prediction of separation and location of separation points on gently curved bodies or obstructions are usually obtained more reliably from experiment than from analysis.

The foregoing may now be generalized into the following simple axiom: *Acceleration of real fluids tends to be an efficient process, deceleration an inefficient one.* Accelerated motion, as it occurs along the surface of the front end of a submerged object, is accompanied by a favorable pressure gradient which serves to stabilize the boundary layer and thus minimize energy dissipation. Decelerated motion is accompanied by an adverse pressure gradient which tends to promote separation, instability, eddy formation, and large energy dissipation.

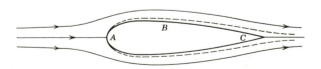

Fig. **7.13**

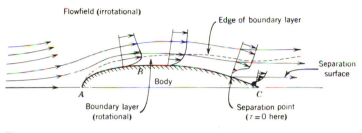

Flowfield (irrotational)

Edge of boundary layer

Separation
surface

Body

A

B

C

Boundary layer
(rotational)

Separation point
($\tau = 0$ here)

Fig. **7.14**

7.6 Secondary Flow

Another consequence of wall friction is the creation of a flow within a flow—a
secondary flow superposed on the main *primary* flow. Figure 7.15*a* shows a cross
section through a river bend where the secondary motion serves to deposit material at
the inside of the bend and to assist in scouring the outer side, thus producing the
well-known meandering characteristic of natural streams. In Fig. 7.15*b* is shown the
horseshoe-shaped vortex that is produced by projections from a boundary surface. Its
origin can be easily seen from the velocity profile along the wall which leads to the
stagnation pressure at *A* being larger than that at *B*. This pressure difference maintains
a downward secondary flow from *A* to *B*, thus inducing a vortex type of motion, the
core of the vortex being swept downstream around the sides of the projection. This
principle is used on the wings of some jet aircraft, the devices (called vortex generators)
being used to draw higher energy fluid down to the wing surface to forestall large-scale
separation.

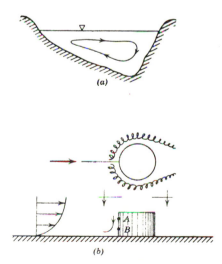

(*a*)

(*b*)

Fig. **7.15**

Because of the complex origin and geometry of secondary flows they have (so far) defied rigorous analysis. They are cited in this elementary text only so that the reader will become aware of them and make due allowance for their possible existence in analyzing new problems. A conservative attitude is to expect secondary flow phenomena wherever there are severe irregularities in boundary geometry.

INTERNAL FLOWS

For the flow of a real fluid in ducts, channels, and pipes many of the concepts developed in earlier chapters can be applied with only minor, but important, changes needed to account for viscous effects. Interestingly, boundary layers, separation, and secondary flows are as important in internal flows as they were in the external flow cases.

7.7 Flow Establishment—Boundary Layers

Just as a boundary layer beginning at the edge of a flat plate in an external flow grows in the downstream direction, viscous effects begin their influence at the entrance to a pipe and establish themselves in internal flows by a process of boundary layer growth (Fig. 7.16). The zone of this growth is the zone of *unestablished* flow and it may be of importance in some engineering problems; it is described here for the simplest case—flow from a large reservoir into a cylindrical passage.

The unestablished flow zone may contain many different flow phenomena. With an unrounded entrance, separation (Section 7.11) will dominate the flow picture—featured by localized eddy formation close to the entrance and followed by decay of the eddies in the final established flow; measurements show that this process extends over a distance of some 50 pipe diameters. However, a rounded (streamlined) entrance is of more engineering interest. For this case, the unestablished flow zone will be dominated by the growth of boundary layers along the walls (accompanied by a diminishing *core* of fluid at the center of the passage); established flow results from the merging of these boundary layers, a phenomenon that occurs in most internal flows.

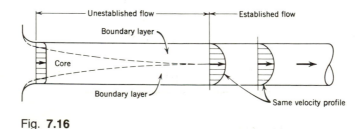

Fig. **7.16**

The mechanism of boundary layer growth in a pipe entrance can be described as follows. As the fluid flows into the entrance, high velocity gradients (dv/dy) develop in the vicinity of the boundary. These gradients are associated with large frictional stresses in the boundary layer which, as in the case of flow over a flat plate, "eat their way" into the flow by slowing down successive fluid elements. Thus the boundary layers steadily thicken until they meet and so envelop the whole flow; downstream from this point the influence of wall friction is felt throughout the flowfield. The flow is, therefore, *established* (there is no further change in the velocity profiles) and is everywhere *rotational*.

Of course, the flow in a boundary layer may be either laminar or turbulent. If the Reynolds number $\mathbf{R} = Vd/\nu$ for the established flow is less than 2 100, it may be safely inferred that the established laminar flow has resulted from the growth of laminar boundary layers. In this case the zone of establishment has a length x given[18] by $x/d \approx \mathbf{R}/20$. For slightly higher $\mathbf{R}$, the flow may be laminar up to and past $x/d \approx \mathbf{R}/20$, with subsequent transition to turbulent flow before the flow is truly established. If $\mathbf{R} \gg 2\ 100$, the boundary layers are ultimately turbulent. For a well-shaped and smooth entrance the boundary layers may be laminar at the upstream end of the zone of unestablished flow, followed by turbulent boundary layers in the downstream portion of this region. In practical cases at high $\mathbf{R}$ the unestablished flow region may extend for 100 diameters or so, but the flow is effectively established beyond $x/d \approx 20$ and the details of the flow are of minor interest.[18] In Section 9.9 the net effect at entrances is treated as a head loss, which is a function of the velocity head in the established flow region.

Comparison of flat plate boundary layers (Section 7.4) with those of the pipe entrance (Fig. 7.16) will reveal certain subtle but critical differences between them. The superficial similarity between Figs. 7.9 and 7.16 is noted immediately; however, for the pipe entrance: (a) the plate has been rolled into a cylinder so it is not flat, (b) the core velocity steadily increases downstream whereas the corresponding free stream velocity of Fig. 7.9 remains essentially constant, and (c) the pressure in the fluid diminishes[19] in a downstream direction whereas for the flat plate there is no such pressure variation.

Although this is not the place for a detailed analysis of the pipe entrance boundary layer problem, certain qualitative facts should be noted (Fig. 7.17). Between the turbulence (of boundary layer or established flow) and solid boundary there exists the viscous sublayer cited in Section 7.3. The velocity of the core flow increases over the region of establishment from a value slightly more than the mean velocity, Q/A, to the centerline velocity of the established flow. The thickening of the laminar boundary layer causes a decrease in velocity gradient and thus a decrease in wall shearing stress in a downstream direction. With the change of velocity profile in the turbulent

---

[18] W. M. Kays and M. E. Crawford, *Convective Heat and Mass Transfer*, 2nd ed., McGraw-Hill, 1980.
[19] This is known as a *favorable pressure gradient*, since it extends the length of the laminar boundary layer and thus reduces frictional effects.

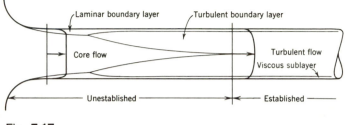

Fig. **7.17**

boundary layer a rather sudden increase of wall shearing stress can be expected after the boundary layer has changed from laminar to turbulent; thereafter this shear stress continues to decrease in a downstream direction. Since the core flow may be treated as frictionless, the fall of pressure may be predicted from the simple Bernoulli equation (without head loss) once the core velocities are known; this has been proved by experiment to be reliable and accurate in the upstream portion of the region of establishment (where the boundary layer is thin) but less accurate in the downstream portion.

7.8   Velocity Distribution and Its Significance

The shearing stresses of laminar and turbulent flow produce velocity distributions characterized by reduced velocities near boundary surfaces. These deviations from the uniform velocity distribution of ideal fluid flow necessitate alterations[20] in the methods for calculation of velocity head and momentun flux. The effect of nonuniform velocity distribution on the computation of flowrate has been indicated in Section 3.5.[21]

The kinetic energy of fluid moving in the differentially small streamtube of Fig. 7.18 is (from Section 4.5) $(dQ)\gamma v^2/2g_n$ or $\rho v^3 \, dA/2$. The momentum flux (Section 6.1) is $(dQ)\rho v$ or $\rho v^2 \, dA$. The total kinetic energy and momentum flux are the respective integrals of these differential quantities; thus

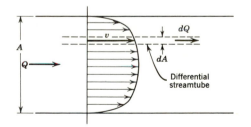

Fig. **7.18**

---

[20] In many practical problems, however, these alterations are so small that they may be neglected.
[21] Section 3.5 should be restudied at this point.

$$\text{Total kinetic energy (J/s or ft} \cdot \text{lb/s)} = \frac{\rho}{2} \int^A v^3 \, dA \qquad (7.8)$$

$$\text{Momentum flux (N or lb)} = \rho \int^A v^2 \, dA \qquad (7.9)$$

Although these quantities are computed as indicated in many engineeering situations, they are also usefully expressed in terms of mean velocity V and total flowrate Q, using the same forms of the equations; thus

$$\text{Total kinetic energy} = \alpha Q \gamma \frac{V^2}{2g_n} = Q\gamma\left(\alpha \frac{V^2}{2g_n}\right) \qquad (7.10)$$

$$\text{Momentum flux} = \beta Q \rho V \qquad (7.11)$$

in which α and β are dimensionless and represent correction factors to the conventional velocity head $V^2/2g_n$ and momentum flux $Q\rho V$, respectively. For a uniform velocity distribution $\alpha = \beta = 1$; for a nonuniform velocity profile $\alpha > \beta > 1$. Expressions for α and β may be derived by equating equations 7.8 and 7.10 and equations 7.9 and 7.11, substituting $\int^A v \, dA$ for Q, and obtaining

$$\alpha = \frac{1}{V^2}\frac{\int^A v^3 \, dA}{\int^A v \, dA} \qquad \text{and} \qquad \beta = \frac{1}{V}\frac{\int^A v^2 \, dA}{\int^A v \, dA} \qquad (7.12)$$

from which it is easily seen that $\alpha = \beta = 1$ for uniform velocity distributions. Comparing the rather pointed (far from uniform) velocity distributions of laminar flow (Fig. 7.6) with the rather flattened (nearer to uniform) ones of turbulent flow (Fig. 7.7), it may be concluded directly that high values of α and β are to be expected for the former, lower ones for the latter.

For the established flow of a real fluid in a prismatic passage the analysis of Section 4.3 applies,[22] and $(p/\gamma + z)$ is a constant throughout any flow cross section. The hydrostatic pressure distribution is not affected by viscosity, but only by the streamline curvature. From this it is immediately seen that $(z + p/\gamma + v^2/2g_n)$ cannot be constant throughout the flow cross section, and this in turn requires some re-examination of the energy line concept. Taking A and C as typical streamlines, it is noted from Fig. 7.19 that each streamline is associated with a different energy line; in other words, the flow along different streamlines possesses different amounts of total energy, and a comprehensive energy line picture would be a "bundle of energy lines"—one for each streamline. However, for flow in a parallel-walled passage, such as a pipe, duct, or open channel, the properties of individual streamlines are seldom of interest, and the

---

[22] Provided that the shearing stresses in the fluid perpendicular to the direction of motion may be neglected. These are safely negligible in engineering problems except for flows dominated by viscous action. Such flows occur at very low Reynolds numbers and low velocity; they are frequently termed *creeping motions*.

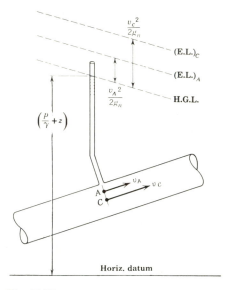

Fig. **7.19**

whole aggregation of streamlines (i.e., the whole flow) is characterized by a single effective energy line a distance $\alpha V^2/2g_n$ above the hydraulic grade line; thus alterations in the Bernoulli equation due to nonuniform velocity distribution are concentrated in the coefficient α alone. However, in problems involving flowfields each streamline will in general be associated with different amounts of total energy, a fact essential to the analysis and interpretation of such problems.

Another consequence of nonuniform velocity distribution can be shown in its simplest aspects by considering a flow through a short constriction (Fig. 7.20) in a passage where the coefficient α changes from section to section. The trend of change of α from section 1 to section 2 may be established by noting that all fluid particles in passing from section 1 to section 2 experience the same change in $(p/\gamma + z)$ or the same drop in the hydraulic grade line. The velocities at section 1 are relatively small, and those at section 2 relatively large; with frictional effects in the constriction relatively small, the increase in the velocity of all particles from section 1 to section 2 will tend to be about the same; the velocity profile at section 2 will therefore be flatter than that at section 1, and α_2 will accordingly be less than α_1. The *exact head loss* between sections 1 and 2 is given by the drop in the energy line between these sections; from Fig. 7.20,

$$\text{Exact } h_{L_{1-2}} = \left(\alpha_1 \frac{V_1^2}{2g_n} + \frac{p_1}{\gamma} + z_1 \right) - \left(\alpha_2 \frac{V_2^2}{2g_n} + \frac{p_2}{\gamma} + z_2 \right)$$

However, the conventional head loss used in engineering computations is defined by ignoring the α terms in the Bernoulli equation and writing

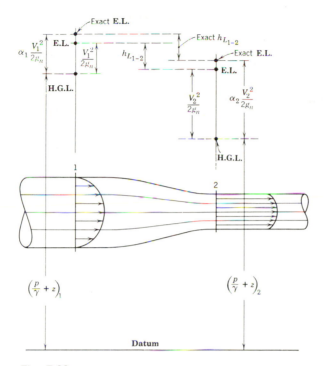

Fig. **7.20**

$$h_{L_{1-2}} = \left(\frac{V_1^2}{2g_n} + \frac{p_1}{\gamma} + z_1\right) - \left(\frac{V_2^2}{2g_n} + \frac{p_2}{\gamma} + z_2\right)$$

Comparison of the two equations leads to

$$h_{L_{1-2}} = \text{Exact } h_{L_{1-2}} + (\alpha_2 - 1)\frac{V_2^2}{2g_n} - (\alpha_1 - 1)\frac{V_1^2}{2g_n} \qquad (7.13)$$

from which it is seen that the conventional head loss does not equal exact head loss unless $(\alpha_2 - 1)V_2^2/2g_n = (\alpha_1 - 1)V_1^2/2g_n$. However, constriction of the passage causes V_2 to be greater than V_1, while the flattening of the velocity profile causes $(\alpha_2 - 1)$ to be less than $(\alpha_1 - 1)$; thus there are some compensating features in equation 7.13 which tend to make exact head loss not very different from conventional head loss in this simple example. Although in other cases these conventional and exact head losses may differ considerably, it should be noted that this is no serious obstacle in most engineering problems; no matter how these losses may be defined or related to each other, there can be (for a given flowrate) only one value for the change of $(p/\gamma + z)$ between sections 1 and 2; if calculations are made with this fact in mind, reliable predictions can be made from conventional head losses even though these losses are never precisely equal to the exact ones.

―――――――――――――― **Illustrative Problem** ――――――――――――――

Assuming Fig. 7.18 to represent a parabolic velocity profile in a passage bounded by two infinite planes of spacing 2R and maximum velocity v_c, calculate q, α, and β.

Relevant Equations

$$q = VA = V(2R) = \int^A v \, dA \tag{3.15}$$

$$\alpha = \frac{1}{V^2} \frac{\int^A v^3 \, dA}{\int^A v \, dA} \qquad \text{and} \qquad \beta = \frac{1}{V} \frac{\int^A v^2 \, dA}{\int^A v \, dA} \tag{7.12}$$

Solution. Taking r as the distance from centerline of the passage to any local velocity, v, and element of area dA, $dA = dr$. The equation of the parabola is $v = v_c(1 - r^2/R^2)$.

$$q = 2 \int_0^R v_c \left(1 - \frac{r^2}{R^2}\right) dr = \tfrac{2}{3}(2Rv_c) \tag{3.15}$$

Since q also equals $2RV$, $V = 2v_c/3$.

$$\alpha = \frac{2 \int_0^R v_c^3 \left(1 - \dfrac{r^2}{R^2}\right)^3 dr}{\left(\dfrac{2v_c}{3}\right)^2 \tfrac{2}{3}(2Rv_c)} = \frac{54}{35} = 1.54 \tag{7.12}$$

$$\beta = \frac{2 \int_0^R v_c^2 \left(1 - \dfrac{r^2}{R^2}\right)^2 dr}{\left(\dfrac{2v_c}{3}\right) \tfrac{2}{3}(2Rv_c)} = \frac{6}{5} = 1.20 \tag{7.12}$$

The meaning of these figures is that the exact velocity head is more than 54% greater than $V^2/2g_n$, and the exact momentum flux 20% greater than $Q\rho V$. Differences of this magnitude warn the engineer that α and β should be considered when applying the energy and momentum equations to one-dimensional laminar flow problems—unless their effects can be shown to have negligible consequence in the results desired.

―――

7.9 The Energy Equation

The energy equation applied to the flow of real fluids is merely an accounting of the various energy changes within a portion of a flow system and presents few difficulties once these energies have been identified. Consider the reasonably (but not com-

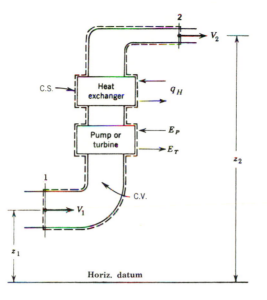

Fig. **7.21**

pletely)[23] general situation of Fig. 7.21. Here the possibility of heat, q_H, being added to or extracted from the flow must be considered; in Fig. 7.21 this action is concentrated in the heat exchanger, but such a flow of heat may also occur through the walls of the streamtube, its direction dependent on the relative temperatures at the surfaces of the tube wall. Such flows of heat are calculated in engineering problems by use of the principles of heat transfer, a large field of engineering which cannot be explored in an elementary fluid mechanics text.

Recall the energy equation 5.5 was [24]

$$\hat{h}_1 + \frac{V_1^2}{2g_n} + z_1 + \frac{q_H}{g_n} + E_P = \hat{h}_2 + \frac{V_2^2}{2g_n} + z_2 + E_T \qquad (5.5)$$

Without heat exchanger, pump, or turbine, this energy equation reduces to

$$\hat{h}_1 + \frac{V_1^2}{2g_n} + z_1 + \frac{q_H}{g_n} = \hat{h}_2 + \frac{V_2^2}{2g_n} + z_2 \qquad (7.14)$$

---

[23] Chemical, electrical, and atomic energies and the kinetic energy of turbulence are excluded from this analysis.

[24] Because of the use of mean velocity, pressure, density, and temperature, correction factors should be applied to the first three terms on each side of this equation to allow for the nonuniform distribution of these quantities across the flow. However, these correction factors are usually close to unity for practical and turbulent flows, and such refinement of the equation is generally not necessary for engineering use; these factors are omitted here.

in which now q_H is the heat energy passing *into*[25] the fluid through the walls of the streamtube. For gases or vapors, equation 7.14 is usually written with neglect of $(z_2 - z_1)$; it then becomes

$$\hat{h}_1 + \frac{V_1^2}{2g_n} + \frac{q_H}{g_n} = \hat{h}_2 + \frac{V_2^2}{2g_n} \tag{7.15}$$

For perfect gases where $g_n(\hat{h}_1 - \hat{h}_2) = h_1 - h_2 = c_p(T_1 - T_2)$, equation 7.15 frequently appears as

$$c_p T_1 + \frac{V_1^2}{2} + q_H = c_p T_2 + \frac{V_2^2}{2} \tag{7.16}$$

For the flow of liquids and gases in the numerous engineering situations where there is negligible change of fluid density, the energy equation is written, in analogy with the original Bernoulli equation, as

$$\frac{p_1}{\gamma} + \frac{V_1^2}{2g_n} + z_1 = \frac{p_2}{\gamma} + \frac{V_2^2}{2g_n} + z_2 + h_{L_{1-2}} \tag{7.17}$$

in which the head loss, $h_{L_{1-2}}$, represents the fall of the energy line between sections 1 and 2. But, now, in terms of the other quantities (through comparisons with equation 7.14 and because $\hat{h} = p/\gamma + ie/g_n$), the head loss is seen to be equal to

$$h_{L_{1-2}} = \frac{(ie_2 - ie_1 - q_H)}{g_n} \tag{7.18}$$

In such problems the separate values of ie_2, ie_1, and q_H are usually not required, and the "packaging" of them into a single term, $h_{L_{1-2}}$, proves highly effective for engineering use. The equation offers proof that head loss is not a loss of total energy but rather a conversion of energy into heat, part of which leaves the fluid, the remainder serving to increase its internal energy. This is the practical case of incompressible flow as it appears in many engineering applications; here head loss is a permissible and useful concept because heat energy leaving the flow and energy converted into internal energy are seldom recoverable and are in effect lost from the useful total of pressure, velocity, and potential energies. For compressible flow this is not generally true, since the useful total of energies will include the internal energy (see Section 9.12).

———————————— **Illustrative Problems** ————————————

A flowrate of 240 lb/s of air occurs in a streamtube similar to that of Fig. 7.21 but without pump or turbine. $A_1 = 4$ ft², $A_2 = 2$ ft², $z_1 = 30$ ft, $z_2 = 80$ ft, $V_1 = 600$ ft/s, $V_2 = 800$ ft/s, $p_1 = 20$ psia, $p_2 = 35$ psia. Calculate the mean temperatures of the air at sections 1 and 2 and the net heat energy added to the fluid between these sections.

---

[25] If heat energy flows *outward* through the walls of the streamtube, q_H will be negative.

Relevant Equations and Given Data

$$\gamma = g_n p / RT \tag{1.3}$$

$$G = A_1 \gamma_1 V_1 = A_2 \gamma_2 V_2 \tag{3.10}$$

$$\hat{h}_1 + \frac{V_1^2}{2g_n} + \frac{q_H}{g_n} = \hat{h}_2 + \frac{V_2^2}{2g_n} \tag{7.15}$$

$$R = 1\,715 \text{ ft} \cdot \text{lb/slug} \cdot {}^\circ R \quad c_p = 6\,000 \text{ ft} \cdot \text{lb/slug} \cdot {}^\circ R$$

Dimensions, and so on, are given in the problem statement.

Solution.

$$240 = 4 \times 600 \times \gamma_1 = 2 \times 800 \times \gamma_2 \tag{3.10}$$

$$\gamma_1 = 0.1 \quad \gamma_2 = 0.15 \text{ lb/ft}^3$$

$$T_1 = \frac{20 \times 144 \times 32.2}{1\,715 \times 0.1} \quad T_2 = \frac{35 \times 144 \times 32.2}{1\,715 \times 0.15} \tag{1.3}$$

$$T_1 = 540^\circ R \quad T_2 = 630^\circ R \blacktriangleleft$$

$$\hat{h}_2 - \hat{h}_1 = \frac{6\,000}{32.2}(630 - 540) = 16\,800 \text{ ft} \cdot \text{lb/lb}$$

$$\frac{(600)^2}{2g_n} + 30 + \frac{q_H}{g_n} = 16\,800 + \frac{(800)^2}{2g_n} + 80 \tag{7.15}$$

$q_H = 877.1$ Btu/slug (1 Btu = 778.2 ft · lb). ◀ Disregarding the z terms (30 and 80), $q_H = 875.0$ Btu/slug, a difference of about 0.2%. The total heat energy added to the fluid represents $240 \times 875.0 \times 778.2/(32.2 \times 550) = 9\,250$ hp.

A flowrate of 1.42 m³/s of water occurs in a streamtube, similar to that of Fig. 7.21, containing a pump (but no heat exchanger) which is delivering 300 kW to the flowing fluid. $A_1 = 0.4$ m², $A_2 = 0.2$ m², $z_1 = 9$ m, $z_2 = 24$ m, $p_1 = 138$ kPa, $p_2 = 69$ kPa. Calculate the head lost between sections 1 and 2.

Relevant Equations and Given Data

$$Q = A_1 V_1 = A_2 V_2 \tag{3.12}$$

$$P = Q \gamma E_P \tag{4.7}$$

$$\hat{h}_1 + \frac{V_1^2}{2g_n} + z_1 + \frac{q_H}{g_n} + E_P = \hat{h}_2 + \frac{V_2^2}{2g_n} + z_2 + E_T \tag{5.5}$$

$$\gamma = 9.81 \text{ kN/m}^3$$

Other data is listed in the problem statement.

Solution.

$$1.42 = 0.4V_1 = 0.2V_2 \quad V_1 = 3.55 \text{ m/s} \quad V_2 = 7.1 \text{ m/s} \tag{3.12}$$

$$E_P = \frac{300 \times 1\,000}{9\,810 \times 1.42} = 21.54 \text{ J/N} \tag{4.7}$$

Applying equation 5.5 with $h_{L_{1-2}} = (ie_2 - ie_1 - q_H)/g_n$ and $E_T = 0$, that is, applying equation 7.17 with E_P included,

$$\frac{p_1}{\gamma} + \frac{V_1^2}{2g_n} + z_1 + E_P = \frac{p_2}{\gamma} + \frac{V_2^2}{2g_n} + z_2 + h_{L_{1-2}} \tag{7.17}$$

$$\frac{138\,000}{9\,810} + \frac{(3.55)^2}{2g_n} + 9 + 21.54 = \frac{69\,000}{9810} + \frac{(7.1)^2}{2g_n} + 24 + h_{L_{1-2}}$$

$$h_{L_{1-2}} = 11.7 \text{ m} \blacktriangleleft$$

7.10   Resistance Force and Energy Dissipation

Although the energy equations of Section 7.9 are essential to engineering analysis, they contain no information about the basic resistance forces that cause energy dissipation in fluid flow. These forces may be isolated, identified, and related to the energy equations through a one-dimensional analysis of the (compressible or incompressible) flow in an element of a cylindrical passage (Fig. 7.22). The wall stress τ_o is the basic resistance (shear) stress to be investigated and will produce a force on the periphery of the streamtube opposing the direction of fluid motion. The other variables shown on the figure will be recognized from previous treatments (e.g., Section 4.1). An impulse-momentum analysis applied *along the direction of the streamtube* to the control volume bounded by sections 1, 2, and the streamtube boundary yields (c.f., equation 6.3)

$$pA - (p + dp)A - \tau_o P \, dl - \left(\gamma + \frac{d\gamma}{2}\right)A \, dl\left(\frac{dz}{dl}\right)$$

$$= \beta_2(V + dV)^2 A(\rho + d\rho) - \beta_1 V^2 A\rho$$

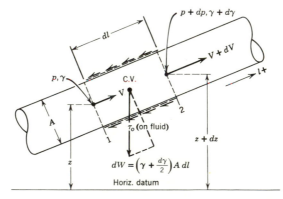

Fig. 7.22

in which P is the perimeter of the streamtube and $dW(dz/dl)$ is the component of the fluid's weight along the streamtube. By assuming $\beta_1 = \beta_2 = 1$, dividing the equation by $A\gamma$, and neglecting terms containing products of differential quantities, the equation may be reduced to[26]

$$\frac{dp}{\gamma} + d\left(\frac{V^2}{2g_n}\right) + dz = -\frac{\tau_o\,dl}{\gamma R_h} \tag{7.19}$$

and compared with the differential form of (energy) equation 7.14,

$$d\left(\frac{p}{\gamma}\right) + d\left(\frac{V^2}{2g_n}\right) + dz = d\left(\frac{q_H}{g_n}\right) - d\left(\frac{ie}{g_n}\right)$$

Recall that in a compressible flow, V can vary even though A remains constant. Since $d(p/\gamma) = dp/\gamma + p\,d(1/\gamma)$, comparison of these equations yields

$$\frac{\tau_o\,dl}{\gamma R_h} = d\left(\frac{ie}{g_n}\right) - d\left(\frac{q_H}{g_n}\right) + p\,d\left(\frac{1}{\gamma}\right) \tag{7.20}$$

which gives a basic relationship between the resistance stress, τ_o, and various terms of the energy equation. Its integration may be indicated,

$$\frac{1}{R_h}\int_1^2 \frac{\tau_o\,dl}{\gamma} = \frac{(ie_2 - ie_1 - q_H)}{g_n} + \int_1^2 p\,d\left(\frac{1}{\gamma}\right)$$

but cannot be performed for compressible flow without information on the thermodynamic process, which will be affected by the amount of heat transfer q_H. For compressible flow, τ_o will also depend upon l, and if the former is considered constant the result will be the mean value $\bar\tau_o$, over the length $(l_2 - l_1)$. Another observation to be made here is that thermodynamics shows (Section 5.1) that for frictionless or isentropic processes

$$d\left[\frac{(ie - q_H)}{g_n}\right] + p\,d\left(\frac{1}{\gamma}\right) = 0$$

and that this is also obtained from equation 7.20 when $\tau_o = 0$.

For established incompressible flow, τ_o is a function of l, γ is constant, and $d(1/\gamma) = 0$. Integration then gives

$$\frac{\tau_o(l_2 - l_1)}{\gamma R_h} = \frac{(ie_2 - ie_1 - q_H)}{g_n}$$

However, the right-hand side of this equation has been defined (equation 7.18) as fall of energy line, or head loss $h_{L_{1-2}}$. Thus for incompressible flow

$$h_{L_{1-2}} = \frac{\tau_o(l_2 - l_1)}{\gamma R_h} \tag{7.21}$$

---

[26] The ratio A/P is known as the *hydraulic radius* R_h; further discussion of it will be found in Section 9.7.

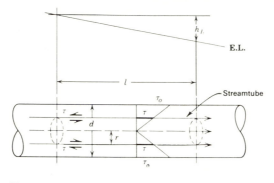

Fig. **7.23**

giving a simple relation between resistance stress τ_o and the head loss caused by the action of resistance.

It should be apparent that the foregoing analysis may be similarly applied to any streamtube of the flow; it is useful to do this for a streamtube of radius r and concentric with the axis of a cylindrical pipe (Fig. 7.23). For such a streamtube the frictional stress, τ, will be that exerted on the outermost fluid layer of the streamtube by the adjacent (more slowly moving) fluid. Without repeating the foregoing development, the following substitutions may be made in equation 7.21: τ for τ_o, $r/2$ for R_h, h_L for $h_{L_{1-2}}$, and l for $l_2 - l_1$; this yields

$$\tau = \left(\frac{\gamma h_L}{2l}\right)r \tag{7.22}$$

and shows that, in established pipe flow, the shear stress τ in the fluid varies linearly with distance from the centerline of the pipe. The relationships of equations 7.21 and 7.22 have been developed without regard to the flow regime; *thus it follows that they are equally applicable to both laminar and turbulent flow in pipes*. These equations are used later as the first step in the analytical treatment of these problems.

──────────────── **Illustrative Problems** ────────────────

Water flows in a 0.9 m by 0.6 m rectangular conduit. The head lost in 60 m of this conduit is determined (experimentally) to be 10 m. Calculate the resistance stress exerted between fluid and conduit walls.

Relevant Equation and Given Data

$$h_{L_{1-2}} = \frac{\tau_o(l_2 - l_1)}{\gamma R_h} \tag{7.21}$$

$$R_h = A/P$$

$$\gamma = 9.8 \text{ kN/m}^3 \qquad h_{L_{1-2}} = 10 \text{ m} \qquad l_2 - l_1 = 60 \text{ m}$$
$$A = 0.9 \times 0.6 = 0.54 \text{ m}^2 \qquad P = 2(0.9 + 0.6) = 3 \text{ m}$$

Solution.

$$\tau_o = \frac{10 \times 9.8 \times 10^3 \times 0.54/3}{60} = 0.29 \text{ kPa} \blacktriangleleft \qquad (7.21)$$

Since this flow is *not* axisymmetric, τ_o must be presumed to be the *mean* shear stress on the perimeter of the conduit.

If the results cited in the preceding problem are obtained for water flowing in a cylindrical pipe 0.6 m in diameter, what shear stress is to be expected (*a*) between fluid and pipe wall, and (*b*) in the fluid at a point 200 mm from the wall?

Relevant Equation and Given Data

$$\tau = \left(\frac{\gamma h_L}{2l} \right) r \qquad (7.22)$$

$$d = 0.6 \text{ m} \qquad r = 0.1 \text{ m}$$

See previous problem for equation 7.21, γ, $h_{L_{1-2}} = h_L$, $l = l_2 - l_1$, R_h, etc.

Solution.   Determine $R_h = d/4$; then

$$\tau_o = \left(\frac{10 \times 9.8 \times 10^3}{60} \right) 0.15 = 0.25 \text{ kPa} \blacktriangleleft \qquad (7.21)$$

From the linear variation of τ with r (equation 7.22),

$$\tau = \frac{100}{300}(0.25 \times 10^3) = 83.3 \text{ Pa} = 0.083 \, 3 \text{ kPa} \blacktriangleleft$$

7.11   Separation

Separation has great impact on design and performance of certain familiar internal flow systems. For example, separation of the flow from the blades of a pump impeller or turbine runner can cause serious degradation of efficiency and often damage to the system. Careful design can prevent separation in these cases; but for flow through a valve or metering orifice or a sudden expansion, separation is inevitable. Figure 7.24 shows both classical ideal fluid and real fluid flows for three cases (see the discussion in Section 7.5 also).

A classic internal flow in which separation is often a major problem is that in a *diffuser* (Fig. 7.25). Here the engineering objective is to provide an expanding passage of proper shape and minimum length that will yield minimum head loss or maximum pressure rise for reduction of mean velocity from V_1 to V_2. A short passage with sharp wall curvatures will prove inefficient because of flow separation and large energy dissipation; a longer one will be efficient but too space-consuming and will produce too

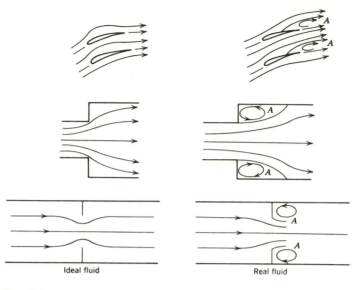

Ideal fluid   Real fluid

Fig. **7.24**

much boundary resistance. The optimum design lies between the extremes but (as yet) cannot be determined wholly by analytical methods; again experiments must be used to supplement and confirm analytical attacks on problems where separation is involved. The optimum included angle, α, in simple diffusers lies between 6° and 8° and depends on the velocity distribution in section 1 at the entrance to the diffuser. In cases in which wide-angle diffusers must be used, their performance can often be markedly improved by use of fixed flow-guide-vanes mounted in the diffuser body or by suction along the diffuser walls to remove slow moving fluid. Both efforts serve to suppress separation in the diffuser.

To reiterate a point made earlier: *acceleration of real fluids tends to be an efficient process, deceleration an inefficient one.* Accelerating motion through a convergent nozzle, for example, is accompanied by a favorable pressure gradient which stabilizes the boundary layer and thus minimizes energy dissipation. Decelerating motion is accompanied by an adverse pressure gradient which promotes separation, instability, eddy formation, and large energy dissipation. This axiom may be extended to explain

Fig. **7.25**

the difference between the maximum efficiencies obtained in comparable complex machines such as hydraulic turbines and centrifugal pumps. In turbines the flow passages are predominantly convergent and the flow is accelerated; in the pump the opposite situation obtains. Hydraulic turbine efficiencies have been obtained up to 94%, but maximum centrifugal pump efficiencies are around 87%. The striking difference between these figures can be attributed primarily to the inherent efficiency and inefficiency of the acceleration and deceleration flow processes, respectively.

7.12 Secondary Flow

The secondary flow that develops in internal flow of real fluids is typified by the double spiral motion produced by a gradual bend in a closed passage. Consider the circular pipe bend of Fig. 7.26 and assume that its curves are gentle enough that separation is not to be expected. For an ideal fluid flowing under these conditions it has been shown (Section 4.8) that a pressure gradient develops across the bend due to the centrifugal forces of fluid particles as they move through the bend. Stability occurs in the ideal fluid when this pressure gradient brings about a balance between centrifugal and centripetal forces on the fluid particles. In a real fluid this stability is disrupted by the velocity reduction toward the walls. The reduction of velocity at the outer part, A, of the bend reduces the centrifugal force of particles moving near the wall, causing the pressure at the wall to be below that which would be maintained in an ideal fluid. However, the velocities of fluid particles toward the center of the bend are about the same as those of the ideal fluid, and the pressure gradient developed by their centrifugal forces is about the same. The "weakening" of the pressure gradient at the outer wall will cause a flow to be set up from the center of the pipe toward the wall which will develop into the twin eddy motion shown, and this secondary motion added to the main flow will cause a double spiral motion. As shown in Section 9.9, a combination of the double-spiral flow and separation leads to increased head losses in bends, a large part of which may be caused by increased wall shear stress in and downstream from the bend because of the redistribution of the flow streamlines by the secondary flow.

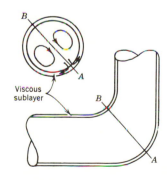

Fig. **7.26**

References

Batchelor, G. K. 1953. *The theory of homogeneous turbulence*. Cambridge: Cambridge Univ. Press.

Hinze, J. O. 1975. *Turbulence*. 2nd ed. New York: McGraw-Hill.

McDowell, D. M., and Jackson, J. D., Eds. 1970. *Osborne Reynolds and engineering science today*. New York: Manchester University Press (Barnes & Noble).

Schlichting, H. 1979. *Boundary layer theory*. 7th ed. New York: McGraw-Hill.

Tennekes, H., and Lumley, J. L. 1972. *A first course in turbulence*. Cambridge: The MIT Press.

Townsend, A. A. 1976. *The structure of turbulent shear flow*. 2nd ed. Cambridge: Cambridge Univ. Press.

Films

Abernathy, F. H. Fundamentals of boundary layers. NCFMF/EDC Film No. 21623, Encyclopaedia Britannica Educ. Corp.

Rouse, H. Mechanics of Fluids: characteristics of laminar and turbulent flow. Film No. U56159, Media Library, Audiovisual Center, Univ. of Iowa.

Stewart, R. W. Turbulence. NCFMF/EDC Film No. 21626, Encyclopaedia Britannica Educ. Corp.

Taylor, E. S. Secondary flow. NCFMF/EDC Film No. 21612, Encyclopaedia Britannica Educ. Corp.

Problems

7.1. When 0.001 9 m^3/s of water flow in a 76 mm pipeline at 21°C, is the flow laminar or turbulent?

7.2. Glycerin flows in a 25 mm (1 in.) pipe at a mean velocity of 0.3 m/s (1 ft/s) and temperature 25°C (77°F). Is the flow laminar or turbulent?

7.3. Carbon dioxide flows in a 50 mm pipe at a velocity of 1.5 m/s, temperature 66°C, and absolute pressure 380 kPa. Is the flow laminar or turbulent?

7.4. What is the maximum flowrate of air that may occur at laminar condition in a 100 mm (4 in.) pipe at an absolute pressure of 200 kPa (30 psia) and 40°C (104°F)?

7.5. What is the smallest diameter of pipeline that may be used to carry 6.3 l/s (100 gpm) of jet fuel (JP-4) at 15°C (59°F) if the flow is to be laminar?

7.6. Derive an expression for pipeline Reynolds number in terms of Q, d, and v.

7.7. A fluid flows in a 75 mm pipe which discharges into a 150 mm line. What is the Reynolds number in the 150 mm pipe if that in the 75 mm pipe is 20 000?

7.8. If water (20°C or 68°F) flows at constant depth (i.e., uniformly) in a wide open channel, below what flowrate q may the regime be expected to be laminar?

7.9. What is the maximum speed at which a spherical sand grain of diameter 0.254 mm (0.01 in.) may move through water (20°C or 68°F) and the flow regime be laminar?

7.10. In the laminar flow of an oil of viscosity 1 Pa · s the velocity at the center of a 0.3 m pipe is 4.5 m/s and the velocity distribution is parabolic. Calculate the shear stress at the pipe wall and within the fluid 75 mm from the pipe wall.

7.11. If the turbulent velocity profile in a pipe 0.6 m in diameter may be approximated by $v = 3.56 \, y^{1/7}$ (v m/s, y m) and the shearing stress in the fluid 0.15 m from the pipe wall is 6.22 Pa, calculate the eddy viscosity, mixing length, and turbulence constant at this point. Specific gravity of the fluid is 0.90.

7.12. A turbulent flow in a boundary layer has a velocity profile $v = (v_*/\kappa) \ln y + C$, where κ is the Kármán constant and the *friction velocity* v_* is defined as $\sqrt{\tau_o/\rho}$ and τ_o is the wall shear stress. Find expressions for the eddy viscosity ε and the shear stress $\tau(y)$ if the mixing length relationship $l = \kappa y$ is assumed valid.

7.13. Given that $\tau = \tau_o$ across the entire boundary layer in a turbulent flow, where τ_o is the wall shear stress, use equation 7.6 to derive the expected velocity profile. *Hint:* express the result in terms of the *friction velocity* $v_* = \sqrt{\tau_o/\rho}$ and (*a*) $y_o > 0$, which is the value of y at which the velocity goes to zero near the wall, or (*b*) δ which is the boundary layer thickness, that is, the y-value at which the velocity is equal to the undisturbed (or free stream) velocity V_o.

7.14. In a pipe flow the shear stress varies linearly with distance from the wall so $\tau = \tau_o(1 - y/R)$, where R is the pipe radius. Also, if v_c is the centerline velocity in the pipe, the typical observed velocity profile is accurately approximated by $(v - v_c)/v_* = (1/\kappa) \ln (y/R)$. Find and plot the variations of the eddy viscosity and the mixing length. Show that $l \sim \kappa y$ for small y/R.

7.15. The thickness of the viscous sublayer (Fig. 7.7) is usually defined by the intersection of a laminar velocity profile (adjacent to the wall) with a turbulent velocity profile. If the former may be described by $v = c_1 y$ and the latter by $v = c_2 y^{1/7}$, derive an expression for the thickness of the film in terms of c_1 and c_2.

7.16. For flow over smooth surfaces, a large number of experimental data can be represented by $v/v_* = v_* y/\nu$ in the viscous sublayer and $v/v_* = 2.5 \ln (v_* y/\nu) + 5$ in the fully turbulent region. Plot this composite "law of the wall" as v/v_* versus $y_+ = v_* y/\nu$ and locate the nominal thickness δ_v of the viscous sublayer at the intersection of the given curves. Recall $v_* = \sqrt{\tau_o/\rho}$. Discuss how δ_v depends on τ_o, ρ, and μ.

7.17. When oil (kinematic viscosity 1×10^{-4} m²/s, specific gravity 0.92) flows at a mean velocity of 1.5 m/s through a 50 mm pipeline, the head lost in 30 m of pipe is 5.4 m. What will be the head loss when the velocity is increased to 3 m/s?

7.18. If in the preceding problem the oil has kinematic viscosity 1×10^{-5} m²/s, the head lost will be 15 m if the roughness is very great. What head loss will occur when the mean velocity is increased to 3 m/s?

7.19. For flow over a smooth plate, what approximately is the maximum length of the laminar boundary layer if $V_o = 9.1$ m/s (30 ft/s) in the irrotational uniform flow and the fluid is air? Water?

7.20. Estimate the maximum laminar boundary layer thicknesses in the preceding problem.

7.21. A model of a thin streamlined body is placed in a flow for testing. The body is 0.9 m (3 ft) long and the flow velocity is 0.6 m/s (2 ft/s). What ν is needed to ensure that the boundary layer on the body is laminar?

7.22. Separation is found to occur 0.76 m (2.5 ft) from the nose of the body in the preceding problem when the fluid is water. Is the boundary layer flow laminar or turbulent at separation?

7.23 From Section 7.7 data, calculate the allowable maximum velocity at the centerline of the established flow of water in a 10 mm or 0.4 in. diameter smooth pipe if the boundary layer must remain laminar. How many pipe diameters from the smooth entrance does established flow begin?

7.24. If a zone of unestablished flow may be idealized to the extent shown and the centerline may be treated as a streamline in an ideal fluid, calculate the drag force exerted by the sidewalls (between sections 1 and 2) on the fluid if the flow is: (*a*) two-dimensional and 0.3 m wide normal to the paper and (*b*) axisymmetric. The fluid flowing has specific gravity 0.90.

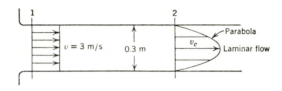

Problem **7.24**

7.25. Calculate α, β, and the momentum flux for the velocity profiles of problem 3.16. Assume that the passage is 0.6 m or 2 ft high, 0.3 m or 1 ft wide normal to the paper, and that water is flowing.

7.26. Calculate α, β, and the momentum flux for the velocity profiles of problem 3.17. Assume that the passage is 1 m or 3 ft in diameter and that water is flowing.

7.27. Calculate α and β for the velocity profile of problem 3.18.

7.28. Just downstream from the nozzle tip the velocity distribution is as shown. Calculate the flowrate past section 1, α, β, and the momentum flux. Assume water is flowing.

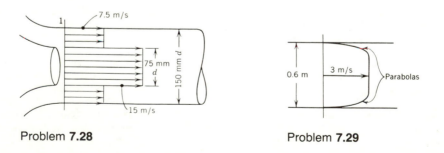

Problem **7.28**   Problem **7.29**

7.29. Calculate α and β for the flow in this two-dimensional passage if q is 1.5 m³/s·m.

7.30. If the velocity profile in a two-dimensional open channel may be approximated by the *parabola* shown, calculate the flowrate and the coefficients α and β.

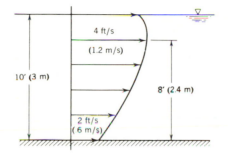

Problem **7.30**

7.31. A horizontal nozzle having a cylindrical tip of 75 mm diameter attached to a 150 mm water pipe discharges 0.05 m³/s. In the pipe just upstream from the nozzle the pressure is 62.6 kPa and α is 1.05. In the issuing jet α is 1.01. Calculate the conventional and exact head losses in the nozzle.

7.32. At a section just downstream from a sluice gate in a wide horizontal open channel the water depth is observed to be 0.20 ft, $q = 2$ cfs/ft, and the velocity distribution uniform. At another section further downstream the depth is 0.30 ft and the velocity profile follows the "seventh-root law" $v/v_s = (y/0.30)^{1/7}$ in which v_s is the surface velocity and y the distance from the channel bottom. Calculate the total drag force (per foot of width) exerted by the water on the length of channel bottom between the two sections.

7.33. These phenomena are observed in a laboratory channel 1 m wide, and the indicated measurements taken. Many velocity measurements at sections 1 and 2 allow calculation of β_1 and β_2, which turn out to be 1.02 and 1.07, respectively. Calculate the total drag force exerted by the water on the walls and bottom of the length of channel between sections 1 and 2.

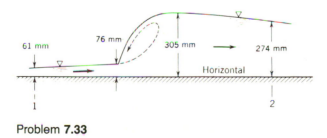

Problem **7.33**

7.34. Calculate the horizontal component of force exerted by the fluid on the upstream half of the semicylinder of problem 4.128.

7.35.   If the velocity profiles at the upstream and downstream ends of the mixing zone of a jet pump may be approximated as shown, and wall friction may be neglected, calculate the rise of pressure from section 1 to section 2, and the power lost in the mixing process. Water is flowing.

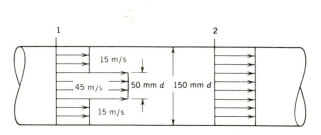

Problem **7.35**

7.36.   Calculate the pressure rise in the preceding problem if the velocity profile at the downstream end of the mixing zone is defined by $v = C(R - r)^{1/7}$.

7.37.   The sketch depicts a simplified flow situation in a two-dimensional wind- or water-tunnel in which the velocity distributions have been measured upstream and downstream from the strut. If wall friction is negligible, determine the drag of the strut in terms of the density of the fluid.

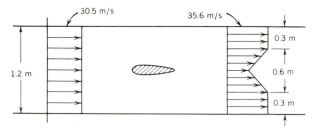

Problem **7.37**

7.38.   A long straight thin-walled tube of 25 mm or 1 in. diameter is towed through water (20°C or 68°F) at a speed of 0.15 m/s or 0.5 ft/s. The mean velocity of the water *through* the tube is observed to be 0.06 m/s or 0.2 ft/s (relative to the tube) with a parabolic velocity distribution in the downstream region of the tube. Calculate the total drag force exerted on the *inside* of the tube.

7.39.   This nozzle discharges ideal liquid of specific weight 9.80 kN/m³ to the atmosphere. The velocity in the pipe is 3.0 m/s. The velocity through the jet may be assumed to vary linearly from top to bottom. The nozzle is of such shape that the weight of liquid contained therein is 135.0 N, with center of gravity 100 mm to the right of section 1. Calculate accurately the magnitude, direction, and location of the resultant force exerted by the liquid on the nozzle.

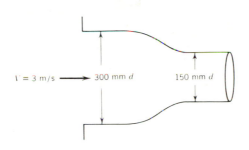

$\text{V} = 3 \text{ m/s} \longrightarrow$ 300 mm d 150 mm d

Problem **7.39**

7.40. In a reach of cylindrical pipe in which air is flowing the mean velocity is observed to increase from 400 to 500 ft/s and the mean temperature to rise from 100°F to 120°F. How much heat is being added to each pound of air in this reach of pipe?

7.41. In a reach of cylindrical pipe wrapped with perfect insulation ($q_H = 0$) the absolute pressure is observed to drop from 690 to 345 kPa and the mean velocity to increase from 91 to 175 m/s. Predict the temperatures in the pipe at each end of the reach if air is flowing in the pipe.

7.42. When water flows at a mean velocity of 3 m/s in a 300 mm pipe, the head loss in 100 m of pipe is 3 m. Estimate the rise of mean temperature of the water if the pipe is wrapped with perfect insulation ($q_H = 0$). If the pipe is not insulated, how much heat must be extracted from the water to hold its temperature constant?

7.43. Use equation 7.22 to derive the velocity profile and calculate the flowrate for laminar pipe flow.

7.44. When fluid of specific weight 50 lb/ft³ flows in a 6 in. pipeline, the frictional stress between fluid and pipe is 0.5 psf. Calculate the head lost per foot of pipe. If the flow rate is 2.0 cfs, how much power is lost per foot of pipe?

7.45. If the head lost in 30 m (100 ft) of 75 mm (3 in.) pipe is 7.6 m (25 ft) when a certain quantity of water flows therein, what is the total dragging force exerted by the water on this reach of pipe?

7.46. Air ($\gamma = 12.6$ N/m³) flows through a horizontal 0.3 m by 0.6 m rectangular duct at a rate of 15 N/s. Find the mean shear stress at the wall of the duct if the pressure drop in a 100 m length is 160 Pa. Compute the power lost per metre of duct length.

7.47. If the velocity near a pipe wall has the form $v = A \ln y + B$, while equation 7.22 is valid in a pipe of radius R, find how the mixing length l in equation 7.3 varies with y. Show that this variation agrees with the discussion of Section 7.2 for small y and propose a method to determine κ.

7.48. Calculate the minimum length of an optimum vaneless diffuser connecting the 1 m or 3.3 ft diameter test section of a wind tunnel to the 3 m or 10 ft diameter main circulation pipe.

8
SIMILITUDE AND DIMENSIONAL ANALYSIS

Most real fluid flow problems can be solved, at best, only approximately by analytical or numerical methods. Thus, experiments play a crucial role in verifying solutions, in suggesting which approximations are valid, or ultimately in providing results that cannot be obtained by theoretical analysis or numerical simulation. Unfortunately, many, if not most, real flow situations are either far too large or far too small for convenient experiment at their true size or, alternatively, the field conditions are so uncontrolled as to make systematic study tedious, if not impossible. When testing the real thing (that is, the *prototype*) is not feasible, a model (that is to say, a scaled version of the prototype) can be constructed and the performance of the prototype simulated in the model.

Near the latter part of the last century, models began to be used to study flow phenomena that could not be solved by analytical methods or by means of available experimental results. Over the era of modern engineering, the use of models and confidence in model studies have steadily increased: the aeronautical engineer obtains data from model tests in wind tunnels; the naval architect tests ship models in towing basins; the mechanical engineer tests models of turbines and pumps and predicts the performance of the full-scale machines from these tests; the civil engineer works with models of hydraulic structures, rivers, and estuaries to obtain more reliable solutions to design problems. The justifications for models include *economics*—a model, being small typically, costs little compared to the prototype from which it is built, and its results may lead to savings of many times its cost—and *practicality*—in a model test environmental and flow conditions can be rigorously controlled. On the other hand, the flow in a river or the wind that carries pollution over a city can neither be controlled nor (often) be predicted. Models are not always smaller than the prototype. Recent models of the blood flow in the human brain, whose goal was to assess the impact of clots on the flow, are an example of models that are much larger than the prototype. In this chapter the laws of similitude are described and provide a basis for interpretation of model results.

Surprisingly, when one is faced with planning or interpreting a set of experiments in a model or prototype situation, it is rapidly discovered that to carry each pertinent flow variable or physical parameter through its appropriate range involves obtaining a prohibitive number of individual experimental data points. Indeed, for a simple incompressible flow where the goal is to find the drag force on a body, Fox and McDonald[1]

[1] R. W. Fox and A. T. McDonald, *Introduction to Fluid Mechanics*, 2nd ed. Wiley, 1978.

320

and White[2] both estimate that 10^4 separate data points are needed to describe the drag force D as a function of the characteristic body size d and the fluid velocity V, density ρ, and viscosity μ. The second major topic of this chapter, namely, dimensional analysis, can be applied to this problem and the results leavened with the knowledge gained from the laws of similitude, to learn that the drag can be characterized completely by a relationship between only *two groups* of variables. In fact, the drag coefficient $C_D = D/\rho V^2 d^2$ is functionally related to the Reynolds number $\mathbf{R} = V\, d\rho/\mu$ so

$$C_D = f(\mathbf{R}) \tag{8.1}$$

Just as Reynolds found (see Section 7.1) that the single number $\mathbf{R}$ defined the limits of laminar and turbulent flow for all fluids in all pipes, it is now found (without recourse to any experiment) that, for an incompressible flow and a given set of bodies with the same *shape* (say, a prototype and any models without regard for their size), their drag is characterized by $\mathbf{R}$ for all body sizes and all fluids. Therefore, using one conveniently sized body and one fluid, one needs to obtain only 10 to 20 data points at different velocities (e.g., see Section 13.6) to define the desired relationship among D, V, d, ρ, and μ. This assistance provided by the laws of similitude and by dimensional analysis in finding the important groups of variables in each flow situation is an indispensable aid in planning and interpreting experiments. In addition, an important understanding is often gained about the physical phenomena being studied, and clues on what to look for in the experiment are often revealed.

8.1 Similitude and Models

Although the basic theory for the interpretation of model tests is quite simple, it is seldom possible to design and operate a model of a fluid phenomenon from theory alone; here the *art* of engineering must be practiced with experience, judgment, ingenuity, and patience if useful results are to be obtained, correctly interpreted, and prototype performance predicted from results.

Similitude of flow phenomena not only occurs between a prototype and its model but also may exist between various natural phenomena if certain laws of similarity are satisfied. Similarity thus becomes a means of correlating the apparently divergent results obtained from similar fluid phenomena and as such becomes a valuable tool of modern fluid mechanics; the application of the laws of similitude will be found to lead to more comprehensive solutions and, therefore, to a better understanding of fluid phenomena in general.

There are three basic types of similitude; all must be obtained if complete similarity is to exist between fluid phenomena.[3] The first and simplest of these is the familiar *geometrical similarity*, which states that the flowfields and boundary geometry of

---

[2] F. M. White, *Fluid Mechanics*, McGraw-Hill, 1979.
[3] However, in some special cases effective and useful similarity is obtained without satisfying this condition.

model and prototype have the same shape[4] and, therefore, that the ratios between corresponding lengths in model and prototype are the same. In the model and prototype of Fig. 8.1, for example,

$$\frac{d_p}{d_m} = \frac{l_p}{l_m}$$

Corollaries of geometric similarity are that corresponding areas vary with the squares of their linear dimensions,

$$\frac{A_p}{A_m} = \left(\frac{d_p}{d_m}\right)^2 = \left(\frac{l_p}{l_m}\right)^2$$

and that volumes vary with the cubes of their linear dimensions.

Consider now the flows past the geometrically similar objects of Fig. 8.1; if these are flowfields of the same shape, and if ratios of *corresponding* velocities and accelerations are the same throughout the flow,[5] the two flows are said to be *kinematically similar*. Thus, flows with geometrically similar streamlines are kinematically similar.

In order to maintain geometric and kinematic similarity between flowfields, the forces acting on *corresponding* fluid masses must be related by ratios similar to those above; this similarity is known as *dynamic similarity*. With the forces shown acting on the corresponding fluid masses M_p and M_m of Fig. 8.1, the vector polygons may be drawn, and from the geometric similarity of these polygons and from Newton's second law (which, of course, is operative in both model and prototype)

$$\frac{\mathbf{F}_{1p}}{\mathbf{F}_{1m}} = \frac{\mathbf{F}_{2p}}{\mathbf{F}_{2m}} = \frac{\mathbf{F}_{3p}}{\mathbf{F}_{3m}} = \frac{M_p \mathbf{a}_{4p}}{M_m \mathbf{a}_{4m}} \qquad (8.2)$$

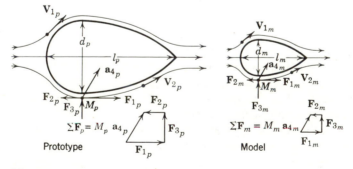

Fig. **8.1**

<hr />

[4] Departure from geometric similarity (resulting in a *distorted* model) is frequently used (for economic and physical reasons) in models of rivers, harbors, estuaries, and so forth; however, most models are geometrically similar to their prototypes.

[5] For example, $\mathbf{V}_{1p}/\mathbf{V}_{1m} = \mathbf{V}_{2p}/\mathbf{V}_{2m}$ and $\mathbf{a}_{3p}/\mathbf{a}_{3m} = \mathbf{a}_{4p}/\mathbf{a}_{4m}$.

For dynamic similarity these force ratios must be maintained on all the corresponding fluid masses throughout the flowfields; thus it is evident that they can be governed only by relations between the dynamic and kinematic properties of the flows and by the physical properties of the fluids involved.

Complete similarity requires simultaneous satisfaction of geometric, kinematic, and dynamic similarity. Kinematically similar flows must be geometrically similar, of course. But, if the *mass distributions* in flows are similar (i.e., if the density ratios for corresponding fluid masses are the same for all masses), then from equation 8.2 it is seen that kinematically similar flows are automatically completely similar. As an example, for constant-density flows, kinematic similarity guarantees complete similarity, while dynamic plus geometric similarity guarantees kinematic similarity.

Referring again to Fig. 8.1, it is apparent that

$$\mathbf{F}_{1p} + \mathbf{F}_{2p} + \mathbf{F}_{3p} = M_p \mathbf{a}_{4p}$$

and

$$\mathbf{F}_{1m} + \mathbf{F}_{2m} + \mathbf{F}_{3m} = M_m \mathbf{a}_{4m}$$

and from this it may be concluded that, if the ratios between three of the four corresponding terms in these equations are the same, the ratio between the corresponding fourth terms must be the same as that between the other three. Thus one of the ratios of equation 8.2 is redundant, and dynamic similarity is characterized by an equality of force ratios numbering one less than the forces.[6] When the first force ratio[7] is eliminated from equation 8.2, the equation may be rewritten in various ways. It is usually written in the form of simultaneous equations,

$$\frac{M_p \mathbf{a}_{4p}}{\mathbf{F}_{2p}} = \frac{M_m \mathbf{a}_{4m}}{\mathbf{F}_{2m}} \tag{8.3}$$

$$\frac{M_p \mathbf{a}_{4p}}{\mathbf{F}_{3p}} = \frac{M_m \mathbf{a}_{4m}}{\mathbf{F}_{3m}} \tag{8.4}$$

The scalar magnitudes of forces that *may* affect a flowfield include:[8] pressure, F_P; inertia, F_I; gravity, F_G; viscosity, F_V; elasticity, F_E; and surface tension, F_T. Since these forces are taken to be those on or of any fluid mass, they may be generalized by the following fundamental relationships.

$$F_P = (\Delta p)A = (\Delta p)l^2$$

$$F_I = Ma = \rho l^3 \left(\frac{V^2}{l}\right) = \rho V^2 l^2$$

---

[6] The force, $M\mathbf{a}$ is the "inertia force."

[7] Any force ratio may be selected for elimination, depending on the quantities that are desired in the equations.

[8] This discussion excludes, for example, the Coriolis force of rotating systems (yields a Rossby number) and the buoyancy forces in stratified flow (yields a Richardson number).

$$F_G = Mg_n = \rho l^3 g_n$$

$$F_V = \mu\left(\frac{dv}{dy}\right)A = \mu\left(\frac{V}{l}\right)l^2 = \mu Vl$$

$$F_E = EA = El^2$$

$$F_T = \sigma l$$

In each case these fundamental forces are obtained by replacing derivatives and other terms in the actual force expressions obtained in earlier chapters by difference expressions and characteristic lengths, velocities, pressures, and so forth, of the flow. Here l and V are a characteristic or typical length and velocity for the system, while ρ, μ, E, and σ are fluid properties and g_n is the acceleration due to gravitational acceleration.

To obtain dynamic similarity between two flowfields when all these forces act, all corresponding force ratios must be the same in model and prototype; thus dynamic similarity between two flowfields when all possible forces are acting may be expressed (after the pattern of equations 8.3 and 8.4) by the following five simultaneous equations:

$$\left(\frac{F_I}{F_P}\right)_p = \left(\frac{F_I}{F_P}\right)_m = \left(\frac{\rho V^2}{\Delta p}\right)_p = \left(\frac{\rho V^2}{\Delta p}\right)_m \qquad \mathbf{E}_p = \mathbf{E}_m \qquad (8.5)$$

$$\left(\frac{F_I}{F_V}\right)_p = \left(\frac{F_I}{F_V}\right)_m = \left(\frac{Vl\rho}{\mu}\right)_p = \left(\frac{Vl\rho}{\mu}\right)_m \qquad \mathbf{R}_p = \mathbf{R}_m \qquad (8.6)$$

$$\left(\frac{F_I}{F_G}\right)_p = \left(\frac{F_I}{F_G}\right)_m = \left(\frac{V^2}{lg_n}\right)_p = \left(\frac{V^2}{lg_n}\right)_m \qquad \mathbf{F}_p = \mathbf{F}_m \qquad (8.7)$$

$$\left(\frac{F_I}{F_E}\right)_p = \left(\frac{F_I}{F_E}\right)_m = \left(\frac{\rho V^2}{E}\right)_p = \left(\frac{\rho V^2}{E}\right)_m \qquad \mathbf{M}_p = \mathbf{M}_m \qquad (8.8)$$

$$\left(\frac{F_I}{F_T}\right)_p = \left(\frac{F_I}{F_T}\right)_m = \left(\frac{\rho l V^2}{\sigma}\right)_p = \left(\frac{\rho l V^2}{\sigma}\right)_m \qquad \mathbf{W}_p = \mathbf{W}_m \qquad (8.9)$$

Over the history of fluid mechanics a set of dimensionless numbers of dynamic similarity have come to be well known, and each is associated (typically) with the name of the person who introduced it; these ubiquitous numbers are:

$$\text{Euler number, } \mathbf{E} = V\sqrt{\frac{\rho}{2\Delta p}} \qquad (8.10)$$

$$\text{Reynolds number, } \mathbf{R} = \frac{Vl}{\nu} \qquad (8.11)$$

$$\text{Froude number, } \mathbf{F} = \frac{V}{\sqrt{lg_n}} \qquad (8.12)$$

$$\text{Cauchy}^9 \text{ number, } \mathbf{C} = \frac{\rho V^2}{E} \tag{8.13}$$

$$\text{Weber number, } \mathbf{W} = \frac{\rho l V^2}{\sigma} \tag{8.14}$$

Each number has a physical interpretation as a force ratio, being the ratio of the inertia force in a flow to the particular force represented by the number; for example, $\mathbf{R}$ is the ratio of the inertia to the viscous force. It is apparent now that the foregoing force-ratio equations may be written in terms of the dimensionless numbers as indicated above. Following the argument leading to equations 8.3 and 8.4, it will be noted that only four of these equations are independent; thus, if any four of them are simultaneously satisfied (e.g., 8.6 to 8.9), dynamic similarity will be ensured and the fifth equation (8.5) will be satisfied automatically.

Fortunately, in most engineering problems four simultaneous equations are not necessary, since some of the forces stated above (1) may not act, (2) may be of negligible magnitude, or (3) may oppose other forces in such a way that the effect of both is reduced. In each new problem of similitude a good understanding of fluid phenomena is necessary to determine how the problem may be satisfactorily simplified by the elimination of the irrelevant, negligible, or compensating forces. The reasoning involved in such analyses is best illustrated by citing certain simple and recurring engineering examples.

Reynolds Similarity

In the classical low-speed submerged body problem typified by the conventional airfoil of Fig. 8.2 there are no surface tension phenomena, negligible compressibility (elastic) effects, and gravity does not affect the flowfield. Thus three of the four equations (8.6 to 8.9), which are typically selected to be satisfied simultaneously, are not relevant to the problem, and dynamic similarity is obtained between model and prototype when the Reynolds numbers or ratio of inertia to viscous forces are the same, that is, when

$$\left(\frac{Vl}{\nu}\right)_p = \mathbf{R}_p = \mathbf{R}_m = \left(\frac{Vl}{\nu}\right)_m \tag{8.6}$$

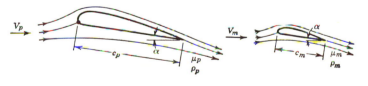

Fig. **8.2**

[9] Because the sonic velocity $a = \sqrt{E/\rho}$ (Section 1.4), the Mach number, $\mathbf{M}$, and Cauchy number, $\mathbf{C}$, are related by $\mathbf{C} = \mathbf{M}^2$.

providing that model and prototype are geometrically similar and are similarly oriented to their oncoming flows. Since the equation places no restriction on the fluids of the model and prototype, the latter could move through air and the former be tested in water; if the Reynolds numbers of model and prototype could be made the same, dynamic similitude would result. If for practical reasons the same fluid is used in model and prototype, the product (Vl) must be the same in both; this means that the velocities around the model will be *larger* than the corresponding ones around the prototype. In aeronautical research the model is frequently tested with compressed air in a *variable density* (or *pressure*) *wind tunnel*; here $\nu_m < \nu_p$ and large velocities past the model are not required. Once equality of Reynolds numbers is obtained in model and prototype, it follows that the ratio of any corresponding forces (such as lift or drag) will be equal to the ratio of any other relevant corresponding forces. Thus (for drag force)

$$\left(\frac{D}{\rho V^2 l^2}\right)_p = \left(\frac{D}{\rho V^2 l^2}\right)_m \tag{8.15}$$

from which the drag of the prototype may be predicted directly from drag and velocity measurements in the model; no corrections for "scale effect" are needed if the Reynolds numbers are the same in model and prototype.

Much of the foregoing reasoning may also be applied to the flow of incompressible fluids through closed passages. Consider, for example, the flows through prototype and model of the contraction of Fig. 8.3. For geometric similarity $(d_2/d_1)_p = (d_2/d_1)_m$, $(l/d_1)_p = (l/d_1)_m$, $(x_1/d_1)_p = (x_1/d_1)_m$, $(x_2/d_2)_p = (x_2/d_2)_m$; *and the roughness pattern of the two passages must be similar in every detail*. Surface tension and elastic effects are nonexistent and gravity does not affect the flowfields. Accordingly dynamic similarity results when equation 8.6 is satisfied, that is, when

$$\left(\frac{V d_1}{\nu}\right)_p = \mathbf{R}_p = \mathbf{R}_m = \left(\frac{V d_1}{\nu}\right)_m \tag{8.6}$$

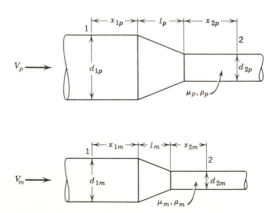

Fig. **8.3**

from which it follows that equation 8.5 is satisfied automatically, so

$$\left(\frac{p_1 - p_2}{\rho V^2}\right)_p = \left(\frac{p_1 - p_2}{\rho V^2}\right)_m \tag{8.5}$$

allowing (for example) prediction of prototype pressure drop ($p_1 - p_2$) from model measurements. In addition, the ratio between the model and prototype flowrates can be deduced because d_{1p}, d_{1m}, V_p, and V_m are known. Clearly, $Q_p = (\pi/4)d_{1p}^2 V_p$ and $Q_m = (\pi/4)d_{1m}^2 V_m$; therefore, $Q_m/Q_p = (d_{1m}^2/d_{1p}^2)(V_m/V_p)$. If $d_{1m}/d_{1p} = 1/5$ and water is used in both model and prototype, then by virtue of equation 8.6, $V_m/V_p = 5$ and $Q_m/Q_p = 1/5$. Here again it is immaterial whether the fluids are the same, dynamic similarity being ensured by the equality[10] of the Reynolds numbers in model and prototype.

Froude Similarity

Another example of wide engineering interest is the modelling of the flowfield about an object (such as the ship of Fig. 8.4) moving on the surface of a liquid. Here geometric similarity is obtained by a carefully dimensioned model suitably weighted so that $(d/l)_p = (d/l)_m$. Compressibility of the liquid is of no consequence in such problems, and surface tension may also be ignored if the model is not too small. However, the motion of the ship through the liquid generates water waves on the liquid surface and these waves move under the influence of gravity. Thus, the drag of the ship is the sum of the effects of the energy dissipated in wave generation (which results in a clearly visible and characteristic wave pattern) and of the frictional action of the liquid on the hull. *If frictional effects are assumed to be negligible*,[11] dynamic similitude is characterized by

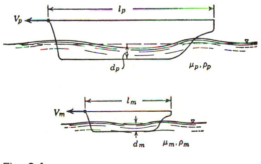

Fig. **8.4**

$$\left(\frac{V}{\sqrt{lg_n}}\right)_p = \mathbf{F}_p = \mathbf{F}_m = \left(\frac{V}{\sqrt{lg_n}}\right)_m \tag{8.7}$$

From this it follows (as for the submerged object of Fig. 8.2) that

$$\left(\frac{D}{\rho V^2 l^2}\right)_p = \left(\frac{D}{\rho V^2 l^2}\right)_m \tag{8.15}$$

which means that, if the model is tested with $\mathbf{F}_p = \mathbf{F}_m$, and the drag measured, the drag of the prototype may be predicted for the corresponding speed; the latter is (from equation 8.7) $V_m \sqrt{l_p/l_m}$.

For ship hulls of good design the contribution of wave pattern and frictional action to the drag are of the same order, and neither can be ignored. Here the phenomena will be associated with both the Froude and the Reynolds numbers, requiring for dynamic similitude the two simultaneous equations

$$\left(\frac{V}{\sqrt{g_n l}}\right)_p = \mathbf{F}_p = \mathbf{F}_m = \left(\frac{V}{\sqrt{g_n l}}\right)_m \tag{8.7}$$

$$\left(\frac{Vl}{\nu}\right)_p = \mathbf{R}_p = \mathbf{R}_m = \left(\frac{Vl}{\nu}\right)_m \tag{8.6}$$

Taking g_n the same for model and prototype and solving these equations (by eliminating V) yields $\nu_p/\nu_m = (l_p/l_m)^{3/2}$, indicating that a specific relation between the viscosities of the liquids is required once the model scale is selected. This means (1) that a liquid of appropriate viscosity must be found for the model test, or (2) if the same liquid is used for model and prototype, the model must be as large as the prototype! Since liquids of appropriate viscosity may not exist and full-scale models are obviously impractical, the engineer is forced to choose between the two equations, since both cannot be satisfied simultaneously with the same liquid for model and prototype. This is done by operating the model so that $\mathbf{F}_p = \mathbf{F}_m$ (resulting in $\mathbf{R}_p \gg \mathbf{R}_m$) and then correcting[12] the test results by experimental data dependent on Reynolds number. William Froude originated this technique for ship model testing in England around 1870, but the same equations and principles apply to any flowfield controlled by the combined action of gravity and viscous action; models of rivers, harbors, hydraulic structures, and open-flow problems in general are good examples. However, such models are considerably more difficult to operate and interpret than ship models because of the less well-defined frictional resistance caused by variations of surface roughness and complex boundary geometry.

Mach Similarity

An example of similitude in compressible fluid flow is that of the projectile of Fig. 8.5. Here gravity and surface tension do not affect the flowfield, and similitude will result

---

[12] This is known as correcting for *scale effect* and is a correction necessitated by incomplete similitude; there would be no scale effect if equations 8.6 and 8.7 could both be satisfied.

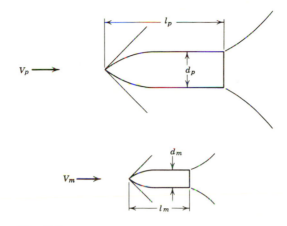

Fig. **8.5**

from the actions of resistance and elasticity (compressibility) characterized by

$$\left(\frac{Vl}{\nu}\right)_p = \mathbf{R}_p = \mathbf{R}_m = \left(\frac{Vl}{\nu}\right)_m \tag{8.6}$$

$$\left(\frac{V}{a}\right)_p = \mathbf{M}_p = \mathbf{M}_m = \left(\frac{V}{a}\right)_m \tag{8.8}$$

which may be solved to yield $l_p/l_m = (\nu_p/\nu_m)(a_m/a_p)$, showing that a relation must exist between model scale and the viscosities and sonic speeds in the gases used in model and prototype if dynamic similarity is to be complete. In this situation (unlike the analogous one for the ship model), gases are available that will allow the equation to be satisfied.

Euler Similarity—The Cavitation Number

When prototype cavitation is to be modeled, the foregoing equations of similitude are inadequate, since they do not include vapor pressure p_v, the attainment of which is the unique feature of the cavitation.[13] Consider the cavitating hydrofoil prototype and model of Fig. 8.6, the model fixed in a water tunnel where p_o and V_o can be controlled. Geometric similarity and the same orientation to the oncoming flow are assumed, and complete dynamic similitude is desired. Here gravity will have small effect on the flowfield, compressibility of the fluid will be insignificant, and surface tension may be neglected.[14] Accordingly, dynamic similitude is obtained when the Reynolds number and the appropriate form of the Euler number are used, that is, when

---

[13] See Appendix 7.
[14] Surface tension may have critical influence in such problems if the cavitation is incipient; the gas content of the liquid is also important. Both of these points are ignored here to simplify the problem.

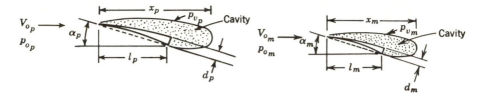

Fig. **8.6**

$$\left(\frac{V_o l}{\nu}\right)_p = \mathbf{R}_p = \mathbf{R}_m = \left(\frac{V_o l}{\nu}\right)_m \tag{8.6}$$

$$\left(\frac{p_o - p_v}{\rho V_o^2}\right)_p = \sigma_p = \sigma_m = \left(\frac{p_o - p_v}{\rho V_o^2}\right)_m \tag{8.16}$$

When these equations are satisfied the cavity shapes in model and prototype will be the same; that is, $(x/l)_p = (x/l)_m$. The second of the two simultaneous equations represents an equality of special Euler numbers (known as the *cavitation number σ*) containing the (absolute) pressures, p_o and p_v, peculiar to cavitation. If these equations can be simultaneously satisfied by the use of appropriate liquid or adjustment of tunnel velocities and pressures or both, then (as for the submerged body without cavitation)

$$\left(\frac{D}{\rho V_o^2 l^2}\right)_p = \left(\frac{D}{\rho V_o^2 l^2}\right)_m \tag{8.15}$$

and

$$\left(\frac{\Delta p}{\rho V_o^2}\right)_p = \left(\frac{\Delta p}{\rho V_o^2}\right)_m \tag{8.5}$$

in which Δp in model and prototype represents the pressure changes between any two *corresponding* points in the flowfields of model or prototype. Here (as for the ship model) it is virtually impossible to satisfy both of the simultaneous equations with essentially the same liquid in model and prototype. Again the engineer must select and satisfy the more important of the two equations and correct the test results to allow for the unsatisfied equation. In this problem it is clear that the existence and extent of the cavitation zone dominate the flowfield and frictional aspects are of comparatively small importance; accordingly the tests would be carried out with the cavitation numbers the same in model and prototype and empirical adjustments made for the frictional phenomena.

Comment

It is a rare occurrence in engineering problems when pure theory alone leads to a complete answer. This is particularly true in the construction, operation, and interpretation of models where compromises of the type above are almost invariably necessary; there is no substitute for experience and judgment in effecting such compromises.

A major aid to the engineer are the equations which govern a fluid process. They serve as a guide to key dimensionless parameters. This topic has been developed in depth by Kline.[15] Simple examples to whet the appetite are seen from the analyses of the hydraulic jump and shock waves in Sections 6.5 to 6.8. The hydraulic jump equation 6.9 shows that the ratio of the depths before and after the jump y_1/y_2 is a function of only the Froude number $F^2 = V_1^2/g_ny_1$. In the normal shock wave case the results in equations 6.18 and 6.19 demonstrate that the proper pressure ratio to be analyzed is $(p_2 - p_1)/p_1$ not p_2/p_1, and that both the pressure ratio and the velocity ratio of values ahead of and behind the jump are functions of only the upstream Mach number M and the ratio of specific heats k.

The foregoing treatment of similitude is intended as an introduction to the theory and to the numerous practical applications cited in the literature at the end of the chapter. Also it is now clear that certain dimensionless groups play a key and recurring role in fluid mechanics problems.

Illustrative Problems

Water (0°C) flows in a 75 mm horizontal pipeline at a mean velocity of 3 m/s. The pressure drop in 10 m of this pipe is 14 kPa. With what velocity must gasoline (20°C) flow in a geometrically similar 25 mm pipeline for the flows to be dynamically similar, and what pressure drop is to be expected in $3\frac{1}{3}$ m of the 25 mm pipe?

Relevant Equations and Given Data

$$\left(\frac{\Delta p}{\rho V_o^2}\right)_p = \left(\frac{\Delta p}{\rho V_o^2}\right)_m \tag{8.5}$$

$$\left(\frac{Vd}{\nu}\right)_p = \mathbf{R}_p = \mathbf{R}_m = \left(\frac{Vd}{\nu}\right)_m \tag{8.6}$$

$V_p = 3$ m/s $\Delta p_p = 14$ kPa $d_p = 75$ mm $\mu_p = 1.781 \times 10^{-3}$ m²/s

$\rho_p = 998.8$ kg/m³ $d_m = 25$ mm $\rho_m = 998.2 \times 0.68 = 680.3$ kg/m³

$\mu_m = 2.9 \times 10^{-4}$ m²/s

Solution.

$$\frac{3 \times (0.075) \times 999.8}{0.001\ 781} = \frac{V_m \times (0.025) \times 680.3}{0.000\ 29} \qquad V_m = 2.16 \text{ m/s} \blacktriangleleft$$

$$\tag{8.6}$$

$$\frac{14}{[999.8 \times (3)^2]} = \frac{\Delta p_m}{[680.3 \times (2.16)^2]} \qquad \Delta p_m = 4.94 \text{ kPa} \blacktriangleleft \tag{8.5}$$

A ship of 400 ft length is to be tested by a model 10 ft long. If the ship travels at 30

[15] S. J. Kline, *Similitude and approximation theory*, McGraw-Hill, 1965.

knots, at what speed must the model be towed for dynamic similitude between model and prototype? If the drag of the model is 2 lb, what prototype drag is to be expected?

Relevant Equations and Given Data

$$\left(\frac{V}{\sqrt{lg_n}}\right)_p = F_p = F_m = \left(\frac{V}{\sqrt{lg_n}}\right)_m \tag{8.7}$$

$$\left(\frac{D}{\rho V^2 l^2}\right)_p = \left(\frac{D}{\rho V^2 l^2}\right)_m \tag{8.15}$$

$$l_p = 400 \text{ ft} \qquad l_m = 10 \text{ ft} \qquad V_p = 30 \text{ knot}$$

$$D_m = 2 \text{ lb} \qquad g_n = 32.2 \text{ ft/s}^2 \qquad \rho_m = \rho_p$$

Solution. Determine the model speed from equality of the Froude numbers. A knot is a nautical mile (6 080 ft) per hour; therefore, $V_p = 50.7$ ft/s and

$$\frac{(50.7)^2}{[400 \times 32.2]} = \frac{V_m^2}{[10 \times 32.2]} \qquad V_m = 8.0 \text{ ft/s} \blacktriangleleft \tag{8.7}$$

Assuming the same liquid for model and prototype and neglecting frictional effects,

$$\frac{2}{[(8)^2 \times (10)^2]} = \frac{D_p}{[(50.7)^2 \times (400)^2]} \qquad D_p = 128\ 525 \text{ lb} \blacktriangleleft \tag{8.15}$$

With no information on hull form, frictional effects cannot be included in this problem; however, if they were included the predicted drag of the prototype would be somewhat *less* than 128 525 lb. See the Illustrative Problem of Section 13.5.

A short smooth hydraulic overflow structure passes a flowrate of 600 m³/s. What flowrate should be used with a 1 : 15 model of this structure to obtain dynamic similitude if friction may be neglected?

Given Data

$$Q_p = 600 \text{ m}^3/\text{s} \qquad l_m/l_p = 1/15$$

Solution. Because of the predominance of gravitational action, similitude will result when equation 8.7 is satisfied. Writing V as Q/A or Q/l^2, the equality of Froude numbers may be expressed as

$$\left(\frac{Q}{l^2\sqrt{g_n l}}\right)_p = \left(\frac{Q}{l^2\sqrt{g_n l}}\right)_m \tag{8.7}$$

whence

$$Q_m = 600 \times \left(\frac{1}{15}\right)^{5/2} = 0.69 \text{ m}^3/\text{s} \blacktriangleleft$$

A 1 : 10 model of a jet plane traveling at a velocity of 900 m/s at 6 km altitude in the U.S. Standard Atmosphere (Appendix 4) is to be tested by firing the model like a projectile in a tank of carbon dioxide at 20°C. Calculate the required velocity of the model and the pressure required in the tank for dynamic similitude between model and prototype.

Relevant Equations and Given Data

$$\rho = p/RT \tag{1.3}$$

$$a = \sqrt{kRT} \tag{1.11}$$

$$V_p = 900 \text{ m/s} \qquad l_m/l_p = 1/10 \qquad alt = 6 \text{ km} \qquad T_m = 293 \text{ K}$$

Solution. For dynamic similarity $\mathbf{R}_p = \mathbf{R}_m$ and $\mathbf{M}_p = \mathbf{M}_m$. Using

$$\left(\frac{V}{a}\right)_p = \mathbf{M}_p = \mathbf{M}_m = \left(\frac{V}{a}\right)_m \tag{8.8}$$

gives $V_m = 900(a_m/a_p)$. From Appendix 4, $T_p = 249$ K, while from Table 2, $R_p = 286.8$ J/kg·K, $k_p = 1.4$, $R_m = 187.8$ J/kg·K, and $k_m = 1.28$. Calculating a_m and a_p from equation 1.11, $a_m = 265.4$ m/s and $a_p = 316.2$ m/s, yielding $V_m = 755.4$ m/s. ◀ Now, using

$$\left(\frac{Vl}{\nu}\right)_p = \mathbf{R}_p = \mathbf{R}_m = \left(\frac{Vl}{\nu}\right)_m \tag{8.6}$$

(with numerical values from Fig. 1.6, Table 2, and Appendix 4),

$$\frac{900 \times 10 \times 0.660}{1.595} = \frac{755.4 \times 1 \times \rho_m}{1.47} \qquad \rho_m = 7.25 \text{ kg/m}^3 \tag{8.6}$$

Finally,

$$7.25 = \frac{p_m}{187.8 \times 293} \qquad p_m\text{(abs)} = 399 \text{ kPa} \blacktriangleleft \tag{1.3}$$

8.2 Dimensional Analysis

Another useful tool of modern fluid mechanics, and one closely related to the laws of similitude, is *dimensional analysis*—the mathematics of the dimensions of quantities. The methods of dimensional analysis are built on Fourier's *principle of dimensional homogeneity* (1882), which states that an equation expressing a physical relationship between quantities must be dimensionally homogeneous; that is, the dimensions of each side of the equation must be the same. This principle was utilized in Chapter 1 in obtaining the dimensions of density and kinematic viscosity, and it was recommended as a valuable means of checking engineering calculations. Further investigation of the principle will reveal that it affords a means of ascertaining the forms of physical equations from knowledge of relevant variables and their dimensions. Although dimensional manipulations cannot be expected to produce analytical solutions to physical problems, dimensional analysis proves a powerful tool in formulating problems which defy analytical solution and must be solved experimentally. Here dimensional analysis comes into its own by pointing the way toward a maximum of information from a minimum of experiment. It accomplishes this by the formation of dimensionless groups, some of which are identical with the force ratios developed with the laws of similitude.

In the SI and FSS unit systems, there are four basic dimensions that are directly relevant to fluid mechanics, namely, length (L), mass (M), time (t), and thermodynamic temperature (T). Only the first three have been included in the following discussion, but should temperature be needed in an analysis, its presence does not change the principles or process. A summary of the fundamental quantities of fluid mechanics and their dimensions and units is given in Appendix 1, the conventional system of capital letters being followed to indicate the dimensions of quantities. An exception is made with respect to the time dimension (t), to avoid confusion with thermodynamic temperature. In problems of mechanics, the basic relation between force and mass is given by the Newtonian law; force or weight = mass × acceleration; and, therefore, dimensionally,

$$F = \frac{ML}{t^2}$$

through which the dimension of force can be expressed in terms of the basic dimensions. Thus, there are only three *independent* fundamental dimensions here.

To illustrate the mathematical steps in a simple dimensional problem, suppose that it is known that the power, P, which can be derived from a hydraulic turbine is dependent on the rate of flow through the machine, Q, the specific weight of the fluid, γ, and the unit mechanical energy, E_T, which is given up by every unit weight of fluid as it passes through the machine. Suppose that the relation between these four variables is unknown but it is known that these are the only variables involved in the problem.[16] With this meager knowledge the following mathematical statement may be made:

$$P = f(Q, \gamma, E_T)$$

From the principle of dimensional homogeneity it is apparent that the quantities involved cannot be added or subtracted since their dimensions are different. This principle limits the equation to a combination of products of powers of the quantities involved, which may be expressed in the general form

$$P = CQ^a\gamma^bE^c$$

in which C is a dimensionless constant which may exist in the equation but cannot, of course, be obtained by dimensional methods and a, b, c are unknowns. The equation can be written dimensionally as

(Dimensions of P) = (Dimensions of Q)^a(Dimensions of γ)^b(Dimensions of E_T)^c

---

[16] Note that experience and analytical ability in determining the relevant variables are necessary before the methods of dimensional analysis can be successfully applied.

or

$$\frac{ML^2}{t^3} = \left(\frac{L^3}{t}\right)^a \left(\frac{M}{L^2t^2}\right)^b (L)^c$$

The principle of dimensional homogeneity being applied, the exponent of *each* of the fundamental dimensions is the same on each side of the equation, giving the following equations in the exponents of the dimensions:

$$M: \quad 1 = b$$

$$L: \quad 2 = 3a - 2b + c$$

$$t: -3 = -a - 2b$$

Solving for a, b, and c yields

$$a = 1 \qquad b = 1 \qquad c = 1$$

and resubstituting these values in the equation above for P gives

$$P = CQ\gamma E_T$$

The form of the equation (confirmed by equations 4.6 and 4.7) has, therefore, been derived, without physical analysis, solely from consideration of the dimensions of the quantities which were known to enter the problem. The magnitude of C must be obtained either (1) from a physical analysis of the problem or (2) from experimental measurements of P, Q, γ, and E_T.

From the foregoing problem it appears that in dimensional analysis (of problems in mechanics) only three equations can be written since there are only three independent fundamental dimensions: M, L, and t. This fact limits the completeness with which a problem with more than three unknowns may be solved, but it does not limit the utility of dimensional analysis in obtaining the form of the terms of the equation. This point may be fruitfully illustrated and a more general analysis procedure developed by re-examining the ship model problem previously treated (Section 8.1) by the laws of similitude. Consider the ship of Fig. 8.4 having a certain *shape*[17] and draft, d. The drag of the ship will depend[18] on the size of the ship [characterized by its length (l)], the viscosity (μ) and density of the fluid (ρ), the velocity (V) of the ship, and acceleration (g_n) due to gravity [which dominates the surface wave pattern].

While Lord Rayleigh put forth the early bases of a dimensional analysis method,

---

[17] *Shape* is emphasized here so that it cannot be confused with *size*. Size of the ship is *not* fixed by *algebraic* dimensions such as l, d, and so forth. With shape constant and size variable, a series of geometrically similar objects is implied.

[18] Selection of the relevant variables is the crucial first step in any problem of dimensional analysis, and obviously this cannot be done without some experience and "feel" for the problem; the mathematical processes of dimensional analysis cannot overcome the selection of irrelevant variables!

Buckingham[19] provided a broad generalization known as the Π-theorem. The modern version of the Buckingham theorem can be stated as follows (see, e.g., Kline[20]):

1. If n variables are functions of each other (such as D, l, ρ, μ, V, and g_n of the present example), then k equations of their exponents (a, b, c, etc.) can be written (where k is the largest number of variables among the n variables which *cannot* be combined into a dimensionless group).
2. In most cases, k is equal to the number m of independent dimensions (equal to three here, that is, M, L, and t). Generally, $k \leqq m$.
3. Application of dimensional analysis allows expression now of the functional relationship in terms of $(n - k)$ distinct dimensionless groups. In the present case, $n = 6$, $k = m = 3$, so three groups are expected. Indeed, the previous similitude analysis yielded $D/\rho l^2 V^2$, $\mathbf{F}$, and $\mathbf{R}$!

Buckingham designated the dimensionless groups by the Greek (capital) letter Π. The important advantage of the Π-theorem is that it shows in advance of the analysis how many groups are to be expected and allows the engineer flexibility in formulating them (particularly as it is already known that certain groups, e.g., force ratios, are relevant).

For the present problem of drag on a ship, the functional relationship among the variables can be written in the form

$$f(D, l, \rho, \mu, V, g_n) = 0$$

by transferring all terms to the left side of the equation. The number of Π's to be expected is $n - m = 6 - 3 = 3$, so the relation between dimensionless groups should be expressible as

$$f'(\Pi_1, \Pi_2, \Pi_3) = 0$$

Two crucial points are, first, that one should check to see if $k < m$ and, second, that there are a sizable number of ways to combine six variables into three dimensionless groups. A rational approach is needed. Experience with similitude suggests that V, l, and ρ appear in most dimensionless groups because they all appear in the inertia force of a similitude analysis. Variables such as D, μ, σ, and E appear only in the unique group describing the ratio of the inertia force to the force related to the variable. Therefore, a useful procedure might be:

1. Find the largest number of variables which do not form a dimensionless Π-group. In the present case the number of independent dimensions is $m = 3$ and V, ρ, and l cannot be formed into a Π-group, so $k = m = 3$.
2. Determine the number of Π-groups to be formed. Here, $n = 6$, $k = m = 3$, and thus three Π's are required.
3. Combine sequentially the variables that cannot be formed into a dimensionless

---

[19] E. Buckingham, "Model experiments and the forms of empirical equations," *Trans. A. S. M. E.* 37, 1915, pp. 263ff.
[20] S. J. Kline, *Similitude and approximation theory*, McGraw-Hill, 1965.

group, with each of the remaining variables to form the requisite Π-groups. Here then the three Π's would be

$$\Pi_1 = f_1(D, \rho, V, l)$$

$$\Pi_2 = f_2(\mu, \rho, V, l)$$

$$\Pi_3 = f_3(g_n, \rho, V, l)$$

Thus it remains only to determine the detailed form of these dimensionless groups. Using Π_1 as an example to illustrate the method,

$$\Pi_1 = D^a \rho^b V^c l^d$$

Writing the equation dimensionally,

$$M^0 L^0 t^0 = \left(\frac{ML}{t^2}\right)^a \left(\frac{M}{L^3}\right)^b \left(\frac{L}{t}\right)^c (L)^d$$

The following equations in the exponents of the dimensions are obtained:

$$M: 0 = a + b$$

$$L: 0 = a - 3b + c + d$$

$$t: 0 = 2a + c$$

Solving these equations in terms of a, b, c, or d (a will be used here),

$$b = -a \qquad c = -2a \qquad d = -2a$$

and substituting them in the equation above for Π_1,

$$\Pi_1 = \left(\frac{D}{\rho l^2 V^2}\right)^a$$

Since a dimensionless group raised to a power is of no more significance than the group itself, the exponent may be taken as any convenient number other than zero, and Π_1 as $D/\rho V^2 l^2$. Similarly Π_2 is obtained as $Vl\rho/\mu$ and Π_3 as $V/\sqrt{lg_n}$. But

$$\mathbf{F} = \frac{V}{\sqrt{lg_n}} \qquad \text{and} \qquad \mathbf{R} = \frac{Vl\rho}{\mu}$$

allowing the equation to be written in the more general form

$$f'\left(\frac{D}{\rho l^2 V^2}, \mathbf{R}, \mathbf{F}\right) = 0$$

or

$$\frac{D}{\rho l^2 V^2} = f''(\mathbf{F}, \mathbf{R})$$

showing without experiment, but from dimensional analysis alone, that $D/\rho l^2 V^2$ depends only on the Froude and Reynolds numbers. This sort of result is the main objective of dimensional analysis in engineering problems and may always be obtained by application of the technique used here. Although it gives no clue to the functional relationship among $D/\rho l^2 V^2$, $\mathbf{F}$, and $\mathbf{R}$, it has arranged the numerous original variables into a relation between a smaller number of dimensionless groups of variables and has thus indicated how test results should be processed for concise presentation. Use of the Π-theorem in dimensional analysis is thus seen to be highly efficient in formulating dimensionless groups which may be easily interpreted in terms of those of geometric, kinematic, and dynamic similitude.

The laws of similitude showed that the flowfield about a prototype surface ship is dynamically similar to that of its model if (with geometric similarity) $\mathbf{F}_p = \mathbf{F}_m$ and $\mathbf{R}_p = \mathbf{R}_m$—and that satisfaction of these conditions leads to $(D/\rho l^2 V^2)_p = (D/\rho l^2 V^2)_m$. The result of dimensional analysis is the same since it has demonstrated the existence of a unique functional relationship among $D/\rho l^2 V^2$, $\mathbf{F}$, and $\mathbf{R}$ for a ship of one shape. Thus (whatever this relationship) for the same $\mathbf{F}$ and same $\mathbf{R}$ in model and prototype, $(D/\rho l^2 V^2)_p = (D/\rho l^2 V^2)_m$.

Illustrative Problem

A smooth symmetric body of cross-sectional area A moves through a compressible fluid of density ρ, viscosity μ, and modulus of elasticity E (recall that the sonic or acoustic velocity $a = \sqrt{E/\rho}$) and with a velocity V_0. If the drag force exerted on the body is D, find the functional dependence of D on the other variables.

Given Data

The relation to be analyzed is

$$f(D, A, \rho, \mu, V_0, E) = 0$$

In the compressible flow about an immersed body, the effect of gravity is generally negligible, so g_n is not included here.

Solution. Follow the steps outlined above.

1. Here A, ρ, and V_0 cannot be formed into a dimensionless group and $m = 3(M, L, t)$ so $k = m = 3$.
2. There are $n = 6$ variables, so $n - k = 3$.
3. The Π-groups are formed as

$$\Pi_1 = f_1(D, A, \rho, V_0)$$
$$\Pi_2 = f_2(\mu, A, \rho, V_0)$$
$$\Pi_3 = f_3(E, A, \rho, V_0)$$

For Π_1,

$$\frac{M^0 L^0}{t^0} = \left(\frac{ML}{t^2}\right)^a (L^2)^b \left(\frac{M}{L^3}\right)^c \left(\frac{L}{t}\right)^d$$

$$M: \quad 0 = a + c$$

$$L: \quad 0 = a + 2b - 3c + d$$

$$t: \quad 0 = -2a - d$$

Solving in terms of c: $a = -c$, $b = c$, and $d = 2c$, so

$$\Pi_1 = \left(\frac{A\rho V_0^2}{D}\right)^c \quad \text{or} \quad \text{(letting } c = -1) \, \Pi_1 = \frac{D}{A\rho V_0^2}$$

For Π_2,

$$\frac{M^0 L^0}{t^0} = \left(\frac{M}{Lt}\right)^a (L^2)^b \left(\frac{M}{L^3}\right)^c \left(\frac{L}{t}\right)^d$$

which yields in terms of d: $a = -d$, $b = \frac{1}{2}d$, and $c = d$, so

$$\Pi_2 = \left(\frac{\sqrt{A}\rho V_0}{\mu}\right)^d \quad \text{or} \quad \text{(for } d = 1) \, \Pi_2 = \frac{\sqrt{A}\rho V_0}{\mu}$$

For Π_3,

$$\frac{M^0 L^0}{t^0} = \left(\frac{M}{t^2 L}\right)^a (L^2)^b \left(\frac{M}{L^3}\right)^c \left(\frac{L}{t}\right)^d$$

which yields in terms of a: $b = 0$, $c = -a$, and $d = -2a$, so

$$\Pi_3 = \left(\frac{E}{\rho V_0^2}\right)^a \quad \text{or} \quad \text{(for } a = -\frac{1}{2}) \, \Pi_3 = \frac{V_0}{\sqrt{E/\rho}}$$

As $\sqrt{E/\rho} = a$, $\Pi_3 = V_0/a$. Accordingly,

$$f'\left(\frac{D}{A\rho V_0^2}, \frac{\sqrt{A}\rho V_0}{\mu}, \frac{V_0}{a}\right) = 0$$

or defining $\mathbf{R} = \sqrt{A}\rho V_0/\mu$ and $\mathbf{M} = V_0/a$,

$$\frac{D}{A\rho V_0^2} = f''(\mathbf{R}, \mathbf{M}) \blacktriangleleft$$

that is, the drag depends on the Reynolds and Mach numbers (see Section 13.2).

8.3 Analysis and Characteristics of Turbomachines

Turbomachines were defined and the applicable momentum principles discussed in Sections 6.10 to 6.12. These principles provided the means for examining specific flow patterns, power transfers from or to blades, and so forth. However, because of the complexity of most turbomachines and often their great size, it is common practice to model such machines. In addition, it is useful to be able to make generalizations and predictions about common characteristics or differences between generic types of machines (e.g., centrifugal and axial-flow pumps or impulse and reaction turbines). Dimensional analysis is very valuable in such matters.

The first step is to select the relevant variables. For the class of turbomachines in which no combustion or significant heat transfer occurs, the power, P, required by or delivered by geometrically similar turbomachines will depend on: the runner or impeller diameter, D (a characteristic length); the rotative speed, N; the volume flowrate, Q; the energy, H, added or subtracted, respectively, from each unit mass[21] of fluid passing through the machine; and the fluid characteristics, namely, viscosity, μ, density, ρ, and elasticity, E. From Section 8.2 and equations 4.6 or 4.7 it is already known that $P \propto \rho Q H$; however, it is very instructive to step back and begin afresh with a new dimensional analysis. Doing this leads directly to significant design coefficients and to a deeper understanding of the relationships among the variables relevant to turbomachines. At the appropriate point in what follows the relation $P \propto \rho Q H$ is used to coalesce the results to a final form from which the nature of the proportionality factor is explicitly seen.

Accordingly, for the variables listed above, it is expected that the general relation

$$f(P, D, N, Q, H, \mu, \rho, E) = 0$$

is valid. There are eight variables and three independent dimensions[22] so $8 - 3 = 5$ Π-terms can be constructed. A Buckingham Π-analysis requires now that μ, E, P, Q, and H be successively combined with the remaining three variables ρ, D, and N to obtain the required five Π-terms (see the problems to find the consequence of using D, ρ, Q or H, ρ, Q as a basis).

Because N has the dimension t^{-1}, ρ, D, N, and μ will form a Reynolds number **R**. Set

$$\Pi_1 = \mu^a \rho^b D^c N^d$$

Solving the dimensional equations gives

$$\Pi_1 = \frac{\rho N D^2}{\mu} = \mathbf{R}$$

Similarly, using E, ρ, D, and N yields

$$\Pi_2 = \frac{\rho N^2 D^2}{E} = \frac{N^2 D^2}{a^2} = \mathbf{M}^2$$

where **M** is a Mach number and $a^2 = E/\rho$.

Now it remains to generate three Π-terms for P, Q, and H, respectively. Taking

$$\Pi_3 = P^a \rho^b D^c N^d$$

---

[21] From Section 4.5, E_P or E_T appear as abrupt rises or falls in the energy line across the machine and have the units J/N or ft · lb/lb. Often E_P or E_T are called the *head* on the unit. Because g_n is not a relevant parameter in this analysis and because it is the mass of fluid that possesses energy, here the *head*, $H = E_{(T \text{ or } P)} g_n$, is defined as the energy per unit mass; (the dimensions of H) $= L^2/T^2$.

[22] Clearly, none of the sets of terms (ρ, D, N), (D, ρ, Q) or (H, ρ, Q) can be combined into a Π-group; hence, $k = m = 3$.

$$\Pi_4 = Q^a \rho^b D^c N^d$$

$$\Pi_5 = H^a \rho^b D^c N^d$$

and solving the dimensional equations gives

$$\Pi_3 = \frac{P}{\rho N^3 D^5} = Power\ Coefficient\ \mathbf{C}_P$$

$$\Pi_4 = \frac{Q}{ND^3} = Capacity\ Coefficient\ \mathbf{C}_Q$$

$$\Pi_5 = \frac{H}{N^2 D^2} = Head\ Coefficient\ \mathbf{C}_H$$

In summary, it has been possible to obtain the required five Π-terms by a combination of the methods and concepts developed earlier in the chapter. The results can be put in the following alternative forms:

$$\mathbf{C}_P = \frac{P}{\rho N^3 D^5} = f'(\mathbf{C}_Q, \mathbf{C}_H, \mathbf{R}, \mathbf{M})$$

$$\mathbf{C}_Q = \frac{Q}{ND^3} = f''(\mathbf{C}_P, \mathbf{C}_H, \mathbf{R}, \mathbf{M})$$

$$\mathbf{C}_H = \frac{H}{N^2 D^2} = f'''(\mathbf{C}_P, \mathbf{C}_Q, \mathbf{R}, \mathbf{M})$$

However, as noted above there is a specific relationship among, P, Q, H, and ρ, that is, $P \propto \rho Q H$. Accordingly, the dimensionless Π-term

$$\Pi_3' = \frac{P}{\rho Q H} = \frac{\mathbf{C}_P}{\mathbf{C}_Q \cdot \mathbf{C}_H} = f^{IV}(\mathbf{C}_Q, \mathbf{C}_H, \mathbf{R}, \mathbf{M}) \qquad (8.17)$$

can replace $\Pi_3 = \mathbf{C}_P$

If now $\mathbf{C}_Q$ and $\mathbf{C}_H$ are held constant for a set of similar machines, all the hydraulic losses in head due to flow friction, eddying, separation, and so forth, are embodied by the Reynolds number effect. However, other losses in power occur in turbomachines (in bearings, etc.) that do not show up as head losses. In the particular case of incompressible flow, $\mathbf{M}$ is not a relevant parameter. Then, equation 8.17 becomes (with $\mathbf{C}_Q$ and $\mathbf{C}_H$ held constant)

$$\frac{P}{\rho Q H} = f^V(\mathbf{R}) = \eta_H$$

where η_H is the hydraulic efficiency of the machine. If the mechanical efficiency η_M (which is a dimensionless number) is taken as a parameter to measure nonhydraulic losses, then, in general,

$$\frac{P}{\rho Q H} = f^{VI}(\eta_M, \mathbf{R}) = \eta$$

where η is the total efficiency of the machine. Clearly, then, under the constraint that C_Q and C_H are held constant in a set of similar machines, the proportionality between P and ρQH depends only on the hydraulic and mechanical efficiencies plus Mach number effects (if any).

Shepherd[23] discusses η and indicates that differences in η between model and prototype can be represented, for example, for hydraulic reaction turbines, by an equation of the form

$$\frac{1 - \eta_m}{1 - \eta_p} = \left(\frac{D_p}{D_m}\right)^{1/5} \tag{8.18}$$

One reason for taking this approach is that the gross Reynolds number defined herein is not a good measure of the similarity of the detailed flow inside a turbomachine. In models, relative roughness effects (caused by not being able to achieve detailed geometric similarity of roughness elements) and scale differences cause transition from laminar to turbulent flow and separation or eddy formation to take place in relatively different places in model and prototype. Equation 8.18 gives a measure of these "scale-effects." Thus, gross dynamic and geometric similarity *do not* guarantee detailed kinematic similarity.

Examination of $\Pi_4(C_Q)$ suggests that similarity can be gaged in terms of kinematic similarity, that is, similarity of velocity patterns, in otherwise geometrically similar machines. The product $ND \propto u$ which is the peripheral velocity of the runner or impeller, while $Q/D^2 \propto V_r$ which is the radial velocity through the runner or impeller. Thus,

$$\Pi_4 = \frac{Q}{ND^3} \propto \frac{V_r}{u} = \Pi_4'$$

Holding Π_4' (and Π_4) constant in a set of geometrically similar machines implies similarity of velocity triangles in the machines (see Section 6.12). Geometrically similar machines having similar velocity triangles are called *homologous machines*. The performance characteristics of any machine in a homologous set are obtained from knowledge of the characteristics of one machine in the set, for example, from data obtained in a model test.

Some variation in efficiency η among homologous machines may have to be accounted for (see equation 8.18) because it is sometimes not possible to keep **R** constant for all machines. Holding **R** constant for a model and prototype requires $(ND^2)_m = (ND^2)_p$; that is, $N_m = N_p(D_p/D_m)^2$. Thus, a one-fifth scale model of a turbine would have to be run at 25 times prototype speed. This may not be feasible; hence, smaller machines usually have lower **R** and proportionately larger friction effects.

Because the relationship between P and ρ, Q and H is known and η is expected to handle both **R** and mechanical effects, a new dimensional analysis for an *incompressible* fluid flowing in a *hydraulic* turbine or pump might consider only

[23] D. G. Shepherd, *Principles of Turbomachinery*, Macmillan, 1956.

$$f(D, Q, H, N, \eta) = 0$$

Two new Π-terms are expected because η is already dimensionless while $k = 2$ (why?). The previous analysis suggests that, for example,

$$\frac{H}{N^2 D^2} = f'\left(\frac{Q}{ND^3}, \eta\right) \qquad \text{or} \qquad C_H = f'(C_Q, \eta)$$

As Shepherd notes, a plot of experimental results for a set of geometrically similar turbomachines of several sizes takes the form shown in Fig. 8.7. The data points lie in a band as shown because η varies not only with D but also with changes in **R** and in mechanical losses in the same machine as its speed changes. If η remains essentially constant, the set of geometrically similar machines represented by point 1 on Fig. 8.7 has the following characteristics

$$H \propto D^2$$

$$Q \propto D^3$$

Thus, a 10% decrease in D causes a 20% head reduction and a 30% reduction in flowrate.

While the above analysis gave considerable insight to turbomachines, two other dimensionless terms are very widely used in engineering practice because through them it is possible to characterize various classes of pumps, compressors, and turbines without regard for their size (D is excluded in forming the terms). For pumps and compressors, the flowrate Q, head H, speed N, and maximum efficiency η_{max} are key performance parameters (η_{max} includes ρ and μ effects) for machines of all sizes. Two Π-terms, Π_4 and Π_5, are obtained with these variables; eliminating D by taking the ratio of $(\Pi_4)^{1/2}$ to $(\Pi_5)^{3/4}$ gives

$$N_S = \frac{NQ^{1/2}}{H^{3/4}} \tag{8.19}$$

where this traditional combination N_S is known as the *specific speed*. (N_S equals the actual rotative speed N (in rad/s) when the machine operates under unit head H and unit flow Q.) Figure 8.8 is a plot of η versus N_S for various types of pumps; the lines

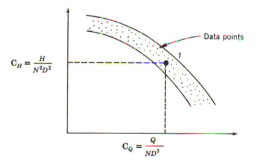

Fig. **8.7**

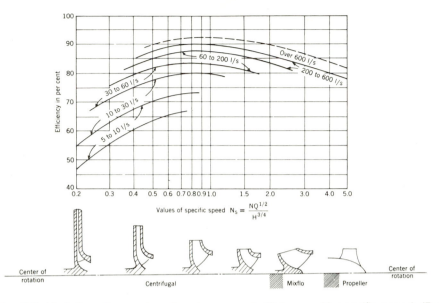

Fig. 8.8 Variation of pump impeller shapes and efficiency with specific speed. (By permission of Worthington Pump International, Inc.)

are the locus of maximum efficiency η_{max} data points. Here η is the total efficiency of the pump and includes mechanical and hydraulic losses. Thus, the various locations of the curves (shown here to be related to flow and by implication to machine size) can be interpreted as representing variations in **R** and η_M (see above).

By using N_S one can gain a feeling for expected pump characteristics because N_S is a constant for a homologous set of machines. For example, for small N_S at a given N, a relatively low flow and high head are needed. Because $H \propto E_P \propto D^2$ for fixed N, a typical low N_S pump has essentially a radial flow, for example, a centrifugal pump, and a low Q to limit velocities within the pump passages (this holds down head losses caused by friction effects). For a large N_S, large Q and small H are appropriate, these conditions being satisfied by propeller or axial flow pumps in which the large flow moves through relatively large passages at reasonable speeds without causing excessive head loss due to friction. If one attempts to operate a centrifugal pump at high N_S, the large Q required will drive the efficiency down (due to increased friction-induced head loss) and it will be much lower than the efficiency of a pump (e.g., axial flow) whose shape suits the flow requirements better (review Fig. 8.8).

For hydraulic turbines the relevant performance and fluid variables are P, H, N, ρ and η_{max}. Using the ratio of $(\Pi_3)^{1/2}$ to $(\Pi_5)^{5/4}$ leads to a specific speed for turbines

$$N_S = \frac{NP^{1/2}}{\rho^{1/2}H^{5/4}} \tag{8.20}$$

Figure 8.9 is a representative plot of η_{max} versus N_S for turbines. Again N_S is an indicator of the performance characteristics of turbines. For fixed N, low N_S corresponds to relatively high head and so to large D coupled with low power (hence low

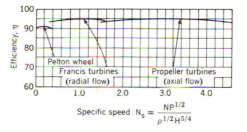

Fig. **8.9** Typical peak efficiencies as a function of specific speed for different types of hydraulic turbine.

Q). In this range the Pelton wheel and Francis reaction turbine (see Sections 6.11 and 6.12) have the best efficiencies. At high N_S, the low head, relatively small D, and large P (hence large Q) describe axial flow machines.

An interesting development in hydropower is the expanding use of pumped storage to provide peak power generation by storage of water pumped to a reservoir during off-peak hours. A typical installation uses a combination pump/turbine-motor/generator set to pump the water and by reverse flow in the same machine to generate power. Figure 8.10 shows a runner for the Coo Trois Ponts project in Belgium. The machines are Francis-type reversible pump/turbines. When generating at rated turbine capacity (145 MW), the specific speed of the pump/turbine is 0.62. When pumping at rated

Fig. **8.10** Coo Trois Ponts (Belgium) pumped-storage pump turbine runner (Francis-type). (Courtesy of Allis-Chalmers.)

discharge of 46 m³/s, the specific speed is 0.60. Where do these numbers[24] fall on Figs. 8.8 and 8.9?

Finally, the performance of hydraulic pumps and turbines is often degraded by cavitation. An energy-line analysis (see Section 4.4) of the flow leading from a reservoir to a pump or from a turbine to the reservoir into which the turbine discharges will show that the minimum head typically occurs in these machines before energy input to or after energy withdrawal from the flowing fluid, that is, at the inlet (or suction side) of a pump and at the outlet (discharge side) of a turbine. The total useful head at these points (i.e., absolute pressure head plus velocity head less vapor-pressure head) is usually called the *net positive suction head*, NPSH. In the present notation (where $H = g_n E_P$, for example), $H_{ps} = g_n \cdot$ NPSH; then

$$H_{ps} = g_n \cdot \text{NPSH} = \frac{p_{o,i} - p_v}{\rho} + \frac{V_{o,i}^2}{2}$$

when $p_{o,i}$ is the absolute pressure and $V_{o,i}$ is the velocity at pump inlet (i) or turbine outlet (o) and p_v is the vapor pressure of the liquid. The reader should reconcile this definition with the cavitation examples in Section 4.4 to see that once NPSH is known the relationship between the physical elevation of a pump (or turbine) and the reservoir levels from which the pump draws (or to which the turbine discharges) is fixed. This parallels the situation in which the elevation of a constriction in a pipeline determines whether or not cavitation will occur there. Clearly, cavitation will occur when $H_{ps} \leq V_{o,i}^2/2$, but a better criterion is needed. As the specific speeds include a head term, it is natural to define *suction specific speeds*

$$\text{Pumps:} \quad N_{SS} = \frac{NQ^{1/2}}{H_{ps}^{3/4}} \tag{8.21}$$

$$\text{Turbines:} \ N_{SS} = \frac{NP^{1/2}}{\rho^{1/2} H_{ps}^{5/4}} \tag{8.22}$$

The data of Wislicenus[25] show, according to Dixon,[26] that cavitation begins at essentially the same value of N_{SS} for all pumps and at a different, but constant, value for all turbines whose passages have been designed to minimize cavitation. Thus, to avoid cavitation in the pump inlet, $N_{SS} \leq 3$, and in the turbine discharge, $N_{SS} \leq 4$. These numbers are useful then in preliminary design and prior to tests of a specific unit because they provide a means of obtaining estimates of the minimum net positive suction head NPSH required for any installation.

---

[24] The reader should not be surprised that the specific speeds for the operation of the same machine as a pump or as a turbine are almost equal. If $P = \rho QH$ (as in an ideal machine where $\eta = 1$), the N_S quoted for pumps (equation 8.19) and turbines (equation 8.20) are just alternate forms of the same equation. Here the efficiencies are greater than 90% and about equal, so the machine has essentially the same specific speed as a pump or as a turbine.

[25] G. F. Wislicenus, *Fluid Mechanics of Turbomachinery*, 2nd ed., McGraw-Hill, 1965.

[26] S. L. Dixon, *Fluid Mechanics, Thermodynamics of Turbomachinery*, 3rd ed., Pergamon, 1978.

References

Allen, J. 1947. *Scale models in hydraulic engineering*. London: Longmans, Green.

A.S.C.E. 1942. Hydraulic models. *A.S.C.E. Manual of Engineering Practice*, No. 25.

Bridgman, P. W. 1931. *Dimensional analysis*. Rev. ed. New Haven: Yale University Press.

Buckingham, E. 1915. Model experiments and the forms of empirical equations. *Trans. A.S.M.E.* 37: 263.

Dixon, S. L. 1978. *Fluid mechanics, thermodynamics of turbomachinery*. 3rd ed. Oxford: Pergamon Press.

Duncan, W. J. 1953. *Physical similarity and dimensional analysis*. London: Edward Arnold.

Hudson, R. Y., Herrmann, F. A., Sager, R. A., Whalin,R. W., Keulegan, G. H., Chatham, C. E., Jr., and Hales, L. Z. 1979. "Coastal hydraulic models." U.S. Army Corps of Engineers *Coastal Engineering Research Center Spec. Rep. No. 5*, May.

Ippen, A. T. 1970. Hydraulic scale models. In *Osborne Reynolds and engineering science today*, Eds. McDowell and Jackson. New York: Barnes and Noble.

Ipsen, D. C. 1960. *Units, dimensions, and dimensionless numbers*. New York: McGraw-Hill.

Kline, S. J. 1965. *Similitude and approximation theory*. New York: McGraw-Hill.

Langhaar, H. L. 1951. *Dimensional analysis and theory of models*. New York: Wiley.

Sedov, L. I. 1959. *Similarity and dimensional methods in mechanics*. New York: Academic Press.

Shepherd, D. G. 1956. *Principles of turbomachinery*. New York: Macmillan.

Taylor, E. S. 1974. *Dimensional analysis for engineers*. London: Oxford University Press.

White, F. M. 1979. *Fluid mechanics*. New York: McGraw-Hill. Chapters 5 and 11.

Films

Eisenberg, P. Cavitation. NCFMF/EDC Film No. 21620, Encyclopaedia Britannica Educ. Corp.

Shapiro, A. H. The fluid dynamics of drag. NCFMF/EDC Films No. 21601–21604, Encyclopaedia Britannica Educ. Corp.

Problems

8.1. An airplane wing of 3 m chord length moves through still air at 15°C and 101.3 kPa at a speed of 320 km/h. A 1 : 20 scale model of this wing is placed in a wind tunnel, and dynamic similarity between model and prototype is desired. (*a*) What velocity is necessary in a tunnel where the air has the same pressure and temperature as that in flight? (*b*) What velocity is necessary in a variable-density wind tunnel where absolute pressure is 1 400 kPa and temperature is 15°C? (*c*) At what speed must the model move through water (15°C) for dynamic similarity?

8.2. A body of 1 m (3.3 ft) length moving through air at 60 m/s (200 ft/s) is to be studied by use of a 1 : 6 model in a water tunnel in which the pressure level is maintained high enough to prevent cavitation. Using air and water properties from Appendixes 4 (altitude zero) and 2 respectively, what velocity should be provided in the test section of the water tunnel for dynamic similarity between model and prototype? What would be the ratio of the prototype drag force to the model drag force? Assume the water temperature 15°C (59°F).

8.3. In a water (20°C) tunnel test the velocity approaching the model is 24 m/s. The pressure head difference between this point and a point on the object is 60 m of water. Calculate the corresponding pressure difference (kPa) on a 12 : 1 prototype in an airstream (101.3 kPa and 15°C) if model and prototype are tested under conditions of dynamic similarity.

8.4. A flat plate 1.5 m long and 0.3 m wide is towed at 3 m/s in a towing basin containing water at 20°C, and the drag force is observed to be 14 N. Calculate the dimensions of a similar plate which will yield dynamically similar conditions in an airstream (101.4 kPa and 15°C) having a velocity of 18 m/s. What drag force may be expected on this plate?

8.5. It is desired to obtain dynamic similarity between 2 cfs of water at 50°F flowing in a 6 in. pipe and crude oil flowing at a velocity of 30 ft/s at 90°F. What size of pipe is necessary for the crude oil?

8.6. A large Venturi meter (Section 11.11) for air flow measurement has d_1 1.5 m and d_2 0.9 m. It is to be calibrated using a 1 : 12 model with water the flowing fluid. When 0.07 m³/s pass through the model, the drop in pressure from section 1 to section 2 is 172 kPa. Calculate the corresponding flowrate and pressure drop in the prototype. Use densities and viscosities for air and water given in Appendixes 4 (altitude zero) and 2. Assume that these properties do not change. Assume the water temperature 15°C.

8.7. This 1 : 12 pump model (using water at 15°C) simulates a prototype for pumping oil of specific gravity 0.90. The input to the model is 0.522 kW. Calculate the viscosity of the oil and the prototype power for complete dynamic similarity between model and prototype.

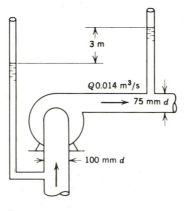

Problem **8.7**

8.8. When crude oil flows at 60°F in a 2 in. horizontal pipeline at 10 ft/s, a pressure drop of 480 psi occurs in 200 ft of pipe. Calculate the pressure drop in the corresponding length of 1 in. pipe when gasoline at 100°F flows therein at the same Reynolds number.

8.9. A flowrate of 0.18 m³/s of water (20°C) discharges from a 0.3 m pipe through a 0.15 m nozzle into the atmosphere. The axial force component exerted by water on nozzle is 3 kN. If frictional effects may be ignored, what corresponding force will be exerted on a 4 : 1 prototype of nozzle and pipe discharging 1.13 m³/s of air (101.4 kPa and 15°C) to the atmosphere? If frictional effects are included the axial force component is 3.56 kN. What flowrate of air is then required for dynamic similarity? What is the corresponding force on the nozzle discharging air?

8.10. The pressure drop in a certain length of 0.3 m horizontal waterline is 68.95 kPa when the mean velocity is 4.5 m/s and the water temperature 20°C. If a 1 : 6 model of this pipeline using air as the working fluid is to produce a pressure drop of 55.2 kPa in the corresponding length when the mean velocity is 30 m/s, calculate the air pressure and temperature required for dynamic similarity between model and prototype.

8.11. A model of this disk rotating in a casing is to be constructed with air as the working fluid. What should be the corresponding dimensions and speed of the model for the torques of model and prototype to be the same? The air is at 59°F and 5.0 psia.

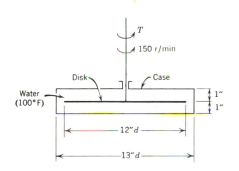

Problem **8.11**

8.12. A force of 9 N is required to tow a 1 : 50 ship model at 4.8 km/h. Assuming the same water in towing basin and sea, calculate the corresponding speed and force in the prototype if the flow is dominated by: (*a*) density and gravity, (*b*) density and surface tension, and (*c*) density and viscosity.

8.13. A tanker 300 m (980 ft) long is to be tested by a 1 : 50 scale model. If the ship is to travel at 46 km/h (25 knots), at what speed must the model be towed to obtain dynamic similarity (neglecting friction) with its prototype?

8.14. A ship model 1 m long (with negligible skin friction) is tested in a towing basin at a speed of 0.6 m/s. To what ship velocity does this correspond if the ship is 60 m long? A force of 4.45 N is required to tow the model; what propulsive force does this represent in the prototype?

8.15. A seaplane is to take off at 130 km/h (80 mph). If the maximum speed available for testing its model is 4.5 m/s (15 ft/s), what is the largest model scale which can be used?

8.16. A ship 120 m (400 ft) long moves through freshwater at 15°C (59°F) at 32 km/h (17.4 knots). A 1 : 100 model of this ship is to be tested in a towing basin containing a liquid of specific gravity 0.92. What viscosity must this liquid have for both Reynolds' and Froude's laws to be satisfied? At what velocity must the model be towed? What propulsive force on the ship corresponds to a towing force of 9 N (2 lb) in the model?

8.17. A perfect fluid discharges from an orifice under a static head. For this orifice and a geometrically similar model of the same, what are the ratios between velocities, flowrates, and jet powers in model and prototype in terms of the model scale?

8.18. The flowrate from a 1.2 in. diameter orifice under a 12 ft static head is 0.133 cfs of water at 68°F. A 1 : 12 model of this setup is to be operated at conditions completely dynamically similar to those of the prototype. The liquid used in the model has a specific gravity of 1.40 and surface-tension effects may be ignored. What flowrate is required in the model? What viscosity should the liquid have?

8.19. The discharge of oil from a tank through an orifice is to be modeled using water as the flowing fluid. The kinematic viscosity of the oil is eight times that of the water. The specific gravity of the oil is 0.90. What oil flowrate is represented by 0.002 2 m³/s in the model? If the force exerted on the model tank bottom is 210 N, what is the corresponding force in the prototype?

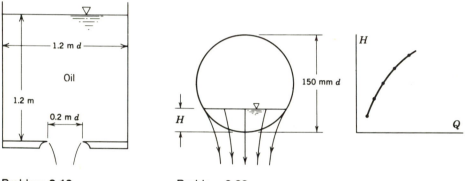

Problem **8.19**   Problem **8.23**

8.20. An overflow structure 480 m long is designed to pass a flood flow of 3 400 m³/s. A 1 : 20 model of the *cross section* of the structure is built in a laboratory channel 0.3 m wide. Calculate the required laboratory flowrate if the actions of viscosity and surface tension may be neglected. When the model is tested at this flowrate, the pressure at a point on the model is observed to be 50 mm of mercury vacuum; how should this be interpreted for the prototype?

8.21. Water at 15°C or 59°F (see Appendix 2) flows over a model of a spillway structure 0.3 m or 1 ft high. If a second model is built one-half the size of the first, what should be the corresponding properties of the liquid (for use in the smaller model) to secure complete dynamic similarity between the two models?

8.22. A 1 : 30 scale model of a cavitating overflow structure is to be tested in a vacuum tank wherein the pressure is maintained at 2.0 psia. The prototype liquid is water at 70°F. The barometric pressure on the prototype is 14.5 psia. If the liquid to be used in the model has a vapor pressure of 1.50 psia, what values of density, viscosity, and surface tension must it have for complete dynamic similarity between model and prototype?

8.23. The plot shows the calibration curve for a sharp-crested circular weir of 150 mm diameter discharging from an infinite reservoir. Show how a calibration curve for a similar weir of 225 mm diameter could be prepared from these data alone. Sketch this new curve on the plot.

8.24. A hydraulic jump from 0.6 m to 1.5 m (2 ft to 5 ft) is to be modeled in a laboratory channel at a scale of 1 : 10. What (two-dimensional) flowrate should be used in the laboratory channel? What are the Froude numbers upstream and downstream from the jump in model and prototype?

8.25. If a $1:2\,000$ tidal model is operated to satisfy Froude's law, what length of time in the model represents a day in the prototype?

8.26. An open cylindrical tank of 1.2 m (4 ft) diameter contains water to a depth of 1.2 m (4 ft). The tank is rotated at 100 r/min about its axis (which is vertical). A half-size model of this tank is to be made with mercury as the fluid. At what speed must the model be rotated for similarity? What is the ratio of the pressures at corresponding points in the liquids? What model law is operative in this situation?

8.27. A laboratory model test is to be made of an ocean breakwater. The ocean wave periods average about 10 seconds (between the arrival of wave crests); however, the laboratory wave generator can only make waves with a 1 second period. What must be the model scale?

8.28. When a sphere of 0.25 mm diameter and specific gravity 5.54 is dropped in water at 25°C it will attain a constant velocity of 0.07 m/s. What specific gravity must a 2.5 mm sphere have so that when it is dropped in crude oil (25°C) the two flows will be dynamically similar when the terminal velocity is attained?

8.29. The flow about a 150 mm artillery projectile which travels at 600 m/s through still air at 30°C and absolute pressure 101.4 kPa is to be modeled in a high-speed wind tunnel with a $1:6$ model. If the wind tunnel air has a temperature of -18°C and absolute pressure of 68.9 kPa, what velocity is required? If the drag force on the model is 35 N, what is the drag force on the prototype if skin friction may be neglected?

8.30. A cavitation zone is expected on an overflow structure when the flowrate is 140 m³/s, atmospheric pressure 101.3 kPa, and water temperature 5°C. The cavitation is to be reproduced on a $1:20$ model of the structure operating in a vacuum tank with water at 50°C. Disregarding frictional and surface-tension effects, determine the flowrate and absolute pressure (kPa) to be used in the tank for dynamic similarity.

8.31. A model is to be built of a flow phenomenon which is dominated by the action of gravity and surface tension forces. Derive an expression for the model scale in terms of the physical properties of the fluid.

8.32. A liquid rises a certain distance in a capillary tube. If this phenomenon is to be modeled, derive a force-ratio expression which must be equal in model and prototype. Compare this with equation 1.16.

8.33. For small models and (small) prototypes of surface ships and overflow structures, the actions of gravity, viscosity, and surface tension may be of equal importance. For dynamic similarity between model and prototype, what relation must exist between viscosity, surface tension, and model scale?

8.34. Prove by dimensional analysis that centrifugal force $= CMV^2/r$.

8.35. Prove by dimensional analysis that $\dot{m} = CA\rho V$.

8.36. Assume that the velocity acquired by a body falling from rest (without resistance) depends on weight of body, acceleration due to gravity, and distance of fall. Prove by dimensional analysis that $V = C\sqrt{g_n h}$ and is thus independent of the weight of the body.

8.37. If $C = f(V, \rho, E)$, prove the expression for Cauchy number by dimensional analysis.

8.38. If $\mathbf{R} = f(V, l, \rho, \mu)$, prove the expression for Reynolds number by dimensional analysis.

8.39.   If $\mathbf{W} = f(V, l, \rho, \sigma)$, prove the expression for Weber number by dimensional analysis.

8.40.   A physical problem is characterized by a relationship among length, velocity, density, viscosity, and surface tension. Derive all possible dimensionless groups significant to this problem.

8.41.   Derive by dimensional analysis an expression for the local velocity in established pipe flow through a smooth pipe if this velocity depends only on mean velocity, pipe diameter, distance from pipe wall, and density and viscosity of the fluid.

8.42.   Derive by dimensional analysis a general expression for the capillary rise of liquids in small tubes if this depends on tube diameter, and the specific weight and surface tension of the liquid. Determine the final functional form by comparison with equation 1.16.

8.43.   The speed of shallow water waves in the ocean (e.g., seismic sea waves or *tsunamis*) depends only on the still water depth and the acceleration due to gravity; derive an expression for wave speed.

8.44.   If the velocity of deep water waves depends only on wave length and acceleration due to gravity, derive an expression for wave velocity.

8.45.   Derive an expression for the velocity of very small ripples on the surface of a liquid if this velocity depends only on ripple length and density and surface tension of the liquid.

8.46.   Derive an expression for the axial thrust exerted by a propeller if the thrust depends only on forward speed, angular speed, size, and viscosity and density of the fluid. How would the expression change if g_n were a relevant variable in the case of a ship propeller?

8.47.   Derive an expression for drag force on a smooth submerged object moving through incompressible fluid if this force depends only on speed and size of object and viscosity and density of the fluid. Discuss results in terms of Fig. 13.13.

8.48.   Derive an expression for the head lost in an established incompressible flow in a smooth pipe if this loss of head depends only on diameter and length of pipe, density, viscosity, and mean velocity of the fluid, and acceleration due to gravity.

8.49.   Derive an expression for the drag force on a smooth object moving through compressible fluid if this force depends only on speed and size of object, and viscosity, density, and modulus of elasticity of the fluid. Discuss in terms of Fig. 13.5 by converting the result to relevant dimensions for a sphere.

8.50.   Derive an expression for the velocity of a jet of viscous liquid issuing from an orifice under static head if this velocity depends only on head, orifice size, acceleration due to gravity, and viscosity and density of the fluid.

8.51.   Derive an expression for the flowrate over an overflow structure if this flowrate depends only on size of structure, head on the structure, acceleration due to gravity, and viscosity, density, and surface tension of the liquid flowing.

8.52.   Derive an expression for terminal velocity of smooth solid spheres falling through incompressible fluids if this velocity depends only on size and density of sphere, acceleration due to gravity, and density and viscosity of the fluid.

8.53.   A circular disk of diameter d and of negligible thickness is rotated at a constant angular speed, ω, in a cylindrical casing filled with a liquid of viscosity μ and density ρ. The casing has an internal diameter D, and there is a clearance y between the surfaces of disk and casing. Derive an expression for the torque required to maintain this speed if it depends only on the foregoing variables.

8.54. Two cylinders are concentric, the outer one fixed and the inner one movable. A viscous incompressible fluid fills the gap between them. Derive an expression for the torque required to maintain constant-speed rotation of the inner cylinder if this torque depends only on the diameters and lengths of the cylinders, the viscosity and density of the fluid, and the angular speed of the inner cylinder.

8.55. Derive an expression for the frictional torque exerted on the journal of a bearing if this torque depends only on the diameters of journal and bearing, their axial lengths (these are the same), viscosity of the lubricant, angular speed of the journal, and the transverse load (force) on the bearing.

8.56. For a hydraulic jump (Fig. 10.25), derive by dimensional analysis an expression for y_2 if y_2 depends only on q, y_1, g_n, μ, and ρ. Compare the resulting expression with equation 10.36.

8.57. Derive by dimensional analysis an expression for the pressure drop, Δp, over the length, X, of unestablished flow if Δp depends only on X, pipe diameter, flowrate, and density and viscosity of the fluid. See Fig. 7.16.

8.58. Fluid flows horizontally through a bed of uniform spherical sand grains. Derive by dimensional analysis an expression for the pressure drop along the flow in terms of velocity, distance, size of grains, and density and viscosity of the fluid.

8.59. The force, F, exerted by the flowing liquid on this two-dimensional sluice gate is to be studied by dimensional analysis. Assuming the flow frictionless, derive an expression for this force in terms of the other variables relevant to the problem.

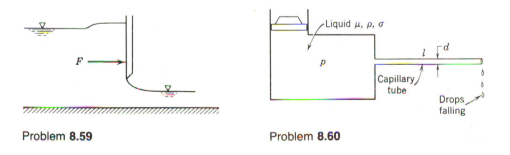

Problem **8.59**   Problem **8.60**

8.60. The time of formation of liquid drops is to be studied experimentally with the apparatus shown. The variables indicated are thought to be the relevant and independent ones. Make a dimensional analysis of the problem assuming "formation time" to be the dependent variable. Show the result in terms of dimensionless groups.

8.61. Tests on the established flow of six different liquids in smooth pipes of various sizes yield the following data:

SI Units

Diameter mm	Velocity m/s	Viscosity mPa·s	Density kg/m³	Wall Shear Pa
300	2.26	862.0	1 247	51.2
250	2.47	431.0	1 031	33.5
150	1.22	84.3	907	5.41
100	1.39	44.0	938	9.67
50	0.20	1.5	861	0.162
25	0.36	1.0	1 000	0.517

FSS Units

Diameter in.	Velocity ft/s	Viscosity × 10⁵ lb·s/ft²	Density slug/ft³	Wall Shear lb/ft²
12	7.43	1 800	2.42	1.070
10	8.10	900	2.00	0.700
6	4.00	176	1.76	0.113
4	4.56	92	1.82	0.202
2	0.67	3.11	1.67	0.003 38
1	1.17	2.10	1.94	0.010 8

Make a dimensional analysis of this problem and a plot of the resulting dimensionless numbers as ordinate and abscissa. What conclusions may be drawn from the plot?

8.62. A long cylinder of diameter d rotates coaxially at angular speed ω within another long cylinder of diameter D. Between the cylinders is a fluid of density ρ and viscosity μ. Make a dimensional analysis of this problem taking as the dependent variable the torque T (per metre of cylinder length) to maintain steady state conditions. *For laminar flow* a physical analysis of this problem yields the equation $T = \pi\mu\omega \, d^3/2(D-d)$, if $(D-d)$ is small. With this equation as a guide, show on a sketch plot the general trend of the family of curves to be expected in the general solution of this problem.

8.63. In a hypothetical problem the following variables are thought to be significant: linear velocity, length, pressure difference, modulus of elasticity, and absolute viscosity. How many different dimensionless numbers may be derived from these? How many will be needed for a general solution of the problem?

8.64. In a hypothetical problem for dimensional analysis, flowrate (m³/s or ft³/s) is thought to be a function of size, acceleration due to gravity, and kinematic viscosity. Derive (or write down from experience) all of the dimensionless numbers obtainable from this aggregation of four variables. How many of these will be significant for a complete solution of the problem?

8.65. Derive an expression for modeling the thrust of geometrically similar screw propellers moving through incompressible fluids if this thrust depends only on propeller diameter, velocity of propeller through the fluid, rotative speed, the density and viscosity of the fluid, and the acceleration due to gravity.

8.66. Find the resulting Π-groups when (a) D, ρ and Q or (b) H, ρ, and Q are the repeating variables in the analysis of a turbomachine where the relevant variables are P, D, N, Q, H, μ, ρ, and E (see paragraph 3 of Section 8.3). Discuss how to interpret each Π obtained.

8.67. Derive an expression for the modeling of the performance of geometrically similar centrifugal compressors. The relevant variables are flowrate, the outlet to inlet pressure ratio, rotative speed, impeller diameter, and the viscosity, density, elasticity, and specific heat ratio of the fluid.

8.68. Derive an expression for the power of hydraulic machines if this power depends only on the angular speed, size and surface roughness of the rotating element of the machine, flowrate, and the density and viscosity of the fluid flowing.

8.69.[27] A 1 : 15 scale model of a hydraulic turbine of 4.5 m (15 ft) diameter is found to develop 2.4 kW (3.21 hp), with a flowrate of 280 l/s (10.0 ft³/s) of water, under a head of 1.14 m (3.75 ft) while rotating at 500 r/min. Calculate the corresponding power, flowrate, head, and speed of the prototype, assuming the same efficiency in model and prototype.

8.70.[27] A centrifugal pump has a 250 mm diameter impeller which rotates at 1 750 r/min and delivers a flowrate of 0.07 m³/s of water at a head of 9 m. Calculate the corresponding quantities for a half-scale model of the pump if the efficiencies and fluid of model and prototype are the same.

8.71. A 0.3 m (1 ft) diameter model of a ship propeller rotates at 500 r/min. The velocity well upstream from the model is 1.5 m/s (5 ft/s). The thrust produced is observed to be 133 N (30 lb). If the prototype (of 1.8 m or 6 ft diameter) drives a ship through still water at a speed of 3 m/s (10 ft/s), what must its thrust and rotational speed be for its operation to be dynamically similar to that of its model? Assume the same efficiency and same (fresh) water in model and prototype.

8.72. In a set of geometrically similar pumps the manufacturer changes the impeller diameter by a factor of two between two models. What change in N is required to make H change by a factor of two also? What is the ratio of flowrates under these conditions?

8.73. A Francis turbine is to operate under a head of 46 m and deliver 18.6 MW while running at 150 r/min. The runner diameter is 4 m. A 1 m diameter model is operated in a laboratory under the same head. Find the model speed, power, and flowrate. If model efficiency is 90%, what is prototype efficiency?

8.74. A pump is being built to deliver 120 m (400 ft) head and 5.7 m³/s (200 ft³/s) at 200 r/min. A geometrically similar model is to be run. The diameter of the prototype impeller is 1.8 m (6 ft) while 0.57 m³/s (20 ft³/s) is the available laboratory flowrate. What should be the model size, speed, and head if the model and prototype efficiencies are the same?

8.75. A Kaplan (propeller with variable pitchblades) turbine with a rated capacity of 83 MW at a head of 24 m and 86 r/min was one of 14 units installed at the McNary project on the

---

[27] When head is expressed in metres, g_n becomes a relevant parameter and Froude number equality must hold; as $V \propto ND$, $(N^2D/g_n)_m = (N^2D/g_n)_p$.

Columbia River. The characteristic runner diameter is 7 m. If a 6 m head is available in the laboratory, what should be the model scale, flowrate, and r/min?

8.76. A turbine model that is 0.3 m in diameter develops 150 kW under a 15 m head at 2 000 r/min. Assuming no change in efficiency, what power would a 4.5 m diameter prototype, operating at 100 r/min under a 225 m head, develop?

8.77. What type of pump should be selected for problem 8.74? What maximum efficiency is expected?

8.78. A pump is designed to operate at 1 800 r/min with 83.7% efficiency, and deliver 250 l/s (4 000 gpm) with a power consumption of 141 kW (189.5 hp). What is its specific speed and type?

8.79. A manufacturer states that it can deliver a pump operating at 690 r/min that delivers 285 l/s or 4 500 gpm at 78% efficiency with a power consumption of about 5.2 kW or 7 hp under a head of 1.5 m or 5 ft. What type of pump is it? Is the efficiency claim reasonable, too high, or too low?

8.80. Determine N_S for both model and prototype turbines in problem 8.75.

8.81. A small turbine must be run at 250 r/min to match generator characteristics and must deliver 745.7 kW from a 6 m head of water. Compute the specific speed of the unit, select a turbine type to be used, and compute the flowrate through a maximum efficiency turbine.

8.82. The Francis turbines for the Shasta plant in California are rated at 76.8 MW (103 000 hp) at 138.5 r/min. If $N_S = 0.74$, find the rated head and expected efficiency of the units.

8.83. A propeller turbine at the Bonneville plant on the Columbia River has $N_S = 2.8$ and a rated head of 18 m (60 ft); it operates at 75 r/min. How much power will the turbine deliver and at what maximum efficiency?

8.84. Reversible pump-turbines are used to provide pumped storage of water and peaking electric power. One such unit operates at 600 r/min and is rated for 125 MW at 376 m head as a turbine and for 29 m³/s at 404 m head as a pump. Calculate the unit's specific speeds; what type of unit is it?

8.85. An impulse turbine delivers 69 000 kW from 750 m head. At what speed should the runner rotate for maximum efficiency?

8.86. Calculate the minimum required NPSH (in metres or feet of water) to avoid cavitation in the pumps of the following problems: (a) problem 8.78, (b) problem 8.79, and (c) problem 8.84 (when operated as a pump).

8.87. Calculate the minimum required NPSH (in metres or feet of water) to avoid cavitation in the turbines of the following problems: (a) problem 8.81, (b) problem 8.82, (c) problem 8.83, and (d) problem 8.84 (when operated as a turbine).

8.88. If the model turbine in problem 8.76 requires a NPSH of 18 m to avoid cavitation, what minimum NPSH is required in the prototype?

9

FLUID FLOW IN PIPES

The problem of fluid flow in pipelines—the prediction of flowrate through pipes of given characteristics, the calculation of energy conversions therein, and so forth—is encountered in many areas of engineering practice; they afford an opportunity of applying many of the foregoing principles to (essentially one-dimensional) fluid flows of a comparatively simple and controlled nature. The subject of pipe flow embraces only those problems in which pipes flow completely full; pipes that flow partially full, such as sewer lines and culverts, are treated as open channels and are discussed in the next chapter.

The solution of practical pipe flow problems results from application of the energy principle, the equation of continuity, and the principles and equations of fluid resistance. Resistance to flow in pipes is offered not only by long reaches of pipe but also by pipe fittings, such as bends and valves, which dissipate energy by producing relatively large-scale turbulence.

INCOMPRESSIBLE FLOW

9.1 Fundamental Equations

The Bernoulli equation for incompressible fluid motion in pipes is, recalling Section 7.9, (if h_L replaces $h_{L_{1-2}}$)

$$\frac{p_1}{\gamma} + \alpha_1 \frac{V_1^2}{2g_n} + z_1 = \frac{p_2}{\gamma} + \alpha_2 \frac{V_2^2}{2g_n} + z_2 + h_L \tag{9.1}$$

However, in most problems of pipe flow the α terms may be omitted for several reasons. (1) Most engineering pipe flow problems involve turbulent flow in which α is only slightly more than unity. (2) In laminar flow where α is large, velocity heads are usually negligible when compared to the other Bernoulli terms. (3) The velocity heads in most pipe flows are usually so small compared to the other terms that inclusion of α has little effect on the final result. (4) The effect of α tends to cancel since it appears on both sides of the equation. (5) Engineering answers are not usually required to an accuracy which would justify the inclusion of α in the equation. Application of equation 9.1 to practical problems thus depends (the Bernoulli terms having been presented in Chapters 4, 5, and 7) primarily on an understanding of the factors which affect the head loss, h_L, and the methods available for calculating this quantity.

Early experiments (circa 1850) on the flow of water in long, straight, cylindrical pipes (Fig. 9.1) indicated that head loss varied (approximately) directly with velocity

357

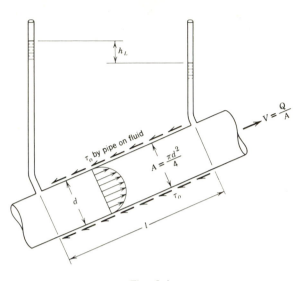

Fig. **9.1**

head and pipe length, and inversely with pipe diameter. Using a dimensionless coefficient of proportionality, f, called the *friction factor*, Darcy, Weisbach, and others proposed equations of the form

$$h_L = f \frac{l}{d} \frac{V^2}{2g_n} \tag{9.2}$$

Observations indicated that the friction factor f depended primarily on pipe roughness but also on velocity and pipe diameter; more recently it was observed that the friction factor also depended on the viscosity of the fluid flowing. This equation, usually called the *Darcy-Weisbach equation,* is still the basic equation for head loss caused by established pipe friction (not pipe fittings) in long, straight, uniform pipes.

Equations 9.2 and 7.21 may now be combined to give a basic relation between frictional stress, τ_o, and friction factor, f; this is

$$\tau_o = \frac{f \rho V^2}{8} \tag{9.3}$$

In this fundamental equation relating wall shear to friction factor, density, and mean velocity, it is apparent that, with f dimensionless, $\sqrt{\tau_o/\rho}$ must have the dimensions of velocity; this is known as the *friction velocity, $v_\star$* which (from equation 9.3) is related to the friction factor and the mean velocity by

$$v_\star = \sqrt{\frac{\tau_o}{\rho}} = V \sqrt{\frac{f}{8}} \tag{9.4}$$

However, the physical meaning of the friction velocity is not revealed by this algebraic definition; since it is a velocity which embodies only wall shear and fluid density, it is defined by the same equation whatever the flow regime (laminar or turbulent) or whatever the boundary texture (rough or smooth). For this reason it is a useful generalization that finds wide application in further developments.

——————————— **Illustrative Problem** ———————————

Water flows in a 150 mm diameter pipeline at a mean velocity of 4.5 m/s. The head lost in 30 m of this pipe is measured experimentally and found to be $5\frac{1}{3}$ m. Calculate the friction velocity.

Relevant Equations and Given Data

$$h_L = f\frac{l}{d}\frac{V^2}{2g_n}$$ (9.2)

$$v_* = \sqrt{\frac{\tau_o}{\rho}} = V\sqrt{\frac{f}{8}}$$ (9.4)

$d = 150$ mm $l = 30$ m $h_L = 5.33$ m $V = 4.5$ m/s

Solution.

$$5.33 = f\frac{30}{0.15}\frac{(4.5)^2}{2g_n} \qquad f = 0.026$$ (9.2)

$$v_* = 4.5\sqrt{\frac{0.026}{8}} = 0.26 \text{ m/s} \blacktriangleleft$$ (9.4)

9.2 Laminar Flow

Although the facts of fluid flow can usually be established by experiment, an analytical approach to the problem is also necessary to an understanding of the mechanics of the flow. For the mechanics of a real fluid, this consists of the application of basic physical laws which have themselves been verified by experiment; the use of "pure theory" alone is seldom possible in this field.

Analysis of laminar flow in a pipeline (Fig. 9.2) may be begun with the following established facts: (1) symmetrical distribution of shear stress and velocity, (2) maximum velocity at the center of the pipe and no velocity at the wall (the no-slip condition), (3) linear shear stress distribution in the fluid given by equation 7.22 (from application of the impulse momentum principle), and (4) shear stress in the fluid given by equation 1.12 (appropriate for laminar flow). In Fig. 9.2 it is clear that $r = R - y$ and $dr = -dy$; thus, equating the expressions for τ yields

$$\tau = \left(\frac{\gamma h_L}{2l}\right)r = \mu\frac{dv}{dy} = -\mu\frac{dv}{dr}$$

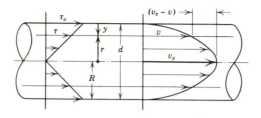

Fig. **9.2**

At $r = R(y = 0)$: $\tau = \gamma h_L R / 2l = \tau_o$, while at $r = 0(y = R)$: $\tau = 0$, and because $\tau = \mu \, dv/dy$, $dv/dy = 0$. Writing the above equation then as

$$\frac{dv}{dr} = -\frac{\tau_o}{\mu R} r$$

allows direct integration to

$$v = -\frac{\tau_o}{\mu R} \left(\frac{r^2}{2}\right) + C$$

The no-slip boundary condition at $r = R$ requires

$$0 = -\frac{\tau_o}{\mu R} \left(\frac{R^2}{2}\right) + C$$

so $C = \tau_o R^2 / 2\mu R$ and the velocity profile is a parabola of the form:

$$v = \frac{\tau_o}{2\mu R} (R^2 - r^2) = v_c \left(1 - \frac{r^2}{R^2}\right) \tag{9.5}$$

because the centerline velocity v_c (at $r = 0$) must be $v_c = \tau_o R / 2\mu$. Recalling that $v_*^2 = \tau_o / \rho$ allows writing equation 9.5 as (with $\nu = \mu/\rho$)

$$\frac{v}{v_*} = \frac{v_*}{2\nu R} (R^2 - r^2) \tag{9.6}$$

or [with $r^2 = (R - y)^2$]

$$\frac{v}{v_*} = \frac{v_*}{\nu} \left(y - \frac{y^2}{2R}\right) \tag{9.6}$$

Thus, the profile can be expressed in terms of distance from the wall and v_* is seen to be a characteristic velocity of the laminar profile (v_* will appear again as an important characteristic velocity of turbulent profiles).

For small values of y, the term $y^2/2R$ is negligible compared to y and near the wall the velocity profile is *linear*, that is, in dimensionless form,

$$\frac{v}{v_*} = \frac{v_* y}{\nu} \qquad y \ll R \tag{9.7}$$

The flowrate Q in a circular pipe is found by integrating the profile equation 9.5 as

$$Q = \int_0^R v(2\pi r \, dr) = \frac{\pi\tau_o}{\mu R} \int_0^R (R^2 - r^2) \, dr = \frac{\pi\tau_o R^3}{4\mu}$$

Since $(1/4)\pi d^2 V = \pi R^2 V = Q$, the mean velocity

$$V = \tau_o R / 4\mu = (1/2)v_c$$

as was shown in the Illustrative Problems of Section 3.5.

Now using $\tau_o = \gamma h_L R / 2l$ yields

$$Q = \pi R^4 \gamma h_L / 8\mu l = \frac{\pi d^4 \gamma h_L}{128\mu l} \tag{9.8}$$

and

$$V = \frac{\gamma R^2 h_L}{8\mu l} = \frac{\gamma d^2 h_L}{32\mu l}$$

Thus,

$$h_L = \frac{32\mu l V}{\gamma d^2} \tag{9.9}$$

Equation 9.8 shows that in laminar flow the flowrate, Q, which will occur in a circular pipe, varies directly with the head loss and with the fourth power of the diameter but inversely with the length of pipe and viscosity of the fluid flowing. These facts of laminar flow were established experimentally, independently, and almost simultaneously by Hagen (1839) and Poiseuille (1840), and thus the law of laminar flow expressed by the equations above is termed the *Hagen-Poiseuille* law. The experimental verification (by Hagen, Poiseuille, and later investigators) of the derivations above serves to confirm the assumptions (1) that there is no velocity adjacent to a solid boundary (i.e., no "slip" between fluid and pipe wall), and (2) that in laminar flow the shear stress is given by $\tau = \mu \, dv/dy$, which were taken for granted in the foregoing derivations.

Equation 9.9 shows that, *in a laminar flow, head loss varies with the first power of the velocity*. Equating the Darcy-Weisbach equation 9.2 for head loss to equation 9.9 yields an expression for the friction factor; it is

$$f = \frac{64\mu}{Vd\rho} = \frac{64}{\mathbf{R}} \tag{9.10}$$

Thus, in laminar flow the friction factor depends *only* on the Reynolds number.

——————————— **Illustrative Problems** ———————————

One hundred gallons of oil (s.g. = 0.90, and μ = 0.001 2 lb·s/ft²) flow per minute through a pipeline of 3 in. diameter. Calculate the centerline velocity, the lost head in 1 000 ft of this pipe, and the shear stress and velocity at a point 1 in. from the centerline.

Relevant Equations and Given Data

$$Q = AV \tag{3.11}$$

$$\tau = (\gamma h_L / 2l)r \tag{7.22}$$

$$\mathbf{R} = Vd / \nu \tag{8.11}$$

$$h_L = f \frac{l}{d} \frac{V^2}{2g_n} \tag{9.2}$$

$$v = v_c \left(1 - \frac{r^2}{R^2}\right) \tag{9.5}$$

$$f = \frac{64}{\mathbf{R}} \tag{9.10}$$

$Q = 100$ gal/min   s.g. $= 0.90$ $\mu = 0.001\ 2$ lb·s/ft²

$d = 3$ in. $h_L = 1\ 000$ ft $\rho_{water} = 1.936$ slug/ft³

Solution.

$$V = \frac{100}{60 \times 7.48} \frac{1}{(\pi/4) (\frac{3}{12})^2} = 4.53 \text{ ft/s} \tag{3.11}$$

$$\mathbf{R} = \frac{4.53 \times (\frac{3}{12}) \times (0.90 \times 1.936)}{0.001\ 2} = 1\ 645 \tag{8.11}$$

Since $\mathbf{R} < 2\ 100$, laminar flow exists, and

$$v_c = 2V = 9.06 \text{ ft/s} \blacktriangleleft$$

Then, combining equations 9.2 and 9.10, yields

$$h_L = \frac{64}{1\ 645} \times \frac{1\ 000}{\frac{3}{12}} \times \frac{(4.53)^2}{2g_n} = 49.6 \text{ ft of oil} \blacktriangleleft \tag{9.2 \& 9.10}$$

(The reader can confirm that use of equation 9.9 yields precisely the same result.)

$$v_1 = 9.06 \left[1 - \left(\frac{1}{1.5}\right)^2\right] = 5.04 \text{ ft/s} \blacktriangleleft \tag{9.5}$$

$$\tau_1 = \left(\frac{62.4 \times 0.90 \times 49.6}{2 \times 1\ 000}\right) \frac{1}{12} = 0.116 \text{ lb/ft}^2 \blacktriangleleft \tag{7.22}$$

A fluid is forced from a large tank through a 100 mm, 4 mm diameter tube. In 600 s, 13×10^{-4} m³ of fluid fall from the tube into a measuring cup. If the head loss in the tube is 1 m, calculate ν. Is the flow laminar?

Relevant Equations and Given Data

$$Q = \pi d^4 \gamma h_L / 128 \mu l \tag{9.8}$$

See the previous problem for equations 3.11 and 8.11.

$d = 4$ mm $\quad l = 100$ mm $\quad$ Vol $= 13 \times 10^{-4}$ m^3 in time $= 600$ s $\quad h_L = 1$ m

Solution. From equation 9.8, since $Q = 13 \times 10^{-4}/600$ and $\gamma = \rho g_n$,

$$\nu = \frac{3.14 \times (0.004)^4 \times 9.81 \times 1}{128 \times (13 \times 10^{-4}/600) \times 0.1} = 0.000\,28 \text{ m}^2/\text{s} \blacktriangleleft \tag{9.8}$$

$$V = \frac{13 \times 10^{-4}}{600} \frac{1}{(\pi/4)(0.004)^2} = 0.17 \text{ m/s} \tag{3.11}$$

$$R = \frac{0.17 \times 0.004}{0.000\,28} = 2.4 < 2\,100 \tag{8.11}$$

The flow is clearly laminar. ◀

9.3 Turbulent Flow—Smooth Pipes

Pipe flow with friction effects has been seen to be a viscosity-inertia phenomenon and, thus, to be characterized by a Reynolds number (Sections 7.1 and 8.1). For smooth pipes, the discussion about flow past solid boundaries (Section 7.3) is relevant and strongly suggests the existence of a viscous sublayer near the pipe walls.

Drawing from the analysis of Section 7.2, the total shear stress can be written as

$$\tau = \left(\mu \frac{dv}{dy} - \rho \,\overline{v_x v_y} \right) \tag{9.11}$$

where $\mu \, dv/dy$ is the viscous stress and $-\rho \,\overline{v_x v_y}$ is the turbulent (Reynolds) stress. It has already been shown (Section 7.10) that the total stress is a linear function of radius r in a pipe. Experiments show (see the typical result sketched in Fig. 9.3) that over most of the flow, the turbulent stress dominates, but the total stress equals the

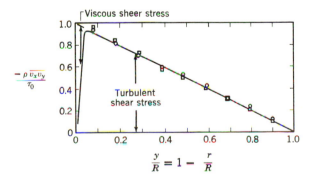

Fig. 9.3 Typical variation of shear stress in a turbulent pipe flow (to scale for $R \gg 2\,100$). Data ($R \sim 5 \times 10^4$ – o; $R \sim 5 \times 10^5$ – □) from J. Laufer, N.A.C.A. Report 1174, 1954.

viscous stress at the wall. Attempts to represent the Reynolds stress have only been partly successful; this area represents one of the classic unsolved problems of fluid flow. As derived in Section 7.2 Prandtl's mixing length theory is a plausible and practical, although not theoretically rigorous, approach. It is used here to derive the velocity distribution and other quantities because the results obtained are close to experimental data and the process gives insight to the physics of the flow.

The analytical treatment is begun by assuming that the viscous stress in equation 9.11 is negligible over most of the flow, employing the Prandtl relationship (equation 7.3) for the turbulent shear stress, and equating this to the linear total shear stress relation (equation 7.22) for pipes. The result is (recalling from Fig. 9.2 that $r = R - y$ and $dr = -dy$)

$$\tau_o\left(1 - \frac{y}{R}\right) = \rho l^2(dv/dy)^2 \tag{9.12}$$

where y is measured from the pipe wall. Now in Section 7.2 it was argued that near the wall ($y \ll R$), $l = \kappa y$. However, here on the centerline ($y = R$), $\tau = 0$, so either $l = 0$ or $dv/dy = 0$ there (or both equal zero). Thus, some insight is needed to proceed further; it comes from experiment. Nikuradse's systematic and comprehensive measurements[1] of velocity profiles in smooth pipes ($5 \times 10^5 < \mathbf{R} < 3 \times 10^6$) and in pipes with a uniform sand grain roughness showed that *all* velocity profiles could be characterized by the single equation

$$\frac{v_c - v}{v_\star} = -2.5 \ln \frac{y}{R} \tag{9.13}$$

Therefore, it must be that $dv/dy \propto 1/y$ and, from equation 9.12, $l \propto y\sqrt{1 - (y/R)}$. Near the wall, $l = \kappa y$; it follows that, in a pipe flow,

$$l = \kappa y\left(1 - \frac{y}{R}\right)^{1/2} \tag{9.14}$$

Equation 9.12 now becomes

$$\tau_o/\rho\kappa^2 y^2 = (dv/dy)^2$$

or

$$dv/dy = \frac{\sqrt{\tau_o/\rho}}{\kappa y} = \frac{v_\star}{\kappa y} \tag{9.15}$$

which illustrates again the pervasive presence of the friction velocity $v_\star$. Integrating equation 9.15 produces

---

[1] J. Nikuradse, "Strömungsgesetze in rauhen Rohren," *VDI-Forschungsheft*, 361, 1933. Translation available N.A.C.A., *Tech Mem.* 1292.

$$v = \frac{v_*}{\kappa} \ln y + C \qquad (9.16)$$

There are two ways to interpret this unknown constant C.

First, if $v = v_c$ at $y = R$ (on the centerline), then

$$v_c = \frac{v_*}{\kappa} \ln R + C$$

and $C = v_c - (v_*/\kappa) \ln R$. The result is

$$v = \frac{v_*}{\kappa} \ln y - \frac{v_*}{\kappa} \ln R + v_c$$

or

$$\frac{v_c - v}{v_*} = -\frac{1}{\kappa} \ln \frac{y}{R} \qquad (9.13)$$

For the usual value $\kappa = 0.40$ this is the experimentally derived equation 9.13.

Second, one can try to find C in terms of the no-slip condition ($v = 0$) at $y = 0$. But at $y = 0$, $v = -\infty$, according to equation 9.16. This is at least unrealistic (and not unexpected because the neglected viscous stresses dominate at the boundary)! It must be that the "turbulent" profile is replaced by a viscous-dominated profile near the wall (see Section 7.3) and that there is some appropriate distance from the wall which marks the onset of this process. One approach is to assume that $v = 0$ (or it is at least negligibly small) at $y = y'$. From equation 9.16, $C = -(v_*/\kappa) \ln y'$ so the profile is

$$\frac{v}{v_*} = \frac{1}{\kappa} \ln \frac{y}{y'}$$

Now, very near the wall in a laminar flow the velocity profile is linear (Section 9.2) and the dimensionless equation 9.7 shows that y is scaled by the length v/v_*. Suppose that, since a viscous-dominated region is being approached as y decreases, $y' \propto v/v_*$ so, say, $y' = \beta(v/v_*)$; then the above equation becomes

$$\frac{v}{v_*} = \frac{1}{\kappa}(\ln y - \ln y') = \frac{1}{\kappa}(\ln y - \ln \beta v/v_*)$$

$$= \frac{1}{\kappa} \ln \frac{v_* y}{v} - \frac{1}{\kappa} \ln \beta$$

or

$$\frac{v}{v_*} = \frac{1}{\kappa} \ln \frac{v_* y}{v} + D$$

where D is an unknown constant which must be determined from experimental data. Nikuradse's data for smooth pipes establishes $D = 5.5$; thus, for $\kappa = 0.40$,

$$\frac{v}{v_\star} = 2.5 \ln \frac{v_\star y}{v} + 5.5 \tag{9.17}$$

or in terms of common logarithms,

$$\frac{v}{v_\star} = 5.75 \log \frac{v_\star y}{v} + 5.5 \tag{9.17}$$

which is a general equation of the velocity distribution for turbulent flow in smooth pipes.

Using equation 9.17 and the fact that a viscous sublayer must exist near the smooth wall, it is possible to describe the structure of the flow in some detail. In the viscous sublayer, the laminar shear stress relation holds; accordingly, the profile result obtained near the wall is applicable. From Section 9.2 (equation 9.7) the dimensionless profile is

$$\frac{v}{v_\star} = \frac{v_\star y}{v} \tag{9.7}$$

The nominal extent of the viscous sublayer is obtained by finding the intersection of the viscous profile 9.7 with the turbulent profile 9.17. At the intersection, y_i, the velocities are the same so

$$\frac{v_\star y_i}{v} = 5.75 \log \frac{v_\star y_i}{v} + 5.5$$

This equation is satisfied by $v_\star v_i / v = 11.6$ and the nominal sublayer thickness[2] δ_v is taken to be represented by y_i; therefore,

$$\frac{v_\star \delta_v}{v} = 11.6 \qquad \text{or} \qquad \delta_v = 11.6(v/v_\star) \tag{9.18}$$

These results are illustrated, together with some experimental data, in Fig. 9.4.

The flowrate Q in a turbulent pipe flow is given by

$$Q = \int_0^R v(2\pi r \; dr) = 2\pi v_\star \int_0^R \left(5.75 \log \frac{v_\star (R - r)}{v} + 5.5\right) r \; dr$$

if the contribution of the viscous sublayer and the imperfection of the profile there and at the centerline are ignored. (At the centerline the derivative dv/dy must be zero, but the present equation 9.17 does not yield this result.) After integration

$$\frac{Q}{\pi R^2 v_\star} = 5.75 \log \frac{v_\star R}{v} + 1.75$$

---

[2] There is no precise boundary of the sublayer and no actual intersection of the two velocity distribution curves; a gradual transition from laminar to turbulent action occurs through a "buffer zone" extending from $3.5 < v_\star y/v < 30$. The velocity profile of Fig. 9.4 is taken to be a universal one for the turbulent flow of fluids over smooth surfaces, providing there is no transfer of heat between fluid and surface.

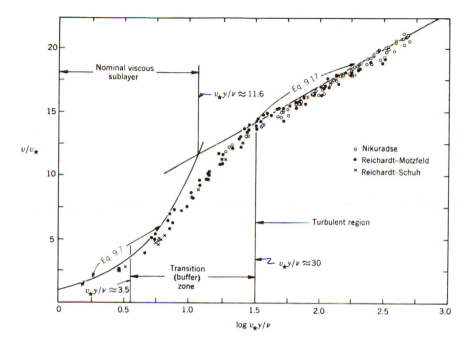

Fig. **9.4** Velocity distribution near a smooth wall. (Adapted from N.A.C.A. Technical Memorandum 1047.)

If the mean velocity $V = Q/\pi R^2$, then

$$\frac{V}{v_\star} = 5.75 \log \frac{v_\star R}{\nu} + 1.75 \qquad (9.19)$$

From equation 9.17 evaluated at $y = R$, where $v = v_c$,

$$\frac{v_c}{v_\star} = 5.75 \log \frac{v_\star R}{\nu} + 5.5$$

Subtracting equation 9.19 from this equation produces

$$\frac{v_c - V}{v_\star} = 3.75$$

or

$$v_c = V + 3.75 v_\star = V(1 + 3.75\sqrt{f/8})$$

because $v_\star = V\sqrt{f/8}$ (equation 9.4). To make this result for v_c conform to experiments, the constant 3.75 is adjusted to 4.07, giving

$$\frac{v_c}{V} = 1 + 4.07\sqrt{f/8} \qquad (9.20)$$

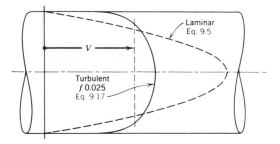

Fig. **9.5**

Figure 9.5 shows a comparison (for the same flowrate and mean velocity) of typical velocity profiles for laminar and turbulent pipe flow.

Next, introducing equation 9.4 into equation 9.19 yields an equation for the friction factor f in the Darcy-Weisbach expression 9.2, namely (replace R with $d/2$ also),

$$\frac{1}{\sqrt{f}} = 2.03 \log (\mathbf{R}\sqrt{f}) - 0.91$$

where $\mathbf{R} = Vd/\nu$ is the pipe Reynolds number. While this result is the correct form of the relationship, it fits the experimental data better if 2.03 and 0.91 are replaced by 2.0 and 0.8 because of the imperfections in the velocity profile as noted above. For turbulent flow in smooth pipes then the friction factor is given by

$$\frac{1}{\sqrt{f}} = 2.0 \log (\mathbf{R}\sqrt{f}) - 0.8 \tag{9.21}$$

Since the wall shear stress is governed by the viscous relation 1.12, that is, $\tau_o = \mu(dv/dy)$ at the wall, the thickness of the viscous sublayer must be an important parameter in relation to friction effects. The dimensionless ratio δ_v/d is, from equation 9.18,

$$\frac{\delta_v}{d} = \frac{11.6\nu}{v_* d} = \frac{11.6\nu}{V\sqrt{f/8}\,d} = \frac{32.8}{\mathbf{R}\sqrt{f}} \tag{9.22}$$

where $v_* = V\sqrt{f/8}$ has been used. From this equation the decrease of sublayer thickness with increasing Reynolds number is readily seen. Furthermore, writing equation 9.22 in the form

$$\mathbf{R}\sqrt{f} = 32.8/(\delta_v/d)$$

and substituting it in the right-hand side of equation 9.21, yields the surprising result that

$$\frac{1}{\sqrt{f}} = 2.0 \log [32.8/(\delta_v/d)] - 0.8 \tag{9.23}$$

that is, for turbulent flow over smooth walls, the friction factor is a function *only* of the ratio of the sublayer thickness to the pipe diameter.

Illustrative Problem

Water (at 20°C) flows in a 75 mm diameter smooth brass pipeline. According to a wall shear meter (Section 11.9), $\tau_o = 3.68$ N/m². Calculate the thickness of the viscous sublayer, the friction factor, the mean velocity and flowrate, the centerline velocity, the shear stress and velocity 25 mm from the pipe centerline, and the head lost in 1 000 m of this pipe.

Relevant Equations and Given Data

$$\tau = \tau_o r / R \qquad v_\star = \sqrt{\tau_o / \rho}$$

$$h_L = f \frac{l}{d} \frac{V^2}{2g_n} \tag{9.2}$$

$$\frac{v}{v_\star} = 5.75 \log \frac{v_\star y}{\nu} + 5.5 \tag{9.17}$$

$$\frac{v_\star \delta_v}{\nu} = 11.6 \qquad \delta_v = 11.6 \, (\nu / v_\star) \tag{9.18}$$

$$\frac{V}{v_\star} = 5.75 \log \frac{v_\star R}{\nu} + 1.75 \tag{9.19}$$

$$Q = \pi R^2 V \qquad \frac{V_c}{V} = 1 + 4.07 \sqrt{f/8} \tag{9.20}$$

$$\frac{1}{\sqrt{f}} = 2.0 \log [32.8/(\delta_v / d)] - 0.8 \tag{9.23}$$

$d = 75$ mm $\qquad R = 37.5$ mm $\qquad \tau_o = 3.68$ N/m² $\qquad l = 1\ 000$ m

$\nu = 1 \times 10^{-6}$ m²/s $\qquad \rho = 998$ kg/m³

Solution.

$$v_\star = (3.68/998)^{1/2} = 0.061 \text{ m/s} \blacktriangleleft$$

$$\delta_v = 11.6(1 \times 10^{-6}/0.061) = 0.19 \text{ mm} \blacktriangleleft \tag{9.18}$$

$$\frac{1}{\sqrt{f}} = 2.0 \log [32.8/(0.19/75)] - 0.8 \tag{9.23}$$

$$V = 0.061 \left(5.75 \log \frac{0.061 \times 37.5 \times 10^{-3}}{1 \times 10^{-6}} + 1.75 \right) \tag{9.19}$$

$$V = 1.29 \text{ m/s} \blacktriangleleft$$

$$V_c = 1.29(1 + 4.07 \sqrt{0.018/8}) = 1.54 \text{ m/s} \blacktriangleleft \tag{9.20}$$

From the linear variation of shear stress with radial distance, $\tau_{25} = 3.68 \times 25/37.5 = 2.45\ N/m^2$. ◀ And because $y_{25} = 37.5 - 25 = 12.5$ mm,

$$v_{25} = 0.061\left(5.75 \log \frac{0.061 \times 12.5 \times 10^{-3}}{1 \times 10^{-6}} + 5.5\right) = 1.35\ m/s \blacktriangleleft \tag{9.17}$$

Finally,

$$h_L = 0.018 \times \frac{1\ 000}{0.075} \times \frac{(1.29)^2}{2g_n} = 20.4\ m \blacktriangleleft \tag{9.2}$$

Before the development of the foregoing generalizations by Prandtl, von Kármán, and Nikuradse, a pioneering effort was made by Blasius to relate velocity profile, wall shear, and friction factor for turbulent flow in smooth pipes. Although Blasius' work has been superseded by these generalizations, it is still of some importance in engineering (in spite of its limited scope and empiricism) because of a mathematical simplicity which allows easy visualization and leads directly to useful (but approximate) results.

First, Blasius[3] showed that the curve representing the friction factor (for $3\ 000 < \mathbf{R} < 100\ 000$) could be closely approximated by the equation

$$f = \frac{0.316}{\mathbf{R}^{0.25}} \tag{9.24}$$

When this is substituted in the Darcy-Weisbach equation 9.2, it will be noted that $h_L \propto V^{1.75}$ for the turbulent flow in smooth pipes with $\mathbf{R} < 10^5$.

Substituting equation 9.24 into equation 9.3 produces

$$\tau_o = \frac{0.316}{(2RV\rho/\mu)^{0.25}} \frac{\rho V^2}{8} = 0.033\ 2\mu^{1/4}\mathbf{R}^{-1/4}V^{7/4}\rho^{3/4} \tag{9.25}$$

Blasius then assumed that the turbulent velocity profile could be approximated by (see Fig. 9.6)

$$\frac{v}{v_c} = \left(\frac{y}{R}\right)^m$$

For this equation the mean velocity V may be related to the center velocity v_c by applying equation 3.15:

$$V\pi R^2 = \int_R^0 v_c\left(\frac{y}{R}\right)^m 2\pi(R - y)(-dy)$$

which gives

---

[3] H. Blasius, *Forschungsarbeiten auf dem Gebiete des Ingenieurwesens*, 131, 1913.

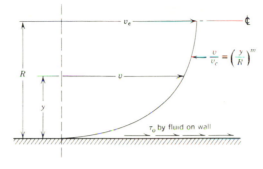

Fig. **9.6**

$$\frac{V}{v_c} = \frac{2}{(m + 1)(m + 2)} \tag{9.26}$$

from which, by substitution into the profile equation, we derive

$$V = \frac{2}{(m + 1)(m + 2)} v \left(\frac{R}{y}\right)^m$$

Substituting this expression for V into equation 9.25,

$$\tau_o = 0.033\,2 \left[\frac{2}{(m + 1)(m + 2)}\right]^{7/4} \mu^{1/4} R^{-1/4 + (7m/4)} v^{7/4} y^{-(7m/4)} \rho^{3/4}$$

However (Blasius reasoned), wall shear τ_o could depend only on the form of the velocity profile and the physical properties of the fluid but could not be affected by pipe size R; thus the exponent of R must be zero, and from this the exponent m must be equal to $\frac{1}{7}$. Validation of this depended on experimental measurements of turbulent velocity profiles which were found to agree quite well with the hypothesis above; thus the so-called *seventh-root law* for turbulent velocity distribution has been widely accepted. It is written

$$\frac{v}{v_c} = \left(\frac{y}{R}\right)^{1/7} \tag{9.27}$$

A useful corollary of this law is an equation for wall shear τ_o in terms of v_c and R. Using $m = \frac{1}{7}$ in equation 9.26 yields $V/v_c = 49/60$, and, substituting $49v_c/60$ into equation 9.25 for V and rearranging,

$$\tau_o = 0.046\,4 \left(\frac{\mu}{v_c \rho R}\right)^{1/4} \frac{\rho v_c^2}{2} \tag{9.28}$$

which will be used later in the approximate analysis of turbulent boundary layers.

Illustrative Problem

For the conditions of the previous Illustrative Problem, viz., $V = 1.29$ m/s, $d = 75$ mm, $\nu = 1 \times 10^{-6}$ m²/s, $\rho = 998$ kg/m³, calculate the friction factor, wall shear stress, centerline velocity, and the velocity 25 mm from the pipe centerline, using the seventh-root law.

Relevant Equations and Given Data

$$f = 0.316/\mathbf{R}^{0.25} \tag{9.24}$$

$$\frac{V}{V_c} = \frac{49}{60} \tag{9.26}$$

$$\frac{V}{V_c} = \left(\frac{y}{R}\right)^{1/7} \tag{9.27}$$

$$\tau_o = 0.046\,4 \left(\frac{\nu}{v_c R}\right)^{1/4} \frac{\rho v_c^2}{2} \tag{9.28}$$

$$V = 1.29 \text{ m/s} \qquad d = 75 \text{ mm} \qquad R = 37.5 \text{ mm}$$
$$\nu = 1 \times 10^{-6} \text{ m}^2/\text{s} \qquad \rho = 998 \text{ kg/m}^3$$

Solution.   Check to see if Blasius' work is applicable, that is, $3\,000 < \mathbf{R} < 10^5$.

$$\mathbf{R} = 1.29 \times 0.075/1 \times 10^{-6} = 96\,750$$

Therefore,

$$f = 0.316/(96\,750)^{1/4} = 0.018 \blacktriangleleft \tag{9.24}$$

$$v_c = 60 \times 1.29/49 = 1.58 \text{ m/s} \blacktriangleleft \tag{9.26}$$

$$\frac{v_{25}}{1.58} = \left(\frac{12.5}{37.5}\right)^{1/7} \qquad v_{25} = 1.35 \text{ m/s} \blacktriangleleft \tag{9.27}$$

$$\tau_o = 0.046\,4 \left(\frac{1 \times 10^{-6}}{1.58 \times 0.037\,5}\right)^{1/4} \frac{998 \times (1.58)^2}{2} \tag{9.28}$$

$$\tau_o = 3.70 \text{ Pa} \blacktriangleleft$$

Compare results with those of the previous Illustrative Problem.

9.4   Turbulent Flow—Rough Pipes

Pipe friction in rough pipes at high Reynolds numbers will be governed primarily by the size and pattern of the roughness, since disruption of the viscous sublayer will render viscous action negligible. However, experiments show, as noted above, that the logarithmic velocity profile given by equation 9.13 is applicable for smooth or rough flow. An equation of this form was derived by use of mixing length theory in Section 9.3, the result being that

$$\frac{v}{v_\star} = \frac{1}{\kappa} \ln \frac{y}{y'}$$

where y' is the distance from the wall at which $v \approx 0$. For smooth walls this distance was characterized by the length scale $v/v_\star$. For a rough wall pipe y' is more plausibly proportional to the mean height e of the roughness projections. Following the previous analysis, the assumption that $y' = \alpha e$ and substitution in the above equation yield (for $\kappa = 0.40$)

$$\frac{v}{v_\star} = 2.5 \ln \frac{y}{\alpha e} = 2.5 \ln \frac{y}{e} - 2.5 \ln \alpha$$

or

$$\frac{v}{v_\star} = 2.5 \ln \frac{y}{e} + B = 5.75 \log \frac{y}{e} + B$$

where B is an unknown constant to be established by comparison with experiments. Nikuradse's experiments, which employed a coating of uniform sand grains on the pipe wall (thus giving an easily measurable index for the roughness, e being the diameter of the sand grain), establish that $B = 8.5$. The general equation of the velocity distribution for turbulent flow in rough pipes is then

$$\frac{v}{v_\star} = 5.75 \log \frac{y}{e} + 8.5 \tag{9.29}$$

As before the flowrate Q is found by integration

$$Q = \int_0^R v \,(2\pi r \, dr) = 2\pi v_\star \int_0^R \left[5.75 \log \left(\frac{R-r}{e} \right) + 8.5 \right] r \, dr$$

yielding

$$V = Q/\pi R^2 = v_\star \left[5.75 \log \frac{R}{e} + 4.75 \right]$$

Thus,

$$\frac{V}{v_\star} = 5.75 \log \frac{R}{e} + 4.75 \tag{9.30}$$

(The reader can use this equation and equation 9.29 to verify that v_c/V is again given by equation 9.20, which is, clearly, then valid for all turbulent flows.) Using $v_\star = V \sqrt{f/8}$ in equation 9.30,

$$\frac{1}{\sqrt{f}} = 2.03 \log \frac{R}{e} + 1.68$$

While the form of this result is correct, comparison with Nikuradse's experimental results suggests that the constants must be adjusted; the final equation is

$$\frac{1}{\sqrt{f}} = 2.0 \log \frac{R}{e} + 1.74 \tag{9.31}$$

or

$$\frac{1}{\sqrt{f}} = 2.0 \log \frac{d}{e} + 1.14 \tag{9.31}$$

For turbulent flow in rough pipes the friction factor is a function *only* of the relative roughness e/R or e/d and is *not* a function of Reynolds number (compare equation 9.31 to equation 9.21). Friction effects are produced in fully rough flow by roughness alone, without dependence on viscous action.

───────── **Illustrative Problem** ─────────

The mean velocity in a 300 mm pipeline is 3 m/s. The relative roughness of the pipe is 0.002, and the kinematic viscosity of water 9×10^{-7} m²/s. Determine the friction factor, centerline velocity, velocity 50 mm from the pipe wall, and the head lost in 300 m of this pipe under the assumption that the pipe is rough.

Relevant Equations and Given Data

$$h_L = f \frac{l}{d} \frac{V^2}{2g_n} \tag{9.2}$$

$$v_* = V \sqrt{f/8} \tag{9.4}$$

$$\frac{v}{v_*} = 5.75 \log \frac{y}{e} + 8.5 \tag{9.29}$$

$$\frac{V}{v_*} = 5.75 \log \frac{R}{e} + 4.75 \tag{9.30}$$

$$\frac{1}{\sqrt{f}} = 2.0 \log \frac{d}{e} + 1.14 \tag{9.31}$$

$d = 0.3$ m $e/d = 0.002$ $V = 3$ m/s $\nu = 9 \times 10^{-7}$ m²/s $\rho = 997$ kg/m³

Solution.

$$\frac{1}{\sqrt{f}} = 2.0 \log \frac{1}{0.002} + 1.14 \quad f = 0.023\ 4 \blacktriangleleft \tag{9.31}$$

$$v_* = 3 \times (0.023\ 4/8)^{1/2} = 0.162 \text{ m/s}$$

Setting $y = R$ in equation 9.29 gives v_c:

$$\frac{v_c}{0.162} = 5.75 \log \frac{0.15}{0.002 \times 0.3} + 8.5 \quad v_c = 3.62 \text{ m/s} \blacktriangleleft \tag{9.29}$$

Setting $y = 0.050$ m in equation 9.29 gives v_{50}:

$$\frac{v_{50}}{0.162} = 5.75 \log \frac{0.050}{0.002 \times 0.3} + 8.5 \qquad v_{50} = 3.17 \text{ m/s} \blacktriangleleft \qquad (9.29)$$

$$h_L = 0.023\,4 \left(\frac{300}{0.3}\right)\frac{(3)^2}{2g_n} = 10.7 \text{ m of water} \blacktriangleleft \qquad (9.2)$$

9.5 Classification of Smoothness and Roughness—Impact on Friction Factor

Turbulent flow in a smooth pipe was found (e.g., equation 9.23) to involve the sublayer thickness δ_v as a characteristic length. Turbulent flow in a rough pipe (e.g., equation 9.31) had the absolute roughness e as a characteristic length. In cases of transition where the pipe acts as neither smooth nor fully rough, e/δ_v must be a significant parameter.

In laminar flow $\delta_v = R$ because the viscous effects dominate the whole flow. In any reasonable pipe e/R is not large and experiments confirm that, for example, the effects of roughness on laminar flow are negligible. Thus, $f = 64/R$ and the other results derived in Section 9.2 are valid for smooth or rough pipes as long as the flow is laminar.

The significant relationship for classification of pipe surfaces as smooth or rough in turbulent flow is derived by expanding e/δ_v in terms of the sublayer thickness definition equation 9.22, namely, by writing

$$\frac{e}{\delta_v} = \frac{e/d}{\delta_v/d} = \frac{e/d}{32.8/R\sqrt{f}} = \left(\frac{e}{d}\right)\frac{R\sqrt{f}}{32.8}$$

so

$$\frac{e}{d}R\sqrt{f} = 32.8\frac{e}{\delta_v} \qquad (9.32)$$

Now the friction factor results obtained in the previous sections can be plotted versus $(e/d)R\sqrt{f}$ and analyzed in terms of experimental results.

For fully rough flow,

$$\frac{1}{\sqrt{f}} - 2.0 \log \frac{d}{e} = 1.14 \qquad (9.31)$$

so it is convenient to plot $(1/\sqrt{f}) - 2.0 \log (d/e)$ versus $(e/d) R\sqrt{f}$ because the rough flow region is a horizontal line. For smooth flow,

$$\frac{1}{\sqrt{f}} = 2.0 \log (R\sqrt{f}) - 0.8 \qquad (9.21)$$

Adding $-2.0 \log (d/e)$ to both sides,

$$\frac{1}{\sqrt{f}} - 2.0 \log \frac{d}{e} = 2.0 \log \left(\frac{e}{d}R\sqrt{f}\right) - 0.8 \qquad (9.33)$$

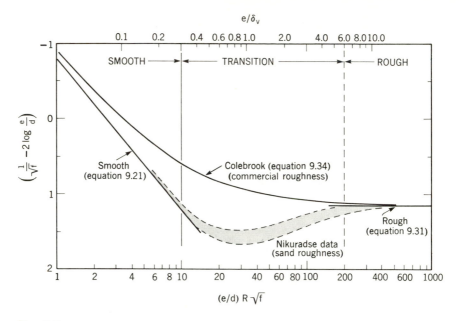

Fig. **9.7**

Equations 9.31 and 9.33 are plotted on Fig. 9.7. The shaded band represents the uniform sand-grain roughness data of Nikuradse. The flow is easily classified now. In particular,

For smooth flow: $\quad \dfrac{e}{d} \mathbf{R} \sqrt{f} \leq 10$

For transition flow: $\quad 10 < \dfrac{e}{d} \mathbf{R} \sqrt{f} < 200$

For rough flow: $\quad 200 \leq \dfrac{e}{d} \mathbf{R} \sqrt{f}$

From equation 9.32, it follows that the pipe acts as a smooth pipe when $e / \delta_v \leq 0.3$, while the pipe acts rough when $6 \leq e / \delta_v$. The e / δ_v scale is plotted along the upper boundary of Fig. 9.7. The general conclusion to be drawn is that the absolute size is not a measure of the effect of surface roughness on fluid flow; the effect is dependent on the size of the roughness *relative to the thickness of the viscous sublayer*.

Unfortunately, the excellent results of Nikuradse cannot be applied directly to engineering problems because the roughness patterns of commercial pipes are entirely different, much more variable, and much less definable than the artificial roughnesses used by Nikuradse. However, Colebrook[4] has shown how these results may be applied

[4] C. F. Colebrook, "Turbulent Flow in Pipes, with Particular Reference to the Transition Region between the Smooth and Rough Pipe Laws," *Jour. Inst. Civil Engrs.*, London, p. 133, February 1939.

toward a quantitative measure of commercial pipe roughness. He found that any pipe of commercial roughness when tested to a high enough Reynolds number gave a friction factor which no longer varied with Reynolds number. This allowed comparison with equation 9.31 and, from measurement of the friction factor, permitted an *equivalent* sand-grain size, e, to be computed. Obtaining e in this manner and plotting test results on many pipes, Colebrook found results from all pipes to be closely clustered about a single line having the equation

$$\frac{1}{\sqrt{f}} - 2 \log \frac{d}{e} = 1.14 - 2 \log \left[1 + \frac{9.28}{R(e/d)\sqrt{f}} \right] \tag{9.34}$$

This result is also shown on Fig. 9.7. Because of the variability of commercial roughness patterns, the previously observed distinctions between smooth, transition, and rough flow are not present for commercial pipes (see Section 9.6).

Another means of classifying roughness effects is to use the velocity profiles directly. For smooth flow, equation 9.13 can be written (for $\kappa = 0.40$) as

$$\frac{v_c - v}{v_\star} = 5.75 \log \frac{R}{y} \tag{9.35}$$

On the other hand, using equation 9.29 for rough flow in the form

$$\frac{v_c}{v_\star} = 5.75 \log \frac{R}{e} + 8.5$$

and subtracting equation 9.29 from this gives

$$\frac{v_c - v}{v_\star} = 5.75 \left(\log \frac{R}{e} - \log \frac{y}{e} \right)$$

or

$$\frac{v_c - v}{v_\star} = 5.75 \log \frac{R}{y}$$

That is, equation 9.35 is valid for both smooth and rough flow (see equation 9.13). Writing equation 9.35 as

$$\frac{v}{v_\star} = \frac{v_c}{v_\star} + 5.75 \log \frac{y}{R}$$

$$= \frac{v_c}{v_\star} + 5.75 \log \frac{e}{R} + 5.75 \log \frac{y}{e}$$

$$= A + 5.75 \log \frac{y}{e}$$

suggests that the term

$$A = \frac{v_c}{v_\star} + 5.75 \log \frac{e}{R} \tag{9.36}$$

is a constant in a fully rough flow. In fact $A = 8.5$ according to experimental data. For smooth flow (equation 9.17)

$$\frac{v}{v_\star} = 5.5 + 5.75 \log \frac{v_\star y}{v} \tag{9.17}$$

$$= 5.5 + 5.75 \log \frac{v_\star y}{v} \left(\frac{e}{e} \right)$$

$$= 5.5 + 5.75 \log \frac{v_\star e}{v} + 5.75 \log \frac{y}{e}$$

$$= A + 5.75 \log \frac{y}{e}$$

but now

$$A = 5.50 + \log \frac{v_\star e}{v} \tag{9.37}$$

and $v_\star e / v$ is the *Roughness Reynolds Number*. Figure 9.8 is a plot of A versus $v_\star e / v$ for Nikuradse's sand roughness data.[5] The results suggest that

For smooth flow: $v_\star e / v \leq 3.5$

For transition flow: $3.5 < v_\star e / v < 70$

For wholly rough flow: $70 \leq v_\star e / v$

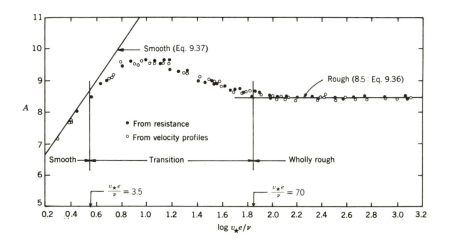

Fig. **9.8**

[5] Figure 9.8 is adapted from N.A.C.A. *Tech Mem.* 1292, which is a translation of J. Nikuradse, "Strömungsgesetze in rauhen Rohren," *VDI-Forschungsheft*, 361, 1933.

If for smooth flow $v_*e/\nu \leq 3.5$ and, from equation 9.18, $v_*\delta_\nu/\nu = 11.6$, it follows that $(11.6/\delta_\nu)e \leq 3.5$ or $e/\delta_\nu \leq 0.3$, which was obtained from the friction factor analysis. Likewise, if $70 \leq v_*e/\nu$, $70 \leq (11.6/\delta_\nu)e$ or $6 \leq e/\delta_\nu$ for rough flow. Thus, both means of classifying roughness effects are consistent.

───────────── **Illustrative Problem** ─────────────

Check the Illustrative Problem of the previous section to see if the flow is truly rough as assumed.

Relevant Equation and Given Data

Rough: $70 \leq v_*e/\nu$

$v_* = 0.162$ m/s $e/d = 0.002$ $d = 0.3$ m $\nu = 9 \times 10^{-7}$ m²/s

Solution. $v_*e/\nu = 0.162 \times 0.002 \times 0.3/9 \times 10^{-7} = 108$. Therefore, the flow is fully rough as assumed. ◀

9.6   Pipe Friction Factors

Application of the methods of similitude and dimensional analysis provides a frame-work on which a rational description of the pipe friction factor can now be constructed. From the analyses of the previous sections, it is known that the wall shear stress τ_o depends on the mean velocity V, the pipe diameter d, the mean height e of the roughness projections, and the fluid density ρ and viscosity μ. Thus,

$$f(\tau_o, V, d, e, \rho, \mu) = 0$$

Using the Π-theorem (Section 8.2), with V, d, and ρ as the repeating variables, gives $\Pi_1 = f_1(\tau_o, V, d, \rho)$, $\Pi_2 = f_2(e, V, d, \rho)$ and $\Pi_3 = f_3(\mu, V, d, \rho)$, and $\Pi_1 = \tau_o/\rho V^2$, $\Pi_2 = e/d$, and $\Pi_3 = Vd\rho/\mu$. Accordingly, on isolating Π_1

$$\frac{\tau_o}{\rho V^2} = f'(Vd\rho/\mu, e/d)$$

or, by comparison with equation 9.3,

$$f = f''(\mathbf{R}, e/d) \tag{9.38}$$

Thus, as confirmed by the experimental results and previous analyses the friction factor may depend only on the Reynolds number of the flow and/or on the *relative roughness* e/d of the pipe. The physical significance of equation 9.38 may be stated briefly: the friction factors of pipes will be the same if their Reynolds numbers, roughness patterns, and relative roughnesses are the same. When this is interpreted by the laws of simili-tude, its basic meaning is: the friction factors of pipes are the same if their flow pictures in every detail are geometrically and dynamically similar.

The relationships of equation 9.38 indicate a convenient means of presenting experimental data on the friction factor. This was used by Stanton[6] (1914) and consists of a logarithmic plot of friction factor against Reynolds number, with surface roughness the parameter as in Fig. 9.9. From such a plot, complete data on the friction factor can be obtained for laminar and turbulent flow of any fluid in smooth or rough pipes.

The results of the previously cited systematic tests by Nikuradse[7] on turbulent flow in smooth and rough pipes demonstrate perfectly the relationship among f, $\mathbf{R}$, and relative roughness. In these tests geometric similarity of the roughness pattern was obtained by fixing a coating of uniform sand grains to a smooth pipe wall; thus the diameter of the sand grain e became the easily measurable index for the roughness. Although the sand-grain roughness of Nikuradse is quite different from that of commercial pipes (see Fig. 9.7), the former is an easily measured, definite quantity which provides a reliable basis for quantitative measurement of roughness effects; Nikuradse's results are generally accepted today as the basic standard for this measurement. The test results when plotted logarithmically in Fig. 9.9 illustrate the following important fundamentals:[8]

1. The physical difference between the laminar and turbulent flow regimes is indicated by the change in the relationship of f to $\mathbf{R}$ near the critical Reynolds number of 2 100.

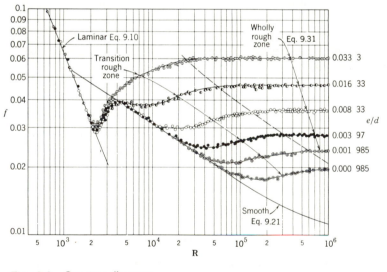

Fig. **9.9**   Stanton diagram.

[6] It is generally known as a *Stanton diagram*.

[7] J. Nikuradse, "Strömungsgesetze in rauhen Rohren," *VDI-Forschungsheft*, 361, 1933. Translation available in N.A.C.A., *Tech. Mem.* 1291.

[8] Figure 9.9 is a summary of Nikuradse's test results. Curves drawn through the points have the equations indicated.

2. The laminar regime is characterized by a single curve, given by the equation $f = 64/\mathbf{R}$ for all surface roughnesses. This confirms that head loss in laminar flow is independent of surface roughness and that $h_L \propto V$.

3. In turbulent flow a curve of f versus $\mathbf{R}$ exists for every relative roughness, e/d, and the horizontal aspect of the curves confirms that for rough pipes the roughness is more important than the Reynolds number in determining the magnitude of the friction factor.

4. At high Reynolds numbers the friction factors of rough pipes become constant, dependent wholly on the roughness of the pipe, and thus independent of the Reynolds number. From the Darcy-Weisbach equation if may be concluded that $h_L \propto V^2$ for completely turbulent flow over rough surfaces.

5. Although the lowest curve was obtained from tests on hydraulically smooth pipes, many of Nikuradse's rough pipe test results coincide with it for $5\,000 < \mathbf{R} < 50\,000$. Here the roughness is submerged in the viscous sublayer (Sections 9.3 and 9.5) and can have no effect on friction factor and head loss, which depend on viscosity effects alone. From the Darcy-Weisbach equation $h \propto V^{1.75}$ for turbulent flow in smooth pipes where the friction factor is given by the Blasius expression (equation 9.24).

6. The series of curves for the rough pipes diverges from the smooth pipe curve as the Reynolds number increases. In other words, pipes that are smooth at low values of $\mathbf{R}$ become rough at high values of $\mathbf{R}$. This is explained by the thickness of the viscous sublayer decreasing (Sections 9.3 and 9.5) as the Reynolds number increases, thus exposing smaller roughness protuberances to the turbulent region and causing the pipe to exhibit the properties of a rough pipe.

Illustrative Problem

Water at 100°F flows in a 3 in. pipe at Reynolds number 80 000. If the pipe contains a uniform sand roughness of grain size 0.006 in. diameter, how much head loss is to be expected in 1 000 ft of the pipe? How much head loss would be expected if this pipe were smooth?

Relevant Equations and Given Data

$$\mathbf{R} = Vd/\nu \tag{8.11}$$

$$h_L = f\frac{l}{d}\frac{V^2}{2g_n} \tag{9.2}$$

$$f = 0.316/\mathbf{R}^{0.25} \tag{9.24}$$

$d = 3$ in. $= 0.25$ ft $l = 1\,000$ ft $e = 0.006$ in. $\mathbf{R} = 80\,000$

Solution.   From Fig. 9.9, with $\mathbf{R} = 80\,000$ and $e/d = 0.006/3 = 0.002$,

$$f \cong 0.021$$

From Appendix 2,

$$\nu = 0.739 \times 10^{-5} \text{ ft}^2/\text{s}$$

$$80\ 000 = \frac{V \times 0.25}{0.739 \times 10^{-5}} \qquad V = 2.36 \text{ ft/s} \tag{8.11}$$

$$h_L \cong 0.021 \times \frac{1\ 000}{0.25} \times \frac{(2.36)^2}{2g_n} = 7.3 \text{ ft} \blacktriangleleft \tag{9.2}$$

For a smooth pipe, since $3\ 000 < \mathbf{R} < 100\ 000$,

$$f = \frac{0.316}{(80\ 000)^{0.25}} = 0.018\ 8 \tag{9.6}$$

which is confirmed by the "smooth" line of Fig 9.9. Thus, by proportion,

$$h_L \cong \left(\frac{0.018\ 8}{0.021}\right)7.3 = 6.53 \text{ ft} \blacktriangleleft$$

Even though Colebrook (see Section 9.5) showed that Nikuradse's results were not applicable to engineering problems with commercial pipes, his equation 9.34 (while conveniently summarizing the data on pipes of commerical roughness) is not easily applicable to engineering problems either. Moody[9] replotted the equation in the form of the Stanton diagram of Fig. 9.10. Moody's plot (now known as the *Moody diagram*) can be used directly in problem solution. Moody extended the work of Colebrook by including information on commercial pipes of other materials and summarized these data in the form of a relative roughness chart (see Fig. 9.11). The roughnesses of such materials are more difficult to classify because of variations in design, workmanship, and age—and are seen to vary between wide limits. This is one of the practical considerations of pipe flow calculations which poses a difficult problem for the designer in predicting friction factors which will be realized after a pipeline is constructed. Obviously extensive practical experience is needed for accuracy in such calculations.

The accuracy of pipe friction calculations is also lessened by the somewhat unpredictable changes in the roughness and friction factor due to the accumulation of dirt and corrosion on the pipe walls. This accumulation not only increases surface roughness but also reduces the effective pipe diameter and may lead to an extremely large increase in the friction factor after the pipe has been in service over a long period. Pipeline designers have different methods to allow for these effects, but no attempt is made to summarize them here.

[9] L. F. Moody, "Friction Factors for Pipe Flow," *Trans. A.S.M.E.*, vol. 66, 1944. Figures 9.10 and 9.11 are adapted from this publication with permission.

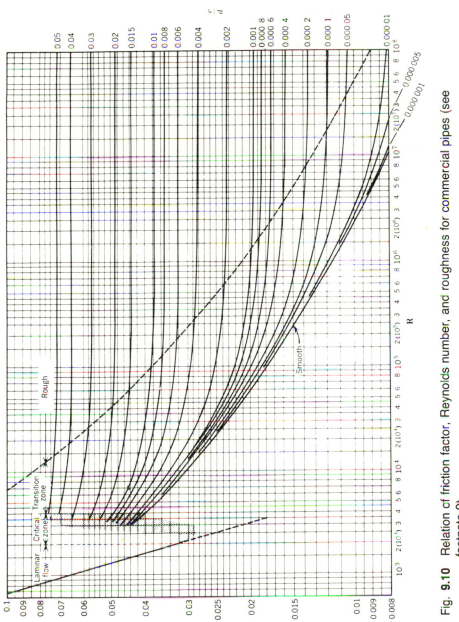

Fig. **9.10** Relation of friction factor, Reynolds number, and roughness for commercial pipes (see footnote 9).

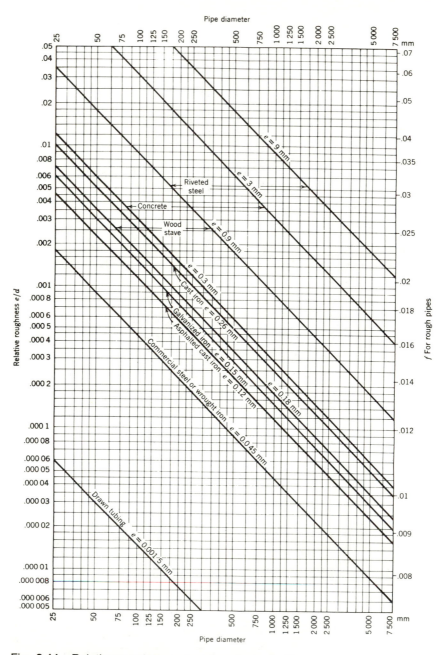

Fig. **9.11** Relative roughness chart (see footnote 9).

Illustrative Problems

Water at 100°F flows in a 3 in. pipe at Reynolds number 80 000. This is a commercial pipe of equivalent sand-grain roughness 0.006 in. What head loss is to be expected in 1 000 ft of this pipe?

Relevant Equation and Given Data

$$h_L = f\frac{l}{d}\frac{V^2}{2g_n} \tag{9.2}$$

$d = 3$ in. $= 0.25$ ft $l = 1\ 000$ ft $e = 0.006$ in. $R = 80\ 000$

Solution. From the Moody diagram of Fig 9.10 with $R = 80\ 000$ and $e/d = 0.006/3 = 0.002$, $f \cong 0.025\ 5$. From the previous Illustrative Problem of this section, $V = 2.36$ ft/s. Therefore,

$$h_L \cong 0.025\ 5\frac{1\ 000}{0.25} \times \frac{(2.36)^2}{2g_n} = 8.8\ \text{ft} \blacktriangleleft \tag{9.2}$$

Crude oil at 20°C flows in a riveted steel pipe of 1 m diameter. The mean velocity is 2 m/s. What is the range of head loss to be expected in 1 km of pipeline?

Relevant Equations and Given Data

$$R = Vd/\nu \tag{8.11}$$

$$h_L = f\frac{l}{d}\frac{V^2}{2g_n} \tag{9.2}$$

$d = 1$ m $V = 2$ m/s $l = 1\ 000$ m

s.g. $= 0.86$ $\mu = 0.007\ 18$ Pa·s $\rho_{water} = 998.2$ kg/m³

Solution.

$$R = \frac{2 \times 1 \times 0.86 \times 998.2}{0.007\ 18} = 2.4 \times 10^5 \tag{8.11}$$

$$\frac{e}{d} = 0.000\ 9 \text{ to } 0.009$$

Therefore $f \cong 0.020\ 5$ to $0.036\ 5$ (in the rough region) and

$$h_L \cong 0.020\ 5\frac{1\ 000}{1} \times \frac{(2)^2}{2 \times 9.81} = 4.2\ \text{m} \tag{9.2}$$

or

$$h_L \cong 0.036\ 5\frac{1\ 000}{1} \times \frac{(2)^2}{2 \times 9.81} = 7.4\ \text{m} \tag{9.2}$$

The range of head loss is from 4.2 to 7.4 metres. ◀

9.7 Pipe Friction in Noncircular Pipes—The Hydraulic Radius

Although the majority of pipes used in engineering practice are of circular cross section, occasions arise when calculations must be carried out on head loss in rectangular passages and other conduits of noncircular form. The foregoing equations for circular pipes may be adapted to these special problems through use of the *hydraulic radius* concept.

The hydraulic radius, R_h, is defined as the area, A, of the flow cross section divided by its "wetted perimeter," P (see Section 7.10). In a circular pipe of diameter d,

$$R_h = \frac{d}{4} \quad \text{or} \quad d = 4R_h \tag{9.39}$$

This value may be substituted into the Darcy-Weisbach equation for head loss and into the expression for the Reynolds number with the following results:

$$h_L = \frac{f}{4R_h} \frac{l}{2g_n} V^2 \tag{9.40}$$

and

$$\mathbf{R} = \frac{V(4R_h)\rho}{\mu} \tag{9.41}$$

from which the head loss in many conduits of noncircular cross section may be calculated with the aid of the Moody diagram of Fig. 9.10.

The calculation of lost head in noncircular conduits thus involves the calculation of the hydraulic radius of the flow cross section and the use of the friction factor obtained for an *equivalent* circular pipe having a diameter d equal to $4R_h$. In view of the complexities of viscous sublayers, turbulence, roughness, shear stress, and so forth, it seems surprising at first that a circular pipe equivalent to a noncircular conduit may be obtained so easily, and it would, therefore, be expected that the method might be subject to certain limitations. The method gives satisfactory results when the problem is one of turbulent flow but, if it is used for laminar flow, large errors are introduced.

The foregoing facts may be justified analytically by examining further the structure of equation 9.40 in which h_L varies with $1/R_h$. From the definition of the hydraulic radius, its reciprocal is the *wetted perimeter per unit of flow cross section* and is, therefore, an index of the extent of the boundary surface in contact with the flowing fluid. The hydraulic radius may be safely used in the equation above when resistance to flow and head loss are primarily dependent on the extent of the boundary surface, as for turbulent flow in which pipe friction phenomena are confined to a thin region adjacent to the boundary surface and thus vary with the size of this surface. However, severe deviation from a circular flow cross section will prevent the hydraulic radius from accounting for changes in head loss, even for cases of turbulent flow. Tests on turbulent flow through annular passages,[10] for example, show a large increase in friction factor with increase of the ratio of core diameter to pipe diameter.

In laminar flow, friction phenomena result from the action of viscosity throughout the whole body of the flow, are independent of roughness, and are not primarily associated with the region close to the boundary walls. In view of these facts the hydraulic radius technique cannot be expected to give reliable conversions from circular to noncircular passages in laminar flow. This expectation is borne out both by experiment and by analytical solutions of laminar flow in noncircular passages.

———————————— **Illustrative Problem** ————————————

Calculate the loss of head and the pressure drop when air at an absolute pressure of 101.3 kPa and 15°C flows through 600 m of a 450 mm by 300 mm smooth rectangular duct with a mean velocity of 3 m/s.

Relevant Equations and Given Data

$$h_L = \frac{f}{4} \frac{l}{R_h} \frac{V^2}{2g_n} \tag{9.40}$$

$$R = \frac{V(4R_h)\rho}{\mu} \tag{9.41}$$

$A = 0.45 \times 0.30 \text{ m}^2$ $P = (2 \times 0.45 + 2 \times 0.30) \text{ m}$ $l = 600 \text{ m}$

$V = 3 \text{ m/s}$ $\rho = 1.225 \text{ kg/m}^3$ $\mu = 1.789 \times 10^{-5} \text{ Pa} \cdot \text{s}$

Solution.

$$R_h = \frac{0.45 \times 0.30}{2 \times 0.45 + 2 \times 0.30} = 0.09 \text{ m}$$

$$R = \frac{3 \times 4 \times 0.09 \times 1.225}{1.789 \times 10^{-5}} = 73\,950$$

From the Moody diagram of Fig. 9.10, for a smooth surface,

$$f \cong 0.019$$

$$h_L \cong 0.019 \times \frac{600}{4 \times 0.09} \times \frac{(3)^2}{2g_n} \cong 14.5 \text{ m of air} \blacktriangleleft \tag{9.40}$$

$$\Delta p \cong 1.225 \times 9.81 \times 14.5 \cong 174 \text{ Pa} \blacktriangleleft$$

[10] See W. M. Owen, "Experimental Study of Water Flow in Annular Pipes," *Trans. A. S. C. E.*, vol. 117, 1952; and J. E. Walker, G. A. Whan, and R. R. Rothfus, "Fluid Friction in Noncircular Ducts," *Jl.A.I. Ch.E.*, vol. 3, 1957.

9.8 Pipe Friction—An Empirical Formulation

The Darcy-Weisbach equation 9.2 has provided a rational basis for the analysis and computation of head loss. However, historically, a number of empirical formulas have been used for pipe friction calculations in engineering practice. One of the most lasting and widely used of these has been the Hazen-Williams[11] formulation. According to their experimental results the mean velocity V in a pipe is given by

$$\text{(FSS units):} \quad V = 1.318 \, C_{hw} R_h^{0.63} S^{0.54} \tag{9.42a}$$

$$\text{(SI units):} \quad V = 0.849 \, C_{hw} R_h^{0.63} S^{0.54} \tag{9.42b}$$

where R_h is the hydraulic radius, S is the head lost per unit length of pipe, and C_{hw} is a roughness (or smoothness) coefficient, typical values of which are given in Table 3.

Table 3 Hazen-Williams Coefficient, C_{hw}

Pipes extremely straight and smooth	140
Pipes very smooth	130
Smooth wood, smooth masonry	120
New riveted steel, vitrified clay	110
Old cast iron, ordinary brick	100
Old riveted steel	95
Old iron in bad condition	60–80

In view of the shape of the formula and the rather indefinite descriptions to be interpreted for selection of coefficients, it is difficult to judge its validity or meaning without wide experience with its application. It is not evident whether C_{hw} is a measure of absolute or relative roughness, whether there is any effect of Reynolds number in the formula, whether it applies only to the rough zone of boundary roughness, etc. Such questions can be answered fairly conclusively if the equation is rewritten so that the more basic yardstick of Fig. 9.10 may be applied. Substituting $d/4$ for R_h, h/l for S, solving for h_L, and multiplying by $2g_n/2g_n$ produces in FSS units:

$$h_L = \left[\frac{194}{C_{hw}^{1.85}} \frac{1}{(Vd)^{0.15} \, d^{0.015}} \right] \frac{l}{d} \frac{V^2}{2g_n}$$

If $(Vd)^{0.15}$ is multiplied by $(\nu/\nu)^{0.15}$, a Reynolds number appears and the equation may be rewritten as

$$h_L = \left[\frac{194}{\nu^{0.15} C_{hw}^{1.85} d^{0.015} \, \mathbf{R}^{0.15}} \right] \frac{l}{d} \frac{V^2}{2g_n}$$

Comparing this with the Darcy-Weisbach equation 9.2 and taking a nominal value of ν for water of, say, 0.000 01 ft²/s, yields

---

[11] G. S. Williams and A. H. Hazen, *Hydraulic Tables,* 3rd ed., John Wiley & Sons, 1933.

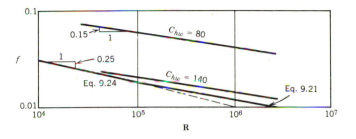

Fig. **9.12**

$$f = \frac{1\ 090}{C_{hw}^{1.85}\,d^{0.015}\,\mathbf{R}^{0.15}}$$

which may be plotted and compared with Fig. 9.10. The appearance of the curves is indicated on Fig. 9.12, and they compare roughly (but favorably) with the curves for the rough pipes of Fig. 9.10. In view of this favorable comparison, it may be concluded that C_{hw} is an index of relative roughness rather than absolute roughness.

The advantages and disadvantages of the Hazen-Williams method are evident from the formula and foregoing discussion. Among its advantages are: (1) the coefficient is a rough measure of relative roughness, (2) the effect of Reynolds number is included in the formula, and (3) the effect of roughness and other variables on the velocity are given directly. Its disadvantages are: (1) its empirical nature, and (2) the impossibility of applying it to all fluids under all conditions. Although the formula appears cumbersome with its fractional exponents, it is very convenient in computer programs because the head loss is given by an explicit expression for any given pipe, whereas the friction factor of Fig. 9.10 changes with $\mathbf{R}$ in a manner not easily reduced to an explicit expression. For example, note that the Colebrook equation 9.34 is an implicit equation for f.

———————— **Illustrative Problem** ————————

If 90 gal/min of water flow through a smooth 3 in. pipeline, calculate the head loss in 3 000 ft of this pipe.

Relevant Equations and Given Data

$$Q = AV \qquad\qquad (3.11)$$

$$R_h = d/4 \qquad\qquad (9.39)$$

$$V = 1.318\,C_{hw}R_h^{0.63}S^{0.54} \qquad\qquad (9.42)$$

$$Q = 90 \text{ gal/min} \qquad d = 3 \text{ in.} = 0.25 \text{ ft} \qquad l = 3\ 000 \text{ ft}$$

Solution.

$$V = \frac{90}{60 \times 7.48} \times \frac{1}{(\pi/4)(\frac{3}{12})^2} = 4.08 \text{ ft/s}$$

$$R_h = (\tfrac{3}{12})/4 = 0.062\ 5 \text{ ft}$$

From Table 3,

$$C_{hw} = 140$$

$$4.08 = 1.318 \times 140 \times (0.062\ 5)^{0.63} S^{0.54}$$

$$S = 0.021\ 8 = h_L/3\ 000 \qquad h_L = 65.5 \text{ ft of water} \blacktriangleleft$$

9.9 Minor Losses in Pipelines

Into the category of minor losses in pipelines fall those losses incurred by change of section, bends, elbows, valves, and fittings of all types. Although in long pipelines these are distinctly "minor" losses and can often be neglected without serious error, in shorter pipelines an accurate knowledge of their effects must be known for correct engineering calculations.

  The general aspects of minor losses in pipelines may be obtained from a study of the flow phenomena about an abrupt obstruction placed in a pipeline (Fig. 9.13), which creates flow conditions typical of those which dissipate energy and cause minor losses. Minor losses usually result from rather abrupt changes (in magnitude or direction) of velocity; in general, increase of velocity (acceleration) is associated with small head loss but decrease of velocity (deceleration) causes large head loss because of the production of large-scale turbulence (Section 7.11). In Fig. 9.13 useful energy is extracted in the creation of eddies as the fluid decelerates between sections 2 and 3, and

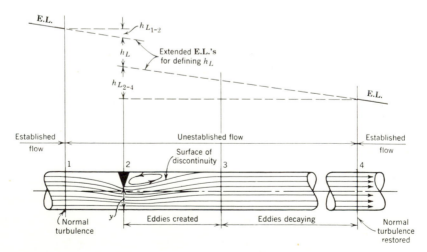

Fig. **9.13**

this energy is dissipated in heat as the eddies decay between sections 3 and 4. Minor losses in pipe flow are, therefore, accomplished in the pipe downstream from the source of the eddies, and the pipe friction processes in this length of pipe are complicated by the superposition of large-scale turbulence on the normal turbulence pattern. To make minor loss calculations possible it is necessary to assume separate action of the normal turbulence and large-scale turbulence, although in reality a complex combination of the two processes exists. Assuming the processes independent allows calculation of the losses due to established pipe friction, $h_{L_{1-2}}$ and $h_{L_{2-4}}$, and also permits the loss h_L, due to the obstruction alone, to be assumed concentrated at section 2. This is a great convenience for engineering calculations since the total lost head in a pipeline may be obtained by simple addition of established pipe friction and minor losses without detailed consideration of the above-mentioned complications.

Early experiments with water (at high Reynolds number) indicated that minor losses vary approximately with the square of velocity and led to the proposal of the basic equation

$$h_L = K_L \frac{V^2}{2g_n} \tag{9.43}$$

in which K_L, the *loss coefficient,* is, for a given flow geometry, practically constant at high Reynolds number; the loss coefficient tends to increase with increasing roughness and decreasing Reynolds number,[12] but these variations are usually of minor importance in turbulent flow. The magnitude of the loss coefficient is determined primarily by the flow geometry, that is, by the shape of the obstruction or pipe fitting.

When an *abrupt enlargement* of section (Fig. 9.14) occurs in a pipeline, a rapid deceleration takes place, accompanied by characteristic large-scale turbulence, which may persist in the larger pipe for a distance of 50 diameters or more before the normal turbulence pattern of established flow is restored. Simultaneous application of the

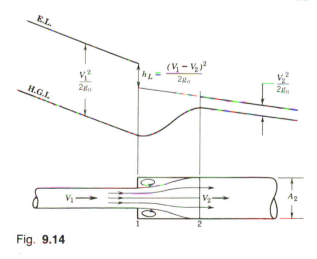

Fig. **9.14**

[12] Note that this is the same trend followed by the friction factor, f.

continuity, Bernoulli, and momentum principles to this problem has shown (Section 6.4) that (with certain simplifying assumptions)

$$h_L = K_L \frac{(V_1 - V_2)^2}{2g_n} \qquad (9.44)$$

in which $K_L \cong 1$. Experimental determinations of K_L confirm this value within a few percent, making it quite adequate for engineering use. A special case of an abrupt enlargement exists when a (relatively small) pipe discharges into a (relatively large) tank or reservoir. Here the velocity downstream from the enlargement may be taken to be zero, and when the lost head (called the exit loss) is calculated from equation 9.44 it is found to be the velocity head in the pipe.

The loss of head due to *gradual enlargement* is, of course, dependent on the shape of the enlargement. Tests have been carried out by Gibson on the losses in conical enlargements, and the results are expressed by equation 9.44, in which K_L is primarily dependent on the cone angle but is also a function of the area ratio, as shown in Fig. 9.15. Because of the large surface of the conical enlargement which contacts the fluid, the coefficient K_L embodies the effects of wall friction as well as those of large-scale turbulence. In an enlargement of small central angle, K_L will result almost wholly from surface friction; but, as the angle increases and the enlargement becomes more abrupt, not only is the surface area reduced but also separation occurs, producing large eddies, and here the energy dissipated in the eddies determines the magnitude of K_L. From the plot it may be observed that (1) there is an optimum cone angle of about 7° where the combination of the effects of surface friction and eddying turbulence is a minimum; (2) it is better to use a sudden enlargement than one of cone angle around 60°, since K_L is smaller for the former.

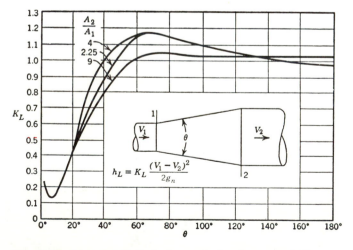

Fig. **9.15** Loss coefficients for conical enlargements.[13]

[13] A. H. Gibson, *Hydraulics and Its Applications*, 4th ed., p. 93, D. Van Nostrand Co., 1930. See also *Engineering Data Book*, Hydraulic Institute, Cleveland, 1979.

Gradual enlargements in passages (termed *diffusers*) of various forms are widely used in engineering practice for *pressure recovery,* that is, pressure rise in the direction of flow. No attempt is made here to review the extensive literature[14] on diffusers, but it should be realized that Gibson's results cannot give a reliable solution to this problem. His tests were made, as are all such tests for minor losses, with long straight lengths of pipe upstream and downstream from the enlargement (see Fig. 9.13). The designer, however, is frequently more interested in the pressure rise through the diffuser and substitution of a short nozzle for the upstream pipe length. The tests of Gibson and others have shown that the pressure will continue to rise for a few pipe diameters downstream from section 2, owing primarily to readjustment of velocity distribution from a rather pointed one caused by deceleration through the diffuser to the flatter one of turbulent flow.[15] From this it may be concluded that pressure rise through the diffuser, computed from application of the Bernoulli equation with data from Fig. 9.15, will be larger than that which will actually be realized. Substitution of a nozzle for the upstream pipe length will alter the inlet velocity distribution from the standard one of turbulent flow to a practically uniform one with a thin boundary layer (see Fig. 7.16). The effect is to reduce the losses by stabilizing the flow and delaying separation; not only do smaller loss coefficients result, but the cone angle for minimum losses is larger, thus allowing a shorter diffuser for the same area ratio.

Illustrative Problem

A 300 mm horizontal water line enlarges to a 600 mm line through a 20° conical enlargement. When 0.3 m³/s flow through this line the pressure in the smaller pipe is 140 kPa. Calculate the pressure in the larger pipe, neglecting pipe friction.

Relevant Equations and Given Data

$$Q = A_1 V_1 = A_2 V_2 \tag{3.12}$$

$$\frac{p_3}{\gamma} + \frac{V_3^2}{2g_n} + z_3 = \frac{p_6}{\gamma} + \frac{V_6^2}{2g_n} + z_6 + h_L \tag{7.17}$$

$$h_L = K_L \frac{(V_3^2 - V_6^2)}{2g_n} \tag{9.44}$$

$Q = 0.3$ m³/s $d_3 = 300$ mm $d_6 = 600$ mm $p_3 = 140$ kPa $z_3 = z_6$ $\theta = 20°$

Solution.

$$V_3 = \frac{0.3}{(\pi/4)(0.3)^2} = 4.24 \text{ m/s} \qquad V_6 = \frac{4.24}{4} = 1.06 \text{ m/s} \tag{3.12}$$

[14] See, for example, E. G. Reid, "Performance Characteristics of Plane-Wall Two-Dimensional Diffusers," *N.A.C.A., Tech. Note* 2888, 1953; S. J. Kline, D. E. Abbott, and R. W. Fox, "Optimum Design of Straight-Walled Diffusers," *Trans. A.S.M.E.,* vol. 81, 1959; and J. M. Robertson and H. R. Fraser, "Separation Prediction in Conical Diffusers," *Trans. A.S.M.E.* (*Series D*), vol. 82, no. 1, 1960.
[15] Compare this with the opposite situation discussed in Section 7.8.

From the plot of Fig. 9.15,

$$K_L = 0.43$$

$$\frac{140 \times 10^3}{9.8 \times 10^3} + \frac{(4.24)^2}{2g_n} = \frac{p_6}{\gamma} + \frac{(1.06)^2}{2g_n} + 0.43\frac{(4.24 - 1.06)^2}{2g_n} \qquad (7.17, 9.44)$$

$$\frac{p_6}{\gamma} = 14.8 \text{ m} \qquad p_6 = 145 \text{ kPa} \blacktriangleleft$$

This is the pressure to be expected a metre or so downstream from the end of the enlargement.

Flow through an *abrupt contraction* is shown in Fig. 9.16 and is featured by the formation of a vena contracta (Section 4.8) and subsequent deceleration and re-expansion of the live stream of flowing fluid.

Experimental measurements of K_L are somewhat conflicting in magnitude although they exhibit a well-established trend from 0.5 for $A_2/A_1 = 0$ to 0 for $A_2/A_1 = 1$. In view of this it is entirely adequate for engineering practice to use a synthesis of analytical approaches and generally accepted experimental information.[16] The result is given in Table 4 where $C_c = A_c/A_2$.

Table 4

A_2/A_1	0	0.1	0.2	0.3	0.4	0.5	0.6	0.7	0.8	0.9	1.0
C_c	0.617	0.624	0.632	0.643	0.659	0.681	0.712	0.755	0.813	0.892	1.00
K_L	0.50	0.46	0.41	0.36	0.30	0.24	0.18	0.12	0.06	0.02	0

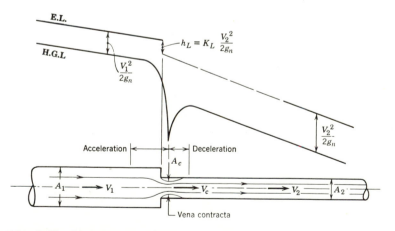

Fig. 9.16   Abrupt contraction.

[16] J. Weisbach, *Die experimental Hydraulik,* 1855.

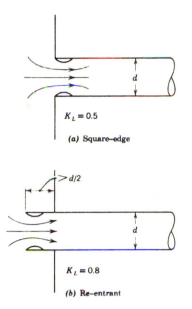

$K_L = 0.5$

(a) Square–edge

$K_L = 0.8$

(b) Re–entrant

Fig. **9.17** Pipe entrances.

A square-edged pipe entrance (Fig. 9.17a) from a large body of fluid is the limiting case of the abrupt contraction, with $A_2/A_1 = 0$ (A_1 is virtually infinite here). The head loss is expressed by equation 9.43 in which K_L is close to 0.5 for highly turbulent flow, as mentioned above.

The entrance of Fig. 9.17b is known as a re-entrant one. If the pipe wall is very thin and if the plane of the opening is more than one pipe diameter upstream from the reservoir wall, the loss coefficient will be close to 0.8, this high value resulting mainly from the small vena contracta and consequent large deceleration loss. For thick-walled pipes the vena contracta can be expected to be larger and the loss coefficient less than 0.8; Harris[17] has shown that pipes of wall thicknesses greater than $0.05d$ (if square-edged) will give a loss coefficient equal to that of the square-edged entrance.

If the edges of a pipe entrance are rounded to produce a streamlined *bell-mouth* (Fig. 9.18) the loss coefficient can be materially reduced. Hamilton[18] has shown that any radius of rounding greater than $0.14d$ will prevent the formation of a vena contracta and thus eliminate the head loss due to flow deceleration. The nominal value of K_L for such an entrance is about 0.1, but its exact magnitude will depend on the detailed geometry of the entrance and structure of the boundary layer (see Fig. 7.16 of Section 7.7).

The head loss caused by short well-streamlined gradual contractions (Fig. 9.19) is so small that it may usually be neglected in engineering problems. However, an

[17] C. W. Harris, "The Influence of Pipe Thickness on Reentrant Intake Losses," *Univ. Wash. Eng. Expt. Sta., Bull.* 48, 1928.
[18] J. B. Hamilton, "The Suppression of Intake Losses by Various Degrees of Rounding," *Univ. Wash. Eng. Expt. Sta., Bull.* 51, 1929.

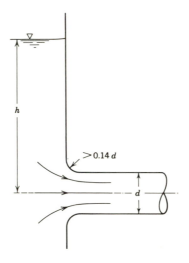

Fig. **9.18** Bell-mouth entrance.

appreciable fall of the hydraulic grade line over such contractions is to be expected; in the pipe downstream from the contraction the hydraulic grade line will be found to slope more steeply than that for established flow because of the change of velocity distribution and boundary layer growth, which cause an increase of α (see Fig. 7.16 of Section 7.7). These effects are known to be small and may be ignored in many problems. A nominal value of K_L (for use in equation 9.43 with $V = V_2$) for short well-streamlined contractions is 0.04; by careful design this figure may be lowered to 0.02, but for long contractions values much larger than 0.04 are to be expected because of extensive wall friction.

Losses of head in *smooth pipe bends* are caused by the combined effects of separation, wall friction, and the twin-eddy secondary flow described in Section 7.12; for bends of large radius of curvature, the last two effects will predominate, whereas, for small radius of curvature, separation and the secondary flow will be the more significant. The loss of head in a bend is expressed by equation 9.43 in which the head loss is the drop of the extended energy lines (see Fig. 9.13) between the entrance and exit of the bend. Reliable and extensive information on this subject will be found in the work of Itō,[19] a small but typical portion of which is shown in Fig. 9.20. Head loss coefficients for smooth pipe bends provide another example of the dependence of K_L

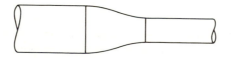

Fig. **9.19** Gradual contraction.

[19] H. Itō, "Pressure Losses in Smooth Pipe Bends," *Trans. A.S.M.E.* (*Series D*), vol. 82, no. 1, 1960.

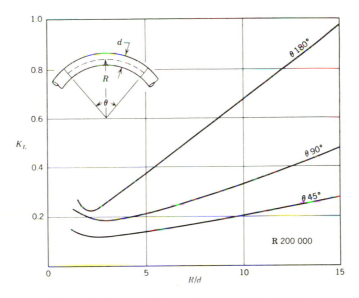

Fig. **9.20** Itõ's loss coefficients for smooth bends ($R = 200\,000$).

on shape of passage (determined by θ and R/d) and Reynolds number; the research of Hofmann[20] provides information on the (expected) dependence of loss coefficient on relative roughness as well. To the engineer the most significant feature of head loss in bends is the minimum value of K_L occurring at certain values of R/d which allows selection of bend shapes for maximum efficiency in pipeline design.

Tests made on bends with $R/d = 0$ have given values of K_L around 1.1. Such bends are known as *miter bends* (Fig. 9.21) and are used widely in large ducts such as wind and water tunnels, where space does not permit a bend of large radius. In these bends, installation of guide vanes materially reduces the head loss and at the same time breaks up the spiral motion and improves the velocity distribution downstream.

The losses of head caused by commercial pipe fittings occur because of their rough and irregular shapes which produce excessive large-scale turbulence. The shapes of commerical pipe fittings are determined more by structural properties, ease in handling, and production methods than by head loss considerations, and it is, therefore, not feasible or economically justifiable to build pipe fittings having completely streamlined

Fig. **9.21** Miter bends.

[20] A. Hofmann, "Loss in 90° Pipe Bends of Constant Circular Cross Section," *Trans. Hydr. Inst. of Munich Techn. Univ., Bull.* 3, A.S.M.E., 1935.

interiors in order to minimize head loss. The loss of head in commercial pipe fittings is usually expressed by equation 9.43 with V the mean velocity in the pipe and K_L a constant (at high Reynolds numbers), the magnitude of which depends on the shape of the fitting. Values of K_L for various common fittings are available in the *Engineering Data Book* of the Hydraulic Institute; typical values are presented in Table 5.

It is generally recognized that when fittings are placed in close proximity the total head loss caused by them is less than their numerical sum obtained by the foregoing methods. Systematic tests have not been made on this subject because a simple numerical sum of losses gives a result in excess of the actual losses and thus produces an error on the conservative side when design calculations of pressures and flowrates are to be made.

9.10   Pipeline Problems—Single Pipes

All steady-flow pipeline problems may be solved by application of the Bernoulli and continuity equations, and the most effective method of doing this is the construction of energy and hydraulic grade lines. From such lines the variations of pressure, velocity, and unit energy can be clearly seen for the whole problem; thus the construction of these lines becomes equivalent to writing numerous equations, but the lines lend a clarity to the solution of the problem which equations alone never can.

In engineering offices, tables, charts, nomograms, and so forth, are employed where numerous pipe-flow problems are to be solved. Although all these methods are different they have their foundations in the Bernoulli principle, usually with certain approximations; no attempt is made here to cover these many methods—the following discussion will be confined to the application of the Bernoulli and continuity principles and the use of certain approximations.

Engineering pipe-flow problems usually consist of (1) calculation of head loss and pressure variation from flowrate and pipeline characteristics, (2) calculation of flowrate from pipeline characteristics and the head which produces flow, and (3) calculation of required pipe diameter to pass a given flowrate between two regions of known pressure difference. The first of these problems can be solved directly, but solution by trial is required for the other two.[22] Trial-and-error solutions are necessitated by the fact that the friction factor, f, and loss coefficients, K_L, depend on the Reynolds number, which in turn depends on flowrate and pipe diameter, the unknowns of problems 2 and 3, respectively. However, many engineering pipeline problems involve flow in rough pipes at high Reynolds numbers. Here trial solutions are seldom required (1) because of the tendency of f and K_L toward constancy in this region, (2) because of the inevitable error[23] in selecting f from Fig. 9.10, and (3) because engineering answers are usually not needed to a precision which warrants trial-and-error solution in the light of the foregoing facts. Construction of energy and hydraulic grade lines for some typical pipeline problems will indicate further approximations which may frequently be used in the solution of engineering problems.

---

[22] Unless special plots are devised for circumventing this.
[23] Due to the inexactness of definition of the roughness.

Table 5 Approximate Loss Coefficient, K_L, for
Commercial Pipe Fittings[21]

Valves, wide open	Screwed		Flanged
Globe	10		5
Gate	0.2		0.1
Swing-check		2	
Angle		2	
Foot		0.8	
Return bend	1.5		0.2
Elbows			
90°—regular	1.5		0.3
—long radius	0.7		0.2
45°—regular	0.4		—
—long radius	—		0.2
Tees			
Line flow	0.9		0.2
Branch flow	2		1

Consider the calculation of flowrate in a pipeline laid between two tanks or reservoirs having a difference of surface elevation H (Fig. 9.22). The energy line must start in one reservoir surface and end in the other; using a gradual drop to represent head loss due to pipe friction, h_{L_f}, and abrupt drops to represent entrance and exit losses, h_{L_e} and h_{L_x}, the energy line is constructed as shown. It is apparent from the energy line that

$$h_{L_e} + h_{L_f} + h_{L_x} = H$$

which is the Bernoulli equation written between the reservoir surfaces. When the appropriate expressions for the head losses are substituted (Sections 9.1 and 9.9),

$$\left(0.5 + f\frac{l}{d} + 1 \right) \frac{V^2}{2g_n} = H$$

If turbulent flow is assumed and an approximate value for f of 0.03 is selected (Fig.

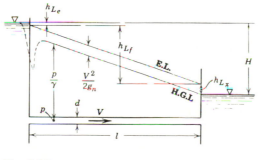

Fig. **9.22**

[21] Adapted with permission of Hydraulic Institute, *Engineering Data Book,* 1979.

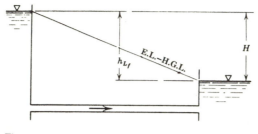

Fig. **9.23**

9.10), the quantity in parentheses becomes 4.5, 31.5, and 301.5 for l/d of 100, 1 000, and 10 000, respectively. The quantities 0.5 and 1.0, which result from inclusion of minor losses, have a decreasing effect on the equation with increasing l/d; if these were omitted entirely, errors of about 18, 2, and 0.3%, respectively, would be produced in the velocity and flowrate. Evidently, the effect of minor losses in pipelines of reasonable length is so small that they may often be neglected entirely and calculations appreciably simplified. Another convenient approximation accompanies the above; increasing l/d also decreases $V^2/2g_n$ and thus brings the energy line and the hydraulic grade line closer together; since $V^2/2g_n$ is of the order of the minor losses, it is consistent to neglect this also, thus making energy and hydraulic grade lines coincident (except near the entrance) and necessitating the construction of only one line. When the single line is drawn for this pipeline problem (Fig. 9.23), the equation

$$h_L = f\frac{l}{d}\frac{V^2}{2g_n} = H$$

may be written, and the velocity and flowrate may be obtained by trial-and-error procedure. The foregoing approximations are convenient in engineering problems but, of course, cannot be applied blindly and without some experience; preliminary calculations similar to those above will usually indicate the effect of such approximations on the accuracy of the result.

——————————— **Illustrative Problems** ———————————

A clean cast iron pipeline of 0.3 m diameter and 300 m long connects two reservoirs having surface elevations 60 and 75. Calculate the flowrate through this line, assuming water at 10°C and a square-edged entrance.

Relevant Equations and Given Data

$$\frac{p_1}{\gamma} + \frac{V_1^2}{2g_n} + z_1 = \frac{p_2}{\gamma} + \frac{V_2^2}{2g_n} + z_2 + h_L \qquad (7.17)$$

$$\mathbf{R} = Vd/\nu \qquad (8.11)$$

$d_1 = 0.3$ m $l = 300$ m $z_1 = 75$ m $z_2 = 60$ m

$\nu = 1.31 \times 10^{-6}$ m²/s   clean cast iron pipe: $e/d = 0.000\ 83$

Solution.

$$R = \frac{V \times 0.3}{(1.31 \times 10^{-6})} = 229\ 000V \tag{8.11}$$

Applying the Bernoulli equation (7.7) with appropriate values of loss coefficients,

$$15 = \left(0.5 + f\frac{300}{0.3} + 1\right)\frac{V^2}{2g_n} \tag{7.17}$$

Assuming $f = 0.030$ gives $V = 3.06$ m/s and $R = 700\ 000$. When these values of f and R are plotted on Fig. 9.10, a point is obtained considerably above the (interpolated) curve for $e/d = 0.000\ 83$. Assuming $f = 0.019$ and repeating the above calculation yields $R = 868\ 000$, which gives (within the accuracy of the plot) a point on the (interpolated) curve. The velocity, V, is then 3.79 m/s, and the flowrate 0.27 m³/s. ◄

A smooth brass pipeline 200 ft long is to carry a flowrate of 0.1 ft³/s between two water tanks whose difference of surface elevation is 5 ft. If a square-edged entrance and water at 50°F are assumed, what diameter of pipe is required?

Relevant Equations and Given Data

$$Q = AV \tag{3.11}$$

See the previous Illustrative Problem for equations 7.17 and 8.11.

$Q = 0.1$ ft³/s $z_1 - z_2 = 5$ ft $l = 200$ ft

$\nu = 1.41 \times 10^{-5}$ ft²/s   smooth pipe

Solution.

$$V = \frac{0.1}{(\pi/4)\ d^2} = \frac{0.127\ 2}{d^2} \tag{3.11}$$

$$R = \frac{0.127\ 2}{d^2} \frac{d}{1.41 \times 10^{-5}} = \frac{9\ 020}{d} \tag{8.11}$$

Applying the Bernoulli equation 7.17 with appropriate values of loss coefficients,

$$5.0 = \left(0.5 + f\frac{200}{d} + 1\right)\frac{V^2}{2g_n} \tag{7.17}$$

Assuming a value of d allows computation of R and f. The solution is obtained when these values give a point on the smooth pipe curve of Fig 9.10. This occurs (within the accuracy of the plot) when $d = 0.187$ ft, $V = 3.63$ ft/s, $R = 48\ 300$, and $f = 0.021\ 4$. A pipe diameter of 2.24 in. is thus theoretically required. Commercial brass pipe of 2.5 in. diameter should be specified. ◄

Energy and hydraulic grade lines are shown in Fig 9.24 for a pipeline terminating in a nozzle. Here, if l/d is large and d_2/d_1 small, minor losses and pipeline velocity head may be neglected. However, the velocity head at the tip of the nozzle cannot be neglected and therefore the energy and hydraulic grade lines, which are coincident over the pipe for the approximate solution, separate as shown over the nozzle. The flowrate may be predicted from the simultaneous solution of the continuity and Bernoulli equations:

$$Q = A_1 V_1 = A_2 V_2 \tag{3.12}$$

$$H = \frac{V_2^2}{2g_n} + f\frac{l}{d}\frac{V_1^2}{2g_n} \tag{7.17}$$

If the jet from the nozzle were to drive an impulse turbine (Section 6.11), the jet power (which is the input to the machine) would be of more interest to engineers than the flowrate; specifically they would be most concerned with providing pipe and nozzle of such relative proportions that this power could be maximized. The general expression for jet power is (from equations 4.6 or 4.7)

$$P = Q\gamma\left(\frac{V_2^2}{2g_n}\right) \tag{9.45}$$

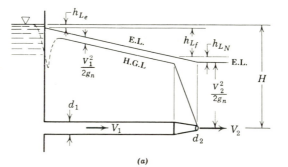

(a)

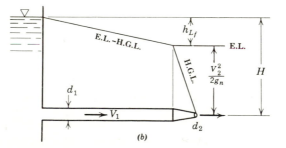

(b)

Fig. **9.24**

However, by substitution of the first simultaneous equation in the second,

$$\frac{V_2^2}{2g_n} = H - \left(\frac{fl}{2g_n d_1 A_1^2}\right) Q^2$$

from which by substitution in equation 9.45 we obtain

$$P = Q\gamma \left[H - \frac{flQ^2}{2g_n d_1 A_1^2} \right]$$

Taking $dP/dQ = 0$ for maximization of the jet power,

$$\frac{flQ^2}{2g_n d_1 A_1^2} = \frac{H}{3}$$

which shows that the maximum jet power may be expected when[24]

$$\frac{flV_1^2}{2g_n d_1} = \frac{H}{3} \qquad \text{and} \qquad \frac{V_2^2}{2g_n} = \frac{2H}{3}$$

from which pipe and nozzle may be sized for any available flowrate.

    When a pipeline runs above its hydraulic grade line, negative pressure in the pipe is indicated (Section 4.4). Sketching pipe and hydraulic grade line to scale indicates regions of negative pressures and critical points which may place limitations on the flowrate. Theoretically the absolute pressure in a pipeline may fall to the vapor pressure of the liquid, at which point cavitation (Appendix 7) sets in; however, this extreme condition is to be avoided in pipelines, and much trouble can be expected before such low pressures are attained. Most engineering liquids contain dissolved gases which will come out of solution well before the cavitation point is reached; since such gases go back into solution very slowly, they move with the liquid as large bubbles, collect in the high points of the line, reduce the flow cross section, and tend to disrupt the flow. In practice, large negative pressures in pipes should be avoided if possible by improvements of design; where such negative pressures cannot be avoided they should be prevented from exceeding about two-thirds of the difference between barometric and vapor pressures.

---------------- **Illustrative Problems** ----------------

How much horsepower must be applied to pump 2.5 ft³/s of water (50°F) from the lower to the upper reservoir? Neglect minor losses and velocity heads. Assume clean cast iron pipe. What is the maximum dependable flow that can be pumped through this system?

[24] In hydropower practice, $V_2^2/2g_n$ will be found to be considerably larger than $(\frac{1}{3})H$, depending on the economic value of the water. The derived result could be used for design only in a situation where the liquid was of no value.

Relevant Equations and Given Data

$$Q = A_8 V_8 = A_6 V_6 \tag{3.12}$$

$$P = Q\gamma E_P / 550 \tag{4.6}$$

$$R = Vd / \nu \tag{8.11}$$

$$h_L = f \frac{l}{d} \frac{V^2}{2g_n} \tag{9.2}$$

$$\gamma = 62.4 \text{ lb/ft}^3 \qquad \nu = 1.41 \times 10^{-5} \text{ ft}^2/\text{s} \qquad Q = 2.5 \text{ ft}^3/\text{s}$$

Solution.

$$V_8 = \frac{2.5}{0.349} = 7.16 \text{ ft/s} \qquad V_6 = \frac{2.5}{0.196\ 4} = 12.72 \text{ ft/s} \tag{3.12}$$

$$R_8 = \frac{7.16 \times 0.667}{1.41 \times 10^{-5}} = 338\ 000$$

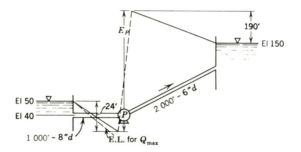

From Fig 9.10,

$$f_8 \cong 0.021$$

$$R_6 = \frac{12.72 \times 0.5}{1.41 \times 10^{-5}} = 450\ 000$$

From Fig. 9.10,

$$f_6 \cong 0.022$$

$$h_{L_8} = 0.021 \frac{1\ 000}{0.667} \frac{(7.16)^2}{2g_n} = 25 \text{ ft} \tag{9.2}$$

$$h_{L_6} \cong 0.022 \frac{2\ 000}{0.5} \frac{(12.72)^2}{2g_n} = 221 \text{ ft} \tag{9.2}$$

From the **E.L.:**

$$E_P = 100 + 25 + 221 = 346 \text{ ft} \cdot \text{lb/lb}$$

Horsepower of pump (100% efficiency) $= \dfrac{2.5 \times 62.4 \times 346}{550} = 98 \text{ hp} \blacktriangleleft \tag{4.6}$

From the energy line the point of maximum negative pressure is at the suction flange of the pump. Allowing this to be a maximum of 20 ft of water places the energy line 20 ft below the suction flange and fixes the head loss in the suction pipe at 30 ft of water. The maximum reliable flow may then be obtained ($f = 0.021$ is approximate) from

$$30 \cong 0.021 \frac{1\ 000\ V_8^2}{0.667\ 2g_n} \qquad V_8 = 7.8\ \text{ft/s} \tag{9.2}$$

$$Q_{max} = 7.8 \times 0.349 = 2.7\ \text{ft}^3/\text{s} \blacktriangleleft \tag{3.12}$$

A centrifugal pump has the characteristics shown in the figure.

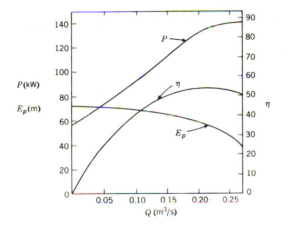

Calculate the actual flowrate as a function of reservoir A's elevation for the system shown. Assume clean commercial steel pipe.

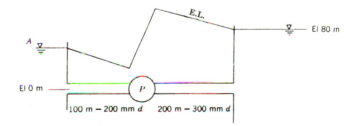

Relevant Equation and Given Data

$$h_L = f \frac{l}{d} \frac{V^2}{2g_n} \tag{9.2}$$

See figures for data; $e = 0.045$ mm.

Solution. By neglecting entrance and exit losses and velocity heads and by assuming rough flow, we obtain from an energy line analysis

$$h_{L\,200} = 0.013 \times \frac{200}{0.3} \times \frac{1}{2g_n} \left(\frac{Q}{\frac{\pi}{4} \times (0.3)^2} \right)^2 = 88.4Q^2 \tag{9.2}$$

$$h_{L\,100} = 0.014 \times \frac{100}{0.2} \times \frac{1}{2g_n} \left(\frac{Q}{\frac{\pi}{4} \times (0.2)^2} \right)^2 = 361.5Q^2 \tag{9.2}$$

$$El_A + E_P - h_{L\,100} = h_{L\,200} + 80$$

The working equation is (it is easiest to specify Q and obtain El_A)

$$El_A = 80 - E_P + (361.5 + 88.4)Q^2$$
$$= 80 - E_P + 449.9Q^2$$

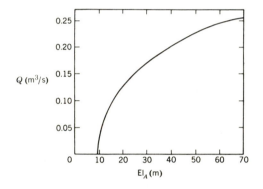

The results are given as a plot of Q versus El_A. The expected increase in flowrate with increasing elevation in reservoir A (i.e., decreasing elevation difference) is clearly illustrated. ◀

9.11 Pipeline Problems—Multiple Pipes

Some of the more complex problems of pipeline design involve the flow of fluids in pipes that intersect. The principles involved in problems of this type may be obtained by a study of pipes that (1) divide and rejoin, and (2) lead from regions of known pressure and elevation and meet at a common point. In such problems, velocity heads, minor losses, and variation of f with $\mathbf{R}$ are usually neglected, and calculations are made on the basis of coincident energy and hydraulic grade lines.

In pipeline practice, *looping* or laying a pipeline B parallel to an existing pipeline A and connected with it (Fig 9.25) is a standard method of increasing the capacity of the line. Here an analogy between fluid flow and the flow of current through a parallel electric circuit is noted if head loss is compared with drop in electric potential and

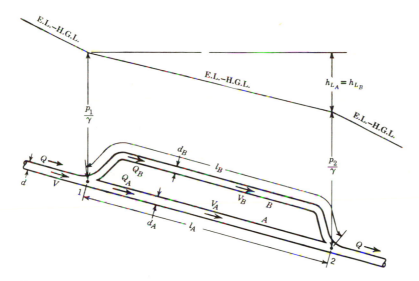

Fig. 9.25

flowrate with electric current. Evidently, the distribution of flow in the branches must be such that *the same head loss occurs in each branch;* if this were not so there would be more than one energy line for the pipe upstream and downstream from the junctions—an obvious impossibility.

Application of the continuity principle shows that the flowrate in the main line is equal to the sum of the flowrates in the branches. Thus the following simultaneous equations may be written:

$$h_{L_A} = h_{L_B}$$

$$Q = Q_A + Q_B$$

Head losses are expressed in terms of flowrate through the Darcy-Weisbach equation (9.2),

$$h_L = f \frac{l}{d} \frac{V^2}{2g_n} = \frac{fl}{2g_n d} \frac{16Q^2}{\pi^2 d^4} = \left(\frac{16fl}{2\pi^2 g_n d^5}\right) Q^2 \qquad (9.46)$$

This equation may be generalized by writing it as

$$h_L = KQ^n \qquad (9.47)$$

Substituting this in the first of the simultaneous equations above,

$$K_A Q_A^n = K_B Q_B^n$$

$$Q = Q_A + Q_B$$

Solution of these simultaneous equations allows prediction of the division of a flowrate

Q into flowrates Q_A and Q_B when the pipe characteristics are known. Application of these principles also allows prediction of the increased flowrate obtainable by looping an existing pipeline.

─────────────── **Illustrative Problem** ───────────────

A 300 mm pipeline 1 500 m long is laid between two reservoirs having a difference of surface elevation of 24 m. The maximum flowrate obtainable through this line (with all valves wide open) is 0.15 m³/s. When this pipe is looped with a 600 m pipe of the same size and material laid parallel and connected to it, what increase of maximum flowrate may be expected?

Relevant Equation and Given Data

$$h_L = KQ^n \tag{9.47}$$

$$d = 300 \text{ mm} \qquad h_L = 24 \text{ m} \qquad l_{common} = 900 \text{ m}$$

Thus for branches:

$$l_A = 600 \text{ m} \qquad l_B = 600 \text{ m}$$

Solution.   For the single pipe, using equation 9.47,

$$24 = K(0.15)^2 \qquad K = 1\ 067$$

Since (see equation 9.46) $K \propto l$, K for each branch of the looped section is $(600/1\ 500) \times 1\ 067 = 427$, and for the unlooped section $(900/1\ 500)\ 1\ 067 = 640$.

For the looped pipeline, the head loss is computed first in the common section plus branch A as

$$24 = 640Q^2 + 427Q_A^2$$

and then in the common section plus branch B as

$$24 = 640Q^2 + 427Q_B^2$$

in which Q_A and Q_B are the flowrates in the parallel branches. Solving these equations by eliminating Q shows that $Q_A = Q_B$ (which is to be expected from symmetry in this problem). Since, from continuity, $Q = Q_A + Q_B$, $Q_A = Q/2$. Substituting this in the first equation yields $Q = 0.18$ m³/s. Thus the gain of capacity by looping the pipe is 0.03 m³/s or 20%. ◀

───

Another engineering example of a multiple pipe system is the classic *three-reservoir problem* of Fig. 9.26 in which pipes lead from (three or more) reservoirs to a common point; this problem may be solved advantageously by the use of the energy line. Here flow may take place (1) from reservoir A into reservoirs B and C, or (2) from

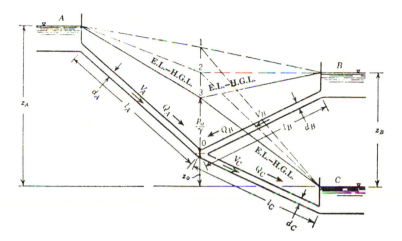

Fig. **9.26**

reservoir A to C without inflow or outflow from reservoir B, or (3) from reservoirs A and B into reservoir C. For situation (1), writing head losses in the form of KQ^n,

$$z_A - K_A Q_A^n = K_C Q_C^n$$

$$z_A - K_A Q_A^n = z_B + K_B Q_B^n$$

$$Q_A = Q_B + Q_C$$

For situation (3),

$$z_A - K_A Q_A^n = K_C Q_C^n$$

$$z_A - K_A Q_A^n = z_B - K_B Q_B^n$$

$$Q_A = Q_C - Q_B$$

For situation (2), $Q_B = 0$ and the sets of equations above become identical. In view of the physical flow picture only one of these sets of equations can be satisfied; this set may be discovered by a preliminary calculation using situation (2) in which Q_A and Q_C may be computed from the first two equations. If $Q_A > Q_C$, it can be seen from the continuity equation that the first set of equations should be used; if $Q_A < Q_C$, the second set will yield a solution. Having identified the set of equations valid for the problem, these may then be solved (by trial) to yield the flowrates Q_A, Q_B, and Q_C from which the pressures at all points in the lines may be predicted.

Multiple pipe systems reach their greatest complexity in the problems of distribution of flow in pipe networks such as those of a city water distribution system. Space does not permit comprehensive treatment of this professional engineering problem here, but one method of analysis is presented to illustrate the basic principles (the reader should consult the references listed at the end of the chapter for a deeper understanding).

A pipe network of a water system is the aggregation of connected pipes used to distribute water to users in a specified area, such as a city or subdivision. The network consists of pipes of various sizes, geometric orientations, and hydraulic characteristics plus pumps, valves and fittings, and so forth. Figure 9.27 shows a simple network in plan view. The pipe junctions are indicated by the capital letters A–H, the individual pipes by the numbers 1–10, and loops (closed circuits of pipes) by the Roman numerals I–III. Flows are assumed positive in a clockwise direction around each loop. Pipes 1, 3, 4, and 2 comprise loop I. Pipes 4, 8, 10, and 7 comprise loop II. Pipe 4 is common to both loops.

The solution of any network problem must satisfy the continuity and Bernoulli principles throughout the network. The continuity principle states that the net flowrate into any pipe junction must be zero. The Bernoulli principle requires that at any junction there be only one position of the energy line, that is, that the net head loss around any single loop (see Fig. 9.27) of the network must be zero. Applying these principles to each junction and loop of the network in Fig 9.27 yields a set of simultaneous equations. The equations for loop I are

$$\sum_A Q = -Q_A + Q_2 - Q_1 = 0$$

$$\sum_F Q = Q_1 + Q_F - Q_3 = 0$$

$$\sum_E Q = Q_3 - Q_4 - Q_8 = 0$$

$$\sum_B Q = -Q_2 + Q_4 + Q_7 + Q_5 = 0$$

$$\sum_I h_L = K_1 Q_1^n + K_3 Q_3^n + K_4 Q_4^n + K_2 Q_2^n = 0$$

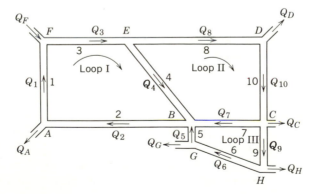

Fig. **9.27**

There are similar equations for the other loops. For the network, flow directions have been *assumed* (these may not be the correct *actual* flow directions). It is also assumed that pipe size, length, and hydraulic characteristics, the inflows and outflows (Q_A, etc.), tentative pump locations and characteristics, and the network layout and elevations (needed if pressures are to be determined) are known. The unknown flowrates, Q_i, $i = 1 - 10$, are to be found.

The solution for the ten unknown flowrates Q_i is obtained by a trial-and-error or iterative process. There are several ways to accomplish this process, the simplest being known as the Hardy Cross method,[25] which is considered here. The essence of the method is to start with a reasonable set of guessed values, Q_{0i}, that satisfy continuity and then to systematically adjust the guessed values, keeping continuity satisfied, until the head loss equations are satisfied to a desired level of accuracy. All the continuity equations at pipe junctions are automatically and continuously satisfied by this approach. Hence only the head loss equations remain and the number of simultaneous equations to be solved is equal to the number of loops.

If the first guesses are reasonable, the true flowrates, Q_i, should be only a small among, Δ_L, different from the guesses, Q_{0i}, in each loop, that is, for a pipe that is only in one loop,

$$Q_i = Q_{0i} \pm \Delta_L \tag{9.48}$$

where the sign ($\pm$) depends on the directions assumed for Q_i and Δ_L. To maintain continuity, the correction, Δ_L, is applied to every pipe in the loop. For example, $Q_3 = Q_{03} + \Delta_I$, $Q_8 = Q_{08} + \Delta_{II}$, but $Q_4 = Q_{04} + \Delta_I - \Delta_{II}$ because the flow Δ_L is always assumed in the clockwise direction around each loop. In general a head loss equation takes the form

$$\sum_L h_L = \sum_L (\pm)_i K_i Q_i^n = 0 \tag{9.49}$$

where the sign $(\pm)_i$ depends on the direction of flow and the Q_i are taken to be the magnitudes of the flow. For the assumed flow directions and loops in Fig. 9.27, the head loss equations 9.49 are (the K_i are positive)

$$K_1 Q_1^n + K_3 Q_3^n + K_4 Q_4^n + K_2 Q_2^n = 0$$

$$-K_4 Q_4^n + K_8 Q_8^n + K_{10} Q_{10}^n + K_7 Q_7^n = 0$$

$$K_5 Q_5^n - K_7 Q_7^n + K_9 Q_9^n + K_6 Q_6^n = 0$$

Introducing equation 9.48 into 9.49,

$$\sum_L h_L = \sum_L (\pm)_i K_i (Q_{0i} \pm \Delta_L)^n = 0$$

[25] H. Cross, "Analysis of Flow in Networks of Conduits and Conductors," *Univ. Illinois Eng. Expt. Sta.,* *Bull.* 286, 1936. See R. W. Jeppson, *Analysis of Flow in Pipe Networks,* Ann Arbor Science, 1976, for the Hardy Cross and other methods plus computer programs.

Expanding this equation by the binomial theorem and neglecting all terms containing products of Δ_L, that is, Δ_L^2, Δ_L^3, and so forth, because Δ_L is presumed small,

$$\sum_L h_L = \sum_L (\pm)_i K_i (Q_{0i}^n \pm n Q_{0i}^{n-1} \Delta_L) = 0$$

Solving now for Δ_L,

$$\Delta_L = -\frac{\displaystyle\sum_L (\pm)_i K_i Q_{0i}^n}{\displaystyle\sum_L \left| n K_i Q_{0i}^{n-1} \right|} \tag{9.50}$$

The absolute value signs in equation 9.50 arise because the $(\pm)$ signs cancel to make all terms in the denominator positive. This equation is used to obtain a correction to guessed flows for each loop in the network.

Because some pipes share loops and because product terms were neglected in arriving at equation 9.50, it does not produce a set of Δ_L that precisely corrects all the Q_{0i} to solution values Q_i. The reader can verify that a value

$$(Q_{0i})_{\text{new}} = (Q_{0i})_{\text{old}} + \sum_L (\pm) \Delta_L \tag{9.51}$$

[where $\sum_L (\pm) \Delta_L$ means that if a pipe is in more than one loop, the correction is the sum of corrections from each loop] is not equal Q_i but is just a better approximation. Hence, the iteration must be continued by using the new (corrected) guesses as the basis for calculating further new corrections. The $(Q_{0i})_{\text{new}}$ also satisfy continuity at each stage. The iteration is continued until *all* the Δ_L's at the last stage are sufficiently small. Then $(Q_{0i})_{\text{new}} \approx (Q_{0i})_{\text{old}}$ and the process is said to have converged. Clearly as Δ_L approaches zero, $\sum_L h_L$ approaches zero also, which is required.

It is very easy to write a computer program to carry out the Hardy Cross analysis for any number of pipes and loops. Once the flowrates are known, actual head losses, pressure changes, and actual pressures can be calculated. For computer and desk calculator analysis it is convenient to modify equations 9.50 and 9.51 for automatic processing. If k represents the number of trials that have been made, it is possible to indicate a sequence of events in the process. The initial guessed flows are $(Q_{0i})_0$, the first correction is $(\Delta_L)_1$, $(Q_{0i})_1 = (Q_{0i})_0 + \sum_L (\pm)(\Delta_L)_1$, or in general

$$(\Delta_L)_{k+1} = -\frac{\displaystyle\sum_L (\pm)_i K_i (Q_{0i}^n)_k}{\displaystyle\sum_L \left| n K_i (Q_{0i}^{n-1})_k \right|} \tag{9.50}$$

and

$$(Q_{0i})_{k+1} = (Q_{0i})_k + \sum_L (\pm)(\Delta_L)_{k+1} \tag{9.51}$$

To add a pump to a pipe in a network an expression representing the head versus capacity curve (see the Illustrative Problems at the end of Section 9.10) is required. It is convenient to fit a polynomial curve through the experimental curve to obtain

$$E_P \approx E_{P_o}(1 + a_1 Q + a_2 Q^2 + \cdots)$$

with as many coefficients, a_i, being used as required for the accuracy of the analysis. If a pump is added to line 8 in loop II of Fig. 9.27, the head loss equation for loop II becomes

$$-K_4 Q_4^n + K_8 Q_8^n - E_{P_o}(1 + a_1 Q_8 + a_2 Q_8^2 + \cdots) + K_{10} Q_{10}^n + K_7 Q_7^n = 0$$

and the analysis proceeds as before.

Illustrative Problem

A parallel commercial steel pipe network was built in two parts. As shown below, section ACD is the original line; the parallel section ABC was added; then section BD completed the job. By accident a valve is left open in the residual connecting link BC. What are the resulting flows in pipes 1 through 5? Neglect minor losses and assume wholly rough flow.

Relevant Equations and Given Data

$$(\Delta_L)_{k+1} = \frac{\displaystyle\sum_L (\pm)_i K_i (Q_{0i}^n)_k}{\displaystyle\sum_L |n K_i (Q_{0i}^{n-1})_k|} \tag{9.50}$$

$$(Q_{0i})_{k+1} = (Q_{0i})_k + \sum_L (\pm)(\Delta_L)_{k+1} \tag{9.51}$$

See the diagram and the table of pipe characteristics for the given data.

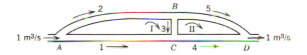

Solution. Construct a table of pipe characteristics; assume the Darcy-Weisbach equation applies so $n = 2$ and $K_i = 16 f_i l_i / 2\pi^2 g_n d_i^5$ (equation 9.46).

Pipe No.	Length (m)	Diameter (m)	e/d	f	K_i (Eq. 9.46)
1	1 000	0.5	9×10^{-5}	0.012	31.7
2	1 000	0.4	1×10^{-4}	0.012	96.8
3	100	0.4	1×10^{-4}	0.012	9.7
4	1 000	0.5	9×10^{-5}	0.012	31.7
5	1 000	0.3	1.4×10^{-4}	0.013	442.0

Form two loops and assume the flow directions shown. Assume an initial set of flowrates that satisfies continuity. The loop equations are

$$(\Delta_I)_{k+1} = -\frac{-K_1(Q_{01}^2)_k + K_2(Q_{02}^2)_k + K_3(Q_{03}^2)_k}{2[\,|\,K_1(Q_{01})_k\,| + |\,K_2(Q_{02})_k\,| + |\,K_3(Q_{03})_k\,|\,]} \qquad (9.50)$$

$$(\Delta_{II})_{k+1} = -\frac{-K_3(Q_{03}^2)_k + K_5(Q_{05}^2)_k - K_4(Q_{04}^2)_k}{2[\,|\,K_3(Q_{03})_k\,| + |\,K_5(Q_{05})_k\,| + |K_4(Q_{04})_k\,|\,]} \qquad (9.50)$$

provided the signs do not change on the assumed flow directions. From equation 9.51,

$$(Q_{01})_{k+1} = (Q_{01})_k - \Delta_I$$

$$(Q_{02})_{k+1} = (Q_{02})_k + \Delta_I$$

$$(Q_{03})_{k+1} = (Q_{03})_k + \Delta_I - \Delta_{II}$$

$$(Q_{04})_{k+1} = (Q_{04})_k - \Delta_{II}$$

$$(Q_{05})_{k+1} = (Q_{05})_k + \Delta_{II}$$

Set up a table to carry out the calculation by desk calculator. The desired results, accurate to two significant figures, are obtained after two trials. The third trial is a check.

k	Q_{01}	Q_{02}	Q_{03}	Q_{04}	Q_{05}	$\dfrac{\text{Num}_I}{\text{Denom}_I}$	Δ_I	$\dfrac{\text{Num}_{II}}{\text{Denom}_{II}}$	Δ_{II}
0	0.5	0.5	0.4	0.9	0.1	17.83/136	−0.13	−22.81/153	0.15
1	0.63	0.37	0.12	0.75	0.25	0.81/114	−0.01	9.65/271	−0.04
2	0.64	0.36	0.15	0.79	0.21	−0.22/114	0.00	−0.51/239	0.00
3	0.64	0.36	0.15	0.79	0.21	—	—	—	—

COMPRESSIBLE FLOW

9.12 Pipe Friction

The preceding calculations are generally valid for flowing liquids. However, the analysis of the flow of gases in pipes often requires an account of compressibility effects. The calculation of pressure, velocity, temperature, and density changes caused by friction and heat transfer during the flow of gases in pipelines is, in general, a rather complex thermodynamic process and such problems cannot be treated exhaustively here. On the other hand, isothermal (constant temperature) and adiabatic flows of gas in pipes have practical applications and serve as a basis for both qualitative and quantitative examples of pipe flow situations with a fluid of varying density. The analysis is limited to conditions under which gases act as perfect gases and to one-dimensional flows.

To begin it is necessary to recall the four basic equations of compressible flow, that is, the fluid equation of state and the continuity, energy, and momentum equations which will be written in differential form now and then integrated. In addition the Second Law of Thermodynamics will be useful.

The equation of state for a perfect gas is

$$\frac{p}{\rho} = RT \tag{1.3}$$

Furthermore, the specific heats c_p and c_v are constant and $k = c_p/c_v$ ($k = 1.4$ for air). The continuity equation 3.9 is

$$A\rho V = \text{constant} \tag{3.9}$$

It follows that the weight flowrate $G = A\gamma V$ is constant as well as the mass flowrate $\dot{m} = A\rho V$. In such a flow the Reynolds number

$$\mathbf{R} = \frac{V d\rho}{\mu} = \frac{G \, d\rho}{A\gamma \, \mu} = \frac{Gd}{\mu g_n A} = \frac{\dot{m}d}{\mu A} \tag{9.52}$$

for a specific flow depends only on temperature because μ is a function of T while $\dot{m}$, d, and A are constants of the flow.

The energy equation 5.4 for steady compressible flow can be written as

$$\frac{d}{dl}(q_H) = \frac{d}{dl}h + \frac{d}{dl}\left(\frac{V^2}{2}\right) \tag{9.53}$$

for changes over a differential distance dl under the assumptions that there is no shear or shaft work and that the elevation terms can be neglected. Recall that q_H is the heat transfer per unit mass of fluid flowing and $h = p/\rho + ie = p/\rho + c_v T = c_p T$. Thus, equation 9.53 can be written as

$$c_p \frac{dT}{dl} + V\frac{dV}{dl} = \frac{d}{dl}(q_H) \tag{9.54}$$

for a perfect gas.

According to equation 7.19 the momentum equation for a differential length of pipe can be written in the form (because $\gamma = \rho g_n$, $R_h = d/4$, and elevation effects are negligible)

$$dp + \rho V \, dV + 4\tau_o \, dl/d = 0$$

Substituting $f\rho V^2/8$ for τ_o produces

$$dp + \rho V \, dV + f\rho V^2 \, dl/2d = 0 \tag{9.55}$$

The Mach number $\mathbf{M} = V/a$ where (see equation 1.10) $a = \sqrt{kp/\rho} = \sqrt{kRT}$. Rewriting equation 9.55 in terms of $\mathbf{M}$ gives

$$\frac{dp}{p} + k\mathbf{M}^2\left(\frac{d\mathbf{M}}{\mathbf{M}} + f\frac{dl}{2d}\right) = 0 \tag{9.56}$$

when a and T are constant (isothermal flow). For fully developed pipe flow Shapiro[26] points out that experiments have shown that f values (cf., Fig. 9.10) for incompressible flow are applicable to compressible flows when $\mathbf{M} < 1$. However, for $\mathbf{M} > 1$ the compressible friction factors are about one-half of the corresponding incompressible value.

According to equation 3.9, ρV = constant in a pipe. Thus, as $\rho = p/RT$, $pV = RT$ for a perfect gas. Again, in the case T = constant, $dp/p = -dV/V = -d\mathbf{M}/\mathbf{M}$ and equation 9.56 becomes

$$\frac{dp}{p} = -\frac{dV}{V} = -\frac{d\mathbf{M}}{\mathbf{M}} = -\frac{k\mathbf{M}^2}{1 - k\mathbf{M}^2}f\frac{dl}{2d} \qquad (9.57)$$

for isothermal flow.

9.13 Isothermal Pipe Flow

To secure isothermal flow in a pipe the heat transferred out of the fluid through the pipe walls and the energy converted into heat by the friction process must be adjusted so that the fluid temperature remains constant. Such an adjustment is approximated naturally in uninsulated pipes where velocities are low (well below sonic) and where temperatures inside and outside the pipe are of the same order; frequently the flow of gases in long pipelines may be treated isothermally.

In an isothermal flow the Reynolds number $\mathbf{R}$ (equation 9.52) is constant regardless of changes in V or ρ because the temperature and hence μ are constant. The friction factor is then constant also. Accordingly, dividing equation 9.55 by ρV^2, replacing ρ by p/RT and V^2 by $\dot{m}^2/A^2\rho^2$ gives

$$\frac{A^2}{\dot{m}^2 RT}p\,dp + \frac{dV}{V} + \frac{f}{2d}\,dl = 0$$

Integrating this equation between the two points 1 and 2 of a control volume in the pipe of Fig 9.28 produces

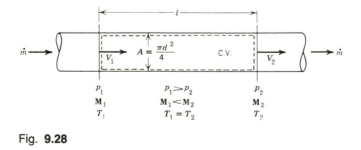

Fig. **9.28**

[26] A. H. Shapiro, *The dynamics and thermodynamics of compressible fluid flow*, vol. I, Ronald Press Co., 1953, pp. 184ff.

$$\frac{A^2}{\dot{m}^2 RT} \int_{p_1}^{p_2} p\, dp + \int_{V_1}^{V_2} \frac{dV}{V} + \frac{f}{2d} \int_{0}^{l} dl = 0$$

or

$$\frac{A^2}{\dot{m}^2 RT} \frac{p_2^2 - p_1^2}{2} + \ln\frac{V_2}{V_1} + f\frac{l}{2d} = 0$$

Thus,

$$p_1^2 - p_2^2 = \frac{\dot{m}^2 RT}{A^2}\left(2\ln\frac{V_2}{V_1} + f\frac{l}{d}\right) \tag{9.58}$$

This equation can be rearranged by use of the continuity equation $A_1\rho_1 V_1 = A_2\rho_2 V_2$ in which $A_1 = A_2$, $\rho_1 = p_1/RT$, and $\rho_2 = p_2/RT$. Thus, by substitution, $p_1 V_1 = p_2 V_2$ and $V_2/V_1 = p_1/p_2$. Hence equation 9.58 may be written

$$p_1^2 - p_2^2 = \frac{\dot{m}^2 RT}{A^2}\left(2\ln\frac{p_1}{p_2} + f\frac{l}{d}\right) \tag{9.59}$$

Generally the solution of this equation for p_2 must be accomplished by trial, but frequently $2\ln p_1/p_2$ is so small (in comparison with fl/d) that it may be neglected, thus allowing a direct solution.

While the actual pressure and velocity at any point in the pipe can be obtained by use of equation 9.59 and a set of prescribed values of p_1, $\mathbf{M}_1$, T_1, A, $\dot{m}$, f, l, and d, the qualitative aspects of the flow are obtained easily by examination of equation 9.57. Thus, if $\mathbf{M} < k^{-1/2}$ the pressure decreases, the velocity increases, the Mach number increases and (as $\rho = p/RT$) the density decreases with increases in l, that is, as point 2 moves downstream (positive l) from point 1. However, if $\mathbf{M}$ becomes greater than $k^{-1/2}$ these trends reverse. In principle then as the value of $\mathbf{M}$ passes $k^{-1/2}$, a reduction of l becomes necessary to achieve a further reduction of p in a subsonic flow. The pressure cannot drop below this point and thus equation 9.59 is applicable only between the pressure p_1 and the limiting value of p_2. Figure 9.29 gives an example of the pressure ratio p_2/p_1 as a function of fl/d for various Mach numbers $\mathbf{M}_1$ for isothermal flow of air ($k = 1.4$). The dotted lines indicate the limits of incompressible flow and $\mathbf{M} < 0.845$.

Equation 9.59 may also be written (in conformance with modern practice in gas dynamics) in terms of pressure ratio and Mach number, $\mathbf{M}$. Divide the equation by p_1^2 and multiply the right-hand side by k/k and substitute $A\rho_1 V_1$ for $\dot{m}$; whereupon $\dot{m}^2 RTk/kp_1^2 A^2$ will be found to be equal to $k\mathbf{M}_1^2$. Since the sonic velocity, a, is constant for isothermal flow, $V_2/V_1 = \mathbf{M}_2/\mathbf{M}_1$ and, since $V_2/V_1 = p_1/p_2$ (see above), $p_2/p_1 = \mathbf{M}_1/\mathbf{M}_2$ and equation 9.59 becomes

$$\frac{\mathbf{M}_1^2}{\mathbf{M}_2^2} = 1 - k\mathbf{M}_1^2\left(2\ln\frac{\mathbf{M}_2}{\mathbf{M}_1} + f\frac{l}{d}\right) \tag{9.60}$$

Considering the flow process in terms of Mach number, $\mathbf{M}_1 \ll 1$ and $\mathbf{M}_2 > \mathbf{M}_1$. From the above-mentioned limitation of the equation the limiting value of $\mathbf{M}_2$ may be found

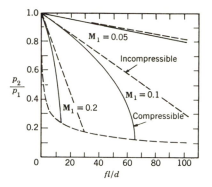

Fig. **9.29** Isothermal pipe flow pressure ratio for perfect gas ($k = 1.4$) as a function of initial conditions and fluid and pipe character.

by differentiating equation 9.60 with respect to l and setting $dp_2/dl = \infty$; the result is $M_2 = \sqrt{1/k}$. Thus, as noted above, the equation is applicable only to that portion of the subsonic flow regime where $M_1 < M_2 \leq \sqrt{1/k}$. Such limitations are of far-reaching importance in engineering problems of gas flow.

--------------------- **Illustrative Problem** ---------------------

If 18 kg/min of air flow isothermally through a smooth 75 mm pipeline at a temperature of 40°C, and the absolute pressure at a point in this line is 350 kPa, calculate the pressure in the line 600 m downstream from this point.

Relevant Equations and Given Data

$$\rho = p/RT \tag{1.3}$$

$$a = \sqrt{kRT} \tag{1.11}$$

$$Q = AV \tag{3.11}$$

$$R = \frac{Vd\rho}{\mu} = \frac{\dot{m}d}{\mu A} \tag{9.52}$$

$$p_1^2 - p_2^2 = \frac{\dot{m}^2 RT}{A^2}\left(2\ln\frac{p_1}{p_2} + f\frac{l}{d}\right) \tag{9.59}$$

$$\frac{M_1^2}{M_2^2} = 1 - kM_1^2\left(2\ln\frac{M_2}{M_1} + f\frac{l}{d}\right) \tag{9.60}$$

$\dot{m} = 18$ kg/min $\qquad d = 75$ mm $\qquad T = 313$ K

$p_1 = 350$ kPa (abs) $\qquad l = 600$ m $\qquad R = 286.8$ J/kg·K

$k = 1.4 \qquad \mu = 1.92 \times 10^{-5}$ Pa·s

Solution.

$$R = \frac{(18/60)(0.075)}{1.92 \times 10^{-5} \times 0.004\ 4} = 266\ 000 \tag{9.52}$$

From the Moody diagram of Fig. 9.10,

$$f \cong 0.015$$

$$\rho_1 = \frac{350 \times 10^3}{286.8 \times 313} = 3.90\ \text{kg/m}^3 \tag{1.3}$$

$$V_1 = \frac{(18/60)}{0.004\ 4 \times 3.9} = 17.5\ \text{m/s} \tag{3.11}$$

$$a = \sqrt{1.4 \times 286.8 \times 313} = 354\ \text{m/s} \tag{1.11}$$

$$M_1 = \frac{17.5}{354} = 0.049$$

The limiting value of M_2 is $\sqrt{1/1.4} = 0.845$. Calculating the limiting value of ℓ for applicability of equation 9.59 from equation 9.60,

$$\left(\frac{0.049}{0.845}\right)^2 = 1 - 1.4(0.049)^2 \left[2 \ln \frac{0.845}{0.049} + 0.015 \frac{\ell}{0.075}\right]$$

$$\ell = 1\ 454\ \text{m}$$

Since $600 < 1\ 454$, equation 9.59 is applicable. Substituting values therein,

$$[(350 \times 10^3)^2 - p_2^2] = \frac{(18/60)^2 \times 286.8 \times 313}{(0.004\ 4)^2} \times \left[2 \ln \frac{350 \times 10^3}{p_2} + 0.015 \frac{600}{0.075}\right]$$

Solving for p_2 by trial, $p_2(\text{abs}) = 269 \times 10^3\ \text{Pa} = 269\ \text{kPa}$. ◀

9.14 Adiabatic Pipe Flow with Friction

In short lengths of pipe or in fully insulated pipes the heat transfer is close to zero, and compressible flow in such pipes is taken to be adiabatic. Because there is no heat transfer and frictional (irreversible) processes are involved, it is clear that the entropy of the fluid must rise as the flow moves downstream. In Section 6.7 it was pointed out that the locus of all states on an enthalpy-entropy diagram for adiabatic flows (they satisfy the continuity and energy equations) is a Fanno line (see Fig. 6.9). From Fig. 6.9, it follows that in adiabatic flows, the Mach number increases downstream in subsonic flow, but decreases downstream in a supersonic adiabatic flow. Using the equations presented in Section 9.12 it is possible to prove these and other conclusions directly and to make useful flow computations.

In the case of adiabatic flow, it is convenient to assemble all the relevant differential equations. Writing equation 9.55 in terms of M gives when T is not constant

$$\frac{dp}{p} + kM^2\left(\frac{dV}{V} + f\frac{dl}{2d}\right) = 0 \tag{9.61}$$

To obtain an expression involving only p, $\mathbf{M}$, and the parameter $f\,dl/d$, it is necessary to utilize relationships derived before. According to equation 3.9 $\rho V = \text{constant}$ so

$$\frac{d\rho}{\rho} = -\frac{dV}{V} \tag{9.62}$$

while from equation 1.3

$$\frac{dp}{p} = \frac{d\rho}{\rho} + \frac{dT}{T} \tag{9.63}$$

For $q_H = 0$ equation 9.54 yields

$$c_p\,dT + V\,dV = 0$$

Dividing by c_pT and using the expressions $c_p = Rk/(k-1)$ and $\mathbf{M}^2 = V^2/kRT$ leads to

$$\frac{dT}{T} = -(k-1)\mathbf{M}^2\frac{dV}{V} \tag{9.64}$$

Introducing equations 9.62 and 9.64 into 9.63 now gives

$$\frac{dp}{p} = -[1 + (k-1)\mathbf{M}^2]\frac{dV}{V} \tag{9.65}$$

This result is used in equation 9.61 to obtain

$$\frac{dp}{p} = -\frac{k\mathbf{M}^2[1 + (k-1)\mathbf{M}^2]}{2(1-\mathbf{M}^2)} \cdot f\frac{dl}{d} \tag{9.66}$$

Then this result is used in equation 9.65 to obtain

$$\frac{dV}{V} = \frac{k\mathbf{M}^2}{2(1-\mathbf{M}^2)} \cdot f\frac{dl}{d} \tag{9.67}$$

As $\mathbf{M}^2 = V^2/kRT$,

$$\frac{d\mathbf{M}^2}{\mathbf{M}^2} = \frac{dV^2}{V^2} - \frac{dT}{T} = \frac{2dV}{V} - \frac{dT}{T}$$

Introducing equation 9.64 produces

$$\frac{d\mathbf{M}^2}{\mathbf{M}^2} = \frac{k\mathbf{M}^2\left[1 + \dfrac{(k-1)}{2}\mathbf{M}^2\right]}{1-\mathbf{M}^2} \cdot f\frac{dl}{d} \tag{9.68}$$

when equation 9.67 is also used.

Noting that $(f/d)\,dl$ is positive in equations 9.66, 9.67, and 9.68 leads to the following conclusions. If $\mathbf{M} < 1$, the pressure decreases, the velocity increases, and

Mach number increases in the downstream direction in adiabatic flow. The flow accelerates in the subsonic case as a result of the friction effects. When $\mathbf{M} > 1$, the trends reverse. Thus the flow decelerates, but the pressure rises in the downstream direction. In adiabatic flow $\mathbf{M}$ tends to unity in either the subsonic or supersonic cases. Clearly $\mathbf{M} = 1$ is the limiting case. Given the conditions at a point in a pipe, the distance l downstream of that point can be increased until $\mathbf{M} = 1$ at the discharge point. No further increase in length can be made without either altering the given conditions or introducing a shock wave because of the reversal of flow trends at $\mathbf{M} = 1$.

To make a flow computation it is necessary to integrate equation 9.68

$$\frac{1}{d}\int_0^l f\,dl = \int_{\mathbf{M}_1^2}^{\mathbf{M}_2^2} \frac{1 - \mathbf{M}^2}{k\mathbf{M}^4\left[1 + \dfrac{(k-1)}{2}\mathbf{M}^2\right]}\,d\mathbf{M}^2 \tag{9.69}$$

On the left side of this equation one can either assume that f is a constant (often an adequate approximation, at high $\mathbf{R}$ in rough pipes for example where $f \approx$ constant) or define an average friction factor

$$\bar{f} = \frac{1}{l}\int_0^l f\,dl$$

In the subsonic flows considered here, $\bar{f}$ is assumed to be equal to f and to the usual incompressible value given by Fig 9.10. Under these conditions equation 9.69 can be integrated to obtain a working relationship for $\mathbf{M}$ as a function of f, l, and d.

To find the maximum length of pipe l_{max} possible in a continuous flow equation 9.69 is integrated with $\mathbf{M}_2 = 1$ and $l = l_{max}$. The result is

$$\bar{f}\frac{l_{max}}{d} = \frac{1 - \mathbf{M}_1^2}{k\mathbf{M}_1^2} + \frac{k+1}{2k}\ln\left\{\frac{(k+1)\mathbf{M}_1^2}{2\left[1 + \dfrac{(k-1)}{2}\mathbf{M}_1^2\right]}\right\} \tag{9.70}$$

Illustrative Problem

Given $\bar{f} = f = 0.02$ find ℓ_{max}/d as a function of $\mathbf{M}_1^2$ for the adiabatic flow of air.

Relevant Equation and Given Data

$$\bar{f}\frac{\ell_{max}}{d} = \frac{1 - \mathbf{M}_1^2}{k\mathbf{M}_1^2} + \frac{k+1}{2k}\ln\left\{\frac{(k+1)\mathbf{M}_1^2}{2\left[1 + \dfrac{(k-1)}{2}\mathbf{M}_1^2\right]}\right\} \tag{9.70}$$

$$\bar{f} = 0.02 \qquad k = 1.4$$

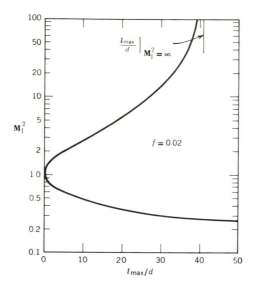

Solution. The working formula is

$$\frac{\ell_{max}}{d} = 50\left\{\frac{1 - M_1^2}{1.4M_1^2} + 0.86 \ln\left[\frac{1.2M_1^2}{1 + 0.2M_1^2}\right]\right\}$$

The results can be plotted as shown.

References

Benedict, R.P. 1980. *Fundamentals of pipe flow*. New York: John Wiley & Sons.

Colebrook, C. F., and White, C. M. 1937. The reduction of carrying capacity of pipes with age. *Jour. Inst. Civil Engrs., London* 7: 99.

Hydraulic Institute. 1979. *Engineering data book*. 1st ed. Cleveland: Hydraulic Institute.

Jeppson, R. W. 1976. *Analysis of flows in pipe networks*. Ann Arbor: Ann Arbor Science Pub. Inc.

Schlichting, H. 1979. *Boundary-layer theory*. 7th ed. New York: McGraw-Hill, Chapter XX.

Shapiro, A. H. 1953. *The dynamics and thermodynamics of compressible fluid flow*, vol. I. New York: Ronald Press, Chapter 6.

Watters, G. Z. 1979. *Modern analysis and control of unsteady flow in pipelines*. Ann Arbor: Ann Arbor Science Pub. Inc.

Films

NCFMF Film Loops. Encyclopaedia Britannica Educ. Corp.

FM-15   Incompressible flow through area contractions and expansions.

FM-16 Flow from a reservoir to a duct.
FM-17 Flow patterns in venturis, nozzles and orifices.
FM-69 Flow through tee-elbow.
FM-134 Laminar and turbulent pipe flow.

Problems

9.1. When 0.3 m³/s of water flows through a 150 mm constriction in a 300 mm horizontal pipeline, the pressure at a point in the pipe is 345 kPa, and the head lost between this point and the constriction is 3 m. Calculate the pressure in the constriction.

9.2. A 50 mm nozzle terminates a vertical 150 mm pipeline in which water flows downward. At a point on the pipeline a pressure gage reads 276 kPa. If this point is 3.6 m above the nozzle tip and the head lost between point and tip is 1.5 m, calculate the flowrate.

9.3. A 12 in. pipe leaves a reservoir of surface elevation 300 at elevation 250 and drops to elevation 150, where it terminates in a 3 in. nozzle. If the head lost through line and nozzle is 30 ft, calculate the flowrate.

9.4. A vertical 150 mm pipe leaves a water tank of surface elevation 24. Between the tank and elevation 12, on the line, 2.4 m of head are lost when 56 l/s flow through the line. If an open piezometer tube is attached to the pipe at elevation 12, what will be the elevation of the water surface in this tube?

9.5. A water pipe gradually changes from 6 in. to 8 in. diameter accompanied by an increase of elevation of 10 ft. If the pressures at the 6 in. and 8 in. sections are 9 psi and 6 psi, respectively, what is the direction of flow: (a) for 3 cfs and (b) for 4 cfs?

9.6. A pump of what power is required to pump 0.56 m³/s of water from a reservoir of surface elevation 30 to one of surface elevation 75, if in the pump and pipeline 12 metres of head are lost?

9.7. Through a hydraulic turbine flow 2.8 m³/s of water. On the 1 m inlet pipe at elevation 43.5, a pressure gage reads 345 kPa. On the 1.5 m discharge pipe at elevation 39, a vacuum gage reads 150 mm of mercury. If the total head lost through pipes and turbines between elevations 43.5 and 39 is 9 m, what power may be expected from the machine?

9.8. In a 225 mm pipeline 0.14 m³/s of water are pumped from a reservoir of surface elevation 30 over a hill of elevation 50. A pump of what power is required to maintain a pressure of 345 kPa on the hilltop if the head lost between reservoir and hilltop is 6 m?

9.9. The pressure drop, Δp, in a pipe is known to be a function of l, d, V, ρ, and a friction factor, f. Recalling that $h_L = \Delta p/\gamma$, use dimensional analysis to derive equation 9.2.

9.10. Derive expressions for wall shearing stress, wall velocity gradient, and friction velocity in terms of V, d, ρ, and μ for laminar flow in a pipe.

9.11. When a horizontal laminar flow occurs between two parallel plates of infinite extent 0.3 m apart, the velocity at the midpoint between the plates is 2.7 m/s. Calculate (a) the flowrate through a cross section 0.9 m wide, (b) the velocity gradient at the surface of the plate, (c) the wall shearing stress if the fluid has viscosity 1.44 Pa · s, (d) the pressure drop in each 30 m along the flow.

9.12. Glycerin (10°C or 50°F) flows in a 50 mm (or 2 in.) pipeline. The center velocity is 2.4 m/s (or 8 ft/s). Calculate the flowrate and the head loss in 3 m (or 10 ft) of pipe.

9.13. In a laminar flow of 0.007 m³/s in a 75 mm pipeline the shearing stress at the pipe wall is known to be 47.9 Pa. Calculate the viscosity of the fluid.

9.14. Oil of viscosity 0.48 Pa · s and specific gravity 0.90 flows with a mean velocity of 1.5 m/s in a 0.3 m pipeline. Calculate shearing stress and velocity 75 mm from the pipe centerline.

9.15. A flowrate of 1 l/min of oil of specific gravity 0.92 exists in this pipeline. Is this flow laminar? What is the viscosity of the oil? For the same flow in the opposite direction, what manometer reading is to be expected?

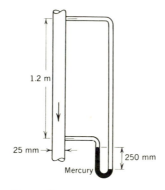

Problem **9.15**

9.16. In a laminar flow in a 12 in. pipe the shear stress at the wall is 1.0 psf and the fluid viscosity 0.002 lb · s/ft². Calculate the velocity gradient 1 in. from the centerline.

9.17. A fluid of specific gravity 0.90 flows at a Reynolds number of 1 500 in a 0.3 m pipeline. The velocity 50 mm from the wall is 3 m/s. Calculate the flowrate and the velocity gradient at the wall.

9.18. Plot the dimensionless mixing length l/R (equation 9.14) versus y/R for pipe flow with $\kappa = 0.40$. Over what range of y/R is $l/R = 0.4(y/R)$ within 10% of the "correct" value?

9.19. In a turbulent flow in a 0.3 m pipe the centerline velocity is 6 m/s, and that 50 mm from the pipe wall 5.2 m/s. Calculate the friction factor and flowrate.

9.20. If the velocity past a smooth surface in turbulent flow depends only on the distance from the surface, viscosity and density of fluid, and wall shear, show by dimensional analysis that $v/v_\star = F(v_\star y/\nu)$. See equation 9.17.

9.21. To determine the frictional stress exerted by fluid on a smooth wall, velocities v_1 and v_2 are measured in the turbulent zone at distances y_1 and y_2 from the wall. Derive an expression for the frictional stress in terms of the four measured quanities and the fluid density.

9.22. Solve problem 9.11 with turbulent flow, smooth plates, and fluid density and viscosity 1 000 kg/m³ and 0.001 4 Pa · s, respectively.

9.23. Solve problem 9.12 assuming water (10°C or 50°F) flowing in a smooth pipe.

9.24. Solve problem 9.13 assuming turbulent flow and a smooth pipe with fluid density 1 000 kg/m^3 and wall shear 4.8 Pa.

9.25. Solve problem 9.14 assuming a viscosity of 0.004 8 Pa · s and the pipe a smooth one.

9.26. Solve problem 9.16 assuming a wall shear of 0.10 psf, viscosity 0.000 02 lb · s/ft^2, and density 1.94 slugs/ft^3. Assume turbulent flow.

9.27. Solve problem 9.17 for a Reynolds number of 150 000 and a smooth pipe.

9.28. Three-tenths of a cubic metre per second of liquid (s.g. 1.25) flows in a smooth brass pipe of 300 mm diameter at Reynolds number 10 000. Predict the velocity where $y = \delta_v$.

9.29. Fluid flows in a very smooth cylindrical pipe at Reynolds number 50 000. Using the velocity distribution of equation 9.27 and the methods of Section 9.3, determine the thickness of the viscous sublayer as a percentage of the pipe radius.

9.30. Fluid of density 1 030 kg/m^3 and kinematic viscosity 1.86×10^{-5} m^2/s flows parallel to a very smooth plane surface. The velocities at 75 mm and 4.4 mm from the wall are measured and found to be 0.3 m/s and 0.08 m/s, respectively. Using *both* of these measurements calculate the shearing stress on the surface.

9.31. Show that the seventh-root law velocity profile gives velocity gradients of $v_c / 7R$ at the center of the pipe and infinity at the wall.

9.32. Fluid flows in a 6 in. or 150 mm smooth pipe at a Reynolds number of 25 000. Compare values of V/v_c computed from equation 9.20 and from the seventh-root law.

9.33. If the $f - \mathbf{R}$ relationship for $10^5 < \mathbf{R} < 10^6$ may be approximated by a straight line (on log-log plot) between $\mathbf{R} = 10^5, f = 0.018\ 0$ and $\mathbf{R} = 10^6, f = 0.011\ 5$, what value of m should be used in the equation $v/v_c = (y/R)^m$ for the velocity profile?

9.34. Calculate a value of $v_* \delta_v / \nu$ that defines the viscous sublayer thickness when the seventh-root law is used for the turbulent velocity profile.

9.35. Solve problem 9.11 for turbulent flow, rough plates with $e = 0.5$ mm, and fluid density and viscosity 1 000 kg/m^3 and 0.001 4 Pa · s, respectively.

9.36. Solve problem 9.12 assuming water (10°C) flowing in a rough pipe with $e = \frac{1}{2}$ mm.

9.37. Solve problem 9.14 assuming turbulent flow and a rough pipe having $e = 0.5$ mm with fluid viscosity 0.000 48 Pa · s.

9.38. Solve problem 9.16 assuming a wall shear of 0.20 psf, rough surface with $e = 0.03$ in., and fluid viscosity and density 0.000 020 lb · s/ft^2 and 1.94 slugs/ft^3, respectively. Assume turbulent flow.

9.39. Solve problem 9.17 for a Reynolds number of 1.5×10^6 and rough surface having $e = 0.5$ mm.

9.40. Water flows in a smooth pipeline at a Reynolds number of 10^6. After many years of use it is observed that half the original flowrate produces the same head loss as for the original flow. Estimate the size of the relative roughness of the deteriorated pipe.

9.41. A horizontal rough pipe of 150 mm diameter carries water at 20°C. It is observed that the fall of pressure along this pipe is 184 kPa per 100 m when the flowrate is 60 l/s. What size of smooth pipe would produce the same pressure drop for the same flowrate?

9.42. In an established flow in a pipe of 0.3 m diameter the centerline velocity is 3.05 m/s and the velocity 75 mm from the centerline 2.73 m/s. Identify: (a) the flow as laminar or turbulent and (b) the pipe wall as smooth or rough.

9.43. If the size of uniform sand-grain roughness in a 10 in. or 254 mm pipe is 0.02 in. or 0.5 mm, below what approximate Reynolds number will the pipe behave as a hydraulically smoooth one? What will be the thickness of the viscous sublayer at this Reynolds number?

9.44. A single layer of steel spheres is stuck to the glass-smooth floor of a two-dimensional open channel. Water of kinematic viscosity 9.3×10^{-7} m²/s flows in the channel at a depth of 0.3 m and surface velocity of $\frac{1}{4}$ m/s. Show that for spheres of 7.2 mm and 0.3 mm diameter that the channel bottom should be classified *rough* and *smooth,* respectively.

9.45. For the velocity measurements shown what is the largest roughness (*e*) which would allow the surface to be classified as smooth?

9.46. Water (20°C or 68°F) flows in a 3 m (10 ft) smooth pipe of $6\frac{1}{4}$ mm (0.25 in.) diameter. Plot head loss against velocity to substantiate the sketch-plot of Fig. 7.3.

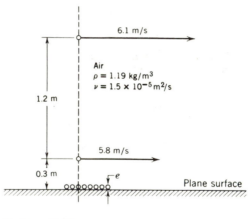

Problem **9.45**

9.47. Water flows through a section of 300 mm pipeline 300 m long running from elevation 90 to elevation 75. A pressure gage at elevation 90 reads 275 kPa, and one at elevation 75 reads 345 kPa. Calculate head loss, direction of flow, and shear stress at the pipe wall and 75 mm from the pipe wall. If the flowrate is 0.14 m³/s, calculate the friction factor and friction velocity.

9.48. When a liquid flows in a horizontal 150 mm (6 in.) pipe, the shear stress at the walls is 100 Pa (2.0 lb/ft²). Calculate the pressure drop in 30 m (100 ft) of this pipeline. What is the shearing stress in the liquid 25 mm (1 in.) from the pipe centerline?

9.49. If the friction factor for a 250 mm (10 in.) waterline is 0.030, and the shearing stress in the water 50 mm (2 in.) from the pipe centerline is 14 Pa (0.3 lb/ft²), what is the flowrate?

9.50. When 0.28 m³/s of water flow in a 0.3 m pipeline, 63 kW are lost in friction in 300 m of pipe. Calculate head loss, friction factor, friction velocity, and shear stress at the pipe wall.

9.51. In a 12 in. pipe, 15 cfs of water flow upward. At a point on the line at elevation 100, a pressure gage reads 130 psi. Calculate the pressure at elevation 150, 2 000 ft up the line, assuming the friction factor to be 0.02.

9.52. A 50 mm water line possesses two pressure connections 15 m apart (along the pipe) having a difference of elevation of 3 m. When water flows upward through the line at a velocity of 3 m/s, a differential manometer attached to the pressure connections and containing mercury and water shows a reading of 254 mm. Calculate the friction factor of the pipe.

9.53. If 680 l/min of water at 20°C flow in a 150 mm pipeline having roughness proturbances of average height 0.75 mm, and if similar roughness having height 0.375 mm exists in a 75 mm pipe, what flowrate of crude oil (40°C) must take place therein for the friction factors of the two pipes to be the same?

9.54. If two cylindrical horizontal pipes are geometrically similar and the flows in them dynamically similar, how will the pressure drops in corresponding lengths vary with density and mean velocity of the fluid?

9.55. Calculate the loss of head in 300 m (1 000 ft) of 75 mm (3 in.) brass pipe when water at 27°C (80°F) flows therein at a mean velocity of 3 m/s (10 ft/s).

9.56. In a 50 mm (2 in.) pipeline, 95 l/min (25 gpm) of glycerin flow at 20°C (68°F). Calculate the loss of head in 50 m (160 ft) of this pipe.

9.57. If 45 kg/min of air flow in a 75 mm galvanized iron pipeline at 1 200 kPa abs. and 27°C, calculate the pressure drop in 90 m of this pipe. Assume the air to be of constant density.

9.58. If 0.34 m³/s of water flows in a 0.3 m riveted steel pipe at 21°C, calculate the smallest loss of head to be expected in 150 m of this pipe.

9.59. A 3 in. smooth brass pipeline 100 ft long carries 100 gpm of crude oil. Calculate the head loss when the oil is at (*a*) 80°F, (*b*) 110°F.

9.60. In a laboratory test, 222 kg/min of water at 15°C flow through a section of 50 mm pipe 9 m long. A differential manometer connected to the ends of this section shows a reading of 480 mm. If the fluid in the bottom of the manometer has a specific gravity of 3.20, calculate the friction factor and the Reynolds number.

9.61. Carbon dioxide flows in a horizontal 100 mm wrought iron pipeline at a velocity of 3 m/s. At a point in the line a pressure gage reads 690 kPa and the temperature is 40°C. What pressure is lost as the result of friction in 30 m of this pipe? Barometric pressure is 101.3 kPa. Assume the fluid is of constant density.

9.62. When water at 20°C (68°F) flows through 6 m (20 ft) of 50 mm (2 in.) smooth brass pipe, the head lost is 0.3 m (1 ft). Calculate the flowrate.

9.63. When glycerin (25°C or 77°F) flows through a 30 m (100 ft) length of 75 mm (3 in.) pipe, the head loss is 36 m (120 ft). Calculate the flowrate.

9.64. A pump of what power is required to pump 40 l/min of crude oil from a tank of surface elevation 12 to one of elevation 18 through 450 m of 75 mm pipe, if the oil is at (*a*) 25°C, (*b*) 40°C?

9.65. If the head lost in 150 m of 75 mm smooth pipe is 21 m when the flowrate is 8.5 l/s, is the flow laminar or turbulent?

9.66. When the flowrate in a certain smooth pipe is 0.14 m³/s, the friction factor is 0.06. What friction factor can be expected for a flowrate of 0.71 m³/s of the same fluid in the same pipe?

9.67. A fluid of kinematic viscosity 4.6×10^{-4} m²/s flows in a certain length of cast iron pipe of 0.3 m diameter with a mean veloctiy of 3 m/s and head loss of 4.5 m. Predict the head loss in this length of pipe when the velocity is increased to 6 m/s.

9.68. When 0.14 m³/s of water (20°C) flow in a smooth 150 mm pipeline, the head lost in a certain length is 4.5 m. What head loss can be expected in the same length for a flowrate of 0.28 m³/s? How is the answer changed if the pipe is concrete?

9.69. Warm oil (s.g. 0.92) flows in a 2 in. or 50 mm smooth brass pipeline at a mean velocity of 8 ft/s or 2.4 m/s and Reynolds number 7 500. Calculate the wall shear stress. As the oil cools, its viscosity increases; what higher viscosity will produce the same shear stress? Neglect variation in specific gravity. The flowrate does not change.

9.70. The same fluid flows through 300 m (1 000 ft) of 75 mm (3 in.) and 300 m (1 000 ft) of 100 mm (4 in.) smooth pipe. The two flows are adjusted so that their Reynolds numbers are the same. What is the ratio between their head losses?

9.71. When 57 l/s of liquid (s.g. 1.27, viscosity 0.012 Pa · s) flow in 150 m of 150 mm pipe, the head lost is 11.2 m. Is this pipe rough or smooth?

9.72. Fluid of specific gravity 0.92 and viscosity 0.096 Pa · s flows in a 50 mm smooth brass pipeline. If $\mathbf{R} = 2\ 100$, calculate the head lost in 30 m of pipe if the flow is (a) laminar, (b) turbulent.

9.73. The head lost in 150 m of 0.3 m pipe having sand-grain roughness projections 2.5 mm high is 12 m for a flowrate of 0.28 m³/s. What head loss can be expected when the flowrate is 0.56 m³/s?

9.74. A liquid of specific gravity 0.85 flows in a 100 mm (4 in.) diameter commercial steel pipe. The flowrate is $4\frac{1}{4}$ l/s (65 gpm). If the pressure drop over a 60 m (200 ft) length of horizontal pipe is $1\frac{3}{4}$ kPa ($\frac{1}{4}$ psi), determine the viscosity of the liquid.

9.75. Water flows in a 100 mm (4 in.) commercial steel pipe at 15°C (59°F). If the center velocity is 1 m/s (3.3 ft/s) what is the flowrate?

9.76. Carbon dioxide flows in a 75 mm wrought iron pipe at an absolute pressure of 345 kPa and 10°C. If the center velocity is 0.6 m/s, calculate the mass flowrate.

9.77. Air at 101.3 kPa and 15.6°C flows in a horizontal triangular smooth duct, having 200 mm sides, at a mean velocity of 3.6 m/s. Calculate the pressure drop per metre of duct. Assume the air has constant density.

9.78. Three-tenths of a cubic metre of water flows per second in a smooth 230 mm square duct at 10°C. Calculate the head lost in 30 m of this duct.

9.79. A concrete conduit of cross-sectional area 10 ft² and wetted perimeter 12 ft carries water at 50°F at a mean veloctiy of 8 ft/s. Calculate the smallest head loss to be expected in 200 ft of this conduit.

9.80. A semicircular concrete conduit of 1.5 m (5 ft) diameter carries water at 20°C (68°F) at a velocity of 3 m/s (10 ft/s). Calculate the smallest loss of head to be expected per metre (foot) of conduit.

9.81. What relative roughness is equivalent to a Hazen-Williams coefficient of 140 for $10^5 < R < 10^6$?

9.82. A new 12 in. riveted steel pipeline carries a flowrate of 2.5 cfs of water. Calculate the head loss of 1 000 ft of this pipe using the Hazen-Williams method. Calculate f and an approximate value of e.

9.83. An 18 in. new riveted steel pipeline 1 000 ft long runs from elevation 150 to elevation 200. If the pressure at elevation 150 is 100 psi and at elevation 200 is 72 psi, what flowrate can be expected through the line?

9.84. Smooth masonry pipe of what diameter is necessary to carry 50 cfs between two reservoirs of surface elevations 250 and 100 if the pipeline is to be 2 miles long?

9.85. What Hazen-Williams coefficient will yield the same head loss as the Darcy-Weisbach equation for a 2 in. smooth pipe with flow at Reynolds number 10^5? Compare the result with the values of Table 3.

9.86. Laboratory tests on cylindrical pipe yield the empirical formula $h_L = 0.002\ 583\ lV^{2.14}\ d^{-0.86}$ with head loss in m, length in m, diameter in m, and velocity in m/s. Water of kinematic viscosity 9.3×10^{-7} m²/s was used in the tests and ranges of d and V were: $0.03 < d < 0.06$ and $0.6 < V < 1.5$. Analyze the formula and comment on its possible validity.

9.87. In the early hydraulic literature there are many empirical head loss formulas of the form $h_L/l = CV^x/d^y$. Show from the shape of the curves on the Moody diagram that $x \leq 2$, $y \geq 1$, and $x + y = 3$.

9.88. If 0.14 m³/s of water flow through a 150 mm horizontal pipe which enlarges abruptly to 300 mm diameter, and if the pressure in the smaller pipe is 138 kPa, calculate the pressure in the 300 mm pipe, neglecting pipe friction.

9.89. The fluid flowing has specific gravity 0.90; $V_{75} = 6$ m/s; $R = 10^5$. Calculate the gage reading.

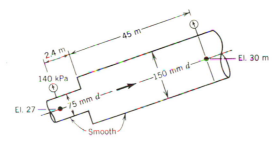

Problem **9.89**

9.90. Water is flowing. Calculate direction and approximate magnitude of the manometer reading.

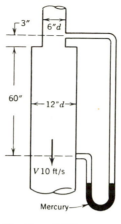

Problem **9.90**

9.91.   Solve problem 9.90 assuming conical enlargements of 70° and 7°.

9.92.   Calculate the magnitude and direction of the manometer reading. Water is flowing.

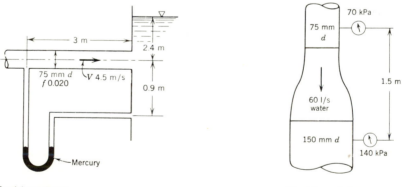

Problem **9.92**   Problem **9.93**

9.93.   Calculate the approximate loss coefficient for this gradual enlargement.

9.94.   Experimental determination of minor losses and loss coefficients are made from measurements of the hydraulic grade lines in zones of established flow. Calculate the head loss and loss coefficient for this gradual enlargement from the data given.

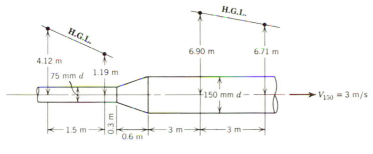

Problem **9.94**

9.95. The mean velocity of water in a 150 mm horizontal pipe is 0.9 m/s. Calculate the loss of head through an abrupt contraction to 50 mm diameter. If the pressure in the 150 mm pipe is 345 kPa, what is the pressure in the 50 mm pipe, neglecting pipe friction?

9.96. A 150 mm horizontal waterline contracts abruptly to 75 mm diameter. A pressure gage 150 mm upstream from the contraction reads 34.5 kPa when the mean velocity in the 150 mm pipe is 1.5 m/s. What will pressure gages read 0.6 m downstream and just downstream from the contraction if the diameter of the vena contracta is 61 mm? Neglect pipe friction.

9.97. Water is flowing. Calculate the gage reading when V_{12} is 8 ft/s.

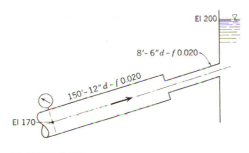

Problem **9.97**

9.98. Calculate the head loss and loss coefficient caused by this restricted contraction.

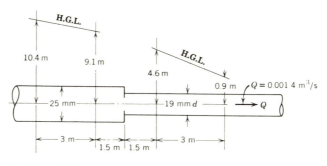

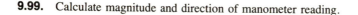

Problem **9.98**

9.99. Calculate magnitude and direction of manometer reading.

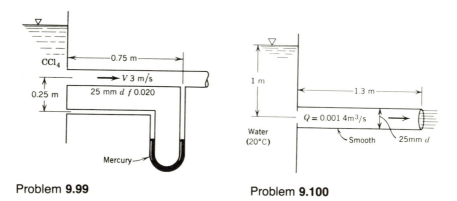

Problem **9.99**   Problem **9.100**

9.100. Determine the head loss and loss coefficient for this restricted pipe entrance.

9.101. The 6 in. suction pipe for a pump extends 10 ft vertically below the free surface of water in a tank. If the mean velocity in the pipe is 20 ft/s, what is the pressure in the pipe at the level of the liquid surface if the entrance is (*a*) rounded, (*b*) re-entrant? Assume the friction factor of the pipe to be 0.020.

9.102. A horizontal pipeline of 150 mm diameter leaves a tank (square-edged entrance) 15 m below its water surface and enters another tank 6 m below its water surface. If the flowrate in the line is 0.11 m³/s, what will gages read on the pipeline a short distance (say 0.6 m) from the tanks? Neglect pipe friction.

9.103. If the length of the unestablished flow zone downstream from a rounded pipe entrance (see Fig. 7.16 of Section 7.7) is 50 pipe diameters when the Reynolds number (Vd/ν) is 1 800, what is the loss coefficient of the entrance if the total head may be assumed to remain constant along the central streamline?

9.104. Solve the preceding problem for a turbulent flow in a smooth pipe at Reynolds number 100 000 if the length of the unestablished flow zone is 25 diameters. Assume that the seventh-root law (Section 9.3) is applicable.

9.105.   A 90° smooth bend in a 6 in. or 150 mm pipeline has a radius of 5 ft or 1.5 m. If the mean velocity through the bend is 10 ft/s or 3 m/s and the Reynolds number 200 000, what head loss is caused by the bend? If the bend were unrolled and established flow assumed to exist in the length of pipe, what percent of the total head loss could be considered due to wall friction?

9.106.   A 90° screwed elbow is installed in a 50 mm (2 in.) pipeline having a friction factor of 0.03. The head lost at the elbow is equivalent to that lost in how many metres (feet) of the pipe? Repeat the calculation for a 25 mm (1 in.) pipe.

9.107.   A 50 mm pipeline 1.5 m long leaves a tank of water and discharges into the atmosphere at a point 3.6 m below the water surface. In the line close to the tank is a valve. What flowrate can be expected when the valve is a (*a*) gate valve, (*b*) globe valve? Assume a square-edged entrance, a friction factor of 0.020, and screwed valves.

9.108.   Calculate the total tension in the bolts. Neglect entrance loss.

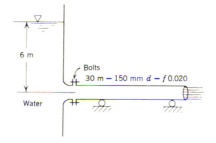

Problem **9.108**

9.109.   Water flows at 10°C from a reservoir through a 25 mm pipeline 600 m long which discharges into the atmosphere at a point 0.3 m below the reservoir surface. Calculate the flowrate, assuming it to be laminar and neglecting minor losses and velocity head in the pipeline. Check the assumption of laminar flow.

9.110.   Glycerin flows through a 2 in. horizontal pipeline leading from a tank and discharging into the atmosphere. If the pipeline leaves the tank 20 ft below the liquid surface and is 100 ft long, calculate the flowrate when the glycerin has a temperature of (*a*) 50°F, (*b*) 70°F. Neglect minor losses and velocity head.

9.111.   A horizontal 50 mm brass pipeline leaves (square-edged entrance) a water tank 3 m below its free surface. At 15 m from the tank, it enlarges abruptly to a 100 mm pipe which runs 30 m horizontally to another tank, entering it 0.6 m below its surface. Calculate the flowrate through the line (water temperature 20°C), including all head losses.

9.112.   Water flows from a tank through 60 m of horizontal 50 mm brass pipe and discharges into the atmosphere. If the water surface in the tank is 1.2 m above the pipe, calculate the flowrate, considering losses due to pipe friction only, when the water temperature is (*a*) 10°C, (*b*) 40°C.

9.113.   A smooth 12 in. pipeline leaves a reservoir of surface elevation 500 at elevation 460. A pressure gage is located on this line at elevation 400 and 1 000 ft from the reservoir (measured along the line). Calculate the gage reading when 10 cfs of water (68°F) flow in the line. Neglect minor losses.

9.114.   One-quarter of a cubic metre per second of liquid (20°C) is to be carried between two tanks having a difference of surface elevation of 9 m. If the pipeline is smooth and 90 m long, what pipe size is required if the liquid is (*a*) crude oil, (*b*) water? Neglect minor losses.

9.115.   A horizontal 50 mm pipeline leaves a water tank 6 m below the water surface. If this line has a square-edged entrance and discharges into the atmosphere, calculate the flowrate, neglecting and considering the entrance loss, if the pipe length is (*a*) 4.5 m, (*b*) 45 m. Assume a friction factor of 0.025.

9.116.   Calculate the flowrate from this water tank if the 6 in. pipeline has a friction factor of 0.020 and is 50 ft long. Is cavitation to be expected in the pipe entrance? The water in the tank is 5 ft deep.

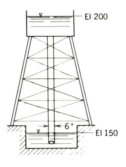

El 200

6″

El 150

Problem **9.116**

9.117.   A 300 mm horizontal pipe 300 m long leaves a reservoir of surface elevation 60 at elevation 54. This line connects (abrupt contraction) to a 150 mm pipe 300 m long running to elevation 30, where it enters a reservoir of surface elevation 39. Assuming friction factors of 0.02, calculate the flowrate through the line.

9.118.   What is the maximum flow which may be theoretically obtained in problem 9.117 when the 150 mm and 300 mm pipes are interchanged?

9.119.   A long 0.3 m pipeline laid between two reservoirs carries a flowrate of 0.14 m³/s of water. A parallel pipe of the same friction factor is laid beside this one. Calculate the approximate diameter of the second pipe if it is to carry 0.28 m³/s.

9.120.   A 6 in. horizontal smooth pipe 1 000 ft long takes oil from a large tank and discharges it into the atmosphere. At the midpoint of the pipe the pressure is 10.0 psi. If the specific gravity and viscosity of the oil are 0.88 and 0.000 5 lb · s/ft², respectively, calculate (*a*) the flowrate and (*b*) the pressure in the tank on the same level as the pipe.

9.121.   There is a leak in a horizontal 0.3 m pipeline having a friction factor of 0.025. Upstream from the leak two gages 600 m apart on the line show a difference of 138 kPa. Downstream from the leak two gages 600 m apart show a difference of 124 kPa. How much water is being lost from the pipe per second?

9.122. The pipe is filled and the plug then removed. Estimate the steady flowrate.

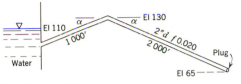

Problem **9.122**

9.123. An irrigation siphon has the dimensions shown and is placed over a dike. Estimate the flowrate to be expected under a head of 0.3 m. Assume a re-entrant entrance, a friction factor of 0.020, and bend loss coefficients of 0.20.

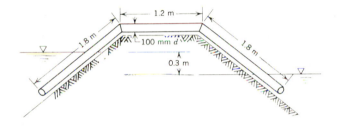

Problem **9.123**

9.124. Calculate the flowrate and the gage reading, neglecting minor losses and velocity heads.

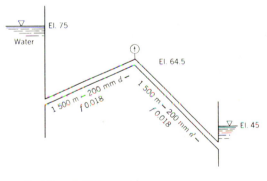

Problem **9.124**

9.125. At least 0.08 m³/s of oil ($\nu = 1.7 \times 10^{-5}$ m²/s) are to flow between two reservoirs having a 15 m difference of free surface elevation. Three hundred millimetre rough steel pipe of equivalent sand-grain size 3 mm and 250 mm smooth pipe are available at the same cost. Which pipe should be used? Provide calculations to justify the choice.

9.126. A 380 mm pipeline having equivalent sand-grain roughness of 6 mm carries 1/3 m³/s of water (20°C). If a smooth liner is installed in the pipe, thereby reducing the diameter to 350 mm, what (percent) reduction of head loss can be expected in the latter pipe for the same flowrate?

9.127. A 6 ft diameter pipeline 4 miles long between two reservoirs of surface elevations 500 and 300 ft carries a flowrate of 250 cfs of water (68°F). It is proposed to increase the flowrate through the line by installing a glass-smooth liner. Above what liner diameter may an increase of flowrate be expected? What is the maximum increase to be expected? Assume the 6 ft diameter to be measured to the midpoint of the roughness projections. Neglect all minor head losses.

9.128. A 0.3 m pipline 450 m long leaves (square-edged entrance) a reservoir of surface elevation 150 at elevation 138 and runs to elevation 117, where it discharges into the atmosphere. Calculate the flowrate and sketch the energy and hydraulic grade lines (assuming that $f = 0.022$) (a) for these conditions, and (b) when a 75 mm nozzle is attached to the end of the line, assuming the lost head caused by the nozzle to be 1.5 m. How much power is available in the jet?

9.129. A 300 mm pipeline ($f = 0.020$) is horizontal and 60 m long and runs between two reservoirs. It leaves the high-level reservoir at a point 36 m below its surface and enters the low-level reservoir 6 m below its surface. (a) Assume standard square-edged entrance and exit and compute the flowrate to be expected. (b) A nozzle of 225 mm tip diameter is now attached to the downstream end of the pipe; neglecting head losses *in* the nozzle, what reduction of flowrate will be expected? (c) The nozzle is now replaced with a diffuser tube of 375 mm exit diameter. Assuming that the diffuser tube flows full and the losses therein may be neglected, what increase of flowrate is to be expected?

9.130. A pipeline of 0.3 m diameter ($f = 0.020$) runs 600 m from a reservoir of surface elevation 150 to a point at elevation 105 where it terminates in a 150 mm diameter nozzle. (a) Neglecting entrance and nozzle losses calculate the flowrate through this pipeline. (b) If a turbine (100% efficiency) is now installed toward the middle of the pipeline, what is the maximum power that may be expected from it? (c) Could a larger power than that of (b) be obtained by changing the nozzle size? If so, calculate the required nozzle diameter. Assume the setting of the turbine low enough so that there are no cavitation considerations in this problem.

9.131. Water flows from a large reservoir through 2 500 ft of 12 in. pipe of constant friction factor 0.030. The pipe terminates in a frictionless nozzle discharging to the atmosphere. The nozzle is at elevation 100, the reservoir surface at elevation 300. What size (diameter) of nozzle should be provided to maximize the jet horsepower? Calculate this maximum horsepower.

9.132. A 0.6 m pipeline 900 m long leaves (square-edged entrance) a reservoir of surface elevation 150 at elevation 135 and runs to a turbine at elevation 60. Water flows from the turbine through a 0.9 m vertical pipe ("draft tube") 6 m long to tail water of surface elevation 56. When 0.85 m³/s flow through pipe and turbine, what power is developed? Take $f = 0.020$; include exit loss; neglect other minor losses and those within the turbine. How much power may be saved by replacing the above draft tube with a 7° conical diffuser of the same length?

9.133. A pump close to a reservoir of surface elevation 100 pumps water through a 6 in. pipeline 1 500 ft long and discharges it at elevation 200 through a 2 in. nozzle. Calculate the pump horsepower necessary to maintain a pressure of 50 psi behind the nozzle, and sketch accurately the energy line, taking $f = 0.020$.

9.134. The horizontal 200 mm suction pipe of a pump is 150 m long and is connected to a reservoir of surface elevation 90 m, 3 m below the water surface. From the pump, the 150 mm discharge pipe runs 600 m to a reservoir of surface elevation 126, which it enters 10 m below the water surface. Taking f to be 0.020 for both pipes, calculate the power required to pump 0.085 m³/s from the lower reservoir. What is the maximum dependable flowrate that may be pumped through this system (a) with the 200 mm suction pipe, and (b) with a 150 mm suction pipe?

9.135. A 0.3 m pipeline 3.2 km long runs on an even grade between reservoirs of surface elevations 150 and 120, entering the reservoirs 10 m below their surfaces. The flowrate through the line is inadequate, and a pump is installed at elevation 125 to increase the capacity of the line. Assuming f to be 0.020, what pump power is required to pump 0.17 m³/s downhill through the line? Sketch accurately the energy line before and after the pump is installed. What is the maximum dependable flowrate that may be obtained through the line?

9.136. Assuming disruption of the flow when the negative pressure head reaches 6 m, to what elevation may the water surface in the tank be lowered by this siphon? Calculate the flowrates when the water surface is at elevation 29.4 and at the point of disruption.

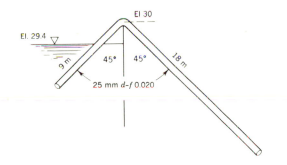

EI 30

EI. 29.4

9 m

45°   45°

18 m

25 mm d–f 0.020

Problem **9.136**

9.137. Calculate the smallest reliable flowrate that can be pumped through this pipeline. Assume atmospheric pressure 14.7 psia.

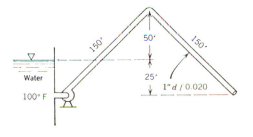

150'

50'

150'

Water

100° F

25'

1" d / 0.020

Problem **9.137**

9.138. The suction side of a pump is connected directly to a tank of water at elevation 30. The water surface in the tank is at elevation 39. The discharge pipe from the pump is horizontal, 300 m long, of 150 mm diameter, and has a friction factor of 0.025. The pipe runs from west to east and terminates in a 75 mm frictionless nozzle. The nozzle stream is to discharge into a tank whose western upper edge is horizontal and is located at elevation 15 and 30 m east of the tip of the nozzle. What is the minimum power that the pump may supply for the stream to pass into the tank?

9.139.   The pump is required to maintain the flowrate which would have occurred without any friction. What power pump is needed? Neglect minor losses.

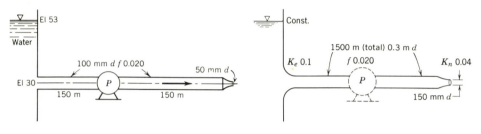

Problem 9.139   **Problem 9.140**

9.140.   The flowrate through this pipeline and nozzle when the pump is not running (and assumed not to impede the flow) is 0.28 m³/s. How much power must be supplied by the pump to produce the same flowrate with a 100 mm nozzle at the end of the line?

9.141.   If the turbine extracts 530 hp from the flow, what flowrate must be passing through the system? What is the maximum power obtainable from the turbine?

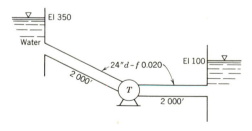

Problem 9.141

9.142.   If there were no pump, 0.14 m³/s of water would flow through this pipe system. Calculate the pump power required to maintain the same flowrate in the opposite direction.

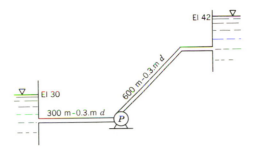

Problem **9.142**

9.143. The flowrate is 0.13 m³/s without the pump. Calculate the approximate pump power required to maintain a flowrate of 0.17 m³/s.

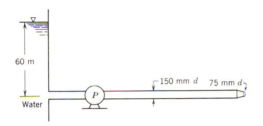

Problem **9.143**

9.144. The pitot tube is on the centerline of the pipe. The flowrate is 2 cfs of water. Calculate the horsepower transferred from pump to fluid.

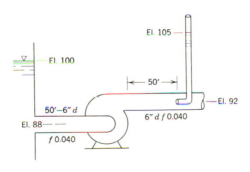

Problem **9.144**

9.145. The pump of the last Illustrative Problem of Section 9.10 is to be installed between two reservoirs that are 2 000 m apart and have an elevation difference of 40 m. Select a pipe material and size so that the pump operates at maximum efficiency.

9.146. The pump of problem 9.145 is mounted in a 0.3 m diameter steel pipe of variable length that discharges to the atmosphere. The pipe is horizontal and its entrance (smooth bellmouth) is located at elevation 50 in a reservoir whose surface elevation is 100. For the range of flow for which the relation between head and flowrate are known for the pump, calculate and plot a relationship between flowrate and the length of the pipe.

9.147. For the situation of problem 9.146 find the length of pipe at which the pump operates at maximum efficiency.

9.148. Five 25 mm diameter brass pipes will conduct 2.4 l/s of water (20°C) between two reservoirs whose respective surface levels are constant. Find the diameter of a single brass pipe that will carry the same flowrate between these reservoirs.

9.149. A 24 in. pipeline branches into a 12 in. and an 18 in. pipe, each of which is 1 mile long, and they rejoin to form a 24 in. pipe. If 30 cfs flow in the main pipe, how will the flow divide? Assume that $f = 0.018$ for both branches.

9.150. For the configuration shown, derive an equation for the velocity in the service line in terms of its diameter, friction factor, and length and of the velocities in the two sections of the venturi meter. Neglect minor losses. Note that in an Arctic setting, flow in the main line produces flow in the smaller home service line, thus keeping the lines from freezing up.

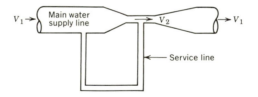

Problem **9.150**

9.151. A 0.6 m pipeline carrying 0.85 m³/s divides into 150 mm, 200 mm, and 300 mm branches, all of which are the same length and enter the same reservoir below its surface. Assuming that $f = 0.020$ for all pipes, how will the flow divide?

9.152. A straight 12 in. pipeline 3 miles long is laid between two reservoirs of surface elevations 500 and 350 entering these reservoirs 30 ft beneath their free surfaces. To increase the capacity of the line, a 12 in. line 1.5 miles long is laid from the original line's midpoint to the lower reservoir. What increase in flowrate is gained by installing the new line? Assume that $f = 0.020$ for all pipes.

9.153. Calculate the nozzle diameter which will maximize the jet power. Calculate this (maximum) jet power. Neglect minor losses. Assume all friction factors 0.020.

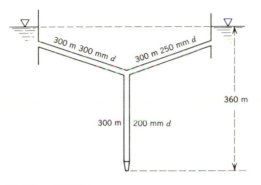

300 m 300 mm d   300 m 250 mm d

360 m

300 m   200 mm d

Problem **9.153**

9.154. Solve problem 4.39 for three 1 000 m (3 300 ft) commercial steel pipes, neglecting entrance losses, but not losses in the tee. State assumptions made.

9.155. Solve problem 4.40 under the conditions set in problem 9.154.

9.156. Both of the pipelines are composed of 1 200 m of 0.3 m pipe terminating in a 0.1 m nozzle. The friction factors for both pipes are 0.020. The head losses in the nozzles are negligible. Points A and B are the midpoints of the lines. There is a short length of pipe between A and B containing a closed gate valve. When this valve is opened, will the total flowrate from the nozzles increase or decrease?

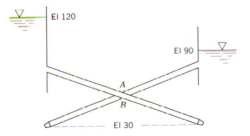

El 120

El 90

A

B

El 30

Problem **9.156**

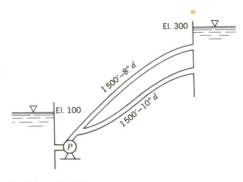

El. 300

El. 100

1 500'–8" d

1 500'–10" d

P

Problem **9.157**

9.157. If the pump supplies 300 hp to the water, what flowrate will occur in the pipes? Assume a friction factor of 0.020 for both pipes.

9.158. A 0.9 m pipe divides into three 0.45 m pipes at elevation 120. The 0.45 m pipes run to reservoirs which have surface elevations 90, 60, and 30, these pipes having respective lengths of 3.2, 4.8, and 6.8 kilometres. When 1.4 m³/s flows in the 0.9 m line, how will the flow divide? Assume that $f = 0.017$ for all pipes.

9.159. Reservoirs A, B, and C have surface elevations 500, 400, and 300, respectively. A 12 in. pipe 1 mile long leaves reservoir A and runs to point O at elevation 450. Here the pipe divides and an 8 in. pipe, 1 mile long, runs from O to B and a 6 in pipe, 1.5 miles long, runs from O to C. Assuming that $f = 0.020$, calculate the flowrates in the lines.

9.160. Three pipes join at a common point at elevation 105. One, a 0.3 m line 600 m long, goes to a reservoir of surface elevation 120; another, a 150 mm line 900 m long, goes to a reservoir of surface elevation 150; the third (150 mm) runs 300 m to elevation 75, where it discharges into the atmosphere. Assuming that $f = 0.020$, calculate the flowrate in each line. Calculate these flowrates when a 50 mm nozzle is attached to the end of the third pipe.

9.161. A 0.3 m pipeline 600 m long leaves a reservoir of surface elevation 150 and runs to elevation 120, where it divides into two 150 mm lines each 300 m long, both of which discharge into the atmosphere, one at elevation 135, the other at elevation 105. Calculate the flowrates in the three pipes if all friction factors are 0.025.

9.162. The pump is to deliver 110 l/s to the outlet at elevation 165 and 220 l/s to the upper reservoir. Calculate pump power and the required diameter of the 300 m pipe.

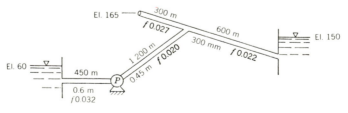

El. 165 — 300 m
f 0.027
600 m
300 mm f 0.022
El. 150
1 200 m f 0.020
300 mm
El. 60
450 m
0.45 m
P
0.6 m
f 0.032

Problem **9.162**

9.163. Water is flowing. For $Q_8 = 4.0$ cfs, calculate Q_6, Q_{12}, and pump horsepower.

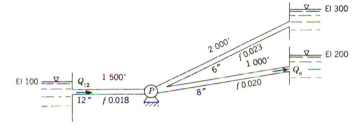

Problem **9.163**

9.164. If a turbine replaces the pump in the pipe system of problem 9.163, what is the maximum horsepower it can produce?

9.165. An analogy to the laminar flow of fluids in pipes may be made with the flow of electric current through a resistance. If head loss is analogous to voltage drop, and flowrate to current, what combination of quantities in the fluid problem is analogous to electric resistance? Applying this idea to problem 9.166 determine the flowrates in the pipes if the kinematic viscosity of the fluid flowing is 2.8×10^{-4} m^2/s.

9.166. Calculate the flowrates in the pipes of this loop if all friction factors are 0.020.

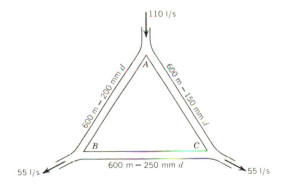

Problem **9.166**

9.167. Calculate the flowrates in the pipes and pressures at the junctions of this network. Assume rough flow and concrete pipes. Neglect minor losses and losses in the short pipe from the reservoir. The network is at elevation 10; the reservoir at 70.

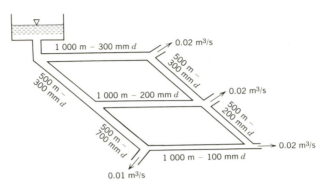

Problem **9.167**

9.168. Calculate the flowrates in the pipes of this network. The length, diameter, and f value is shown for each pipe.

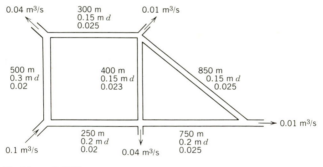

Problem **9.168**

9.169. Calculate the flowrates in the pipes of this network. The length, diameter, and f value is shown for each pipe.

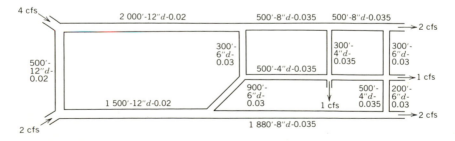

Problem **9.169**

9.170. Suppose that the head versus flowrate characteristic of a pump is approximately $E_P = 70(1 - 2Q)$. Calculate the flowrates in the pipes in this network with and without the pump. Let $f = 0.015$ in all pipes.

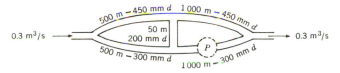

Problem **9.170**

9.171. Air flows isothermally in a 75 mm pipeline at a mean velocity of 3 m/s, absolute pressure of 276 kPa, and a temperature of 10°C. If pressure is lost by friction, calculate the mean velocity where the absolute pressure is 207 kPa.

9.172. Through a horizontal section of 6 in. cast iron pipe 1 000 ft long in which the temperature is 60°F, 200 lb/min of air flow isothermally. If the pressure at the upstream end of this length is 30 psia, calculate the pressure at the downstream end and the mean velocities at these two points.

9.173. Carbon dioxide flows isothermally in a 50 mm horizontal wrought iron pipe, and at a certain point the mean velocity, absolute pressure, and temperature are 18 m/s, 400 kPa, and 25°C, respectively. Calculate the pressure and mean velocity 150 m downstream from this point.

9.174. The Mach number, V/a, at a point in an isothermal (38°C) flow of air in a 100 mm pipe is 0.20, and the absolute pressure at this point is 345 kPa. If the friction factor of the pipe is 0.020, what pressure and Mach number may be expected 60 m downstream from this point?

9.175. What are the limiting Mach numbers for isothermal flow of helium, carbon dioxide, and methane?

9.176. Construct a figure comparable to Fig 9.29 for helium or carbon dioxide.

9.177. Nitrogen is being pumped adiabatically in an insulated 100 mm (4 in.) pipeline in which $f = 0.01$. If the Mach number at a point in the line is 0.3, how far downstream will the Mach number be 0.8?

9.178. What is the maximum possible distance downstream that the flow in problem 9.177 can continue before discharge from the pipe at sonic speed?

9.179. Construct a figure comparable to that accompanying the Illustrative Problem of Section 9.14 for nitrogen, carbon dioxide, or helium. Plot $\bar{f}$ (l_{max}/d) versus $\mathbf{M}_1^2$.

9.180. Carbon dioxide enters a pipe that is fully insulated at $\mathbf{M}_1 = 3.0$. Fifteen pipe diameters downstream $\mathbf{M}_2 = 1.5$. What is the effective $\bar{f}$ for the pipe?

9.181. Resolve problem 9.180 if the gas is helium.

9.182. A supersonic flow of helium gas in an insulated pipe is needed for a test. The pipe is 300 mm (12 in.) diameter. Assuming the flow is rough and that for a supersonic flow $\bar{f} = \frac{1}{2}f$ for any material, plot l_{max} versus f for $\mathbf{M}_1 = 2$. Indicate appropriate pipe materials corresponding to various l_{max} values.

9.183. Use equations 9.66 and 9.68 to obtain

$$\frac{dp/p}{d\mathbf{M}^2/\mathbf{M}^2} = -\frac{1 + (k-1)\mathbf{M}^2}{2\left(1 + \dfrac{k-1}{2}\mathbf{M}^2\right)}$$

Integrate this result between sections where $p = p_1$, $\mathbf{M} = \mathbf{M}_1$, and where $\mathbf{M} = 1$ and $p = p_s$ to prove that

$$\frac{p_1}{p_s} = \frac{1}{\mathbf{M}_1}\left[\frac{k+1}{2\left(1 + \dfrac{k-1}{2}\mathbf{M}_1^2\right)}\right]^{1/2}$$

9.184 Use the results of problem 9.183 to compute the pressure ratio p_2/p_1 for problem 9.180 or 9.181.

10
LIQUID FLOW IN OPEN CHANNELS

Open-channel flow embraces a variety of problems that arise when water flows in natural water courses, regular canals, irrigation ditches, sewer lines, flumes, and so forth—a province of paramount importance to the civil engineer. Although open-channel problems practically always involve the flow of water, and although the experimental results used in these problems were obtained by hydraulic tests, modern fluid mechanics indicates the extent to which these results may be applied to the flow of other liquids in open channels.

10.1 Fundamentals

In the problems of pipe flow, as may be seen from the hydraulic grade line, the pressures in the pipe can vary along the pipe and depend on energy losses and the conditions imposed on the ends of the line. Open flow, however, is characterized by the fact that pressure conditions are determined by the constant pressure, usually atmospheric, existing on the entire surface of the flowing liquid. Usually pressure variations within an open-channel flow can be determined by the principles of hydrostatics since the streamlines are ordinarily straight, parallel, and approximately horizontal (compare Fig. 4.3, Section 4.3); when these conditions are satisfied, the hydraulic grade lines for all the streamtubes are the same and coincide with the liquid surface. There are, of course, many exceptions to the foregoing situation and, in flowfields where streamlines are convergent, divergent, curved, or steeply sloping, each streamline has in general its own hydraulic grade line and these do not lie in the liquid surface.

A typical reach of open-channel flow is shown in Fig. 10.1.[1] The individual streamlines of the flow are very slightly convergent but may safely be assumed to be parallel, and the hydraulic grade line coincides with the liquid surface. The energy line is located one velocity head[2] above the hydraulic grade line, and the head loss is defined, as usual, as the drop in the energy line. Thus the energy line-hydraulic grade

---

[1] The channel slopes in many of the illustrations in this chapter have been exaggerated to emphasize their existence. In open-channel practice, slopes are very seldom encountered which are greater than 1 (vertical) in 100 (horizontal), or 0.01.

[2] Strictly, this distance should be $\alpha V^2/2g_n$ but, in view of the relatively flat velocity profile, α is in most cases only slightly greater than unity.

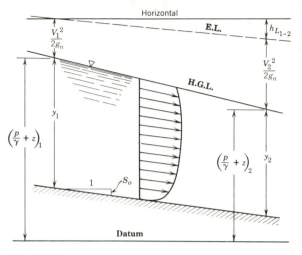

Fig. **10.1**

line approach to one-dimensional flow problems can be expected to find wide application in many problems of open flow.

Open-channel flow may be laminar or turbulent, steady or unsteady, *uniform* or
varied, subcritical or *supercritical*. The complexity of unsteady open-flow problems
forbids their treatment in an elementary text, but the other categories are examined
herein; the emphasis, however, is on steady turbulent flow, which is the problem
generally encountered in practice. The definitions of subcritical and supercritical flow
are presented subsequently, but the significance, causes, and limits of uniform and
varied flow must be examined first. The meaning of these terms and also the fundamentals of open-channel flow may be seen from a comparison of ideal-fluid flow and
real-fluid flow in identical prismatic channels[3] leading from reservoirs of the same
surface elevation (Fig. 10.2). No resistance is encountered by the ideal fluid as it flows
down the channel. Because of this lack of resistance, the ideal fluid continually accelerates under the influence of gravity. Thus the mean velocity continually increases, and
with this increase of velocity a reduction in flow cross section is required by the
continuity principle. Reduction in flow cross section is characterized by a decrease in
depth of flow; since the depth of flow continually *varies* (from section to section, not
with time) this type of fluid motion is termed *varied flow*.[4]

When real fluid flows in the same channel, there are resistance forces due to fluid
viscosity and channel roughness. Analysis of the resistance forces originating from
these same properties in pipes has shown that such forces depend on the velocity of flow
(equation 9.3). Thus, in the upper end of the channel where motion is slow, resistance
forces are small, but the components of gravity forces in the direction of motion are

---

[3] A prismatic channel has an unvarying cross-sectional shape and a constant bottom slope.
[4] The term *nonuniform flow* is also used, but *varied flow* is preferred.

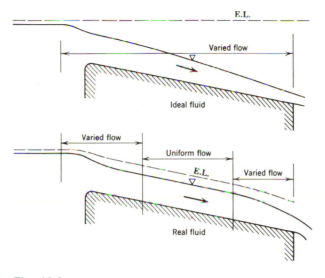

Fig. **10.2**

about the same as for the ideal fluid. The resulting unbalanced forces in the direction of motion bring about acceleration and varied flow in the upper reaches of the channel. However, with an increase of velocity, the forces of resistance increase until they finally balance those caused by gravity. When this force balance occurs, constant-velocity motion is attained; it is characterized by no change of flow cross section and thus no change in the depth of flow; this is *uniform flow*. Toward the lower end of the channel, pressure and gravity forces again exceed resistance forces and varied flow again results.

Obviously, an inequality between the above-mentioned forces is more probable than a balance of these forces, and, hence, varied flow occurs in practice to a far greater extent than uniform flow. In short channels, for example, uniform-flow conditions may never be attained because of the long reach of channel necessary for the establishment of uniform flow; nevertheless, solution of the uniform-flow problem forms the basis for open-channel flow calculations.

10.2 Uniform Flow—The Chezy Equation

The fundamental equation for uniform open-channel flow may be derived readily by equating (since there is no change of momentum) the equal and opposite force components of gravity and resistance and applying some of the fundamental notions of fluid mechanics encountered in the analysis of pipe flow. Consider the uniform flow of a liquid between sections 1 and 2 of the open channel of Fig. 10.3. The forces acting on the control volume of liquid $ABCD$ are (1) the forces of static pressure, F_1 and F_2, acting on the ends of the body; (2) the weight, W, which has a component $W \sin \theta$ in the direction of motion; (3) the pressure forces exerted by the bottom and sides of the

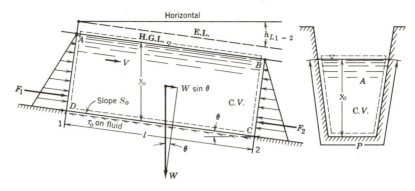

Fig. 10.3

channel;[5] and (4) the force of resistance exerted by the bottom and sides of the channel $Pl\tau_o$. A summation of forces along the direction of motion gives

$$F_1 + W \sin \theta - F_2 - Pl\tau_o = 0$$

since there is no change of momentum between sections 1 and 2. Obviously, $F_1 = F_2$, $W = A\gamma l$, and $\sin \theta = h_L/l$. The slope, S_o, of both channel bed and liquid surface is $\tan \theta$, and for the small slopes encountered in open-channel practice the approximation $\tan \theta = \sin \theta$ may be made. A/P is recognized (Sections 7.10 and 9.7) as the hydraulic radius, R_h. When these substitutions are made in the equation above an expression for the *mean shear stress* results; it is

$$\tau_o = \gamma R_h S_o \tag{10.1}$$

In pipe flow, τ_o was shown (Section 9.1) to be given by $f\rho V^2/8$ in which f, the friction factor, is dependent on surface roughness and Reynolds number, but more dependent on the magnitude of roughness for highly turbulent flow. The mechanism of real-fluid motion is similar in pipes and open channels, and if it is assumed[6] that the hydraulic radius concept will account adequately for the differences in the cross-sectional shapes of circular pipes and open channels, these expressions for τ_o may be equated. Solving for V and replacing γ/ρ by g_n,

$$V = \sqrt{\frac{8g_n}{f}} \sqrt{R_h S_o}$$

or, if $C = \sqrt{8g_n/f}$,

$$V = C\sqrt{R_h S_o} \tag{10.2}$$

---

[5] These are not shown since they do not enter the following equations.
[6] Experience has shown this to be a valid assumption for regular prismatic channels (canals, flumes, etc.) with turbulent flow. It cannot be expected to apply to laminar flow (Section 10.4) or to channels of irregular or distorted cross section (such as natural streams at flood stage).

This is called the *Chezy equation* after the French hydraulician who established this relationship experimentally in 1775. By applying the continuity principle, the equation may be placed in terms of flowrate as

$$Q = CA\sqrt{R_h S_o} \tag{10.3}$$

which is the fundamental equation of uniform flow in open channels.

10.3 The Chezy Coefficient

From the time of Chezy to the present, many experiments in field and laboratory have been performed to determine the magnitude of the Chezy coefficient, C, and its dependence on other variables. The simplest relation and the one most widely used may be derived from the work of Manning,[7] an analysis of experimental data obtained from his own experiments and from those of others. His results may be summarized by the empirical relation

$$\text{(SI units): } C = \frac{R_h^{1/6}}{n}$$

which may be combined with the Chezy equation to give the so-called Chezy-Manning equation,

$$\text{(SI units): } Q = \left(\frac{1}{n}\right) A R_h^{2/3} S_o^{1/2}$$

These last two equations were derived from metric data; hence, the unit of length is the metre. Although n is often supposed to be a characteristic of channel roughness, it is convenient to consider n to be dimensionless.[8] Then, while the English (FSS) and metric (SI) equations differ, the values of n are the same in both systems. Unfortunately, the coefficient $(1/n)$ in the equations has the dimensions $t^{-1}L^{1/3}$; as a consequence, conversion to English (FSS) units requires inclusion of the factor $(3.28)^{1/3} = 1.49$, where 3.28 is the number of feet in a metre. The result is

$$\text{(FSS units): } C = \frac{1.49 R_h^{1/6}}{n}$$

and

$$\text{(FSS units): } Q = \left(\frac{1.49}{n}\right) A R_h^{2/3} S_o^{1/2}$$

Accordingly, the results can be summarized as

[7] *Trans. Inst. Civil Engrs. Ireland*, vol. 20, p. 161, 1890.
[8] See V. T. Chow, *Open-Channel Hydraulics*, McGraw-Hill, 1959, p. 98, for a complete discussion of the dimensional aspects of n.

$$C = \frac{uR_h^{1/6}}{n} \tag{10.4}$$

and

$$Q = \left(\frac{u}{n}\right)AR_h^{2/3}S_o^{1/2} \tag{10.5}$$

where $u = 1$ for SI units and $u = 1.49$ for FSS units.

The Manning n is obtained from a descriptive statement of the channel character; some typical values are given in Table 6. There is no substitute for experience and judgment in the interpretation and selection of values for n. Systematic experiments and analyses, similar to those of Nikuradse and Colebrook in the field of pipe flow, have yet to produce a clear definition of open-channel roughness or a scientific interpretation of the Chezy C or Manning n.

Table 6 Typical Values of n[9]

Partly-full pipes	
Corrugated metal drains	0.024
Concrete culverts (normal)	0.013
Drainage tiles (clay)	0.013
Sewers (normal)	0.013
Man-made channels (lined)	
Steel	0.012
Timber	0.012
Concrete (troweled)	0.013
Concrete (gunite)	0.019–0.022
Good ashlar masonry or brickwork	0.015
Rubble masonry	0.025
Asphalt	0.013–0.016
Earth (clean)	0.022
Earth (with vegetation)	0.027–0.035
Natural channels	
Clean and straight	0.030
Winding with some pools and shoals	0.040
Very weedy, deep pools	0.100
Mountain streams	0.040–0.050
Major streams (width greater than 100 ft or 30 m at flood stage)	0.025–0.100

[9] Table 6 was adapted, with permission of McGraw-Hill Book Company, from V. T. Chow, *Open Channel Hydraulics*, Chapter 5, copyright © 1959 McGraw-Hill Book Co. H. H. Barnes, "Roughness Characteristics of Natural Channels," *U.S. Geol. Survey Wat. Supply Pap.* 1849, 1967, gives a set of typical n values together with matching descriptive data and color photographs of natural channels.

From the relationship $C = \sqrt{8g_n/f}$ and equation 10.4,

$$n = uR_h^{1/6}\sqrt{\frac{f}{8g_n}}$$

Introducing numerical values for g_n in the respective unit systems

$$\text{(FSS units): } n = 0.093f^{1/2}R_h^{1/6}$$

$$\text{(SI units): } n = 0.113f^{1/2}R_h^{1/6}$$

Because $(3.28)^{1/6} = 1.22$ and $0.133/0.093 = 1.22$ the n values obtained are equal irrespective of units. Clearly, n is not an absolute roughness coefficient because it depends in a complex manner on the friction factor, which in turn depends on relative roughness and Reynolds number, and hydraulic radius. Interestingly, increases in R_h are sometimes offset by related decreases in f so that n may remain relatively constant for a given surface.

On the basis of experience and by use of careful judgment, it is possible to estimate n and use the Chezy-Manning equation successfully. When used with care it is simple and reliable; however, it is a dimensionally nonhomogeneous (see Section 8.2), empirical equation. This nonhomogeneity leads to the requirement that n or the constants in the equation have dimensions and makes it impossible to establish a fundamentally sound basis for determining n.

--- **Illustrative Problem** ---

A rectangular channel lined with asphalt is 20 ft wide and laid on a slope of 0.000 1. Calculate the depth of uniform flow in this channel when the flowrate is 400 ft³/s.

Relevant Equation and Given Data

$$Q = \left(\frac{u}{n}\right)AR_h^{2/3}S_o^{1/2} \tag{10.5}$$

$A = By_o$ $B = 20 \text{ ft}$ $Q = 400 \text{ ft}^3/\text{s}$ $S_o = 0.000\ 1$

$u = 1.49$ asphalt surface

Solution. Using y_o as the depth of uniform flow,

$$A = 20y_o \qquad P = 20 + 2y_o \qquad R_h = \frac{20y_o}{(20 + 2y_o)}$$

From Table 6,

$$n = 0.013$$

$$400 = \left(\frac{1.49}{0.013}\right)(20y_o)\left(\frac{20y_o}{20 + 2y_o}\right)^{2/3}(0.000\ 1)^{1/2} \tag{10.5}$$

from which we obtain

$$y_o\left(\frac{20y_o}{20 + 2y_o}\right)^{2/3} = 17.4$$

Solving by trial, $y_o = 6.85$ ft. ◀

If a value of n of 0.016 (23% greater than 0.013) had been selected, the depth would have been 7.95 ft (16% greater than 6.85 ft).

10.4   Uniform Laminar Flow[10]

Laminar flow ($Vy_o/v \gtrsim 500$) in open channels occurs in drainage from streets, airport runways, parking areas, and so forth. Here the flow is in thin sheets of virtually infinite width (i.e., without sidewalls) with resistance only at the bottom of the sheet; thus the flow is essentially a two-dimensional one. A definition sketch for uniform laminar flow is shown in Fig. 10.4. From the pipe flow analysis of Section 9.2, a parabolic velocity profile and linear shear stress profile are to be expected. The hydraulic radius for such a flow may be deduced by considering a section of width b normal to the flow; the flow cross section is thus a rectangle of depth y_o and width b. However, the wetted perimeter (i.e., the boundary offering resistance to the flow) is only the channel bottom, of width b; therefore $R_h = A/P = by_o/b = y_o$. Inserting this in equation 10.1,

$$\tau_o = \gamma y_o S_o$$

However, in laminar flow, τ_o is also given by $\mu(dv/dy)_o$ in which $(dv/dy)_o$ is the velocity gradient at the channel bottom. From the properties of the parabolic velocity profile, $(dv/dy)_o$ may be shown to equal $2v_s/y_o$ and the mean velocity V to be $2v_s/3$. Combining these, equating the two expressions for τ_o, and substituting ρg_n for γ, and v for μ/ρ produce

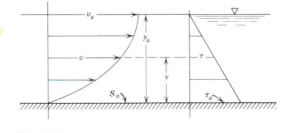

Fig. **10.4**

---

[10] See W. M. Owen, "Laminar to Turbulent Flow in a Wide Open Channel," *Trans. A.S.C.E.*, vol. 119, 1954, and L. G. Straub, E. Silberman, and H. C. Nelson, "Open Channel Flow at Small Reynolds Number," *Trans. A.S.C.E.*, vol. 123, 1957.

$$V = \frac{g_n y_o^2 S_o}{3\nu} \qquad (10.6)$$

which relates mean velocity, slope, and depth when the flow is laminar. As in pipes, the limit of laminar flow is defined by an experimentally determined critical Reynolds number, in this case having a value of around 500 if the Reynolds number is defined as $V y_o / \nu$. From this a specific relationship between Chezy coefficient and Reynolds number may be derived for this laminar flow. Substitute V from equation 10.6 into the definition for Reynolds number to obtain

$$\mathbf{R} = g_n y_o^3 S_o / 3\nu^2 \qquad (10.7)$$

By rearranging equation 10.6 to the Chezy form,

$$V = \frac{g_n \sqrt{S_o} y_o^{3/2}}{3\nu} \sqrt{y_o S_o} \qquad (10.8)$$

and by comparing equations 10.7 and 10.8 we find that

$$C = \sqrt{g_n \mathbf{R}/3} \qquad (10.9)$$

10.5 Hydraulic Radius Considerations

With the hydraulic radius R_h playing a prominent role in the equations of open-channel flow, and with depth variation a basic characteristic of such flows, the variation of hydraulic radius with depth (and other variables) becomes an important consideration. Clearly this is a problem of section geometry that requires no principles of mechanics for solution.

Consider first the variation of hydraulic radius with depth in a rectangular channel (Fig. 10.5a) of width B. Here $R_h = By/(B + 2y)$ and it is immediately evident that: for $y = 0$, $R_h = 0$ and, for $y \rightarrow \infty$, $R_h \rightarrow B/2$; therefore the variation of R_h with y must be as shown. From this comes a useful engineering approximation: for *narrow deep sections* $R_h \cong B/2$; since any (nonrectangular) section when deep and narrow approaches a rectangle this approximation may be used for any deep and narrow section—for which the hydraulic radius may be taken to be one half of the mean width.

Another useful engineering approximation may be discovered by examining the variation of hydraulic radius with channel width (Fig. 10.5b) for a constant depth y. With $R_h = By/(B + 2y)$ it is noted that: for $B = 0$, $R_h = 0$ and, for $B \rightarrow \infty$, $R_h \rightarrow y$; thus the variation of R_h with B is as shown. From this it may be concluded that for wide shallow rectangular sections $R_h \cong y$; for nonrectangular sections the approximation is also valid if the section is wide and shallow—here the hydraulic radius is approximately the mean depth.

A simple, useful, and typical nonrectangular section is the trapezoidal one of Fig. 10.6 for which $R_h = A/P = (By + zy^2)/(B + 2\sqrt{1 + z^2}y)$. The derivative of R_h with respect to y will allow investigation of the form of the relationship between R_h and y.

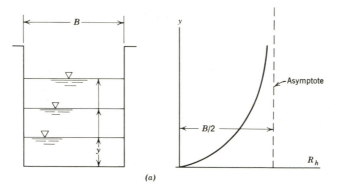

(a)

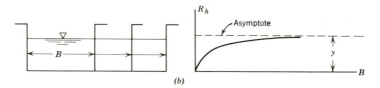

(b)

Fig. **10.5**

Performing this operation yields

$$\frac{dR_h}{dy} = \frac{1 + 2z(y/B)[1 + (y/B)\sqrt{1 + z^2}]}{1 + 4\sqrt{1 + z^2}(y/B)[1 + (y/B)\sqrt{1 + z^2}]} \qquad (10.10)$$

For $0 \le \alpha < 90°$, $0 \le z < \infty$ and $2z < 4\sqrt{1 + z^2}$; therefore, for all values of y the denominator of the fraction is larger than the numerator, $dR_h/dy < 1$, and also dR_h/dy diminishes with increasing y. Thus the form of the variation of R_h with y is as shown on Fig. 10.7; it superficially resembles the curve of Fig. 10.5a but has no vertical asymptote since dR_h/dy does not approach zero as y approaches infinity. These are also general properties of most (nontrapezoidal) channel sections with divergent side walls.

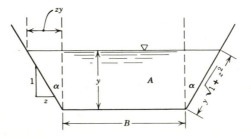

Fig. **10.6**

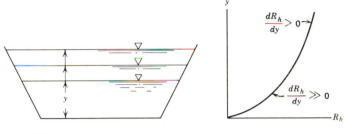

Fig. **10.7**

For a trapezoidal channel with convergent side walls, Fig. 10.8 and equation 10.10 may be used with $-90° < \alpha < 0$, that is, $z < 0$. In this range the possibility of dR_h/dy (equation 10.10) becoming equal to or less than zero is evident. The variation of R_h with y for such a channel is shown in Fig. 10.8. The foregoing form of relation between R_h and y will be found valid for any section with convergent side walls; the circle (Fig. 10.9) is a simple example of this, and it is of wide use in engineering practice for underground open channels such as sewer lines, storm drains, and culverts. Here, for the full circle or semicircle, $R_h = d/4$ and, for $y = 0$, $R_h = 0$; thus a continuous variation of R_h with y may be expected to feature a maximum value as shown. Of more engineering significance in such problems is the variation of $AR_h^{2/3}$ with y, since this combination of A and R_h appears in the Chezy-Manning equation 10.5. With $AR_h^{2/3}$ plotted against y (see Figs. 10.8 and 10.9) another maximum point is noted; from this it may be concluded that for given S_o and n there is a point at which the (uniform) flowrate is also maximum. Engineers usually disregard this in channel design, but it sometimes helps to explain phenomena which occur when channels with convergent walls flow nearly full.

Another important engineering problem of section geometry is the reduction of boundary resistance by minimization of the wetted perimeter for a given area of flow cross section. With A fixed, $R_h = A/P$, and, with P to be minimized, it is apparent that this may be considered a problem of maximization of the hydraulic radius. The desirability of this is evident from the above, but it is to be noted that reduction of

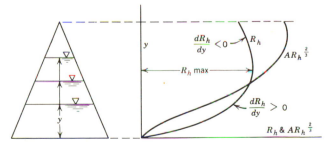

Fig. **10.8**

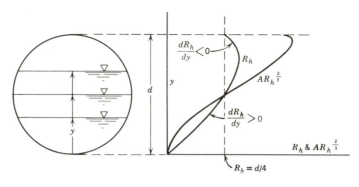

Fig. **10.9**

wetted perimeter also tends to reduce the cost of lining material, grading, and construction. Hence a channel section of maximum hydraulic radius not only results in optimum hydraulic design but also tends toward a section of minimum cost. For this reason a section of maximum hydraulic radius is known as the *most efficient*, or *best, hydraulic cross section*. Since there is little hope of a generalized solution for all possible section shapes, the trapezoidal one (Fig. 10.6) is again studied because of its wide use and good approximation to other sections. As before, $A = By + zy^2$ and $P = B + 2y\sqrt{1 + z^2}$, but here[11] A and z are constants and the relation between B and y for maximum R_h is desired. Eliminating B, R_h may be written as a function of y:

$$R_h = \frac{A}{P} = \frac{A}{A/y - zy + 2y\sqrt{1 + z^2}}$$

Differentiating R_h in respect to y and equating to zero gives

$$A = y^2(2\sqrt{1 + z^2} - z)$$

which, when substituted into the foregoing expression for R_h, yields

$$R_{h_{max}} = y/2 \qquad (10.11)$$

Thus, for best design conditions, a trapezoidal open channel should be proportioned so that its hydraulic radius is close to one-half of the depth of flow; this may also be used as a rough guide for the design of other channel sections that approach trapezoidal shape.

Since a rectangle is a special form of trapezoid, the foregoing results may be applied directly to the rectangular channel. Here again $R_{h_{max}} = y/2$ and, as $z = 0$, $A = 2y^2$. Since A also equals By, it is evident that $B = 2y$ and that the best proportions for a rectangular channel exist when the depth of flow is one-half the width of the channel.

---

[11] The slope of the side walls z is limited in an earth canal by the angle of repose of the soil. If the canal is appropriately lined, z may have any value.

_____ **Illustrative Problem** _____

What are the best dimensions for a rectangular channel that is to carry a uniform flow of 10 m³/s if the channel is lined with gunite concrete and is laid on a slope of 0.000 1?

Relevant Equations and Given Data

$$Q = \left(\frac{u}{n}\right)AR_h^{2/3}S_o^{1/2} \tag{10.5}$$

$$R_{h_{max}} = y/2 \tag{10.11}$$

$Q = 10$ m³/s $S_o = 0.000\ 1$ $u = 1$ gunite concrete surface

Solution. From Table 6,

$$n = 0.019$$

$$A = 2y_o^2 \qquad R_h = y_o/2 \tag{10.11}$$

$$10 = \left(\frac{1}{0.019}\right)2y_o^2\left(\frac{y_o}{2}\right)^{2/3}(0.000\ 1)^{1/2} \tag{10.5}$$

whence

$$y_o^{8/3} = 15.1 \qquad y_o = 2.77 \text{ m} \qquad B = 2y_o = 5.54 \text{ m} \blacktriangleleft$$

A more general method of solution (necessary for nonrectangular sections) is:

$$10 = \left(\frac{1}{0.019}\right)By_o\left(\frac{By_o}{B + 2y_o}\right)^{2/3}(0.000\ 1)^{1/2} \tag{10.5}$$

$$\frac{y_o}{2} = \frac{By_o}{B + 2y_o} \tag{10.11}$$

and, solving these equations simultaneously, $y_o = 2.77$ m, $B = 5.54$ m. $\blacktriangleleft$

_____

For an extended or distorted channel cross section such as that of Fig. 10.10, which represents a simplification of a river section in flood, routine calculation of

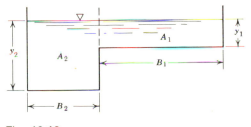

Fig. **10.10**

hydraulic radius from A/P will lead to large errors; such a calculation would imply that the effect of boundary resistance is uniformly distributed through the flow cross section, which is clearly not the case. Furthermore, accurate estimation of the effective value of n is virtually impossible since n for the narrow deep portion of the section may be very different from that of the wide shallow one. A logical (but not necessarily precise) means of treating such problems is to consider them composed of parallel channels separated by the vertical dashed line. The total flowrate is then expressed as $Q_1 + Q_2$, and the separate Q's by the Chezy-Manning equation,

$$Q = \left(\frac{u}{n_1}\right)A_1 R_{h_1}^{2/3} S_o^{1/2} + \left(\frac{u}{n_2}\right)A_2 R_{h_1}^{2/3} S_o^{1/2}$$

in which $A_1 = B_1 y_{o_1}$, $A_2 = B_2 y_{o_2}$, $P_1 = B_1 + y_{o1}$, $P_2 = B_2 + 2y_{o_2} - y_{o_1}$, and for SI units $u = 1$, while for FSS units $u = 1.49$ as before.

10.6 Specific Energy, Critical Depth, and Critical Slope—Wide Rectangular Channels

Many problems of open-channel flow are solved by a special application of the energy principle using the channel bottom as the datum plane. The concept and use of *specific energy*, the distance between channel bottom and energy line, were introduced by Bakhmeteff in 1912 and have proved fruitful in the explanation and analysis of new and old problems of open-channel flow. Today a knowledge of the fundamentals of specific energy is absolutely necessary in coping with the advanced problems of open flow; these fundamentals and a few of their applications are developed in the following paragraphs.

Considering the relative positions of channel bottom, liquid surface, and energy line in the typical reach of open channel shown in Fig. 10.11, it is not possible to predict the character of the change of specific energy between sections 1 and 2. Although the energy line must fall in the direction of the flow, *the specific energy may increase or decrease*, depending on other factors to be investigated.

With the specific energy defined as the vertical distance between channel bottom and energy line,

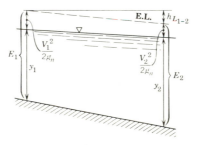

Fig. **10.11**

$$E = y + \frac{V^2}{2g_n} \tag{10.12}$$

or in terms of flowrate, which in steady flow is the same through each cross section,

$$E = y + \frac{1}{2g_n}\left(\frac{Q}{A}\right)^2 \tag{10.13}$$

It is convenient at this point to deal with flow in a wide channel of rectangular cross section to simplify the equations for better illustration of fundamentals; the principles may be applied, of course, to channels of other shapes (Section 10.7), but the resulting equations are considerably more unwieldy. The assumption of a wide rectangular channel allows the use of the two-dimensional approximation and the use of the two-dimensional flowrate, q. Substituting q/y for V in equation 10.12 gives

$$E = y + \frac{1}{2g_n}\left(\frac{q}{y}\right)^2 \quad \text{or} \quad q = \sqrt{2g_n(y^2 E - y^3)} \tag{10.14}$$

This equation gives clear and simple relationships between specific energy, flowrate, and depth. Although a three-dimensional plot of these variables may be visualized, a better understanding of the equation may more easily be acquired by (1) holding q constant and studying the relation between E and y and (2) holding E constant and examining the relation between q and y. Plotting these relations yields, respectively, the *specific energy diagram* and *q-curve* of Fig. 10.12. Since these curves are merely different plots of the same equation, it may be expected (proof will be offered later) that the points of minimum E and maximum q are entirely equivalent. The depth associated with these points is known as the critical depth, y_c, and it is a boundary line between zones of open-channel flow which are very different in physical character. Flows at depths greater than critical are known as *subcritical*[12] flows; flows at depths less than critical are known as *supercritical* flows.

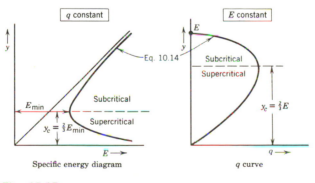

Fig. **10.12**

From the specific energy diagram the question raised by Fig. 10.11 can now be answered; an increase in depth is seen to cause an increase in specific energy in subcritical flow but a decrease in specific energy in supercritical flow; in other words, specific energy may be gained or lost in Fig. 10.11, depending on whether the depths of flow are greater or less than the critical depth. Furthermore, the form of the diagram shows that for a given specific energy two depths (and thus two flow situations) are, in general, possible, one of subcritical flow and one of supercritical flow; these two depths are known as the *alternate depths*.

The importance of the calculation of critical depth as a means of identifying the type of flow is apparent from the previous discussion, and equations for this may be obtained from the specific energy diagram where $dE/dy = 0$, and from the q-curve where $dq/dy = 0$. Performing the differentiations of equations 10.14 yields (from the first equation)

$$q = \sqrt{g_n y_c^3} \quad \text{or} \quad y_c = \sqrt[3]{\frac{q^2}{g_n}} \tag{10.15}$$

and (from the second equation)

$$y_c = \frac{2E_{min}}{3} \quad \text{or} \quad E_{min} = \frac{3y_c}{2} \tag{10.16}$$

Substituting the latter relation into equation 10.14 for q gives $q = \sqrt{g_n y_c^3}$ as before, thus proving that the points of minimum E and maximum q have the same properties. Of great significance is the fact that (equation 10.15) *critical depth is dependent on flowrate only*; the many other variables of open flow are not relevant to the computation of this important parameter. Equation 10.15 also suggests the use of critical flow as a means of flowrate measurement; if critical flow can be created or identified in a channel, the depth may be measured and flowrate determined.

Critical velocity, or velocity at the critical depth, may be obtained from $q = V_c y_c$ and $q = \sqrt{g_n y_c^3}$, from which we obtain

$$V_c = \sqrt{g_n y_c}$$

This equation is similar to another equation of open flow (see Appendix 3),

$$a = \sqrt{g_n y}$$

which gives the velocity of propagation of a small wave on the surface of a liquid of depth y. The similarity of these equations offers a means of identifying subcritical and supercritical flows in the field.

In subcritical and supercritical flows, velocities are, respectively, less than and more than $\sqrt{g_n y}$; thus in subcritical flow small surface waves will progress upstream, but in supercritical flow such waves will be swept downstream (Fig. 10.13) to some angle β with the direction of flow. For such waves to remain stationary their component of propagation velocity (upstream) must be exactly equal and opposite to the flow velocity, V, whence $\sin \beta = \sqrt{g_n y}/V$. From this relationship the engineer in the field

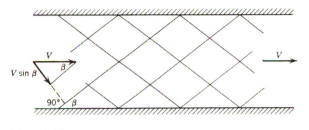

Fig. **10.13**

may estimate[13] velocity (and flowrate) from simple measurements of wave angle and water depth (as well as channel width).

Since wave phenomena are characterized by the interaction of inertia and gravity effects, it is to be expected that critical depth considerations may be written in terms of Froude number (Section 8.1). Defining Froude number by $\mathbf{F} = V/\sqrt{g_n y}$ and applying the foregoing principles: for subcritical flow $\mathbf{F} < 1$, for critical flow $\mathbf{F} = 1$, for supercritical flow $\mathbf{F} > 1$, and $\sin \beta = 1/\mathbf{F}$. The similarity between wave phenomena in open flow and compressible flow (Sections 6.5–6.8) is to be noted here. This analogy has been used by aerodynamicists and nozzle designers in visualizing (hydraulically) the wave patterns in the flow of compressible fluids.

Uniform critical flow (i.e., uniform flow at critical depth) will occur in long open channels if they are laid on the *critical slope*, S_c. For a rectangular channel *of great width* a simple expression for S_c is obtained by equating the flowrates of equations 10.14 and 10.3:

$$q = \sqrt{g_n y_c^3} = C y_c \sqrt{y_c S_c}$$

from which we obtain

$$S_c = \frac{g_n}{C^2}$$

However, from equation 10.4, $C = u y^{1/6}/n$ so that S_c may also be expressed

$$S_c = \frac{g_n n^2}{u^2 y^{1/3}} \tag{10.17}$$

showing critical slope to be a function of depth.[14] For a wide rectangular channel the form of this function is as shown on Fig. 10.14, featuring a relatively small change of S_c with y over a large range of depth. Although the form of this function is different

---

[13] Great accuracy cannot be expected since surface velocity is not equal to mean velocity, nor are these velocities uniformly distributed through the flow cross section. Also it is reemphasized that the application is to small waves only; larger waves produced by gross flow disturbances will stand at larger angles than arcsin $\sqrt{g_n y}/V$. See Section 6.6.

[14] Although this equation shows a general relation between y and S_c, the latter may be computed from it only for $y = y_c$, which (through equation 10.15) implies a certain flowrate.

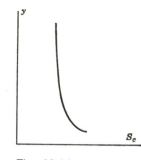

Fig. **10.14**

for other cross sections, this feature is retained for practically all sections except those which are narrow and deep; it justifies the useful rough approximation that over a relatively large range of depths (and flowrates) S_c may be assumed constant. Uniform critical flow is, of course, a rare borderline situation between subcritical and super-critical flows. On the specific energy diagram it is noted that, in the vicinity of the crit-ical depth, the depth may change considerably with little variation of specific energy; physically this means that, since many depths may occur for practically the same specific energy, flow near the critical depth will possess a certain instability (which manifests itself by undulations in the liquid surface). Uniform flow near the critical depth has been observed in both field and laboratory to possess these characteristics, and because of this the designer seeks to prevent flow situations close to the critical and uniform.

Slopes greater than and less than the critical slope, S_c, are known, respectively, as *steep* and *mild* slopes. Evidently, channels of steep slope (if long enough) will produce supercritical uniform flows, and channels of mild slope (if long enough) will produce subcritical uniform flows. Whether or not uniformity is actually realized, the depth of uniform flow may always be computed from flowrate, slope, channel shape, and roughness through the Chezy-Manning equation. Depths computed in this manner are called *normal* or *neutral* depths; although such depths may not occur in short channels, they are, nevertheless, useful parameters of the flow and essential to an understanding of problems of varied flow.

———————————— **Illustrative Problem** ————————————

For a flowrate of 500 cfs in a rectangular channel 40 ft wide, the water depth is 4 ft. Is this flow subcritical or supercritical? If $n = 0.017$, what is the critical slope of this channel for this flowrate? What channel slope should be provided to produce uniform flow at 4 ft depth?

Relevant Equations and Given Data

$$q = Q/B \qquad\qquad (3.14)$$

$$Q = (u/n)AR_h^{2/3}S_o^{1/2} \qquad\qquad (10.5)$$

$$y_c = \sqrt[3]{\frac{q^2}{g_n}} \tag{10.15}$$

$$S_c = \frac{g_n n^2}{u^2 y^{1/3}} \tag{10.17}$$

$Q = 500 \text{ ft}^3/\text{s}$ $B = 40 \text{ ft}$ $y_o = 4 \text{ ft}$ $n = 0.017$ $u = 1.49$

Solution.

$$q = \frac{500}{40} = 12.5 \text{ ft}^3/\text{s/ft} \tag{3.14}$$

$$y_c = \sqrt[3]{\frac{(12.5)^2}{32.2}} = 1.69 \text{ ft} \tag{10.15}$$

Since $4 > 1.69$, this flow is subcritical: ◄

$$S_c = \frac{32.2(0.017)^2}{2.21(1.69)^{1/3}} \cong 0.003\,5 \blacktriangleleft \tag{10.17}$$

The foregoing calculation is approximate since it implies that $y = R_h = 1.69$ ft; actually $y = 1.69$ ft and $R_h = 1.56$ ft. For uniform flow at 4 ft depth, $R_h = 160/48 = 3.33$ ft and $u = 1.49$,

$$500 = \left(\frac{1.49}{0.017}\right) 160(3.33)^{2/3} S_o^{1/2} \qquad S_o = 0.000\,255 \blacktriangleleft \tag{10.5}$$

S_o is a *mild* slope since it is less than S_c.

10.7 Specific Energy, Critical Depth, and Critical Slope—Nonrectangular Channels

The application of the principles of Section 10.6 to channels of nonrectangular cross section (Fig. 10.15) leads to more generalized and complicated mathematical expressions, owing to the more difficult geometrical aspects of such problems. The specific energy equation is written as before,

$$E = y + \frac{1}{2g_n}\left(\frac{Q}{A}\right)^2 \tag{10.13}$$

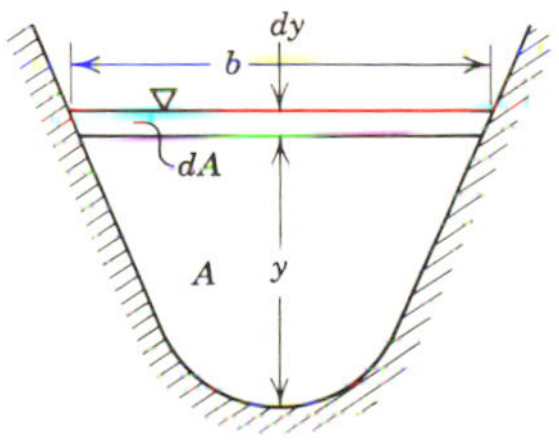

Fig. **10.15**

Here A is some function of y, depending on the form of the channel cross section, and the equation becomes

$$E = y + \frac{1}{2g_n}\left(\frac{Q}{f(y)}\right)^2 \tag{10.18}$$

which is analogous to equation 10.14 and leads to a specific energy diagram and a Q-curve (Fig. 10.16) having the same superficial appearance as those of Fig. 10.12 but for which there are other critical depth relationships. These may be worked out for the generalized channel cross section by differentiating equation 10.13 in respect to y and setting the result equal to zero:

$$\frac{dE}{dy} = 1 + \frac{Q^2}{2g_n}\left(-\frac{2}{A^3}\frac{dA}{dy}\right) = 0$$

From Fig. 10.15, $dA = b\,dy$ in which b is the channel width *at the liquid surface*; dA/dy is thus equal to b, and substitution gives

$$\frac{Q^2}{g_n} = \frac{A^3}{b} \qquad \text{or} \qquad \frac{Q^2 b}{g_n A^3} = 1 \tag{10.19}$$

as the equation which allows calculation of critical depth in nonrectangular channels. Substituting V for Q/A in equation 10.19 and defining a Froude number by

$$\mathbf{F} = \sqrt{\frac{Q^2 b}{g_n A^3}} = \frac{V}{\sqrt{g_n(A/b)}} \tag{10.20}$$

it is evident that (as for the wide rectangular channel): for subcritical flow $\mathbf{F} < 1$, for critical flow $\mathbf{F} = 1$, and for supercritical flow $\mathbf{F} > 1$.

Critical slope for a nonrectangular channel may be derived by following the same procedure as for the wide rectangular one:

$$Q = \sqrt{\frac{g_n A^3}{b}} = \left(\frac{u}{n}\right)AR_h^{2/3}S_c^{1/2}$$

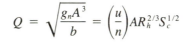

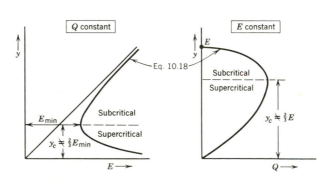

Fig. **10.16**

from which we obtain

$$S_c = \frac{g_n n^2}{u^2}\left(\frac{A}{bR_h^{4/3}}\right) \qquad (10.21)$$

In this expression A, b, and R_h are all functions of depth and dependent on the section shape. Critical slope is thus a function of section shape also, and the form of the variation of S_c with y will usually be similar to that of Fig. 10.14.

─────────────── **Illustrative Problem** ───────────────

A flow of 28 m³/s occurs in an earth-lined trapezoidal canal having base width 3 m, side slopes 1 (vert.) on 2 (horiz.), and $n = 0.022$. Calculate the critical depth and critical slope.

Relevant Equations and Given Data

$$\frac{Q^2}{g_n} = \frac{A^3}{b} \qquad (10.19)$$

$$S_c = \frac{g_n n^2}{u^2}\left(\frac{A}{bR_h^{4/3}}\right) \qquad (10.21)$$

$Q = 28$ m³/s $B = 3$ m $n = 0.022$ slope: 1 on 2 $u = 1$

Solution.

$z = 2$ $A = 3y + 2y^2$ $b = 3 + 4y$ $P = 3 + 2\sqrt{5}y$

$$R_h = A/P = (3y + 2y^2)/(3 + 2\sqrt{5}y)$$

$$\frac{(28)^2}{9.81} = \frac{(3y_o + 2y_c^2)^3}{3 + 4y_c} \qquad (10.19)$$

When this is solved (by trial), $y_c = 1.5$ m. ◀ Thus

$A_c = 9$ m² $b_c = 9$ m $R_{h_c} = 9/9.7 = 0.93$ m

$$S_c = 9.81(0.022)^2\left(\frac{9}{9(0.93)^{4/3}}\right) = 0.005\ 23 ◀ \qquad (10.21)$$

10.8   Occurrence of Critical Depth

The analysis of open-channel flow problems usually begins with the location of points in the channel at which the relationship between water depth and flowrate is known or controllable. These points are known as *controls* since their existence governs, or controls, the liquid depths in the reach of a channel either upstream or downstream from such points. In a broad sense control points usually feature a change from subcritical to supercritical flow. In this context, controls occur at physical barriers, for example, at sluice gates, dams, weirs, drop structures, etc., where critical flow is forced to occur, and at changes in channel slope, for example, from mild to steep, which causes the flow to pass through the critical depth. It is also possible that, where the channel slope changes (e.g., from steep to less steep or to mild), the flow will move from a normal depth upstream to another normal depth or into a hydraulic jump downstream. In this case the slope change point is a control for the downstream reach and the depth at the control is the normal depth for the flow upstream of the slope change. The prediction of control point location is, thus, a powerful tool for "blocking in" the variety of flow phenomena to be expected before making detailed calculations.

The most obvious place where critical depth can be expected is in the situation pictured in Fig. 10.17, where a long channel of mild slope ($S_o < S_c$) is connected to a long channel of steep slope ($S_o > S_c$). Far up the former channel uniform subcritical flow at normal depth, y_{o1}, will occur, and far down the latter a uniform supercritical flow at a smaller normal depth, y_{o2}, can be expected. These two uniform flows will be connected by a reach of varied[15] flow in which at some point the depth must pass through the critical. Experience shows that this point is close to (actually slightly upstream from) the break in slope and may be assumed there for most purposes. It is to be noted, however, that in the immediate vicinity of the change in bottom slope the streamlines are both curved and convergent, and the problem cannot be treated as one-dimensional flow with hydrostatic pressure distribution; the fact that the critical depth is not found precisely at the point of slope change prevents the use of such points in the field for obtaining the exact flowrate by depth measurement and application of equations 10.15 or 10.19; however, the possibility of this method as an approximate means of flowrate measurement should not be overlooked.

When a long channel of steep slope discharges into one of mild slope (Fig. 10.18), normal depths will occur upstream and downstream from the point of slope change, but usually the critical depth will not be found near this point. Under these conditions a *hydraulic jump* (Sections 6.5 and 10.10) will form whose location will be dictated (through varied[16] flow calculations) by the details of slopes, roughness, channel shapes,

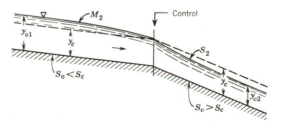

Fig. **10.17**

[15] See Section 10.9 for explanation of the symbols M_2 and S_2.
[16] See Section 10.9 for explanation of symbol M_3.

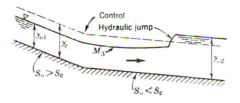

Fig. **10.18**

and so forth, but the critical depth will be found within the hydraulic jump. (See Section 10.10).

The occurrence of critical depth on overflow structures can be proved by examining the flow over the top of a high frictionless broad-crested weir (Fig. 10.19) equipped with a movable sluice gate at the downstream end and discharging from a large reservoir of constant surface elevation. With gate closed (position A), the depth of liquid on the crest will be y_A, and the flowrate will obviously be zero, giving point A on the q-curve. With the gate raised to position B, a flowrate q_B will occur, with a decrease in depth from y_A to y_B. This process will continue until the gate is lifted clear of the flow (C) and can therefore no longer affect it. With the energy line fixed in position at the reservoir surface level and, therefore, giving constant specific energy, it follows that points A, B, and C have outlined the upper portion of the q-curve, that the flow occuring without gates is maximum, and that the depth on the crest is, therefore, the critical depth. For flow over weirs, a relation between head and flowrate is desired; this may be obtained by substituting (for a very high weir) $y_c = 2H/3$ in equation 10.15, which yields

$$q = \sqrt{g_n y_c^3} = \sqrt{g_n \left(\frac{2H}{3}\right)^3} = 0.577 \times \tfrac{2}{3}\sqrt{2g_n}H^{3/2} \qquad (10.22)$$

The last form of this equation is readily compared with the standard weir equation (Section 11.17); here 0.577 is the *weir coefficient*. Because of neglect of friction in the analysis this coefficient of 0.577 is higher than those obtained in experiments; tests on

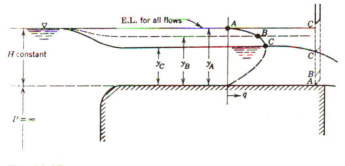

Fig. **10.19**

high, broad-crested weirs give coefficients between 0.50 and 0.57 (depending on the details of weir shape).

The reasoning of the foregoing paragraph may be extended to a *free outfall* (Fig. 10.20)[17] from a long channel of mild slope, to conclude that the depth must pass through the critical in the vicinity of the brink. Rouse[18] has found that for such rectangular channels the critical depth occurs a short distance (3 to $4y_c$) upstream from the brink and that the brink depth (y_b) is 71.5% of the critical depth.[19] Using this figure and equation 10.15, Rouse proposes the free outfall as a simple device for metering the flowrate, which requires only measurement of the brink depth. Free outfalls also provide an opportunity for recognition of the limitations of specific energy theory. Upstream from the point of critical depth, the streamlines are essentially straight and parallel and the pressure distribution hydrostatic; between this point and the brink neither of these conditions obtains, so the one-dimensional theory is invalid in this region. For example, it *cannot* be concluded that the flow at the brink is supercritical merely because the brink depth is less than the critical depth.

Critical depth may be obtained at points in an open channel where the channel bottom is raised by the construction of a low hump or the channel is constricted by moving in the sidewalls. Such contractions (usually containing a rise in the channel bottom) are known generally as Venturi flumes, and specific designs[20] of these flumes are widely used in the measurement of irrigation water. Preliminary analysis of such contractions (which feature accelerated flow) may be made assuming one-dimensional flow and neglecting head losses, but final designs require more refined information either from other designs or from laboratory experiments. The advantages of such flumes are their ability to pass sediment-laden water without deposition and the small net change of water level required between entrance and exit channels.

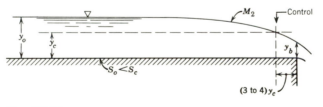

Fig. **10.20**

---

[17] See Section 10.9 for explanation of symbol M_2.

[18] H. Rouse, "Discharge Characteristics of the Free Overfall," *Civil Engr.* vol. 6, no. 4, p. 257, 1936. See also T. Strelkoff and M. S. Moayeri, "Pattern of Potential Flow in a Free Overfall," *Jour. Hydr. Div., A.S.C.E.,* vol. 96, no. HY4, pp. 879–901, April, 1970.

[19] Experiments on free outfalls from *circular conduits* of mild slope show brink depth to be about 75% of critical depth when the brink depth is less than 60% of the diameter of the conduit.

[20] See, for example, P. Ackers, W. R. White, J. A. Perkins, and A. J. M. Harrison, *Weirs and Flumes for Flow Measurement,* John Wiley & Sons, 1978.

——————————————— **Illustrative Problem** ———————————————

Uniform flow at a depth of 5 ft occurs in a long rectangular channel of 10 ft width, having a Manning n of 0.015, and laid on a slope of 0.001. Calculate (a) the minimum height of hump which can be built in the floor of this channel to produce critical depth, and (b) the maximum width of contraction which can produce critical depth.

Relevant Equations and Given Data

$$Q = (u/n)AR_h^{2/3} S_o^{1/2} \qquad (10.5)$$

$$E = y + \frac{1}{2g_n}\left(\frac{Q}{A}\right)^2 \qquad (10.13)$$

$$y_c = \sqrt[3]{\frac{q^2}{g_n}} \qquad (10.15)$$

$$E_{min} = \frac{3y_c}{2} \qquad (10.16)$$

$$y_o = 5 \text{ ft} \qquad B = 10 \text{ ft} \qquad n = 0.015 \qquad S_o = 0.001 \qquad u = 1.49$$

Solution.

$$Q = \left(\frac{1.49}{0.015}\right)50\left(\frac{50}{20}\right)^{2/3}(0.001)^{1/2} = 289 \text{ ft}^3/\text{s} \qquad (10.5)$$

$$E = 5.00 + \left(\frac{1}{2g_n}\right)\left(\frac{289}{50}\right)^2 = 5.52 \text{ ft} \qquad (10.13)$$

(a)

$$q = 289/10 = 28.9 \text{ ft}^3/\text{s/ft}$$

so

$$y_c = \sqrt[3]{\frac{(28.9)^2}{32.2}} = 2.96 \text{ ft} \qquad (10.15)$$

$$E_{min} = 3 \times \frac{2.96}{2} = 4.44 \text{ ft} \qquad (10.16)$$

A hump height smaller than x will lower the water surface over the hump but cannot produce critical depth; a hump height larger than this will produce critical depth, the hump then being a broad-crested weir. The latter condition would, however, raise the energy line in the vicinity of the hump and (this being a subcritical flow) increase the depth upstream from this point. Therefore the minimum height of hump to produce critical depth is that which will do so without raising the energy line. From the sketch, this is seen to be $5.52 - 4.44$, or 1.08 ft. ◀

(b) The width b at the contraction being unknown but critical depth being specified,

$$y_c = \sqrt[3]{\frac{(289/b)^2}{32.2}} \qquad (10.15)$$

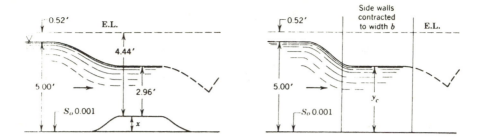

From part (*a*) it may be seen that contraction widths smaller than that required to just produce critical depth will do so by raising the energy line and the water depth upstream from the contraction. Accordingly, the condition sought is that which will produce critical depth in the contraction without raising the energy line. The contraction being of rectangular cross section,

$$y_c = 2 \times \frac{5.52}{3} = 3.68 \text{ ft} \tag{10.16}$$

Equating this to the preceding expression and solving for *b*, *b* = 7.2 ft. ◄

10.9 Varied Flow

For design of open channels and analyses of their performance, the engineer must be able to predict forms and positions of water-surface profiles of varied (nonuniform) flow and to acquire some facility in their calculation. The first objective may be attained by development and study of the differential equation of varied flow, and the second objective either by performing the integration or by using step-by-step calculations.

The differential equation of varied flow may be derived by considering the differential element of Fig. 10.21 in which the bottom slope S_o is small and over the distance

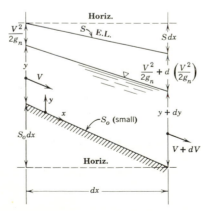

Fig. **10.21**

dx along the channel bottom, the depth is assumed to increase by dy, and the velocity head by $d(V^2/2g_n)$. Taking the slope of the energy line to be S and the slope of channel bottom S_o, the drops of these lines (in the distance dx) will be $S\,dx$ and $S_o\,dx$, respectively. From the sketch may be written the equation

$$S_o\,dx + y + \frac{V^2}{2g_n} = y + dy + \frac{V^2}{2g_n} + d\!\left(\frac{V^2}{2g_n}\right) + S\,dx \qquad (10.23)$$

which, by appropriate cancellations and division by dx, may be reduced to

$$\frac{dy}{dx} + \frac{d}{dx}\!\left(\frac{V^2}{2g_n}\right) = S_o - S$$

Multiplying the second term by dy/dy and solving the equation for dy/dx,

$$\frac{dy}{dx} = \frac{S_o - S}{1 + \dfrac{d}{dy}\!\left(\dfrac{V^2}{2g_n}\right)}$$

The derivative in the denominator of the fraction must be recognized before the equation can be fully explored; this has been already calculated in the derivation of equation 10.19 and shown to be $-Q^2 b/g_n A^3$ in which b is the top width of the flow cross section A. Since $Q^2 b/g_n A^3 = \mathbf{F}^2$ (see Section 10.7), the differential equation of varied flow may be written alternatively as

$$\frac{dy}{dx} = \frac{S_o - S}{1 - Q^2 b/g_n A^3} = \frac{S_o - S}{1 - \mathbf{F}^2} \qquad (10.24)$$

from which the forms of all possible water-surface profiles of varied flow may be deduced.

From the first step (equation 10.23) in the development of this equation two limitations are evident: (1) Since hydrostatic pressure distribution has been assumed in the use of specific energy, application is limited to flows with streamlines essentially straight and parallel, and of small slope S_o. (2) Since the depth y is measured vertically from the channel bottom, the slope of the water surface dy/dx is *relative to this channel bottom*, and thus does not have the same meaning as the conventional dy/dx of analytic geometry; its implications are shown on Fig. 10.22, which is basic to the prediction of surface profiles from analysis of equation 10.24.

Surface profiles are categorized in terms of bottom slope S_o as follows: (a) $S_o > S_c$ (S for *steep*), (b) $S_o = S_c$ (C for *critical*), (c) $S_o < S_c$ (M for *mild*), (d) $S_o = 0$ (H for *horizontal*), and (e) $S_o < 0$ (A for *adverse*). The form of the profile also depends on the water depth relative to both normal and critical depths; the relative positions of the latter are shown in column 2 of Table 7 and show that twelve possible profiles are to be expected. Here the only point not previously discussed involves the values of y_o for $S_o = 0$ and $S_o < 0$; these are readily confirmed by solving the Chezy-Manning equation for y_o.

Analysis of equation 10.24 shows $(S_o - S)$ to depend only on the relative sizes of depth and normal depth—and $(1 - \mathbf{F}^2)$ to depend solely on the relative magnitudes

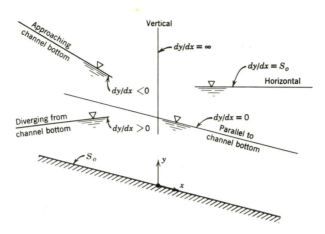

Fig. **10.22**

of depth and critical depth; these statements may be validated in the following manner. From the Chezy-Manning equation it is easily seen that (for a given Q and n) S will decrease with increasing $AR_n^{2/3}$; since the latter term has been shown (Section 10.5) to be a direct (not linear) function of depth for most open channels, S will also decrease with increasing depth. This is to be expected since increased depth implies larger flow cross sections and smaller velocities, with consequent reduction of energy dissipation. Since, when $y = y_o$, $S = S_o$ (uniform flow), the above variations may be applied to conclude that: for $y > y_o$, $S < S_o$, and $(S_o - S) > 0$; for $y < y_o$, $S > S_o$ and $(S_o - S) < 0$. Analysis of the quantity $(1 - \mathbf{F}^2)$ is much easier since it has been shown (Section 10.7) that, for $y > y_c$, $\mathbf{F} < 1$ and, for $y < y_c$, $\mathbf{F} > 1$; hence, for $y > y_c$, $(1 - \mathbf{F}^2) > 0$ and, for $y < y_c$, $(1 - \mathbf{F}^2) < 0$. With this information on $(S_o - S)$ and $(1 - \mathbf{F}^2)$ the signs of dy/dx may be determined as indicated in columns 3, 4, and 5 of Table 7.

   The final forms of the profiles (excepting those for $S_o = S_c$) may be established (from equation 10.24) by observing that (1) for $y \to \infty$: $S \to 0$, $\mathbf{F} \to 0$, and $dy/dx \to S_o$; (2) for $y \to y_o$: $S \to S_o$, and $dy/dx \to 0$; (3) for $y \to y_c$: $\mathbf{F} \to 1$ and $dy/dx \to \infty$. The exception occurs in the case of $S_o = S_c$ since here, as $y \to (y_o = y_c)$, dy/dx cannot approach both zero and infinity. Of the foregoing conclusions: (1) expresses the well-known fact that for large depths (and accompanying small velocities) liquid surfaces approach the horizontal; (2) shows that all surface profiles approach uniform flow asymptotically; (3) shows (if taken literally) that the free surface is vertical when passing through the critical depth! Although the last conclusion is invalid because the original development assumed essentially straight and parallel streamlines, it is nevertheless useful as a guide in the final formulation of the surface profiles, which, to satisfy the conditions above, must be as shown in Table 7. Examples of physical configurations leading to some of these surface profiles are shown in Figs. 10.17, 10.18, 10.20, 10.27, and 10.28.

   Investigation of the interesting (but relatively unimportant) borderline situation

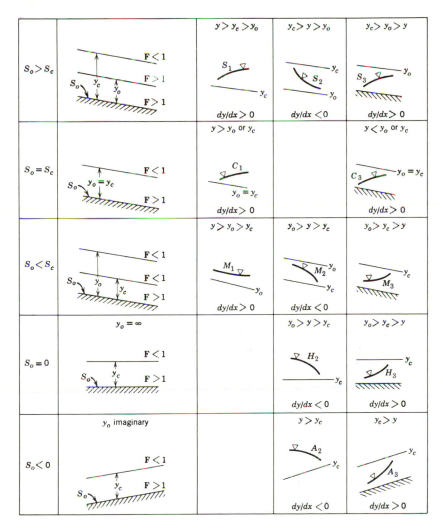

Table 7 Surface Profiles of Varied Flow

where $S_c = S_o$ may be undertaken using a wide rectangular section as an approximation to most other sections.[21] Here equation 10.24 may be rearranged to

$$\frac{dy}{dx} = S_c\left(\frac{1 - S/S_c}{1 - \mathbf{F}^2}\right) \qquad (10.25)$$

For a section of unit width $A = R_h = y$, and from the Chezy-Manning equation,

---

[21] For narrow and deep sections this is obviously a poor approximation.

$$q = \left(\frac{u}{n}\right)y^{5/3}S^{1/2} = \left(\frac{u}{n}\right)y_c^{5/3}S_c^{1/2}$$

whence

$$\frac{S}{S_c} = \left(\frac{y_c}{y}\right)^{10/3} \tag{10.26}$$

However, with $\mathbf{F} = V/\sqrt{g_n y}$ and $q^2 = g_n y_c^3$ (equation 10.15),

$$\mathbf{F}^2 = \frac{V^2}{g_n y} = \frac{q^2}{g_n y^3} = \frac{g_n y_c^3}{g_n y^3} = \left(\frac{y_c}{y}\right)^3 \tag{10.27}$$

By substitution of relations 10.26 and 10.27 into equation 10.25, it becomes

$$\frac{dy}{dx} = S_c\left(\frac{1 - (y_c/y)^{10/3}}{1 - (y_c/y)^3}\right) \tag{10.28}$$

which shows that: (1) for $y > y_c$ or $y < y_c$, $dy/dx > S_c$, and (2) for $y \rightarrow \infty$, $dy/dx \rightarrow S_c$, which justifies the form of the C_1 and C_3 profiles shown in Table 7.

Lack of space precludes a review of the numerous attempts (over the last century) to integrate the differential equation of varied flow to provide the engineer with y as a function of x so that water-surface profiles may be plotted for known channel cross section, flowrate, slope, and Manning n. However, Prasad[22] has produced an efficient numerical technique that is well-suited for both desk calculators and digital computers. The essence of his method is outlined in the following paragraphs for a prismatic channel. In his paper Prasad generalizes this analysis to natural channels with lateral inflow.

The integration is based on equation 10.24 and the assumption that the head loss for varied flow is equal to the head loss for a uniform flow at the same depth and flowrate. This plausible assumption used in the calculation of S has never been precisely confirmed by experiment, but errors arising from it are small compared to those incurred in the selection of n, and over the years it has proved to be a reliable basis for design calculations. The assumption is doubtless more valid for accelerating (depth decreasing downstream) flow than for decelerating flow, but in a prismatic channel the difference between them would be very small. The result is $S = n^2Q^2P^{4/3}/u^2A^{10/3}$ and

$$\dot{y} = \frac{dy}{dx} = \frac{S_o - \dfrac{n^2Q^2P^{4/3}}{u^2A^{10/3}}}{1 - \dfrac{Q^2b}{g_nA^3}} \tag{10.29}$$

where, as before, $u = 1.0$ for SI units and $u = 1.49$ for FSS units. Equation 10.29 is a first-order, nonlinear, ordinary differential equation for y as a function of x. It may

[22] R. Prasad, "Numerical Method of Computing Flow Profiles," *Jour. Hydr. Div.*, A.S.C.E., vol. 96, no. HY1, pp. 75–86, January, 1970.

be integrated by any number of means.[23] Here an iterative technique known as the trapezoidal method is used; through numerical analysis the accuracy of the solution can be assured.

Consider Fig. 10.23 which is essentially the same as the definition sketch of Fig. 10.21. Because y is a function of x,

$$y_{i+1} = y_i + \frac{dy}{dx} \Delta x = y_i + \dot{y} \, \Delta x \tag{10.30}$$

where the subscripts are a measure of distance along the channel (the actual distance depending on Δx of course). For the numerical integration Δx is presumed to be relatively very small so that to a good approximation, in the interval i to $i + 1$, $\dot{y} = \frac{1}{2}(\dot{y}_i + \dot{y}_{i+1})$. Equation 10.30 becomes

$$y_{i+1} = y_i + \tfrac{1}{2}(\dot{y}_1 + \dot{y}_{i+1}) \, \Delta x \tag{10.31}$$

Equations 10.29 and 10.31 contain two unknowns, y_{i+1} and $\dot{y}_{i+1}$, if the channel characteristics $[A(y), n, P(y), b(y), Q]$ and y_i are presumed known. The solution for y_{i+1} and $\dot{y}_{i+1}$ proceeds as follows:

(a) Use equation 10.29 to calculate $\dot{y}_i$ where y_i is known either as the initial point or from a previous cycle of this calculation.
(b) Set $\dot{y}_{i+1} = \dot{y}_i$ as a first trial.
(c) Use current values of $\dot{y}_i$ and $\dot{y}_{i+1}$ to calculate y_{i+1} from equation 10.31 with chosen Δx.
(d) Find $\dot{y}_{i+1}$ from equation 10.29 using y_{i+1} from step c.
(e) If new $\dot{y}_{i+1}$ is not close enough to that previously obtained, repeat steps c, d, and e using $\dot{y}_{i+1}$ found in step d. Otherwise, advance calculation one step to the interval $(i + 1, i + 2)$ and return to step a.
(f) Terminate calculation when desired reach has been covered.

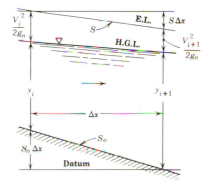

Fig. **10.23**

[23] R. L. Ketter and S. P. Prawel, Jr., *Modern Methods of Engineering Computation*, McGraw-Hill, 1969.

A good check is to recompute the profile with Δx halved; if the two profiles are substantially the same, Δx was small enough. From Table 8, it appears that a computation can, in principle, proceed either upstream or downstream from a point of known depth on a varied flow profile. It has been widely reported[24] that calculations must start at a control and move in the direction in which the control is exercised (e.g., upstream in Fig. 10.20). The works of Prasad[22] and McBean and Perkins[25] show that, from a mathematical perspective, this is *untrue for a sufficiently accurate numerical scheme*. However, it does appear that the accumulation of error is greater in a calculation which moves toward rather than away from a control. As a practical matter, the starting depths are most often known at (or near) controls, so one normally begins the calculation there and steps away from the control.

A flowchart and program for a varied flow calculation are given in Fig. 10.24 and Table 8.[26] The variable names in the program are

EL = Total length of channel under consideration
B = Width of channel bottom
Z = Side slope of channel

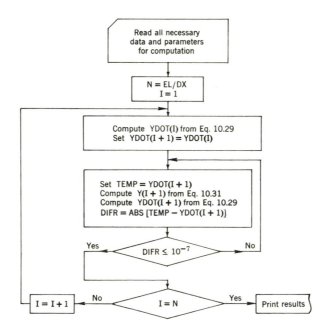

Fig. **10.24** Flowchart for numerical integration.[26]

[24] For example, F. M. Henderson, *Open Channel Flow*, MacMillan Co., 1966, p. 125
[25] E. McBean and F. Perkins, "Numerical errors in water profile computation," *Joue. Hydr. Div., A.S.C.E.,* vol. 101, no. HY11, pp. 1389–1403, November, 1975, and vol. 103, no. HY6, pp. 665–666, June, 1977.
[26] Adapted from R. Prasad, "Numerical Method of Computing Flow Profiles," *Jour. Hydr. Div., A.S.C.E.,* vol. 96, no. HY1, p. 77 and 85, January, 1970.

```
$WATFIV
C     THIS PROGRAM WORKS IN SI UNITS OR FSS UNITS
C     ALFA= 1, SI(U= 1; G= 9.81); ALFA= 2, FSS(U= 1.49; G= 32.17).
C     EL= REACH LENGTH, B= CHANNEL BOTTOM WIDTH, Z= CHANNEL SIDE SLOPE,
C     SO= CHANNEL SLOPE, EN= MANNING'S N, DX= REACH SUBDIVISION UNIT,
C     PLSM= 1 IF COMPUTING DOWNSTREAM AND -1 IF UPSTREAM, Q= DISCHARGE,
C     Y(1)= STARTING DEPTH, AND THE Y(I) ARE DEPTHS OF THE FLOW PROFILE.
C
      DIMENSION Y(1000)
  100 FORMAT(F6.0,2F6.2,2F6.4,4F6.1,5F6.2)
  101 FORMAT(55H OUT*Y IS SMALL OR YDOT IS LARGE(NEAR CRITICAL DEPTH)**)
  102 FORMAT(F10.1,F10.3)
  103 FORMAT(1X,6H     X,6X,7H   Y(X))
C
C.....EQUATION 10.29...........................
C
      YDOTF(B,S,EN,Z,U,G,Q,YY)=(S-((EN*Q)**2*(B+2.*YY*SQRT(1.+Z*Z))**(
     14./3.))/((U*U)*((B+Z*YY)*YY)**(10./3.)))/(1.-((Q*Q*(B+2.*Z*YY))/
     2(G*((B+Z*YY)*YY)**3)))
  104 READ(5,100,END=999)EL,B,Z,SO,EN,ALFA,DX,PLSM,Q,Y(1)
C         SET UNITS
      U=1.
      G=9.81
      IF(ALFA-1.5) 120,110,110
  110 U= 1.49
      G= 32.17
  120 CONTINUE
      WRITE(6,100)EL,B,Z,SO,EN,U,G,DX,PLSM,Q,Y(1)
C         IF Y(1)=0 (NOT GIVEN), Y(1)= Y AT THE PREVIOUS REACH.
      IF(Y(1)) 999,130,140
  130 Y(1)= Y(N+1)
  140 N= EL/DX + 0.1
      N1= N+1
C         DIFFERENTIAL EQUATION IS SOLVED IN THE DO LOOP
      DO 150 I=1,N
      YDOTI=YDOTF(B,SO,EN,Z,U,G,Q,Y(I))*PLSM
      YDOTJ=YDOTI
      DO 160 J=1,15
      TEMP=YDOTJ
      Y(I+1)=Y(I)+(YDOTI+YDOTJ)*DX*0.5
      IF(Y(I+1)) 170,170,180
  180 YDOTJ=YDOTF(B,SO,EN,Z U,G,Q,Y(I+1))*PLSM
      IF(ABS(TEMP-YDOTJ)-.1E-06) 150,150,160
  160 CONTINUE
      GO TO 170
  150 CONTINUE
      GO TO 190
  170 CONTINUE
      XSTEP = -DX
      DO 200 K=1,I
      XSTEP = XSTEP + DX
      STEP = PLSM*XSTEP
      WRITE(6,102) STEP,Y(K)
  200 CONTINUE
      WRITE(6,101)
  999 STOP
  190 CONTINUE
      WRITE(6,103)
      XSTEP = -DX
      DO 210 K=1,N1
      XSTEP = XSTEP + DX
      STEP = PLSM*XSTEP
      WRITE(6,102) STEP,Y(K)
  210 CONTINUE
C         PROCEED TO ANOTHER REACH, A NEW SET OF DATA, OR STOP.
      GO TO 104
      END
$DATA
1800.  6.   0.    .0001 .013  1.     100.   -1.    10.   1.5
$STOP
/*
```

$\quad$ SO $=$ Channel slope

$\quad$ EN $=$ Manning coefficient, n, for channel

ALFA $=$ Selects SI or FSS units

$\quad$ DX $=$ Δx, length of integration step

PLSM $=$ Sign, depends on direction in which computations are desired to proceed, has a value $+1$ when computations proceed downstream and -1 when they proceed upstream

$\quad$ Q $=$ Discharge

Y(1) $=$ Initial depth of water

YDOTI $=$ $\dot{y}_i$ or dy/dx at i

YDOTJ $=$ $\dot{y}_{i+1}$ or dy/dx at $i + 1$

Y(I) $=$ y_i or depth of water at ith point along channel [$i = $ subscript denoting distance along channel, given by $\Delta x(i - 1)$, from starting point]

$\quad$ To handle isolated or occasional problems of varied flow, the engineer may resort to a simple step-by-step method of computation after having sketched in the appropriate surface profiles from Table 7. The definition sketch of Fig. 10.23 is used again; from the sketch

$$S_o \Delta x + y_i + \frac{V_i^2}{2g_n} = y_{i+1} + \frac{V_{i+1}^2}{2g_n} + S \Delta x \tag{10.32}$$

from which, by substitution of E for $y + V^2/2g_n$ and solving for Δx, we obtain

$$\Delta x = \frac{E_i - E_{i+1}}{S - S_o} \tag{10.33}$$

Procedure in using this equation begins in knowing channel shape, Q, S_o, n, and one of the depths (usually at a control section); assumption of the other depth allows calculation of Δx, the horizontal distance between the two cross sections of depths y_i and y_{i+1}—once a suitable method is devised for calculation of S. Making the assumption that the head loss between sections i and $i + 1$ for varied flow is equal to the head loss for a uniform flow of average velocity (V_{avg}) and hydraulic radius ($R_{h\,\text{avg}}$),S may be calculated from the Chezy-Manning equation (10.5),

$$V_{\text{avg}} = \left(\frac{u}{n}\right)(R_{h\,\text{avg}})^{2/3}S^{1/2} \tag{10.34}$$

in which $V_{\text{avg}} = (V_i + V_{i+1})/2$ and $R_{h\,\text{avg}} = (R_{h_i} + R_{h_{i+1}})/2$.

_____ **Illustrative Problem** _____

A flowrate of 10 m³/s occurs in a rectangular channel 6 m wide, lined with concrete (troweled) and laid on a slope of 0.000 1; at a point in this channel the depth is 1.50 m. How far (upstream or downstream) from this point will the depth be 1.65 m?

Relevant Equations and Given Data

$$Q = (u/n)AR_h^{2/3}S_o^{1/2} \tag{10.5}$$

$$y_c = \sqrt[3]{q^2/g_n} \tag{10.15}$$

$$\Delta x = \frac{E_i - E_{i+1}}{S - S_o} \tag{10.33}$$

$$V_{avg} = \left(\frac{u}{n}\right)(R_{h\,avg})^{2/3}S^{1/2} \tag{10.34}$$

$Q = 10\ \text{m}^3/\text{s}$ $B = 6\ \text{m}$ $S_o = 0.000\ 1$ $z = 0$ $u = 1$ $y_{initial} = 1.50\ \text{m}$

$y_{end} = 1.65\ \text{m}$ concrete surface (troweled): $n = 0.013$

Solution.   From solution of equation 10.5 by trial,

$$y_o = 1.94\ \text{m}$$

$$y_c = \sqrt[3]{\frac{(1.67)^2}{9.81}} = 0.66\ \text{m} \tag{10.15}$$

Since $y_o > 1.65 > 1.50 > y_c$, reference to Table 7 indicates an M_2 surface profile with downstream control, so the depth of 1.65 m will be found _upstream_ from that of 1.50 m.

The result according to the computer program of Table 8 is 1 750 m. A table of results for $n = 0.013$, $\Delta x = 100$ m, PLSM $= -1$, and a reach length EL of 1800 m follows.

x	y	x	y
0	1.500	−900	1.589
−100	1.512	−1 000	1.597
−200	1.523	−1 100	1.605
−300	1.533	−1 200	1.612
−400	1.543	−1 300	1.620
−500	1.553	−1 400	1.627
−600	1.563	−1 500	1.634
−700	1.572	−1 600	1.640
−800	1.580	−1 700	1.647
		−1 800	1.653

← 1 750   1.650 ◄

For a calculation by the step method, set up a table for the calculations (and use equations 10.33 and 10.34).

y	A	V (V_{avg})	$V^2/2g_n$	E (ΔE)	R_h $(R_{h\,avg})$	S	$S - S_o$	Δx
1.50	9.0	1.11	0.062 8	1.562 8	1.00			
		(1.095)		(0.046 6)	(1.01)	0.000 200	0.000 100	466
1.55	9.3	1.08	0.059 4	1.609 4	1.02			
		(1.060)		(0.045 7)	(1.03)	0.000 183	0.000 083	551
1.60	9.6	1.04	0.055 1	1.655 1	1.04			
		(1.025)		(0.046 9)	(1.05)	0.000 166	0.000 066	711
1.65	9.9	1.01	0.052 0	1.702 0	1.06			

$$\sum(\Delta x) = 1\ 728$$

The distance requested is 1 728 m according to the step method. ◀ In this case where the surface slope is very small, the rough step method agrees quite well with the more accurate computer based trapezoidal method. Note that the trend of y values for various x shows the curve of y versus x to be convex up, that is, of M_2-type, as expected from Table 7.

10.10 The Hydraulic Jump

When a change from supercritical to subcritical flow occurs in open flow a *hydraulic jump* appears, through which the depth increases abruptly in the direction of flow. In spite of the complex appearance of a hydraulic jump with its turbulence[27] and air entrainment, it may be successfully analyzed by application of the impulse-momentum principle (see Section 6.5) to yield results and relationships which conform closely with experimental observations.

A hydraulic jump in an open channel of small slope is shown in Fig. 10.25.[28] In engineering practice the hydraulic jump frequently appears downstream from overflow structures (spillways) or underflow structures (sluice gates) where velocities are high. It may be used as an effective dissipator of kinetic energy (and thus prevent scour of channel bottom) or as a mixing device in water or sewage treatment designs where chemicals are added to the flow. In design calculations the engineer is concerned mainly with prediction of existence, size, and location of the jump.

In Section 6.5 the basic equation for the jump in a rectangular[29] open channel was derived; it is

---

[27] See H. Rouse, T. T. Siao, and S. Nageratnam, "Turbulence Characteristics of the Hydraulic Jump," *Trans. A.S.C.E.*, vol. 124, 1959.

[28] The jump is not so steep as shown in the figure; the length of the jump is approximately $6y_2$. See U. S. Bureau of Reclamation, "Research studies on stilling basins, energy dissipators, and associated appurtenances," *Hydr. Lab. Rep.*, Hydr 399, 1 June 1955, and V. T. Chow, *Open Channel Hydraulics*, McGraw-Hill, 1959, Chapter 15.

[29] Here the analysis is confined to the rectangular channel for both mathematical simplicity and practical application. The same methods may be applied to channels of nonrectangular cross section.

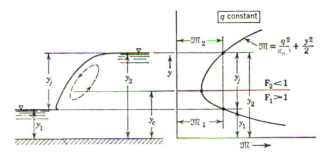

Fig. **10.25**

$$\frac{q^2}{g_n y_1} + \frac{y_1^2}{2} = \frac{q^2}{g_n y_2} + \frac{y_2^2}{2} \tag{10.35}$$

the solution of which may be written

$$\frac{y_2}{y_1} = \frac{1}{2}\left[-1 + \sqrt{\frac{8V_1^2}{g_n y_1}}\right] \tag{10.36}$$

or

$$\frac{y_1}{y_2} = \frac{1}{2}\left[-1 + \sqrt{1 + \frac{8V_2^2}{g_n y_2}}\right] \tag{10.37}$$

in which $V_1^2/g_n y_1$ is recognized as $\mathbf{F}_1^2$ and $\mathbf{F}_2^2 = V_2^2/g_n y_2$; these equations show that $y_2/y_1 > 1$ only when $\mathbf{F}_1 > 1$ and $\mathbf{F}_2 < 1$, thus proving the necessity of supercritical flow for jump formation. Another way of visualizing this is by defining a quantity $\mathfrak{M}$ by

$$\mathfrak{M} = \frac{q^2}{g_n y} + \frac{y^2}{2}$$

and plotting $\mathfrak{M}$ as a function of y (Fig. 10.25) for constant flowrate, whereupon the solution of the equation occurs when $\mathfrak{M}_1 = \mathfrak{M}_2$; the curve obtained features a minimum value of $\mathfrak{M}$ at the critical depth[30] and thus superficially resembles (but must not be confused with) the specific energy diagram. After construction of this curve and with one depth known, the corresponding, or *conjugate*, depth may be found by passing a vertical line through the point of known depth. Since a vertical line is a line of constant $\mathfrak{M}$, the intersection of this line and the other portion of the curve gives a point where $\mathfrak{M}_1$ is equal to $\mathfrak{M}_2$, and allows the conjugate depth and height of jump y_j to be taken directly from the plot. The shape of the $\mathfrak{M}$-curve also shows clearly and convincingly that hydraulic jumps can take place only across the critical depth (i.e., from supercritical to subcritical flow).

Because of eddies (rollers), air entrainment, and flow decelerations in the hydraulic jump, large head losses are to be expected; they may be calculated (as usual) from

---

[30] This may be easily proved by setting $d\mathfrak{M}/dy$ equal to zero.

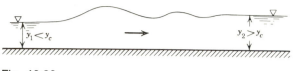

Fig. **10.26**

the fall of the energy line,

$$ h_{L_i} = \left(y_1 + \frac{V_1^2}{2g_n} + z_1 \right) - \left(y_2 + \frac{V_1^2}{2g_n} + z_2 \right) $$

in which $z_1 = z_2$ if the channel bottom is horizontal. For very small jumps, the eddies and air entrainment disappear, the form of the jump changes to that of a smooth standing wave, known as an *undular* jump (Fig. 10.26), with very small head loss. Laboratory tests show that undular jumps are to be expected for $F_1 \gtrsim 1.7$.

A simple problem in hydraulic jump location is shown in Fig. 10.27. Here a rectangular channel of steep slope ($S_o > S_c$) with uniform flow discharges into a channel of mild slope ($S_o < S_c$) of sufficient length to produce uniform flow. For the given flowrate the normal depths y_{o_1} and y_{o_2} may be calculated from the Chezy-Manning equation (10.5), and the critical depth from equation 10.15.

In view of the variety of phenomena produced by combinations of uniform flow, varied flow, and hydraulic jump, the sketch of Fig. 10.27 must first be validated. The depth at the break in slope must be y_{o_1} since the flow in channel 1 is supercritical (with flow velocity larger than wave velocity); accordingly any influence from channel 2 cannot be transmitted upstream into channel 1. The varied flow surface profile over the reach l in channel 2 is identified as M_3 from Table 7 and may be calculated (and plotted) starting with y_{o_1} and using the methods of Section 10.9. The depth y_1 may be obtained from equation 10.37 of the hydraulic jump, with $y_2 = y_{o_2}$. The validity of using $y_2 = y_{o_2}$ may be confirmed by assuming $y_2 > y_{o_2}$ and $y_2 < y_{o_2}$; for either of these conditions no varied flow surface profile would be possible (see Table 7) between y_2 and y_{o_2}. There are two other possible flow situations in this problem: (1) $l = 0$ with the

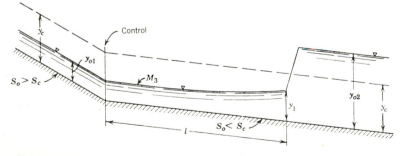

Fig. **10.27**

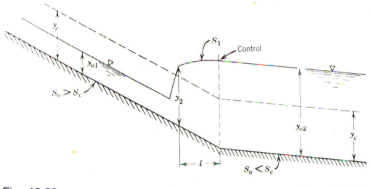

Fig. **10.28**

hydraulic jump at the break in slope; and (2) hydraulic jump in channel 1 (Fig. 10.28). The first of these will obtain when solution of the hydraulic jump equation yields $y_{o_1} = y_1$ and $y_{o_1} = y_2$; the second will result when the hydraulic jump equation is solved (with $y_1 = y_{o_1}$) for y_2, and y_2 found to be less than y_{o_2}. The hydraulic jump will then be found to be followed by an S_1 profile from y_2 to y_{o_2} at the downstream end of channel 1.

────────────── **Illustrative Problem** ──────────────

A hydraulic jump occurs in a V-shaped open channel having sides sloping at 45°. Derive an equation (analogous to equation 10.36) relating the two depths and flowrate.

Relevant Equations and Given Data

$$F = \gamma h_c A \tag{2.12}$$

$$Q = AV \tag{3.11}$$

$$\sum F_x = \oint_{c.s.} V_x(\rho \mathbf{V} \cdot d\mathbf{A}) \tag{6.4}$$

For 45° side slopes: $A = y^2$

Solution.

$$F_1 - F_2 = Q\frac{\gamma}{g_n}(V_2 - V_1) \tag{6.4}$$

$$F_1 + \frac{Q\gamma V_1}{g_n} = F_2 + \frac{Q\gamma V_2}{g_n}$$

$$F = \gamma h_c A = \gamma\frac{y}{3}y^2 \qquad V = \frac{Q}{y^2} \tag{2.12} \ \ (3.11)$$

Substituting these values above and canceling γ,

$$\frac{y_1^3}{3} + \frac{Q^2}{g_n y_1^2} = \frac{y_2^3}{3} + \frac{Q^2}{g_n y_2^2} \blacktriangleleft$$

References

Ackers, P., White, W. R., Perkins, J. A., and Harrison, A. J. M. 1978. *Weirs and flumes for flow measurement*. New York: Wiley.

Chow, V. T. 1959. *Open-channel hydraulics*. New York: McGraw-Hill.

Elevatorski, E. A. 1959. *Hydraulic energy dissipators*. New York: McGraw-Hill.

Henderson, F. M. 1966. *Open channel flow*. New York: Macmillan.

Ippen, A. T. 1950. Channel transitions and controls. Chapter VIII in *Engineering hydraulics*, Ed. H. Rouse. New York: Wiley.

McBean, E., and Perkins, F. 1975. Numerical errors in water profile computation. *Jour. Hydr. Div., A.S.C.E.* 101, HY11: 1389–1403 (see also 1977: 103, HY6: 665–666).

Minton, P. and Sobey, R. J. 1973. Unified nondimensional formulation for open channel flow. *Jour. Hydr. Div., A.S.C.E.* 99, HY1: 1–12.

Posey, C. J. 1950. Gradually varied open channel flow. Chapter IX in *Engineering hydraulics*, Ed. H. Rouse. New York: Wiley.

Prasad, R. 1970. Numerical method of computing flow profiles. *Jour. Hydr. Div., A.S.C.E.* 96, HY1: 75–86.

Films

Rouse, H. Mechanics of fluids: fluid motion in a gravitational field. Film No. U45961, Media Library, Audiovisual Center, Univ. of Iowa.

St. Anthony Falls Hydraulic Laboratory. Some phenomena of open channel flow. Film No. 3, St. Anthony Falls Hydr. Lab., Minneapolis, Minn.

Problems

10.1.   Liquid of specific weight γ flows uniformly down an inclined plane of slope angle θ. Derive an expression for the pressure on the plane in terms of γ, θ, and the depth y (measured *vertically* from the plane to the surface of the liquid).

10.2.   Water flows uniformly at a depth of 1.2 m (4 ft) in a rectangular canal 3 m (10 ft) wide, laid on a slope of 1 m per 1 000 m (1 ft per 1 000 ft). What is the mean shear stress on the sides and bottom of the canal?

10.3.   Calculate the mean shear stress over the wetted perimeter of a circular sewer 3 m in diameter in which the depth of uniform flow is 1 m and whose slope is 0.000 1.

10.4.   What is the mean shear stress over the wetted perimeter of a triangular flume 2.4 m (8 ft) deep and 3 m (10 ft) wide at the top, when the depth of uniform flow is 1.8 m (6 ft)? The slope of the flume is 1 in 200, and water is flowing.

10.5.   Calculate the Chezy coefficient that corresponds to a friction factor of 0.030.

10.6. Uniform flow occurs in a rectangular channel 10 ft wide at a depth of 6 ft. If the Chezy coefficient is 120 ft$^{1/2}$/s, calculate friction factor, Manning n, and approximate height of the roughness projections.

10.7. What uniform flowrate will occur in a rectangular timber flume 1.5 m wide and having a slope of 0.001 when the depth therein is 0.9 m?

10.8. Calculate the uniform flowrate in an earth-lined ($n = 0.020$) trapezoidal canal having bottom width 3 m (10 ft), sides sloping 1 (vert.) on 2 (horiz.), laid on a slope of 0.000 1, and having a depth of 1.8 m (6 ft).

10.9. A steel flume in the form of an equilateral triangle (apex down) of 1.2 m sides is laid on a slope of 0.01. Calculate the uniform flowrate that occurs at a depth of 0.9 m.

10.10. This large (uniform) open channel flow is to be modeled (without geometric distortion) in the hydraulic laboratory at a scale of 1 to 9. What flowrate, bottom slope, and Manning n will be required in the model?

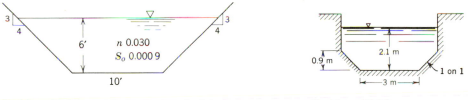

| Problem **10.10** | Problem **10.12** |

10.11. A rectangular open channel of 2.4 m width has a homogeneous roughness estimated to have an effective height of 6.4 mm. To visualize the variation of the various resistance coefficients with depth calculate and tabulate: relative roughness, friction factor, Chezy coefficient, and Manning n for depths of 0.15, 0.3, 0.6, 1.2, 1.8, and 2.4 m.

10.12. What uniform flowrate will occur in this canal cross section if it is laid on a slope of 1 in 2 000 and has $n = 0.017$?

10.13. A semicircular canal of 1.2 m (4 ft) radius is laid on a slope of 0.002. If n is 0.015, what uniform flowrate will exist when the canal is brim full?

10.14. A flume of timber has as its cross section an isosceles triangle (apex down) of 2.4 m base and 1.8 m altitude. At what depth will 5 m^3/s flow uniformly in this flume if it is laid on a slope of 0.01?

10.15. At what depth will 4.25 m^3/s flow uniformly in a rectangular channel 3.6 m wide lined with rubble masonry and laid on a slope of 1 in 4 000?

10.16. At what depth will 400 cfs flow uniformly in an earth-lined ($n = 0.025$) trapezoidal canal of base width 15 ft, having side slopes 1 on 3, if the canal is laid on a slope of 1 in 10 000?

10.17. Calculate the depth of uniform flow for a flowrate of 11 m^3/s in the open channel of problem 10.12.

10.18.   An earth-lined trapezoidal canal of base width 10 ft and side slopes 1 (vert.) on 3 (horiz.) is to carry 100 cfs uniformly at a mean velocity of 2 ft/s. What slope should it have?

10.19.   What slope is necessary to carry 11 m³/s uniformly at a depth of 1.5 m in a rectangular channel 3.6 m wide, having $n = 0.017$?

10.20.   A trapezoidal canal of side slopes 1 (vert.) or 2 (horiz.) and having $n = 0.017$ is to carry a uniform flow of 37 m³/s (1 300 ft³/s) on a slope of 0.005 at a depth of 1.5 m (5 ft). What base width is required?

10.21.   A rectangular channel 1.5 m wide has uniform sand grain roughness of diameter 6 mm. To observe the variation of n with depth, calculate values of n for depths of 0.03, 0.3, 0.6, and 0.9 m.

10.22.   A winding natural channel has a uniform flow of 30 m³/s at a particular depth in the winter. What will be the uniform flowrate in the summer at the same depth when the channel is very weedy?

10.23.   A flowrate of 0.1 cfs of oil (per foot of width) is to flow uniformly down an inclined glass plate at a depth of 0.05 ft. Calculate the required slope. The viscosity and specific gravity of the oil are 0.009 lb · s/ft² and 0.90, respectively.

10.24.   Water (20°C) flows uniformly in a channel at a depth of 0.009 m. Assuming a critical Reynolds number of 500, what is the largest slope on which laminar flow can be maintained? What mean velocity will occur on this slope?

10.25.   What flowrate (per foot of width) may be expected for water (68°F) flowing in a wide rectangular channel at a depth of 0.02 ft if the channel slope is 0.000 1? Assume laminar flow and confirm by calculating Reynolds number.

10.26.   At what depth will a flowrate of water (20°C) of 0.28 l/s · m occur in a wide open channel of slope 0.000 15? Assume laminar flow and confirm by calculating Reynolds number.

10.27.   Plot curves similar to those of Fig 10.8 for an equilateral triangle (apex up). Find the maximum points of the curves mathematically.

10.28.   Plot curves similar to those of Fig 10.8 for a square, laid diagonal vertical. Find the maximum points of the curves mathematically.

10.29.   Derive an equation for the shape of the walls of an open channel that features no variation of hydraulic radius with water depth.

10.30.   The cross section of an open channel is 6 m wide at the water surface and 3 m deep at the center. If its shape may be closely approximated by a parabola calculate the hydraulic radius.

10.31.   This canal cross section is proposed for carrying a flowrate of 1 760 cfs at a velocity of 5 ft/s. With the same area dimension, propose another (trapezoidal) one which will be

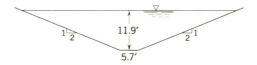

Problem **10.31**

hydraulically *better*. What dimensions should be used (for the same area) for the section to be hydraulically *best*? The side slopes, bottom width, and depth may all be changed.

10.32. What uniform flowrate occurs in a 1.5 m circular brick conduit laid on a slope of 0.001 when the depth of flow is 1.05 m? What is the mean velocity of this flow?

10.33. Calculate the depth at which 0.7 m³/s (25 ft³/s) will flow uniformly in a smooth cement-lined circular conduit 1.8 m (6 ft) in diameter, laid on a slope of 1 in 7 000.

10.34. Rectangular channels of flow cross section 50 ft² have dimensions (width × depth) of (*a*) 25 ft by 2 ft, (*b*) 12.5 ft by 4 ft, (*c*) 10 ft by 5 ft, and (*d*) 5 ft by 10 ft. Calculate the hydraulic radii of these sections.

10.35. A channel flow cross section has an area of 18 m². Calculate its best dimensions if (*a*) rectangular, (*b*) trapezoidal with 1 (vert.) or 2 (horiz.) side slopes, and (*c*) V-shaped.

10.36. Calculate the required width of a rectangular channel to carry 45 m³/s uniformly at best hydraulic conditions on a slope of 0.001 if *n* is 0.035.

10.37. What are the best dimensions for a trapezoidal canal having side slopes 1 (vert.) on 3 (horiz.) and *n* of 0.020 if it is to carry 40 m³/s (1 400 ft³/s) uniformly on a slope of 0.009?

10.38. What is the minimum slope at which 5⅔ m³/s may be carried uniformly in a rectangular channel (having a value of *n* of 0.014) at a mean velocity of 0.9 m³/s.

10.39. What is the minimum slope at which 28 m³/s (1 000 ft³/s) may be carried uniformly at a mean velocity of 0.6 m/s (2 ft/s) in a trapezoidal canal having *n* = 0.025 and sides sloping 1 (vert.) on 4 (horiz.)?

10.40. Prove that the best form for a V-shaped open-channel section is one of vertex angle 90°.

10.41. This flood channel has Manning *n* 0.017 and slope 0.000 9. Estimate the depth of uniform flow for a flowrate of 1 200 cfs.

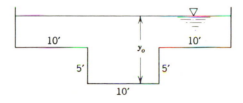

Problem **10.41**

10.42. Calculate the specific energies when 225 cfs flow in a rectangular channel of 10 ft width at depths of (*a*) 1.5 ft, (*b*) 3 ft, and (*c*) 6 ft.

10.43. At what depths may 0.85 m³/s flow in a rectangular channel 1.8 m wide if the specific energy is 1.2 m?

10.44. In a two-dimensional open-channel flow the water depth is 0.9 m (3 ft) and the specific energy 2.4 m (8 ft). For the same flowrate what depths may be expected for a specific energy of 3 m (10 ft)?

10.45. Eight hundred cubic feet per second flow in a rectangular channel of 20 ft width having $n = 0.017$. Plot accurately the specific energy diagram for depths from 0 to 10 ft, using the same scales for y and E. Determine from the diagram (a) the critical depth, (b) the minimum specific energy, (c) the specific energy when the depth of flow is 7 ft, and (d) the depths when the specific energy is 8 ft. What type of flow exists when the depth is (e) 2 ft, (f) 6 ft; what are the channel slopes necessary to maintain these depths? What type of slopes are these, and (g) what is the critical slope, assuming the channel to be of great width?

10.46. Flow occurs in a rectangular channel of 6 m width and has a specific energy of 3 m. Plot accurately the q-curve. Determine from the curve (a) the critical depth, (b) the maximum flowrate, (c) the flowrate at a depth of 2.4 m, and (d) the depths at which a flowrate of 28.3 m³/s may exist, and the flow condition at these depths.

10.47. Five hundred cubic feet per second flow in a rectangular channel 15 ft wide at depth of 4 ft. Is the flow subcritical or supercritical?

10.48. If 11 m³/s flow in a rectangular channel 5.4 m wide with a velocity of 1.5 m/s, is the flow subcritical or supercritical?

10.49. If 8.5 m³/s flow uniformly in a rectangular channel 3.6 m wide having $n = 0.015$ and laid on a slope of 0.005, is the flow subcritical or supercritical? What is the critical slope for this flowrate, assuming the channel to be of great width?

10.50. A uniform flow in a rectangular channel 12 ft wide has a specific energy of 8 ft; the slope of the channel is 0.005 and the Chezy coefficient 120 ft$^{1/2}$/s. Predict all possible depths and flowrates.

10.51. What is the maximum flowrate which may occur in a rectangular channel 2.4 m wide for a specific energy of 1.5 m?

10.52. Thirty cubic metres per second (1 000 ft³/s) flow uniformly in a rectangular channel, 4.5 m (15 ft) wide (n is 0.018, S_o is 0.002), at best hydraulic conditions. Is this flow subcritical or supercritical? What is the critical slope for the flowrate, assuming the channel to be of great width?

10.53. The flowrate in a rectangular channel 5 ft wide is 100 cfs. Calculate the angle of small surface waves with the wall of the channel for depths of 1 ft and 3 ft.

10.54. In a wide rectangular channel n is 0.017 and does not vary over a depth range from 0.3 to 1.5 m. To confirm the trend of Fig. 10.14, calculate values of the critical slope for depths of 0.3, 0.9, and 1.5 m. What flowrates are associated with these depths? What is the average value of the criticial slope for this depth range? What is the maximum (percent) deviation from this average value?

10.55. Assuming the canal entrance frictionless, calculate the two-dimensional flowrate and determine the water depths at the lettered sections.

10.56. This "Venturi flume" has straight walls, horizontal floor, and may be assumed frictionless. For a flowrate of 300 cfs and energy line 10 ft above the floor determine the water depths at the lettered sections. Plot and use a q-curve in the solution of this problem.

10.57. Eleven cubic metres per second are diverted through ports in the bottom of the channel between sections 1 and 2. Neglecting head losses and assuming a horizontal channel, what depth of water is to be expected at section 2? What channel width at section 2 would be required to produce a depth of 2.5 m?

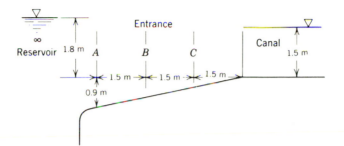

Problem **10.55**

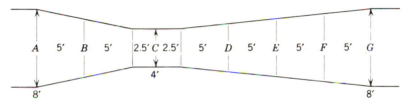

Problem **10.56**

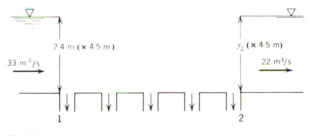

Problem **10.57**

10.58. Derive a general expression for critical slope in terms of b, y, and n for a rectangular channel.

10.59. Calculate the specific energy when 300 cfs flow at a depth of 4 ft in a trapezoidal channel having base width 8 ft and sides sloping at 45°.

10.60. Calculate the specific energy when 2.8 m³/s flow at a depth of 0.9 m in a V-shaped flume if the width at the water surface is 1.2 m.

10.61. Determine the critical depth for a flowrate of 2 m³/s in this square open channel.

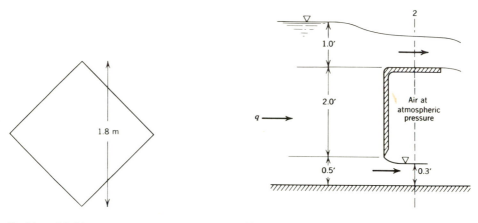

Problem **10.61**   Problem **10.79**

10.62. At what depths may 800 cfs flow in a trapezoidal channel of base width 12 ft and side slopes 1 (vert.) on 3 (horiz.) if the specific energy is 7 ft?

10.63. Eleven cubic metres per second flow in a trapezoidal channel of base width 4.5 m and side slopes 1 (vert.) on 3 (horiz.). Calculate the critical depth and the ratio of critical depth to minimum specific energy. If $n = 0.020$, what is the critical slope?

10.64. Uniform flow occurs in a trapezoidal canal of base width 1.5 m and side slopes 1 (vert.) on 2 (horiz.) laid on a slope of 0.002. The depth is 1.8 m. Is this flow subcritical or supercritical?

10.65. A sewer line of elliptical cross section 10 ft high by 7 ft wide carries 200 cfs of water at a depth of 5 ft. Is this flow subcritical or supercritical?

10.66. What is the critical depth for a flowrate of 14 m³/s in a channel having the cross section of problem 10.12?

10.67. For best hydraulic cross section in a trapezoidal channel of side slopes 2 (vert.) on 1 (horiz.), a uniform flowrate of 1 000 cfs occurs at critical depth. If n is 0.017, calculate the critical slope.

10.68. Plot the variation of critical slope with depth for an open channel whose cross section is an isosceles triange (apex up) of 2.4 m base and 2.4 m altitude.

10.69. Derive an expression for critical depth for an open channel of V-shaped cross section with side slopes of 45°. What is the ratio between critical depth and minimum specific energy for this channel?

10.70. Solve the preceding problem for a channel of parabolic cross section defined by $y = x^2$.

10.71. Calculate more exact values of critical slopes for problems (a) 10.45, (b) 10.49, (c) 10.52 by including the effect of the sidewalls.

10.72. Solve problem 10.54 for a rectangular channel of 4.5 m width.

10.73. What theoretical flowrate will occur over a high broad-crested weir 30 ft long when the head thereon is 2 ft?

10.74. The elevation of the crest of a high broad-crested weir is 100.00 ft. If the length of this weir is 12 ft and the flowrate over it 200 cfs, what is the elevation of the water surface upstream from the weir?

10.75. If the energy line is 0.9 m (3 ft) above the crest of a frictionless broad-crested weir of 1.2 m (4 ft) height, what is the water depth just upstream from the weir?

10.76. When the depth of water just upstream from a frictionless broad-crested weir 0.6 m (2 ft) high is 0.9 m (3 ft), what flowrate per metre (foot) of crest length can be expected?

10.77. A dam 1.2 m high and having a broad horizontal crest is built in a rectangular channel 4.5 m wide. For a depth of water *on* the crest of 0.6 m, calculate the flowrate and the depth of water just upstream from the dam.

10.78. The velocity in a rectangular channel of 3 m width is to be reduced by installing a smooth broad-crested rectangular weir. Before installation the mean velocity is 1.5 m/s and the water depth 0.3 m; after installation these quantities are to be 0.3 m/s and 1.5 m, respectively. What height of weir is required?

10.79. Flow occurs over and under this control gate as shown. Calculate the flowrate (per foot of gate width), assuming streamlines straight and parallel at section 2.

10.80. Assuming frictionless flow, calculate the two-dimensional flowrate over the weir crest and the water depths at the lettered sections.

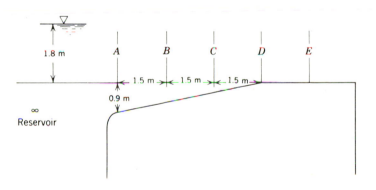

Problem **10.80**

10.81. Four hundred cubic feet per second flow in a long rectangular open channel 12 ft wide which ends in a free outfall. The slope of the channel is 0.000 9 and Manning's n 0.017. A frictionless broad-crested weir is to be installed near the end of the channel to produce uniform flow throughout the length of the channel. What weir height is required?

10.82. In the preceding problem a sluice gate installed near the end of the channel is used to accomplish the same purpose. If the water depth just downstream from the gate is equal to the gate opening, what gate opening is required to produce uniform flow?

10.83.   Show that the freely falling sheet of water well downstream from the brink of Fig 10.20 must have a thickness (measured vertically) of $2y_c/3$.

10.84.   An open rectangular channel 1.5 m wide and laid on a mild slope ends in a free outfall. If the brink depth is measuerd as 0.264 m, what flowrate exists in the channel?

10.85.   Fifty cubic feet per second flow in a rectangular channel 8 ft wide. If this channel is laid on a mild slope and ends in a free outfall, what depth at the brink is to be expected?

10.86.   A horizontal pipe of 0.6 m diameter discharges as a free outfall. The depth of water at the brink of the outfall is 0.15 m. Calculate the flowrate.

10.87.   This rectangular laboratory channel is 1 m wide and has a movable section 30 m long set on frictionless rollers and connected to the fixed channel with material so flexible that it transmits no force. When the indicated steady flow is established in the channel the reading on the spring scales is observed to increase by 75 N. Calculate Manning's n for the channel.

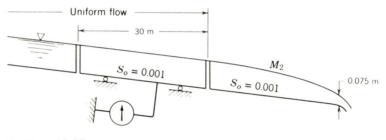

Problem **10.87**

10.88.   Is this water surface profile possible? If not, what profile is to be expected? The channel and constriction are both of rectangular cross section. The flowrate is 7.2 m³/s.

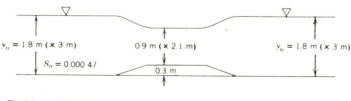

Problem **10.88**

10.89.   Up to what flowrate can critical depth be expected to exist in this constriction? For convenience assume Chezy C for the channel 86 ft$^{1/2}$/s and neglect head losses caused by hump and constriction. Assume the channel very long upstream and downstream from the constriction.

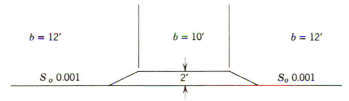

Problem **10.89**

10.90. The critical depth is maintained at a point in a rectangular channel 1.8 m wide by building a gentle hump 0.3 m high in the bottom of the channel. When the depth over the hump is 0.66 m, what water depths are possible just upstream from the hump?

10.91. If 150 cfs flow uniformly in a rectangular channel 10 ft wide, laid on a slope of 0.000 4, and having $n = 0.014$, what is the minimum height of the hump that may be built across this channel to create critical depth over the hump? Sketch the energy line and water surface, showing all vertical dimensions. Neglect head losses caused by the hump.

10.92. Uniform flow in a rectangular channel occurs at a depth of 0.45 m and velocity of 7.5 m/s. When a smooth frictionless hump 0.6 m is built in the floor of the channel, what depth can be expected on the hump? What hump height would be required to produce critical depth on the hump? What would happen if a taller hump than this were installed.

10.93. Solve the preceding problem for a depth of 3 m and velocity of $1\frac{1}{8}$ m/s.

10.94. Uniform flow at 6.00 ft depth occurs in a rectangular open channel 10 ft wide having Manning n of 0.017 and laid on a slope of 0.003 6. There is a flexible hump in the floor of the channel which can be raised or lowered. Neglecting losses caused by the hump: (*a*) How large may the hump height be made without changing the water depth just upstream from the hump? (*b*) How large should the hump height be to make the water depth 7.00 ft just upstream from the hump?

10.95. The water depth over a hump of 0.3 m height in a rectangular open channel is 1.05 m. Just upstream and just downstream from the hump the water depths are 1.5 m. What depths would be expected on and just upstream from the hump when the channel downstream from the hump is removed? There will be no change of flowrate.

10.96. A long rectangular channel 10 ft wide carries a flowrate of 372 cfs uniformly at a depth of 2.00 ft and ends in a free outfall. When a smooth hump 1.50 ft high is installed at the very end of this channel, what depth is to be expected on the hump? How large a hump would cause the depth there to be critical?

10.97. A rectangular channel 3.6 m wide is narrowed to a 1.8 m width to cause critical flow in the contracted section. If the depth in this section is 0.9 m, calculate the flowrate and the depth in the 3.6 m section, neglecting head losses in the transition. Sketch energy line and water surface, showing all pertinent vertical dimensions.

10.98. Four and one-quarter cubic metres per second flow uniformly in a rectangular channel 3 m wide having $n = 0.014$, and laid on a slope of 0.000 4. This channel is to be narrowed to cause critical flow in the contracted section. What is the maximum width of contracted section which will accomplish this? For this width and neglecting head losses, sketch the energy line and water surface, showing vertical dimensions. If the contraction is narrowed to 1.2 m width, what depths are to be expected in and just upstream from the contraction?

10.99. A uniform of 21.2 m³/s occurs in a rectangular channel 4.5 m wide at a depth of 3 m. A hump 0.6 m high is built in the bottom of the channel, and at the same point width is reduced to 3.6 m. When friction is neglected, what is the water depth over the hump?

10.100. A flow of 50 cfs per foot of width occurs in a wide rectangular channel at a depth of 6 ft. A bridge to be constructed across this channel requires piers spaced 20 ft on centers. Assuming that the noses of the piers are well streamlined and frictionless, how thick may they be made without causing "backwater effects" (i.e., deepening of the water) upstream from the bridge?

10.101. A flow of 8.5 m³/s occurs in a long rectangular channel 3 m wide, laid on a slope of 0.001 6, and having $n = 0.018$. There is a smooth gradual constriction in the channel to 1.8 m width. What depths may be expected in and just upstream from the constriction?

10.102. A long rectangular channel 15 ft wide has $n = 0.014$ and is laid on a slope of 0.000 90. In this channel there is a smooth constriction to 13.5 ft width containing a hump 1.50 ft high. Predict the water depths in and just upstream from the constriction for a flowrate of 1 000 cfs.

10.103. In a trapezoidal canal of 3 m base width and sides sloping at 45°, 23 m³/s flow uniformly at a depth of 3 m. The channel is constricted by raising the sides to a vertical position. Calculate the depth of water in the constriction, neglecting local head losses. What is the minimum height of hump which may be installed in the constriction to produce critical depth there?

10.104. A long rectangular channel of 4.5 m width, slope of 0.001, and n of 0.015 reduces to 3.3 m width as it passes through a culvert in a highway embankment. Below what flowrate can depths greater than normal depth be expected upstream from the embankment?

10.105. For a rectangular open channel of 8 ft width, $n = 0.015$, and slope 0.003 5, the following data are given. Approximately how far apart are sections 1 and 2?

Section	Depth (ft)	Velocity (ft/s)	Hydraulic radius (ft)	Specific energy (ft)
1	3.00	15.00	1.715	6.49
2	3.20	14.06	1.775	6.26

10.106. The channels of Fig 10.17 are of 3.6 m width and have $n = 0.020$. Their slopes are 0.005 00 and 0.010 0 and the depth of water at the break in slope is 1.5 m. Considering this depth to be critical depth, how far upstream and downstream from this point will depths of 1.62 and 1.38 m be expected? If the respective slopes were slightly decreased, would the calculated distances be lengthened or shortened?

10.107. This uniform flow occurs in a very long rectangular channel. The sluice gate is lowered to a position so that the opening is 0.6 m. Sketch the new water surface profiles to be expected and identify them by letter and number. Calculate and show all significant depths and longitudinal distances.

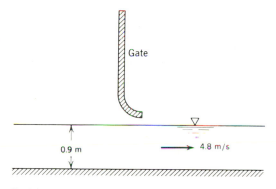

Problem **10.107**

10.108. Solve the preceding problem for the same flowrate but with 1.8 m depth of uniform flow.

10.109. The depth of uniform flow for a flowrate of 300 cfs in an infinitely long rectangular open channel of 10 ft width is 5.00 ft. A single short smooth structure exists in the channel. This structure consists of a hump 2 ft high and a constriction to 8 ft width; hump and constriction are at the same flow cross section. Sketch the expected water surface profiles in the channel upstream and downstream from the structure. Identify any gradually varied flow profiles by letter and number. Calculate and show on the sketch all dimensions.

10.110. Solve the preceding problem for depths of uniform flow of (*a*) 2.50 ft and (*b*) 7.50 ft.

10.111. The length of "Reach 1" of the San Luis Canal in California is $25\frac{1}{4}$ km. The canal has a bottom width of 33 m and side slopes 7 (vert) on 15 (horiz). The bottom slope of this reach of canal is zero. With no flow in the canal, the depth of water will be everywhere 10.0 m, and the water surface coincident with that of the large reservoir (forebay) at the upstream end of the canal. Flow in this canal is produced by pumps at the downstream end which extract a (maximum) flowrate of 371 m^3/s. For this flowrate what drop in the water surface is to be expected at the downstream end of the canal? Assume that the water surface in the forebay remains constant and that n is 0.014 9.

10.112. A uniform flow at 5 ft depth occurs in a long rectangular channel 15 ft wide, having $n = 0.017$, and laid on a slope of 0.001 0. A hump is built across the floor of the channel of such height that the water depth just upstream from the hump is raised to 6.00 ft. Identify the surface profile upstream from the hump and calculate the distance from the hump to where a depth of 5.60 ft is to be expected. Use depth intervals of 0.20 ft for noncomputer solutions.

10.113. Water discharges through a sluice gate into a long rectangular open channel 2.4 m wide having $n = 0.017$. The depth and velocity at the vena contracta are 0.6 m and 7.5 m/s, respectively. Identify the surface profiles if the channel slope is (*a*) favorable, (*b*) zero, (*c*) adverse. If these slopes are (a) 0.002, (b) 0.000, and (c)−0.002, how far downstream from the sluice will a depth of 0.67 m be expected?

10.114. A rectangular channel 3 m wide is laid on a slope of 0.002 and has Manning n of 0.014 9. At a certain point in the channel the water depth is 0.60 m. Sixty metres downstream from this point the depth is 0.67 m. Estimate the flowrate and calculate the water surface profile.

10.115. The flowrate in a trapezoidal channel of base width 10 ft, side slope 2 on 3, and n 0.014 9 is 600 cfs. A very long reach of this channel has a slope of 0.000 3. At a point in the channel the depth is 5.00 ft. Approximately how far (upstream or downstream) may a depth of 6.00 ft be expected?

10.116. Construct the computer solution to the Illustrative Problem of Section 10.9.

10.117. Compute the profile downstream from a point where $y = 1.9$ m in a trapezoidal channel in which $n = 0.02$, $z = 2$, $S_o = 0.002$, $b = 30$ and $Q = 211$ m^3/s. How far can calculation proceed?

10.118. A very wide rectangular channel of bottom slope 0.002 has a constant $C = 100$ ft$^{1/2}$/s. At a point in this channel the water depth is 5.0 ft. Estimate the flowrate per foot of width and identify the type of the water surface profile.

10.119. Eleven cubic metres per second flow uniformly in an infinitely long trapezoidal canal having base width 3 m, side slopes 1 on 1, Manning n 0.021, and depth 0.75 m. The channel is locally constricted by raising the walls to a vertical position. This alteration is short and streamlined and may be assumed to cause no additional head loss. What water depths are to be expected in the constriction, just upstream from there, and just downstream from there? Also find any other depths that can be obtained without use of the equations of gradually varied flow. Show all results on a sketch with any gradually varied flow profiles identified by letter and number.

10.120. Compute all gradually varied flow profiles for problem 10.119.

10.121. Eight hundred cubic feet per second flow in a rectangular channel of 20 ft width. Plot the $\mathfrak{M}$-curve of hydraulic jumps on the specific energy diagram of problem 10.45. From these curves determine (a) the depth after a hydraulic jump has taken place from a depth of 1.5 ft, (b) the height of this jump, (c) the specific energy before the jump, (d) the specific energy after the jump, (e) the loss of head in the jump, and (f) the total horsepower lost in the jump.

10.122. A hydraulic jump occurs in a rectangular open channel. The water depths before and after the jump are 0.6 m and 1.5 m, respectively. Calculate the critical depth.

10.123. For a rectangular open channel, prove that hydraulic jumps can occur only across the critical depth. Prove this for a nonrectangular channel.

10.124. If the maximum F_1 for an undular hydraulic jump is $\sqrt{3}$, what is the maximum value of y_2/y_1 for such jumps? Show that the head lost in a hydraulic jump is given by $h_L/y_1 = (y_2/y_1 - 1)^3/4(y_2/y_1)$. Note that such head losses will be very small (usually negligible) for low values of y_2/y_1.

10.125. A supercritical flow of 100 cfs occurs at 3 ft depth in a V-shaped open channel of side slopes 45°. Calculate the depth just downstream from a hydraulic jump in this flow.

10.126. A hydraulic jump occurs in a V-shaped open channel with sides sloping at 45°. The depths upsream and downstream from the jump are 0.9 m and 1.2 m, respectively. Estimate the flowrate in the channel.

10.127. A hydraulic jump occurs in a horizontal storm sewer of circular cross section 1.2 m in diameter. Before the jump the water depth is 0.6 m and just downstream from the jump the sewer is full with a a gage pressure of 7 kPa at the top. Predict the flowrate.

10.128. A rectangular channel 3 m wide carries a flowrate of 14 m³/s uniformly at 0.6 m depth. The channel is constricted at the end to produce a hydraulic jump in the channel. Calculate the width of constriction required for the jump to be just upstream from the constriction.

10.129. The depths of water upstream and downstream from a hydraulic jump on the horizontal "apron" downstream from a spillway structure are observed to be approximately 3 ft and 8 ft. If the structure is 200 ft long (perpendicular to the direction of flow), about how much horsepower is being dissipated in this jump?

10.130. Calculate y_2, h, and y_3 for this two-dimensional flow picture. State any assumptions clearly.

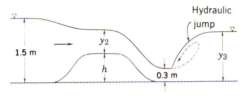

Problem **10.130**

10.131. The situation shown on the sketch is observed (in the field) in a rectangular open channel 5 ft wide and 12 ft deep. The 10 ft and 6 ft depths are easily (although not accurately) measured. The depth at the vena contracta cannot be obtained because of danger and inaccessibility. From the measurements shown estimate the flowrate in the channel.

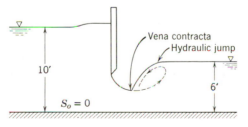

Problem **10.131**

10.132. What shape (y_2/y_1) of hydraulic jump will cause dissipation of one-third of the total energy of the approaching flow?

10.133. Determine the horizontal component of force exerted on the broad-crested weir of problem 11.91 using (a) a conventional impulse-momentum analysis and (b) the $\mathfrak{M}$-diagram of Fig 10.25.

10.134. Determine the force component of problem 6.34 by use of the $\mathfrak{M}$-diagram of Fig. 10.25.

10.135. In which zone $(AB, BC,$ etc.) in this flow is a hydraulic jump to be expected? Show numerical proof for selection of zone. The numbers above the H_2 profile are the water depths for that profile. The numbers below the channel floor are the depths for the H_3 profile. Upstream of section A the flow may be considered frictionless.

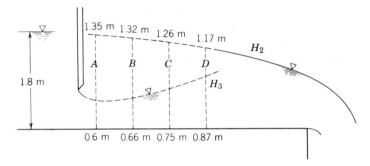

Problem **10.135**

10.136. A very long rectangular channel of 8 ft width and n of 0.017, laid on a slope of 0.03, carries a flowrate of 200 cfs and ends in a free outfall. Across the downstream end of the channel a smooth frictionless hump is constructed. Calculate the depths (a) on the hump, (b) just upstream from the hump, and (c) far up the channel, and sketch and identify the surface profiles. If a hydraulic jump occurs calculate the depths y_1 and y_2 and the varied flow profiles. The hump is 2.5 ft high.

10.137. A flowrate of 254 cfs occurs in a very long rectangular open channel 10 ft wide, having slope and Manning n such that the depth of uniform flow is 6.00 ft. In this channel there is a smooth hump 2 ft high and (at the same section) a constriction to 6 ft width. Predict the water depths; (a) in the constriction, and in the channel, (b) just upstream, and (c) just downstream from the constriction. Calculate and sketch the water-surface profile carefully, identifying any profiles of gradually varied flow by letter and number.

10.138. The flowrate through this Venturi flume is 28.3 m³/s; the flume may be considered frictionless. The slope of the approach channel is such that the normal depth is 2.7 m. Over what depth range at the downstream end of the flume may hydraulic jumps be expected in the flume?

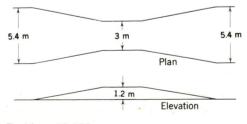

Problem **10.138**

10.139. Neglecting wall, bottom, and hump friction (but not losses in the jump), what height of hump will produce this flow picture?

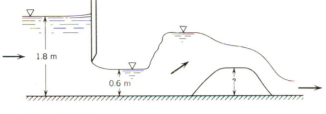

Problem **10.139**

10.140. The sketch shows a plan view of a frictionless Venturi flume of rectangular cross section with horizontal floor. A hydraulic jump from 1 ft to 3 ft occurs at section 3. Calculate the depth of water at sections 1, 2, and 4. Section 1 may be assumed to be infinitely wide.

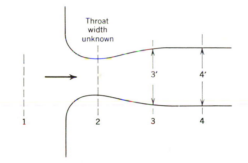

Problem **10.140**

10.141. The flowrate in Fig. 10.27 is 15.4 m³/s. If the channels are 3.6 m wide with $n = 0.017$ and the downstream channel laid on a slope of 0.002 28, what depth must exist in this channel for a hydraulic jump to occur to uniform flow? Calculate the length l if $y_{o_1} = 0.80$ m.

10.142. The flowrate in Fig 10.28 is 15.4 m³/s, the channels are $3\frac{2}{3}$ m wide, and $n = 0.017$. The downstream channel is laid on a slope of 0.001 5. If $y_{o_1} = 0.79$ m, calculate l.

11
FLUID MEASUREMENTS

In engineering and industrial practice one of the fluid mechanics problems most frequently encountered by the engineer is the *measurement* of many of the variables and properties discussed in the foregoing chapters. Efficient and accurate measurements are also absolutely essential for correct conclusions in the various fields of fluid mechanics research. Whether the necessity for precise measurements is economic or scientific, the engineer must be well equipped with a knowledge of the fundamentals and existing methods of measuring various fluid properties and phenomena. It is the purpose of this chapter to indicate only the principles and phenomena of fluid measurements; the reader will find available in the abundant engineering literature[1] the details of installation and operation of the various measuring devices.

11.1 Measurement of Fluid Properties

Of the fluid properties density, viscosity, elasticity, surface tension, and vapor pressure, the engineer is usually called on to measure only the first two. Since measurements of elasticity, surface tension, and vapor pressure are normally made by physicists and chemists, the various experimental techniques for measuring these properties are not reviewed here.

Density measurements of liquids may be made by the following methods, listed in approximate order of their accuracy: (1) weighing a known volume of liquid, (2) hydrostatic weighing, (3) Westphal balance, and (4) hydrometer.

To weigh accurately a known volume of liquid a device called a *pycnometer* is used. This is usually a glass vessel whose weight, volume, and variation of volume with temperature have been accurately determined. If the weight of the empty pycnometer is W_1, and the weight of the pycnometer, when containing a volume V of liquid at temperature t, is W_2, the specific weight of the liquid, γ_t, at this temperature may be calculated directly from

$$\gamma_t V = W_2 - W_1$$

Density determination by hydrostatic weighing consists essentially in weighing a plummet of known volume (1) in air, and (2) in the liquid whose density is to be determined (Fig. 11.1a). If the weight of the plummet in air is W_a, its volume, V, and its weight when suspended in the liquid, W_l, the equilibrium of vertical forces on the plummet gives

[1] See the References in and at the end of this chapter.

502

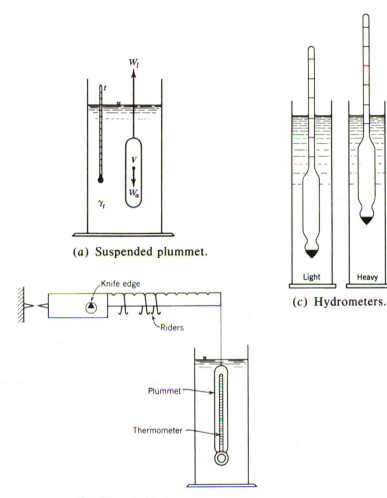

(a) Suspended plummet.

(c) Hydrometers.

(b) Westphal balance.

Fig. **11.1** Devices for density measurements.

$$W_l + \gamma_t V - W_a = 0$$

from which the specific weight, γ_t, at the temperature t, may be calculated directly.

Like the method of hydrostatic weighing, the Westphal balance (Fig. 11.1b) utilizes the buoyant force on a plummet as a measure of specific gravity. Balancing the scale beam with special riders placed at special points allows direct and precise reading of specific gravity.

Probably the most common means of obtaining liquid densities is with the hydrometer (Fig. 11.1c), whose operation is governed by the fact that a weighted tube will float with different immersions in liquids of different densities. To create a great variation of immersion for small density variation, and, thus, to provide a sensitive instrument, changes in the immersion of the hydrometer occur along a slender tube, which is graduated to read the specific gravity of the liquid at the point where the liquid surface intersects the tube.

Viscosity measurements are made with devices known as *viscosimeters* or *viscometers*, the common varieties of which may be classified as *rotational* or *tube* devices. The operation of these viscometers *depends on the existence of laminar flow* (which has been seen in foregoing chapters to be dominated by viscous action) under certain controlled and reproducible conditions. In general, however, these conditions involve too many complexities to allow the constants of the viscometer to be calculated analytically, and they are, therefore, usually obtained by calibration with a liquid of known viscosity. Because of the variation of viscosity with temperature all viscometers *must be immersed in constant-temperature baths* and *provided with thermometers* for taking the temperatures at which the viscosity measurements are made.

Two instruments of the rotational type are the MacMichael and Stormer viscometers, whose essentials are shown diagrammatically in Fig. 11.2. Both consist of two concentric cylinders, with the space between containing the liquid whose viscosity is to be determined. In the MacMichael type, the outer cylinder is rotated at constant speed, and the rotational deflection of the inner cylinder (accomplished against a spring) becomes a measure of the liquid viscosity. In the Stormer instrument, the inner cylinder is rotated by a falling-weight mechanism, and the time necessary for a fixed number of revolutions becomes a measure of the liquid viscosity.

The measurement of viscosity by the above-mentioned variables may be justified by a simplified mechanical analysis, using the dimensions of Fig. 11.2. Assuming ΔR, Δh, and $\Delta R/R$ small, and the peripheral velocity of the moving cylinder to be V, the torque, T, may be calculated from the principles and methods of Section 1.5. The result is

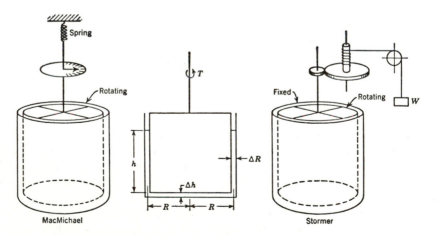

Fig. **11.2**   Rotational viscometers (schematic).

$$T = \frac{2\pi R^2 h\mu V}{\Delta R} + \frac{\pi R^3 \mu V}{2\Delta h}$$

in which the first term represents the torque due to viscous shear in the space between the cylinder walls, and the second term that between the ends of the cylinders. With R, h, ΔR, and Δh constants of the instrument, and the rotational speed (N) proportional to V, the equation may be written

$$T = K\mu N \qquad \text{or} \qquad \mu = \frac{T}{KN}$$

in which the constant K depends on the foregoing factors. For the MacMichael instrument the torque is proportional to the torsional deflection θ ($T = K_1\theta$) with the result

$$\mu = \frac{K_1\theta}{KN}$$

which shows that the viscosity may be obtained from deflection and speed measurements. In the Stormer viscometer the torque is constant since it is proportional to the weight W, which is a constant of the instrument; also, the time t required for a fixed number of revolutions is inversely proportional to N ($t = K_2/N$) with the result

$$\mu = \left(\frac{T}{KK_2}\right)t$$

and the time required for a fixed number of revolutions produced by the same falling weight thus becomes a direct measure of viscosity.

Typical tube-type viscometers are the Ostwald and Saybolt instruments of Fig. 11.3. The former is used typically for low-viscosity fluids, for example, water, while the latter is used for medium- to high-viscosity fluids. Similar to the Ostwald is the

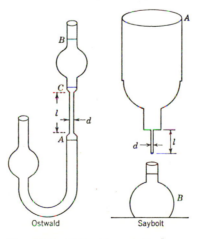

Ostwald   Saybolt

Fig. **11.3** Tube viscometers.

Bingham type, and similar to the Saybolt are the Redwood and Engler viscometers. All these instruments involve the *unsteady laminar flow* of a fixed volume of liquid through a small tube under standard head conditions. The time for the quantity of liquid to pass through the tube becomes a measure of the kinematic viscosity of the liquid.

The Ostwald viscometer is filled to level A, and the meniscus of the liquid in the right-hand tube is drawn up to a point above B and then released. The time for the meniscus to fall from B to C becomes a measure of the kinematic viscosity. In the Saybolt viscometer the outlet is plugged, and the reservoir filled to level A; the plug is then removed, and the time required to collect a fixed quantity of liquid in the vessel B is measured. This time then becomes a direct measure of the kinematic viscosity of the liquid.

The relation between time and kinematic viscosity for the tube-type viscometer may be indicated approximately by applying the Hagen-Poiseuille law for laminar flow in a circular tube (Section 9.2). The approximation involves the application of a law of steady established laminar motion to a condition of unsteady flow in a tube which may be too short for the establishment of laminar flow and, therefore, cannot be expected to give a complete or perfect relationship between efflux time and kinematic viscosity; it serves, however, to indicate elementary principles. From equation 9.8

$$Q = \frac{\pi d^4 \gamma h_L}{128 \mu l} \tag{11.1}$$

for steady laminar flow in a circular tube. But $Q = V/t$, in which V is the volume of liquid collected in time t. Substituting this in equation 11.1 and solving for μ,

$$\mu = \left(\frac{\pi d^4 h_L}{128 V l} \right) \gamma t$$

The head loss, h_L, however, is nearly constant since it is approximately equal to the imposed head which varies between fixed limits. Since d, l, V, and h_L are constants of the instrument, the equation reduces to $\mu \cong K\gamma t = K\rho g_n t$, from which $\mu/\rho = \nu \cong Kg_n t$, and kinematic viscosity is seen to depend almost linearly on measured time. A more exact (but empirical) equation relating ν and t for the Saybolt Universal viscometer is (for $t > 32$ s)

$$\nu(\text{m}^2/\text{s}) = 10^{-4} \left(0.002\ 197t - \frac{1.798}{t} \right)$$

in which t is the time in seconds (called *Saybolt seconds*). This equation approaches the linear one predicted by the approximate analysis for large values of the time, t. The familiar S.A.E. numbers used for motor oils are indices of kinematic viscosity, as shown in Table 9.

Of the tube viscometers, the Saybolt, Engler, and Redwood are built of metal to rigid specifications and hence may be used without calibration. Since the dimensions of the glass viscometers such as the Bingham and Ostwald cannot be so perfectly controlled, these instruments must be calibrated before viscosity measurements are made.

Table 9[a] S.A.E. Viscosities

S.A.E. Viscosity No.	Saybolt seconds at 99°C	ν(mm²/s) at 99°C
20	45 to 58	5.85 to 9.66
30	58 to 70	9.66 to 12.7
40	70 to 85	12.7 to 16.5

[a] Adapted from *S.A.E. Handbook*, Soc. Auto. Engrs., 1959.

MEASUREMENT OF PRESSURE

11.2 Static Pressure

The accurate measurement of pressure in a fluid at rest may be accomplished with comparative ease since it depends only on the accuracy of the gage or manometer used to record this pressure and is independent of the details of the connection between fluid and recording device. To measure the static pressure within a moving fluid with high accuracy is quite another matter, however, and depends on painstaking attention to the details of the connection between flowing fluid and measuring device.

For perfect measurement of static pressure in a flowing fluid a device is required which fits the streamline picture and causes no flow disturbance; it should contain a small smooth hole whose axis is normal to the direction of motion at the point where the static pressure is to be measured; to this opening is connected a manometer or pressure transducer. Meeting all these requirements is a virtual impossibility, but it is evident that the device must be as small as possible and constructed with great care. However, the most troublesome point in measuring the static pressure in a flowfield is the proper orientation or alignment of the device with the flow direction, which usually is not known in advance. Two basic designs (there are many adaptations) have solved this problem successfully. One of these is the thin disk of Fig. 11.4a containing separate piezometer openings in each side which lead to a differential manometer or transducer; this device is inserted in the flowfield and turned (about its stem) until the connected differential manometer reads zero, which shows the pressure on both sides of the disk to be the same and the disk thus aligned with the flow. After alignment is secured either pressure may be separately measured and taken to be the static pressure at the piezometer opening. A second device (for two-dimensional flows) is the cylinder of Fig. 11.4b containing two separate piezometer openings connected to a differential manometer or transducer; the cylinder is turned about its own axis until the manometer balances, showing the direction of flow to be along the bisector of the angle between the openings. At the stagnation point A, $p_A > p_o$ and, at B, $p_B < p_o$ with continuous fall of pressure from p_A to p_B; accordingly, at some point between A and B on the surface of the cylinder, a local pressure equal to the static pressure will be found and

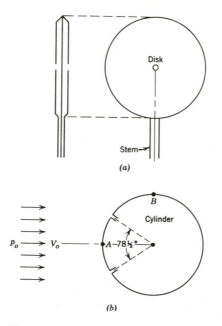

(a)

(b)

Fig. **11.4.**

may be measured on a separate manometer. Fechheimer[2] has shown that the appropriate angle between the piezometer openings should be 78.5° for incompressible flow, and Thrasher and Binder[3] have confirmed this for compressible flows of velocity up to one-fourth of the sonic (at higher velocities a larger angle is required). Once the flow direction has been found and the static pressure recorded, the cylinder may be rotated through the appropriate angle so that one of the piezometer openings is at the stagnation point, at which time the stagnation pressure may be measured and the velocity V_o obtained. Thus this simple and compact device serves a threefold purpose in the measurement of static pressure, flow direction, and velocity. A spherical counterpart of this device has been developed for use in three-dimensional flowfields.

In fields where the flow direction is known to fair accuracy, the *static tube* (Fig. 11.5) is the usual means of measuring static pressure. Such a tube is merely a small

Fig. **11.5**   Static tube.

[2] C. J. Fechheimer, "The Measurement of Static Pressure," *Trans. A.S.M.E.*, vol. 48, p. 965, 1926.
[3] L. W. Thrasher and R. C. Binder, "Influence of Compressibility on Cylindrical Pitot-Tube Measurements," *Trans. A.S.M.E.*, vol. 72, p. 647, 1950.

smooth cylinder with a rounded or pointed upstream end; in the side of the cylinder are piezometer holes or a circumferential slot through which pressure is transmitted to transducer or manometer. Assuming perfect alignment with the flow, the flow past the tube will be symmetrical and of mean velocity slightly larger than V_o; hence the pressure at the piezometer openings may be expected to be slightly less than p_o. This error is minimized by making the tube as small as possible and is usually negligible in engineering work. In experimental work or in the use of the static tube on aircraft, some misalignment of the tube with flow direction is to be expected; when this occurs the pressure on one side of the tube becomes larger than p_o, on the other side less than p_o, and some flow through the tube (from opening to opening) results. With such complexities, the pressure carried to gage or manometer cannot be exactly predicted but will be close to p_o for small angles of misalignment. A static tube which is insensitive to misalignment is desired by the experimentalist since, with larger (and incurable) errors of alignment, accurate values of static pressure may still be obtained.

    The static pressures in the fluid passing over a solid surface (such as a pipe wall or the surface of an object; Fig. 11.6) may be measured successfully by small smooth piezometer holes drilled normal to the surface; such surfaces fit the flow perfectly since (assuming no separation) they are streamlines of the flow. These piezometer openings can measure only the local static pressures at their locations on the solid surface and cannot, in general, measure the pressures at a distance from this surface since such pressures differ from those at the surface owing to flow curvatures and accelerations. Where no flow curvatures exist, as in a straight passage, a wall piezometer opening will allow pressures throughout the cross section of the passage to be predicted.

    In addition to the manometer (see Section 2.3), pressure may be sensed by a range of transducers that convert pressure or pressure differential to an electric output. A common form of transducer is a diaphragm gage (Fig. 11.7), which is essentially a differential pressure device (when port 1 is open to the atmosphere, gage pressure is measured). The diaphragm is an elastic (metal) element. If $p_1 = p_2$, the diaphragm is centered in the cavity; it is deflected to one side or the other depending on how p_1 and p_2 vary. Two types of diaphragm transducer are the strain-gage and capacitance types. If electrical-resistance strain gages are attached to the diaphragm, its deflection can be measured as an electrical output signal that, through a calibration process, can be correlated to the differential pressure. On the other hand, if the metal diaphragm is properly insulated so as to form one plate of a capacitor in an electric circuit, the changing capacitance caused by motion of the diaphragm can be correlated with

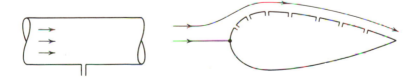

Fig. **11.6**

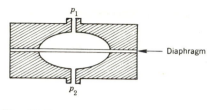

Fig. **11.7**

pressure differential. These and other types of transducer are described by Benedict and Holman (see end-of-chapter References).

11.3 Surface Elevation

The elevation of the surface of a liquid at rest may be determined by manometer, piezometer column, or pressure-gage readings (Sections 2.1–2.3).

The same methods may be applied to flowing liquids if the foregoing precautions for static pressure measurement (Section 11.2) are followed and if the piezometer method is used only where the streamlines of flow are *essentially* straight and parallel. Correct and incorrect measurements of a liquid surface by piezometer openings are illustrated in Fig. 11.8.

Floats are often used in connection with chronographic water-level recorders for measuring liquid-surface elevations. The arrangement of such floats is indicated schematically in Fig. 11.9. As the liquid level varies, the motion of the cable is measured on a scale, plotted automatically on a chronographic record sheet, or recorded electrically for later analysis.

There are several other methods of sensing the location of a liquid surface. Sonic devices can be used; they depend on the precise measurement of the time necessary for a sound pulse to travel to the surface and return to its source. Two very popular and cheap sensors are based on the introduction of either a single insulated wire or two bare conductors into the water as a component of an electrical system. In the first case a cylindrical capacitor is formed with the wire conductor being one plate, the water being the other plate, and the insulation acting as the dielectric between the plates. With the water as the ground terminal, a voltage applied to the capacitor encounters a capacitance proportional to the length of wire immersed in the water. The resistive sensor is created by applying a voltage across two bare wires which are partially immersed in the

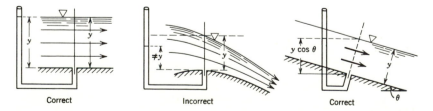

Fig. **11.8** Measurement of surface elevation with piezometer columns.

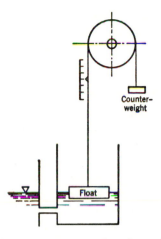

Fig. **11.9** Float for measurement of surface elevation (schematic).

water. The resistance between the terminals is that which is offered by the water so that changes in the depth of immersion are indicated by the change in the resistance.

When these sensors are included in one arm of an electrical bridge of the appropriate type—capacitance or resistance—the changes due to the motion of the water surface are easily measured. For both types of sensors the output bridge voltage will be proportional to the change in depth if the change in the electrical quantity measured is small compared to that in the bridge arm to which the sensor is connected. Figure 11.10 illustrates the basic operational concepts of these sensors.

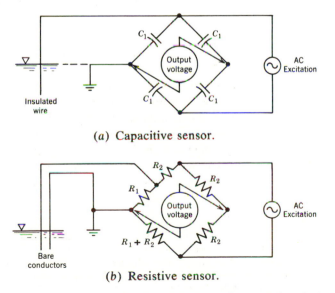

(a) Capacitive sensor.

(b) Resistive sensor.

Fig. **11.10** Schematic diagrams for surface sensing devices.

Staff gages give comparatively crude but direct measurements of liquid-surface elevation. From casual observation, the reader is familiar with their use as tide gages, in the measurement of reservoir levels, and in registering the draft of ships.

11.4 Stagnation Pressure

The stagnation[4] pressure (also called *total pressure*) may be measured accurately by placing in the flow a small solid object having a small piezometer hole at the stagnation point. The piezometer opening may be easily located at the stagnation point if the hole is drilled along the axis of a symmetrical object such as a cylinder, cone, or hemisphere; with the axis of the object properly aligned with the direction of flow, the piezometer opening is automatically located at the stagnation point, and the pressure there may be transferred through the opening to a recording device. Theoretically, the upstream end of solid objects for this purpose may be of any shape, since the shape of the object does not affect the magnitude of the stagnation pressure; in laboratory practice, however, the upstream end is usually made convergent (conical or hemispherical) in order to fix more precisely the location of the stagnation point.

When used on airplanes or in experimental work, misalignments of tube with flow are inevitable, but reliable measurements of stagnation pressure are still required; research[5] led to the development of a tube shielded by a special jacket which gives reliable measurements at angles of misalignment up to 45°.

The early experimental work of Henri Pitot (1732) provided the basis for the measurement of stagnation pressure by showing that a small tube with open end facing upstream (Fig. 11.11) provided a means for measuring velocity. He found that when such tubes (later called *pitot tubes*) were placed in an open flow where the velocity was V_o the liquid in the tube rose above the free surface a distance $V_o^2/2g_n$. From Fig. 11.11 the expression for stagnation pressure,

$$\frac{p_s}{\gamma} = \frac{p_o}{\gamma} + \frac{V_o^2}{2g_n} \qquad \text{or} \qquad p_s = p_o + \tfrac{1}{2}\rho V_o^2 \tag{4.3}$$

may be written directly, thus confirming the equations of Section 4.4.

In the use of pitot tubes for the measurement of velocity profiles in shear flows, experimentalists have noted[6] an error due to the asymmetry of the flow near the tip of the tube. This error is small where the velocity gradient is low but increases near solid boundaries where the velocity gradient is high. The sense of the error is to move the effective center of the tube toward the region of lower velocity gradient; this means that the velocity being measured by stagnation pressure will be found at a point farther from

---

[4] See Sections 4.4 and 5.4.

[5] W. R. Russell, W. Gracey, W. Letko, and P. G. Fournier, "Wind Tunnel Investigation of Six Shielded Total-Pressure Tubes at High Angles of Attack," *N.A.C.A. Tech. Note* 2530, 1951.

[6] See, for example, A. D. Young and J. N. Maas, "The Behavior of a Pitot Tube in a Transverse Pressure Gradient," *Aero. Res. Comm. R. & M.* 1770, 1936; and F. A. McMillan, "Experiments on Pitot Tubes in Shear Flow," *Aero. Res. Comm. R. & M.* 3028, 1956.

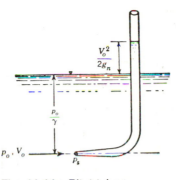

Fig. **11.11**   Pitot tube.

the wall than the axis of the pitot tube; the magnitude of this error also depends on the detailed geometry of the tube.

MEASUREMENT OF VELOCITY

11.5 Pitot-Static Tube in Incompressible Flow

From the stagnation pressure equation for an incompressible fluid may be obtained

$$V_o = \sqrt{\frac{2(p_s - p_o)}{\rho}} = \sqrt{\frac{2g_n(p_s - p_o)}{\gamma}} \tag{11.2}$$

from which it is apparent that velocities may be calculated from measurements of stagnation and static pressures. It has been shown that stagnation pressures may be measured easily and accurately by a pitot tube, and static pressures by various methods such as tubes, flat plates, and wall piezometer openings. Therefore, a pitot tube can be used together with any static pressure device to obtain the necessary pressure difference $p_s - p_o$ from which V_o is deduced. An integral combination of stagnation- and static-pressure-measuring devices into a single device is known usually as a *pitot-static tube* .

Modern practice favors the combined type of pitot-static tube, several of which are illustrated in Fig. 11.12. Here the static tube jackets the stagnation pressure tube, resulting in a compact, efficient, velocity-measuring device. When connected to a differential pressure-measuring instrument, the pressure difference $(p_s - p_o)$, which is seen from equation 11.2 to be a direct measure of the velocity V_o, may be read.

A static tube has been shown to record a pressure slightly less than the true static pressure, owing to the increase in velocity past the tube (Section 11.2). This means that equation 11.2 must be modified by an experimentally determined instrument coefficient, C_l, to

$$V_o = C_l\sqrt{\frac{2(p_s - p_o')}{\rho}} = C_l\sqrt{\frac{2g_n(p_s - p_o')}{\gamma}} \tag{11.3}$$

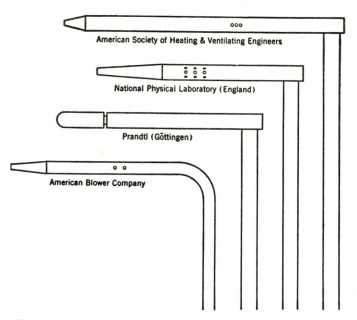

Fig. **11.12**   Pitot-Static tubes (to scale).

in which p_o' is the actual pressure measured by the static tube. Since p_o' is less than p_o, it is evident that C_I will always be less than unity. However, for most engineering problems the value of C_I may be taken as 1.00 for the conventional types of pitot-static tubes (Fig. 11.12), since the differences between p_o and p_o' are very small. Prandtl has designed a pitot-static tube in which the difference between p_o and p_o' is completely eliminated by ingenious location of the static-pressure opening. The opening is so located (Fig. 11.13) that the underpressure caused by the tube is exactly compensated by the overpressure due to the stagnation point on the leading edge of the stem, thus giving the true static pressure at the piezometer opening.

A practical aspect of velocity-measuring devices is their sensitivity to yaw or misalignment with the flow direction. Since perfect alignment is virtually impossible, it is advantageous for a pitot-static tube to produce minimum error when perfect alignment does not exist. Prandtl's pitot-static tube, designed to be insensitive to small angles of yaw, gives a variation of only 1% in its coefficient at an angle of yaw of 19°. For the same percentage variation in coefficient the American Society of Heating and Ventilating Engineers' pitot-static tube may have an angle of yaw of 12°, and that of the National Physical Laboratory only 7°.[7]

---

[7] Data from K. G. Merriam and E. R. Spaulding, "Comparative Tests of Pitot-Static Tubes," *N.A.C.A. Tech. Note* 546, 1935.

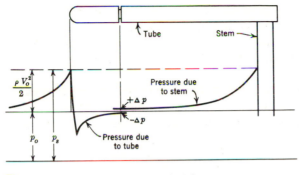

Fig. **11.13** Prandtl's pitot-static tube

---------- **Illustrative Problem** ----------

A pitot-static tube having a coefficient of 0.98 is placed at the center of a pipeline in which benzene is flowing. A manometer attached to the pitot-static tube contains mercury and benzene and shows a reading of 75 mm. Calculate the velocity at the centerline of the pipe.

Relevant Equation and Given Data

$$V_o = C_l \sqrt{\frac{2(p_s - p_o')}{\rho}} = C_l \sqrt{\frac{2g_n(p_s - p_o')}{\gamma}} \qquad (11.3)$$

$$\text{s.g.}_{Hg} = 13.57 \qquad \text{s.g.}_{ben} = 0.88$$

$$MR = 0.075 \text{ m} \qquad C_l = 0.98$$

Solution.   Using manometer principles,

$$\frac{p_s - p_o'}{\gamma} = 0.075 \times \frac{13.57 - 0.88}{0.88} = 1.08 \text{ m of benzene}$$

$$V_c = 0.98\sqrt{2g_n(1.08)} = 4.51 \text{ m/s} \blacktriangleleft \qquad (11.3)$$

11.6 Pitot-Static Tube in Compressible Flow

For velocity measurements in compressible flow, separate measurements of static pressure, stagnation pressure, and stagnation temperature are required. For subsonic flow ($M_o < 1$) equation 5.18 may be used directly:

$$\frac{V_o^2}{2} = c_p T_s \left[1 - \left(\frac{p_o}{p_s}\right)^{(k-1)/k} \right] \qquad (5.18)$$

with p_s obtained by pitot tube, p_o by static tube, and T_s by temperature probe. The temperature probe consists of a small thermocouple surrounded by a jacket with open upstream end and small holes at the rear; a stagnation point exists on the upstream end of the probe, and a temperature close to the stagnation temperature is measured by the thermocouple.[8]

For supersonic flow ($M_o > 1$) a short section of normal shock wave (Section 6.7) will be found upstream from the stagnation point (Fig. 11.14) and velocity calculations are considerably more complicated. Applying equation 6.18 through the shock wave,

$$\frac{p_1}{p_o} = 1 + \frac{2k}{k+1}(M_o^2 - 1) \tag{11.4}$$

Applying equation 5.16 between the downstream side of the shock wave and the stagnation point,

$$\frac{p_s}{p_1} = \left(1 + \frac{k-1}{2}M_1^2\right)^{k/(k-1)} \tag{11.5}$$

Multiplying these two equations together and substituting the relation between M_o and M_1 of equation 6.17 yields

$$\frac{p_s}{p_o} = \frac{k+1}{2}M_o^2\left[\frac{(k+1)^2M_o^2}{4kM_o^2 - 2k + 2}\right]^{1/(k-1)} \tag{11.6}$$

which shows that measurements of p_s and p_o will allow prediction of the Mach number M_o of the undisturbed stream. Using the fact that the stagnation temperature is the same on both sides of the shock wave,

$$V_o^2/2 = c_p(T_s - T_o) \tag{11.7}$$

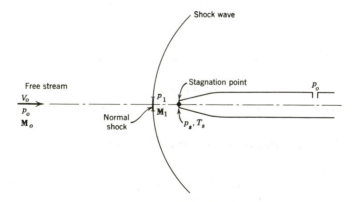

Fig. **11.14**   Pitot tube in a supersonic flow.

[8] The measured temperature is not exactly the stagnation temperature, and calibration of the instrument is required.

However, from $a_o^2 = kRT_o$ and $\mathbf{M}_o = V_o/a_o$, $T_o = V_o^2/kR\,\mathbf{M}_o^2$, which when substituted in the foregoing equation yields

$$\frac{V_o^2}{2}\left(1 + \frac{2c_p}{kR\,\mathbf{M}_o^2}\right) = c_pT_s \qquad (11.8)$$

Use of equations 11.6 and 11.8 allows V_o to be calculated from measurements of T_s, p_s, and p_o.

From measurements of p_s and p_o the experimentalist may easily identify subsonic and supersonic flow and thus select the proper equation for velocity calculation. By inserting $\mathbf{M}_o = 1$ in equation 5.16 or 11.6, we obtain

$$\frac{p_s}{p_o} = \left(\frac{k + 1}{2}\right)^{k/(k-1)} \qquad (11.9)$$

Thus, for $\mathbf{M}_o < 1$, $p_s/p_o < [(k + 1)/2]^{k/(k-1)}$ and, for $\mathbf{M}_o > 1$, $p_s/p_o > [(k + 1)/2]^{k/(k-1)}$.

--------------- **Illustrative Problems** ---------------

A pitot-static tube in an air duct indicates an absolute stagnation pressure of 635 mm of mercury; the absolute static pressure in the duct is 457 mm of mercury; a temperature probe shows a stagnation temperature of 66°C. What is the local velocity just upstream from the pitot-static tube?

Relevant Equations and Given Data

$$\frac{V_o^2}{2} = c_pT_s\left[1 - \left(\frac{p_o}{p_s}\right)^{(k-1)/k}\right] \qquad (5.18)$$

$$\frac{p_s}{p_o} = \left(\frac{k + 1}{2}\right)^{k/(k-1)} \qquad (11.9)$$

$$p_s = 635 \text{ mm Hg} \qquad p_o = 457 \text{ mm Hg} \qquad T_s = 66°C$$

$$c_p = 1\,003 \text{ J/kg} \cdot \text{K} \qquad k = 1.4$$

Solution.

$$\frac{635}{457} = 1.39 < \left(\frac{1.4 + 1}{2}\right)^{3.5} = 1.893 \qquad (11.9)$$

Therefore equation 5.18 should be used for velocity calculation:

$$V_o^2/2 = 1\,003(273 + 66)\left[1 - \left(\frac{457}{635}\right)^{0.286}\right] = 30\,529 \text{ m}^2/\text{s}^2 \text{ of air}$$

$$V_o = 247 \text{ m/s} \blacktriangleleft \qquad (5.18)$$

The instruments of a high-speed airplane flying at high altitude show stagnation pressure 20 in. (508 mm) of mercury abs., static pressure 5 in. (127 mm) of mercury abs., and stagnation temperature 150°F (66°C). Calculate the speed of this airplane.

Relevant Equations and Given Data

$$\frac{p_s}{p_o} = \frac{k+1}{2} M_o^2 \left[\frac{(k+1)^2 M_o^2}{4k M_o^2 - 2k + 2} \right]^{1/(k-1)} \tag{11.6}$$

$$\frac{V_o^2}{2}\left(1 + \frac{2c_p}{kR M_o^2}\right) = c_p T_s \tag{11.8}$$

See equation 11.9 in the problem above.

$$p_s = 20 \text{ in. Hg} \qquad p_o = 5 \text{ in. Hg} \qquad T_s = 150°F$$
$$R = 1\ 715 \text{ ft·lb/slug·°R} \qquad c_p = 6\ 000 \text{ ft·lb/slug·°R} \qquad k = 1.4$$

Solution.

$$\frac{20}{5} = 4 > \left(\frac{1.4+1}{2}\right)^{3.5} = 1.893 \tag{11.9}$$

Therefore equations 11.6 et seq. should be used for velocity calculation:

$$4 = \frac{1.4+1}{2} M_o^2 \left[\frac{(2.4)^2 M_o^2}{5.6 M_o^2 - 2.8 + 2} \right]^{2.5} \tag{11.6}$$

Solving (by trial),

$$M_o^2 = 2.71 \qquad M_o = 1.645$$

$$\frac{V_o^2}{2}\left(1 + \frac{2 \times 6\ 000}{1\ 715 \times 1.4 \times 2.71}\right) = 6\ 000(610) \qquad V_o = 1\ 600 \text{ ft/s} \blacktriangleleft \tag{11.8}$$

11.7 Anemometers and Current Meters

Mechanical devices of similar characteristics are utilized in the measurement of velocity in air and water flow. Those for air are called *anemometers*; those for water, *current meters*. These devices consist essentially of a rotating element whose speed of rotation varies with the local velocity of flow, the relation between these variables being found by calibration. Anemometers and current meters fall into two main classes, depending on the design of the rotating elements; these are the cup type and vane (propeller) type, as illustrated in Fig. 11.15. Anemometers and current meters differ slightly in shape, ruggedness, and appurtenances because of the different conditions under which they are used. The cup-type anemometer for the measurement of wind velocity is usually mounted on a rigid shaft; the vane-type anemometer is held in the hand while readings are taken. The current meters are usually suspended in a river or canal by a cable and hence must have empennages and weights to hold them in fixed positions in the flow.

Cup type *N.Y.U.*   Vane type *N.Y.U.*

Anemometers

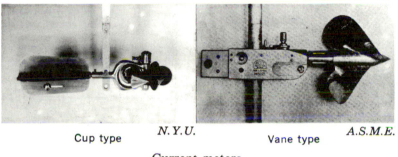

Cup type *N.Y.U.*   Vane type *A.S.M.E.*

Current meters

Fig. **11.15**

Several nonmechanical devices which utilize wave propagation concepts are also available. For example, Kolin[9] developed a nonmechanical velocity-measuring device based on electromagnetic principles which may be used successfully in liquid flows. Here the flowing liquid is utilized as a conductor which develops a voltage as it passes through a magnetic field; after calibration, measurement of this voltage allows computation of the velocity. The device may be used to obtain the mean velocity in a pipe flow or, when built in very small sizes, to obtain local velocities within a flowing liquid. Elrod and Fouse[10] discuss the application of such meters to the flow measurement of liquid metals.

[9] A. Kolin, "Electromagnetic Velometry, I. A Method for the Determination of Fluid Velocity Distribution in Space and Time," *Jour. Appl. Phys.*, vol. 15, no. 2, p. 150, 1944. See also J. A. Shercliff, *Electromagnetic Flow Measurement*, Cambridge University Press, 1962, or for all flow metering devices mentioned, the recent references at the end of the chapter.

[10] H. G. Elrod, Jr., and R. R. Fouse, "An Investigation of Electromagnetic Flowmeters," *Trans. A.S.M.E.*, vol. 74, no. 4, p. 589, 1952. See also V. Cushing, "Induction Flowmeter," *Rev. Sci. Instruments,* vol. 29, 1958, and V. P. Head, "Electromagnetic Flow Meter Primary Elements," *Trans. A.S.M.E.* (*Series D*), December 1959.

Two other devices, based on the effect of the fluid motion on wave propagation, are coming into wide use. These are the acoustic flowmeter[11] and the laser-Doppler-anemometer.[12] Both share the distinct advantage that they do not disturb the fluid flow and have no moving ports.

11.8 Hot-Wire and Hot-Film Anemometers

Another type of anemometer which is used for measuring both mean and instantaneous velocities in gas and liquid flows is the *hot-wire* or *hot-film anemometer*. At present, most measurements of turbulent velocity fluctuations are made with these instruments. However, for accessible flows (those where a light beam can enter and leave the flow through a free surface or transparent walls or ports) the laser-Doppler anemometer is taking over this role. Figure 11.16 shows typical sensing elements and their supports for hot-wire and hot-film operation.

As seen in Fig. 11.16 the sensing element of the anemometer is either a thin wire or a metal film laid over a glass support and coated to protect the film. Because of the supporting rod and coating the hot-film sensor is mechanically superior and usable in contaminated environments. The sensor element is heated electrically by an electronic circuit which allows measurement of the power supplied to the element. The power supplied is related to the fluid velocity *normal to the sensor* through the laws of convective heat transfer between the sensor and the fluid. With the proper electronic control and compensation a thin hot-wire can respond to velocity fluctuations at frequencies up to 500 000 Hz.

The anemometer sensors are usually an electronic part of either a *constant temperature* or *constant current* circuit, the former being the system most often used. In a constant temperature anemometer (Fig. 11.17a) the sensor is one leg of a bridge circuit. The system of resistances is balanced at a no-flow condition by use of the variable resistor so that there is no unbalance-voltage output. Then flow past the sensor will cool it and decrease its resistance. The detector senses the unbalanced condition and changes its output voltage to increase the current flowing in the sensor, thus bringing its temperature back to the original value. The bridge voltage, V (the output voltage of the amplifier), is proportional to the current, I, flowing in the sensor because all resistances are kept constant. But, $P = I^2R$, so the square of the output voltage is proportional to the instantaneous heat transfer from the sensor.

In a constant current anemometer (Fig. 11.17b), the current in the sensor is maintained constant by putting the sensor in series with a very large (relative to the sensor resistance) resistor. When a flow cools the sensor, its resistance decreases, thereby unbalancing the bridge by changing the voltage at point A. This unbalance voltage is the output which must be greatly amplified. Generally, a constant current

---

[11] F. C. Lowell, Jr., and F. Hirschfeld, "Acoustic flowmeters for pipelines," *Mechanical Engineering*, October 1979, pp. 28–35.

[12] F. Durst, A. Melling, and J. H. Whitelaw, *Principles and practice of laser-Doppler anemometry*. Academic Press, 1976.

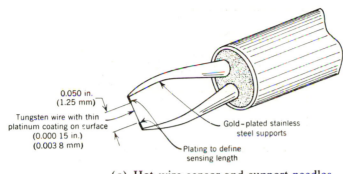

0.050 in.
(1.25 mm)

Tungsten wire with thin
platinum coating on surface
(0.000 15 in.)
(0.003 8 mm)

Gold–plated stainless
steel supports

Plating to define
sensing length

(*a*) **Hot-wire sensor and support needles.**

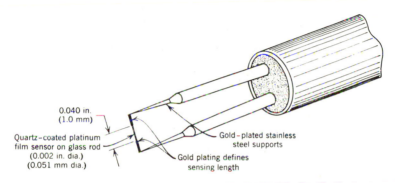

0.040 in.
(1.0 mm)

Quartz–coated platinum
film sensor on glass rod
(0.002 in. dia.)
(0.051 mm dia.)

Gold–plated stainless
steel supports

Gold plating defines
sensing length

(*b*) **Hot-film sensor and support needles.**

Fig. **11.16** Anemometer sensors. (Reproduced from TB5, Thermo-Systems, Inc.,
2500 Cleveland Avenue North, St. Paul, Minnesota, 55113.)

system is more complex to operate and less accurate than a constant temperature
system.

King[13] showed that the power transferred from a wire (or film) sensor can be
expressed as

$$P = I^2 R_{sensor} = (a + b \sqrt{\rho v'}) (T_{sensor} - T_{fluid})$$

where I is the current flowing through the sensor, v' is the instantaneous fluid velocity
normal to the sensor, and a and b are empirical constants obtained by calibration of the
wire (or film). For the constant temperature system, R_{sensor} and T_{sensor} are constant, so
$V^2 \propto I^2$. Thus, provided T_{fluid} is constant,

$$V^2 = (A + B \sqrt{\rho v'}) \tag{11.10}$$

[13] L. V. King, "On the Convection of Heat from Small Cylinders in a Stream of Fluid, with Applications to
Hot-Wire Anemometry," *Phil. Trans. Roy. Soc. London*, vol. 214, no. 14, 1914, pp. 373–432.

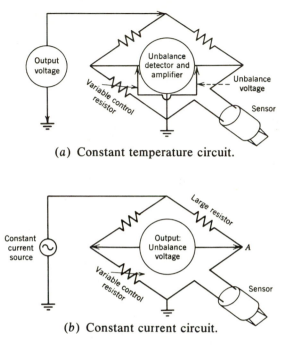

(*a*) Constant temperature circuit.

(*b*) Constant current circuit.

Fig. **11.17**   Schematic of anemometers.

where V is the output voltage of the bridge. Figure 11.18 shows a typical calibration curve for a hot-film sensor.

Figure 7.4 shows a schematic of the variation of one component of the instantaneous velocity in a turbulent flow. Figure 11.19 gives a plot of experimental data, obtained at Stanford, for air flow over a smooth flat plate in a wind tunnel. Here a pair of hot-wires were mounted in an X-array.[14] In this configuration, the sum and difference of the velocities normal to each wire are used to obtain the horizontal and vertical components of the flow, respectively. With mean values removed, the remainder is used to calculate the intensities shown in Fig. 11.19.

MEASUREMENT OF SHEAR

11.9   Shear Measurements—By Inference

No device has yet been invented which is capable of measuring the stress between moving layers of fluid. Shear measurements consist entirely of measurements of wall shear (τ_o) from which the shear between moving layers may be deduced from certain

---

[14] The two wires lie next to each other in vertical parallel planes that are in line with the flow. One wire is inclined at $+45°$ to the horizontal and one at $-45°$.

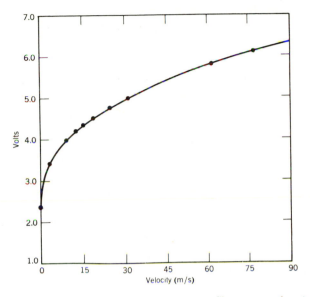

Fig. **11.18** Calibration for a 0.051 mm diameter hot-film sensor in atmospheric air: 0–90 m/s. (Reproduced from TB5, Thermo-Systems, Inc., 2500 Cleveland Avenue North, St. Paul, Minnesota 55113.)

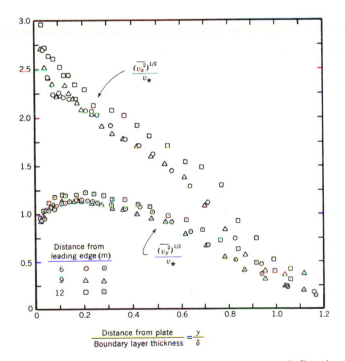

Fig. **11.19** Turbulent velocity intensities in air flow over a smooth flat plate.

equations of fluid mechanics; such deductions may be of high or low accuracy, depending on the equations used and the approximations necessary for solving them.

The wall shear for a cylindrical pipe of uniform roughness and with established flow may be obtained easily and accurately from pressure measurements through the use of equation 7.21, which may be expressed as

$$\tau_o = \frac{\gamma d}{4l}\left[\frac{p_1}{\gamma} + z_1 - \frac{p_2}{\gamma} - z_2\right] \tag{11.11}$$

in which the bracketed term is recognized as the head loss between points 1 and 2. With all details of the flow axisymmetric and no variation of wall roughness, it may be safely assumed that τ_o is the same at all points on the pipe wall and its value deducible from equation 11.11. The same procedure may be applied to any prismatic conduit and the same equation may be used with d replaced by $4R_h$; here, however, the flow is not axisymmetric and the shear stress must be interpreted as the *mean value*. Although on any longitudinal element of such a conduit the wall shear may be presumed constant, the equation provides no information on whether the wall shear at any point is larger or smaller than the mean value.

11.10 Shear Measurements—Wall Probes

Because of the foregoing limitations and for applications to problems of more complex boundary geometry, a more basic type of shear meter has been developed. This consists (Fig. 11.20) in replacing a small section of the wall by a movable plate mounted on elastic columns fastened to a rigid support. The columns are deflected slightly by the shearing force of fluid on plate, the small deflection measured by strain gages, and the shear stress deduced from this deflection.[15] Although the device is basically simple, it is costly, unwieldy, and by no means easy to operate and interpret because of the relatively small shear force to be measured and the relatively large extraneous forces that also contribute to the deflection of the columns.

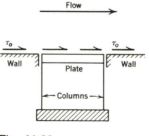

Fig. **11.20**

---

[15] See Section 6.5.2.1 in W. H. Graf, *Hydraulics of Sediment Transport*, McGraw-Hill, 1971.

Wall pitot tubes have been used successfully for the measurement of wall shear. Stanton[16] first used the design shown in Fig. 11.21, the wall forming one side of the pitot tube; calibration in laminar pipe flow and in the viscous sublayer of turbulent pipe flow showed τ_o to be a function of $(p_s - p_o)$. Its use for turbulent flows is thus restricted to measurements in the viscous sublayer covering the wall, where the velocity profile is essentially the same as that in the pipe for the same fluid viscosity and wall shear. Taylor[17] presented a dimensionless calibration for the Stanton tube which may be expressed by $\tau_o h^2/4\rho\nu^2$ as a function of $(p_s - p_o)h^2/4\rho\nu^2$; this relationship is shown in Fig. 11.21, which allows τ_o to be predicted from measurements of $(p_s - p_o)$, ρ, ν, and h, providing that h is smaller than the thickness of the viscous region.

More recently, Preston[18] has applied for foregoing idea to turbulent flow over smooth surfaces with a tube of simpler design. The Preston tube (Fig. 11.22) is not submerged in the viscous sublayer, and its performance depends on the similarity of the velocity profiles through the buffer[19] zone between the viscous sublayer and the turbulent region. The single calibration curve of Fig. 11.22 validates this similarity over the range indicated.

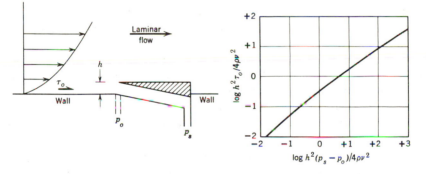

Fig. 11.21 Stanton tube.

[16] T. E. Stanton, D. Marshal, and (Mrs.) C. N. Bryant, "On the Conditions at the Boundary of a Fluid in Turbulent Motion," *Proc. Roy. Soc.*, (A), vol. 97, 1920.

[17] G. I. Taylor, "Measurements with Half-Pitot Tubes," *Proc. Roy. Soc.*, (A), vol. 166, 1938.

[18] J. H. Preston, "The Determination of Turbulent Skin Friction by Means of Pitot Tubes," *Journ. Roy. Aero. Soc.*, vol. 58, 1954. See also E. Y. Hsu, "The Measurement of Local Turbulent Skin Friction by Means of Pitot Tubes," *David Taylor Model Basin* Rept. 957, August 1955, and V. C. Patel, "Calibration of the Preston Tube and Limitations on its Use in Pressure Gradients," *J. Fluid Mech.*, vol. 23, no. 1, 1965, pp. 185–208.

[19] See Fig. 9.4 of Section 9.3.

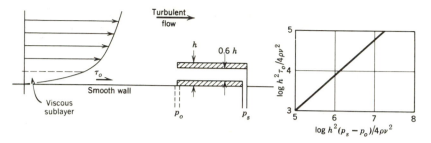

Fig. **11.22** Preston tube.

MEASUREMENT OF FLOWRATE

In some steady flows, the flowrate (Q or G) can be obtained by simple measurement of the total quantity of fluid collected in a measured time. Such collections can be made by weight or by volume and are a primary means of measuring fluid flow. However, they are usually practical only for small flows under laboratory conditions. For gas flows, volumetric measurements must be made under conditions of constant pressure and temperature.

For routine practical measurements there exist a plentiful supply of *flowmeters* and flow-measuring techniques. Among the flowmeters are positive displacement types (e.g., reciprocating piston meters, nutating disk meters, rotary piston and vane meters, etc.) and differential measurement systems (e.g., venturi meters, orifices, nozzles, and elbow meters). For open-channel flows, weirs and the dilution and salt-velocity methods are useful, while multiple current meter (Section 11.7) measurements across a channel section can be integrated to give flowrate. For many flow applications, magnetic and ultrasonic (acoustic) flow measurement devices[20] offer nonintrusive measurements.

No attempt is made here to describe all possible systems. A representative sample is discussed; the appropriate references at the end of the chapter contain detailed information.

11.11 Venturi Meters

A constriction in a streamtube has been seen (Sections 4.4 and 5.8) to produce an accelerated flow and fall of hydraulic grade line or pressure which is directly related to flowrate, and thus is an excellent meter in which rate of flow may be calculated from pressure measurements. Such constrictions used as fluid meters are obtained by Venturi meters, nozzles, and orifices.

A Venturi meter is shown in Fig. 11.23. It consists of a smooth entrance cone of angle about 20°, a short cylindrical section, and a diffuser of 5° to 7° cone angle in order

---

[20] N. P. Cheremisinoff, *Applied Fluid Flow Measurement*, M. Dekker, Inc., 1979.

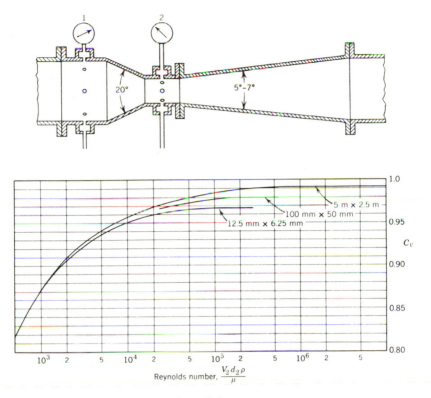

Fig. **11.23** Venturi meter and coefficients.

to minimize head loss (Section 9.9). For satisfactory operation of the meter the flow should be an established one as it passes section 1. To ensure this the meter should be installed downstream from a section of straight and uniform pipe, free from fittings, misalignments, and other sources of large-scale turbulence, and having a length of at least 30 (preferably 50) pipe diameters. Straightening vanes may also be placed upstream from the meter for reduction of rotational motion in the flow.

The pressures at the *base* of the meter (section 1) and at the *throat* (section 2) are obtained by piezometer rings, and the pressure difference is usually measured by differential manometer or pressure transducer.[21] For the metering of gases, separate measurements of pressure and temperature are required at the base of the meter, but for liquids the differential pressure reading alone will allow computation of the flowrate.

For ideal incompressible flow the flowrate may be obtained by solving simultaneous equations 3.12 and 4.1 (see Section 4.4) to yield

---

[21] In industrial practice, this pressure difference is frequently carried to an electromechanical device which calculates and plots a chronographic record of the flowrate.

$$Q = \frac{A_2}{\sqrt{1 - (A_2/A_1)^2}} \sqrt{2g_n \left(\frac{p_1}{\gamma} + z_1 - \frac{p_2}{\gamma} - z_2 \right)} \tag{11.12}$$

However, for real-fluid flow and the same $(p_1/\gamma + z_1 - p_2/\gamma - z_2)$ the flowrate will be expected to be less than that given by equation 11.12 because of frictional effects and consequent head loss between sections 1 and 2; in metering practice it is customary to account for this by insertion of an experimentally determined coefficient C_v in equation 11.12,[22] which then becomes

$$Q = \frac{C_v A_2}{\sqrt{1 - (A_2/A_1)^2}} \sqrt{2g_n \left(\frac{p_1}{\gamma} + z_1 - \frac{p_2}{\gamma} - z_2 \right)} \tag{11.13}$$

Since C_v is merely a convenient means of expressing the head loss $h_{L_{1-2}}$, an exact relation between these variables is to be expected; this may be found by substituting $A_2 V_2$ for Q in the foregoing equation and rearranging it to the form of the conventional Bernoulli equation. The result is

$$h_{L_{1-2}} = \left[\left(\frac{1}{C_v^2} - 1 \right) \left(1 - \left(\frac{A_2}{A_1} \right)^2 \right) \right] \frac{V_2^2}{2g_n} \tag{11.14}$$

The bracketed quantity is the conventional minor loss coefficient, K_L, for the entrance cone of the meter and may be calculated from C_v.[23] Loss coefficients and pipe friction factors have been seen to depend on Reynolds number and to diminish with increasing Reynolds number; from the structure of the bracketed quantity it can be predicted that C_v will increase with increasing Reynolds number. This prediction is borne out by the typical experimental results of Fig. 11.23[24] for Venturi meters of different size but of the same diameter ratio.[25] It should be observed in view of the principles of similitude (Chapter 8) that geometrically similar meters could be expected to give results falling on a single line on the plot, when installed in pipelines with established flow.

───────────── **Illustrative Problem** ─────────────

Air flows through a 150 mm by 75 mm Venturi meter. The gage pressure is 140 kPa and the fluid temperature 15°C at the base of the meter, and the differential manometer shows a reading of 150 mm of mercury. The barometric pressure is 101.3 kPa. Calculate the flowrate.

Relevant Equations and Given Data

$$\rho = p/RT \tag{1.3}$$

---

[22] For compressible flow C_v is inserted in equations 5.23, 5.24, 5.25, and 5.27 in the same way.
[23] Nominal values of K_L and C_v in turbulent flow are 0.04 and 0.98, respectively. See Section 9.9.
[24] Test results for Simplex Valve and Meter Co. and Builders Iron Foundry Venturi meters. *Fluid Meters: Their Theory and Application*, 4th edition, A.S.M.E., 1937.
[25] Experiments at other diameter ratios (d_2/d_1) show a decrease of coefficient with increase of diameter ratio.

$$\dot{m} = \frac{YA_2\rho_1}{\sqrt{1 - \left(\frac{A_2}{A_1}\right)^2}} \sqrt{2g_n \left(\frac{p_1 - p_2}{\gamma_1}\right)} \tag{5.27}$$

$p_1 = 140$ kPa $p_{atm} = 101.3$ kPa $MR = 150$ mm Hg

$T_1 = 15°C = 288$ K $d_1 = 0.150$ m $d_2 = 0.075$ m $R = 286.8$ J/kg·K

Solution.

$$\frac{p_2}{p_1} = \frac{140 + 101.3 - 150(101.3/760)}{140 + 101.3} = 0.917$$

$$\rho_1 = \frac{241.3 \times 10^3}{286.8 \times 288} = 2.92 \text{ kg/m}^3 \tag{1.3}$$

$$\frac{A_2}{A_1} = 0.25 \qquad A_2 = 0.004\ 42 \text{ m}^2$$

From Appendix 8, $Y = 0.949$ and, assuming $C_v = 0.98$,

$$\dot{m} = \frac{0.949 \times 0.98 \times 0.004\ 42 \times 2.92}{\sqrt{1 - (0.25)^2}} \sqrt{2g_n(150 \times (101.3 \times 10^3/760))/2.92\,g_n}$$

$$\dot{m} = 1.45 \text{ kg/s} \tag{5.27}$$

Checking R_2 gives a value of 1 250 000, which (from Fig. 11.23) gives a better value of C_v of 0.981. The true flowrate is, therefore, $\dot{m} = 1.45$ kg/s (0.981/0.980) = 1.45 kg/s. ◀

11.12 Nozzles

Nozzles are used in engineering practice for the creation of jets and streams for all purposes as well as for fluid metering; when placed in or at the end of a pipeline as metering devices they are generally termed *flow nozzles*. Since a thorough study of flow nozzles will develop certain general principles which may be applied to other special problems, only the flow nozzle will be treated here.

A typical flow nozzle is illustrated in Fig. 11.24. Such nozzles are designed to be clamped between the flanges of a pipe, generally possess rather abrupt curvatures of the converging surfaces, terminate in short cylindrical tips, and are essentially Venturi meters with the diffuser cones omitted. Since the diffuser cone exists primarily to minimize the head losses caused by the meter, it is obvious that larger head losses will result from flow nozzles than occur in Venturi meters and that herein lies a disadvantage of the flow nozzle; this disadvantage is somewhat offset, however, by the lower initial cost of the flow nozzle.

Extensive research on flow nozzles, sponsored by the American Society of Mechanical Engineers and the International Standards Association, has resulted in the accumulation of a large amount of reliable data on nozzle installation, specifications,

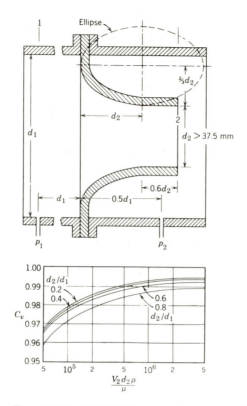

Fig. **11.24**   A.S.M.E. flow nozzle and coefficients.[26]

and experimental coefficients. Only the barest outline of these results can be presented here; the reader is referred to the original papers of these societies for more detailed information.

The A.S.M.E. flow nozzle installation of Fig. 11.24 is typical of those employed in American practice, section 1 being taken one pipe diameter upstream and section 2 at the nozzle tip. It has been found that a pressure representative of that at the latter point may be obtained by a wall piezometer connection which leads, fortunately, to the simplification of the nozzle installation, since a wall piezometer is easier to construct than a direct connection to the tip of the nozzle. Pressures obtained in this manner are not, of course, the exact pressures existing in the live stream of flowing fluid passing section 2, but the slight deviations incurred are of no consequence since they are absorbed in the experimental coefficient, C_v. The variation of C_v with area ratio and Reynolds number is typical of the geometrically similar conditions specified; the con-

[26] Data from *Fluid Meters: Their Theory and Application,* 4th edition, A.S.M.E., 1937; and H. S. Bean, S. R. Beitler, and R. E. Sprenkle, "Discharge Coefficients of Long Radius Flow Nozzles When Used with Pipe Wall Pressure Taps," *Trans. A.S.M.E.,* p. 439, 1941.

stancy of C_v at high Reynolds number and decrease of C_v with decreasing Reynolds number is observed for the flow nozzle as for the Venturi meter. Coefficients for standardized flow nozzle installations at Reynolds numbers below 50 000 are available, too.[27]

The flow nozzle being essentially equivalent to the entrance cone of the Venturi meter, flowrates may be computed by equations 5.27 and 11.13.

Illustrative Problem

An A.S.M.E. flow nozzle of 75 mm diameter is installed in a 150 mm waterline. The attached differential manometer contains mercury and water and shows a reading of 150 mm. Calculate the flowrate through the nozzle and the head loss caused by its installation. The water temperature is 15.6°C.

Relevant Equations and Given Data

$$h_L = K_L \frac{(V_1^2 - V_2^2)^2}{2g_n} \tag{9.44}$$

$$Q = \frac{C_v A_2}{\sqrt{1 - (A_2/A_1)^2}} \sqrt{2g_n \left(\frac{p_1}{\gamma} + z_1 - \frac{p_2}{\gamma} - z_2 \right)} \tag{11.13}$$

$$h_{L\,1-2} = \left[\left(\frac{1}{C_v^2} - 1 \right) \left(1 - \left(\frac{A_2}{A_1} \right)^2 \right) \right] \frac{V_2^2}{2g_n} \tag{11.14}$$

$d_1 = 0.150$ m $\qquad d_2 = 0.075$ m $\qquad T = 15.6°C$

$s.g._{Hg} = 13.57 \qquad s.g._{water} = 1.00 \qquad MR = 0.150$ m

Solution. Selecting tentatively $C_v = 0.99$ from Fig. 11.24 and calculating A_2 as 0.004 56 m^2,

$$Q = \frac{0.99 \times 0.004\ 56}{\sqrt{1 - (0.25)^2}} \sqrt{2g_n (0.15)(13.57 - 1)} = 0.028\ 6\ \text{m}^3/\text{s} \tag{11.13}$$

Calculating R_2 gives 423 000, which yields a better value of $C_v = 0.988$. Using this value in place of 0.99, $Q = (0.988/0.99)0.028\ 6 = 0.028\ 5$ m^3/s. ◀

Precise calculation of head loss caused by the nozzle installation is not possible or necessary, but adequate values may be obtained by computing $h_{L\,1-2}$ from equation 11.14 (giving 0.03 m) and adding it to the head loss caused by the flow deceleration downstream from the nozzle. Treating this as an abrupt enlargement, the head loss may be computed from equation 9.44 as $h_L = (V_2 - V_1)^2/2g_n = 1.13$ m. Thus the total head loss caused by the nozzle installation is about 1.16 m of water. ◀

---

[27] H. S. Bean, ed. *Fluid Meters—Their Theory and Application*, 6th ed., A.S.M.E., 1971.

11.13 Orifices

Like nozzles, orifices serve many purposes in engineering practice other than the metering of fluid flow, but the study of the orifice as a metering device will allow the application of principles to other problems, some of which will be treated subsequently.

The orifice for use as a metering device in a pipeline consists of a concentric square-edged circular hole in a thin plate which is clamped between the flanges of the pipe (Fig. 11.25). The flow characteristics of the orifice differ from those of the nozzle in that the minimum section of the streamtube occurs not within the orifice but downstream from it, owing to the formation of a *vena contracta* at section 2. The cross-sectional area at the vena contracta, A_2, is characterized by a coefficient of contraction,[28] C_c, and given by $C_c A$. Substituting this into equation 11.13,

$$Q = \frac{C_v C_c A}{\sqrt{1 - C_c^2(A/A_1)^2}} \sqrt{2g_n \left(\frac{p_1}{\gamma} + z_1 - \frac{p_2}{\gamma} - z_2 \right)} \tag{11.15}$$

which is customarily written

$$Q = CA \sqrt{2g_n \left(\frac{p_1}{\gamma} + z_1 - \frac{p_2}{\gamma} - z_2 \right)} \tag{11.16}$$

thus defining the orifice coefficient C as

$$C = \frac{C_v C_c}{\sqrt{1 - C_c^2(A/A_1)^2}} \tag{11.17}$$

which is thus dependent not only on C_v and C_c but on the shape of the installation (defined by A/A_1) as well.

In practice it is not feasible to locate the downstream pressure connection at the vena contracta because the location of the vena contracta depends on both Reynolds number and A/A_1. Accordingly, it is frequently located (as for the flow nozzle) at a

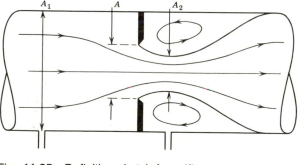

Fig. **11.25** Definition sketch for orifice meter.

[28] The Weisbach values of Section 9.9 (Table 4) may be used as nominal at high Reynolds numbers.

fixed proportion of the pipe diameter downstream from the orifice plate, and the other connection one pipe diameter upstream. Any coefficient, C, will thus be dependent on, and associated with, the particular location of the pressure connections. Values of C over a wide range of Reynolds numbers may be obtained from Fig. 11.26; it is convenient and standard practice to define the Reynolds number on the basis of flowrate and orifice diameter as

$$\mathbf{R} = \frac{Qd}{(\pi d^2/4)\nu} = \frac{4Q}{\pi d\nu}$$

which is used as the abscissa for the plot. The trend of the coefficient with Reynolds number is of interest when compared with that of the Venturi meter and flow nozzle. The constancy of C at high Reynolds number is again noted, reflecting the substantial constancy of both C_v and C_c in this range. At lower Reynolds numbers an increase of C is noted, in spite of the expectation of a decrease of C_v in this region; evidently, increased viscous action not only lowers C_v but also raises C_c (by increasing the size of the vena contracta), and the latter effect is predominant. In the range of very low

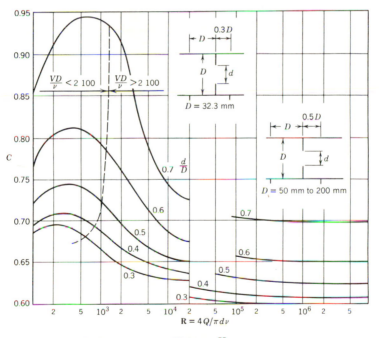

Fig. **11.26** Orifice meter coefficients.[29]

[29] Data from G. L. Tuve and R. E. Sprenkle, "Orifice Discharge Coefficients for Viscous Liquids," *Instruments,* vol. 6, p. 201, 1933; vol. 8, pp. 202, 225, 232, 1935; and *Fluid Meters: Their Theory and Application,* 4th edition, A.S.M.E., 1937.

Reynolds numbers the effect of viscous action on the vena contracta remains at a maximum (with C_c around 1), and the decrease of C with further decrease of Reynolds number reflects the steady decrease of C_v produced by viscous resistance.

In metering the flow of compressible fluids at high pressure ratios, equation 5.23 may be used but it strictly applies only when the downstream pressure connection is at the vena contracta. Values of Y from Appendix 8 may generally be used only as a first approximation; accurate values of Y for various locations of the pressure taps will be found in the A.S.M.E. Fluid Meters report, cited in the References at the end of the chapter.

An extension of the pipeline orifice of Fig. 11.25 is the *submerged orifice* of Fig. 11.27 featured by orifice discharge from one large reservoir into another. Here with A/A_1 virtually zero, C (of equation 11.17) becomes $C_v C_c$. Assuming a perfect fluid and applying the Bernoulli equation between the upstream reservoir and section 2,

$$h_1 = h_2 + \frac{V_2^2}{2g_n} \quad \text{or} \quad V = \sqrt{2g_n(h_1 - h_2)}$$

providing that the pressure distribution may be considered hydrostatic[30] in the downstream reservoir. For the real fluid, frictional effects will prevent the attainment of this velocity and C_v is introduced as before, so that

$$V_2 = C_v \sqrt{2g_n(h_1 - h_2)}$$

The flowrate may be calculated from $A_2 V_2$, in which A_2 is replaced by $C_c A$:

$$Q = A_2 V_2 = C_c C_v A \sqrt{2g_n(h_1 - h_2)} = CA \sqrt{2g_n(h_1 - h_2)} \quad (11.18)$$

in which $C_v C_c$ is defined as the coefficient (of discharge) of the orifice. When the orifice discharges freely into the atmosphere (Fig. 11.28), h_2 becomes zero and the equation reduces to

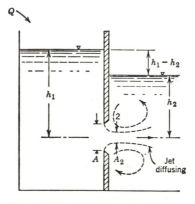

Fig. **11.27** Submerged orifice.

[30] This is a valid assumption if h_2 is large compared to orifice size.

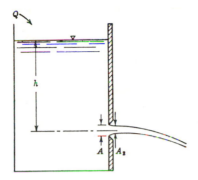

Fig. **11.28** Orifice discharging freely.

$$Q = C_c C_v A \sqrt{2g_n h} = CA \sqrt{2g_n h}$$

The dependence of the various orifice coefficients on shape of orifice is illustrated by Fig. 11.29. The coefficients given are nominal values for large orifices ($d > 1$ in. or 25 mm) operating under comparatively large heads of water ($h > 4$ ft or 1.2 m). Above these limits of head and size, various experiments have shown that the coefficients are practically constant. Coefficients for sharp-edged orifices over a wide range of Reynolds numbers are given in Fig. 11.30, which shows the same trend of values (for the same reasons) as that of Fig. 11.26. The plot of Fig. 11.30, although convenient and applicable to the flow of all fluids, has a certain limitation in orifice size caused by the action of surface tension. Surface-tension effects (although impossible to predict except in idealized situations) will increase with decreasing orifice size; the plotted values are valid only where such effects are negligible and, thus, cannot be applied to very small orifices.

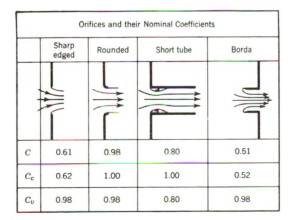

Orifices and their Nominal Coefficients				
	Sharp edged	Rounded	Short tube	Borda
C	0.61	0.98	0.80	0.51
C_c	0.62	1.00	1.00	0.52
C_v	0.98	0.98	0.80	0.98

Fig. **11.29**

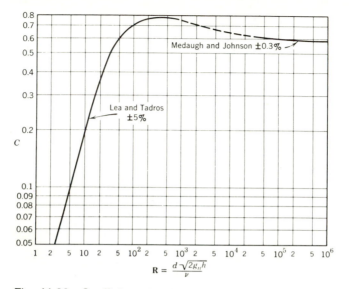

Fig. **11.30**   Coefficients for sharp-edged orifices
under static head[31] ($h/d > 5$).

The head lost between the reservoir and section 2 in an orifice operating under
static head may be calculated from the coefficient of velocity and flowrate by equation
11.14. Since $A_2/A_1 \cong 0$ this equation reduces to

$$h_L = \left(\frac{1}{C_v^2} - 1 \right) \frac{V_2^2}{2g_n}$$

One special problem of orifice flow is that of the two-dimensional sluice gate of
Fig. 11.31 in which jet contraction occurs only on the top of the jet and pressure
distribution in the vena contracta is hydrostatic. Assuming an ideal fluid,

$$y_1 + \frac{V_1^2}{2g_n} = y_2 + \frac{V_2^2}{2g_n}$$

and substituting $V_2 y_2/y_1$ for V_1, and solving for V_2,

$$V_2 = \frac{1}{\sqrt{1 - (y_2/y_1)^2}} \sqrt{2g_n(y_1 - y_2)}$$

The actual velocity (allowing for head loss) is obtained by multiplying the above
by C_v, and the flowrate by multiplying the result by $C_c A$. The flowrate through the
sluice is, therefore,

$$q = \frac{C_v C_c A}{\sqrt{1 - (y_2/y_1)^2}} \sqrt{2g_n(y_1 - y_2)}$$

[31] Data from F. C. Lea, *Hydraulics,* 6th edition, p. 87, Edward Arnold and Co., 1938; and F. W. Medaugh
and G. D. Johnson, *Civil Eng.,* vol. 10, no. 7, p. 422, July, 1940.

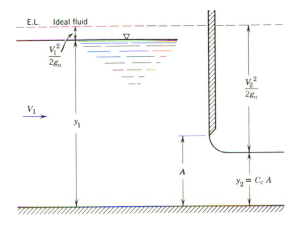

Fig. 11.31 Sluice gate.

from which it is noted that the effective head on the sluice is $(y_1 - y_2)$, that the equation is analogous to equation 11.15, and that it reduces to equation 11.18 as the depth y_1 becomes large compared to y_2.

A second special problem of orifice flow is represented by the *rotameter*[32] which can be calibrated to read velocity or flowrate directly. Basically a rotameter consists of a precisely manufactured, tapered vertical tube through which fluid flows upward. As the tube diameter increases upward, the fluid velocity in the tube, at any fixed flowrate, decreases with distance up the tube. Within the tube is placed a "float" which has a specially designed shape, a density slightly greater than that of the flowing fluid, and (often) spiral grooves which cause the float to spin (and hence to remain roughly centered in the tube) when fluid is flowing. Accordingly, the flow around the float is quite like that through a needle valve or annular orifice.

When there is no flow, the float sits at the bottom of the tapered tube. When fluid is flowing, the float rises until the upward drag and buoyancy forces on it are balanced by its weight. Since the tube is tapered, the velocity past the float and, thus, the drag on it decrease as the float moves up in a constant rate flow. The points of equilibrium can be noted as a function of flowrate and, with a marked glass tube, the level of the float becomes a direct measure of flowrate. Rotameters are widely used in industrial applications where the visual output can be used by operators controlling flowrates in various processes.

11.14 Elbow Meters

The orifice, nozzle, and Venturi meters as applied in the measurement of pipeline flow have been seen to be fundamentally methods of producing a regular and reproducible fall of the hydraulic grade line which is related to flowrate. Another meter of this type is the elbow meter of Fig. 11.32 which utilizes the fall of hydraulic grade line between

---

[32] E. Ower and R. C. Pankhurst, *The Measurement of Air Flow*, 5th ed. (SI units). Pergamon Press, 1977.

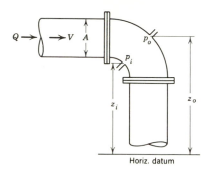

Fig. **11.32**   Elbow meter.

the outside and inside of a regular pipe bend (see Fig. 4.12 of Section 4.8). Analytical solutions of such problems are not feasible, and such devices are calibrated by determining experimentally the relation between fall of hydraulic grade line and flowrate. Lansford[33] has done this for a variety of 90° flanged elbows, and it allows their use as accurate and economical flow meters; for a basic equation he proposes

$$\left(\frac{p}{\gamma} + z\right)_o - \left(\frac{p}{\gamma} + z\right)_i = C_k \frac{V^2}{2g_n}$$

with coefficient C_k ranging between 1.3 and 3.2, the magnitudes depending on the size and shape of the elbow. This equation may be solved for V and multiplied by A to obtain the flowrate, Q. If the coefficient of the meter is then defined as $\sqrt{1/C_k}$ the resulting equation has the same form as that for nozzles and orifices,

$$Q = CA \sqrt{2g_n \left(\frac{p_o}{\gamma} + z_o - \frac{p_i}{\gamma} - z_i\right)}$$

in which C will have values between 0.56 and 0.88.

11.15   Dilution Methods[34]

Dilution methods for measuring flowrate consist essentially of introducing (at a steady rate) a concentrated foreign substance to the flow, measuring the concentration of the substance after thorough mixing has taken place, and calculating from the dilution of the substance the flowrate which has brought about this dilution.

A concentrated salt solution has been used in Europe in applying this method to the calculation of flow in small mountain streams, and in this country to the calculation of the flow through the turbines of hydroelectric power plants. If the rate of flow of salt solution into the unknown flow is Q_s, and the concentration of salt in this solution is

---

[33] W. M. Lansford, "The Use of an Elbow in a Pipe Line for Determining the Flow in the Pipe," *Eng. Exp. Sta. Univ. Ill., Bull.* 289, 1936.

[34] See W. A. Cawley and J. W. Woods, "An Improved Dilution Method for Flow Measurement," *Trans. A.S.C.E.*, vol. 123, 1958.

C_1 (weight/unit volume), the weight per second of salt added to the unknown flow is given by $Q_S C_1$. If the concentration of salt in the unknown flow after mixing has occurred is C_2, the weight of salt flowing in the stream per second is also given by $(Q + Q_S)C_2$. Therefore

$$Q_S C_1 = (Q + Q_S)C_2$$

Since Q_S is so small compared to Q, it is usually neglected on the right-hand side of the equation and the flowrate computed from $Q = Q_S(C_1/C_2)$. Hence by control and measurement of Q_S and with C_1 and C_2 easily obtainable by titration methods, the unknown flowrate Q may be found, providing that complete mixing has been attained.

11.16 Salt-Velocity Method

An ingenious method of flow measurement, which has met with success in the measurement of flowrate in large pipelines, is the salt-velocity method developed by Allen and Taylor.[35] In this method a quantity of concentrated salt solutiion is introduced suddenly to the flow, and the mean velocity is obtained by measuring the velocity of the salt solution as it moves with the flow.

The mechanism for the method (illustrated schematically in Fig. 11.33) is essentially a device for introducing the salt solution and two similar electrodes and circuits. The passage of the salt solution between the plates of an electrode is recorded by a momentary increase in the ammeter reading due to the greater conductivity of the salt solution. By noting the time, t, between the deflections of the two ammeter needles and knowing the distance l between electrodes, the mean velocity in the pipe may be calculated from $V = l/t$ and the flowrate Q from $Q = AV$.

The testing techniques involved in the salt-velocity method are somewhat more complex than the above statement of principles implies, owing primarily to the use of a chronographic device to record automatically (1) the variation of electrical current with the passage of the salt solution, and (2) the time of passage of the solution between electrodes.

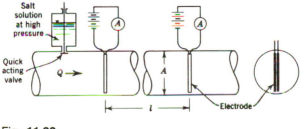

Fig. **11.33**

[35] C.M. Allen and E.A. Taylor, "The Salt Velocity Method of Water Measurement," *Trans. A.S.M.E.*, vol. 45, p. 285, 1923.

11.17   Weirs

For measuring large and small open flows in field or laboratory, the weir finds wide application. A weir may be defined in a general way as "any regular obstruction *over* which flow occurs." Thus, for example, the overflow section (spillway) of a dam is a special type of weir and may be utilized for flow measurement. However, weirs for measuring purposes are usually of more simple and reproducible form, consisting of smooth, vertical, flat plates with upper edges sharpened. Such weirs, called *sharp-crested weirs,* appear in a variety of forms, the most popular of which is the rectangular weir; this type has a straight, horizontal crest and extends over the full width of the channel in which it is placed. The flow picture produced by such a weir is essentially two-dimensional and for this reason it will be used as a basis for the following discussion.

The flow of liquid over a sharp-crested weir is at best an exceedingly complex problem and impossible of rigorous analytical solution.[36] An appreciation for the complexities, however, is necessary to an understanding of experimental results and of the deficiencies of simplified weir formulas. These complexities may be discovered by considering the flow over the sharp-crested weir of Fig. 11.34. Although it is obvious at once that the head, H, on the weir is the primary factor causing the flow (Q) to occur, no simple relationship between these two variables can be derived, for two fundamental reasons: (1) the shape of the flow picture, and (2) the effect of turbulence and frictional processes cannot be calculated. The most important factors which affect the shape of the flow picture are the head on the weir, H, the weir height, P, and the extent of ventilation beneath the nappe. Although the effect of these factors may be found experimentally, there is no simple method of predicting the flow picture from the values of H, P, and pressure beneath the nappe. The effects of turbulence and friction cannot be predicted, nor can they be isolated for experimental measurement. It may be noted, however, that frictional resistance at the sidewalls will affect the flowrate to an increasing extent as the channel becomes narrower and the weir length, b (normal to the

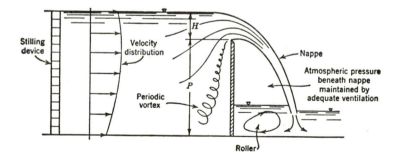

Fig. **11.34**   Weir flow (actual).

---

[36] However, the two-dimensional problem for the ideal fluid has been solved by application of potential theory and complex functions and by numerical finite-difference methods employing digital computers.

paper), smaller. Fluid turbulence and frictional processes at the sides and bottom of the approach channel also contribute to the velocity distribution in an unpredictable way. The effects of velocity distribution on weir flow have been shown by Schoder and Turner[37] to be appreciable, and an effort should be made in all weir installations to provide a good length of approach channel, with stilling devices such as racks and screens for the even distribution of turbulence and the prevention of abnormal velocity distributions. Another influence of frictional processes is the creation of a periodic helical secondary flow in the corners just upstream from the weir plate, resulting in a vortex (Fig. 11.34), which influences the flow in an unpredictable manner. The free liquid surfaces of weir flow also bring surface-tension effects into the problem, and these forces, although small, affect the flow picture appreciably at low heads and small flows.

In the light of the foregoing complexities, the derivation of any simple weir formula obviously requires drastic simplification of the problem which leads to an approximate result; however, by such methods the *form* of the relationship between flowrate and head can be found and an experimental coefficient defined. To derive a simple weir equation, let it be assumed that (1) velocity distribution upstream from the weir is uniform, (2) all fluid particles move horizontally as they pass the weir crest, (3) the pressure in the nappe is zero, and (4) the influence of viscosity, turbulence, secondary flows, and surface tension may be neglected. These assumptions produce the flow picture of Fig. 11.35. Taking section 1 in the approach channel well upstream from the weir and section 2 slightly downstream from the weir crest, Bernoulli's equation may be applied to a typical streamline to find the velocity v_2. Using streamline AB as typical and taking the weir crest as datum,

$$H + \frac{V_1^2}{2g_n} = (H - h) + \frac{v_2^2}{2g_n}$$

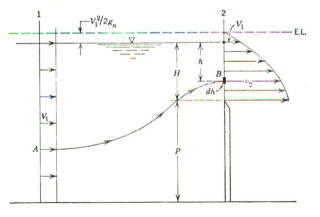

Fig. **11.35** Weir flow (simplified).

[37] E. W. Schoder and K. B. Turner, "Precise Weir Measurements," *Trans. A.S.C.E.*, vol. 93, p. 999, 1929.

from which we obtain

$$v_2 = \sqrt{2g_n\left(h + \frac{V_1^2}{2g_n}\right)}$$

which shows v_2 to be dependent on h. Taking dq to be the two-dimensional flowrate through a strip of height dh, $dq = v_2\, dh$ allowing integration to obtain the flowrate, q,

$$q = \int_0^H v_2\, dh = \sqrt{2g_n}\int_0^H \left(h + \frac{V_1^2}{2g_n}\right)\, dh$$

The result is

$$q = \tfrac{2}{3}\sqrt{2g_n}\left[\left(H + \frac{V_1^2}{2g_n}\right)^{3/2} - \left(\frac{V_1^2}{2g_n}\right)^{3/2}\right]$$

Because of the cumbersome form of this equation[38] and because in many weir problems $P \gg H$ and V_1 is small, $V_1^2/2g_n$ is customarily neglected and the equation further simplified to

$$q = \tfrac{2}{3}\sqrt{2g_n}\,H^{3/2} \tag{11.19}$$

which is the basic equation for rectangular weirs. Into this equation must be inserted an experimentally determined coefficient, C_w, which includes the effects of the many phenomena disregarded in the foregoing development and simplifications. For real weir flow the relation between flowrate and head then becomes

$$q = C_w\tfrac{2}{3}\sqrt{2g_n}\,H^{3/2} \tag{11.20}$$

showing that (to the extent C_w is constant), for rectangular weirs, $q \propto H^{3/2}$. The coefficient C_w is essentially a factor which transforms the simplified weir flow of Fig. 11.35 into the real weir flow of Fig. 11.34 and its magnitude is thus fixed by the most important difference between these flows—the shape of the flowfield. Thus the weir coefficient is primarily a coefficient of contraction which expresses the extent of contraction of the true nappe below that assumed in the simplified analysis. Since the size of the weir coefficient depends primarily on the shape of the flowfield, the effect of other fluid properties and phenomena may usually be discovered by examining their influence on this shape.

Although a dimensional analysis of the weir problem must necessarily be incomplete because of the impossibility of including all the pertinent factors, it provides a rational basis for an understanding of some of the factors affecting weir coefficients. The expressible independent variables entering the two-dimensional weir problem are q, H, P, μ, σ, ρ, and g_n. Application of the Buckingham Π-theorem analysis of Section 8.2 to these variables shows that there are $7 - 3 = 4$ distinct dimensionless Π-groups. Using q, H, and ρ as the repeating variables produces

---

[38] Which (knowing P and H) requires trial-and-error solution for q.

$$\Pi_1 = P/H \qquad \Pi_2 = \rho q/\mu = \mathbf{R} \qquad \Pi_3 = \rho q^2/\sigma H = \mathbf{W}$$

$$\Pi_4 = q/\sqrt{g_n}\,H^{3/2} = \mathbf{F}$$

Noting that, from Π_4, $q \propto \sqrt{g_n H}\,H$ allows Π_2 and Π_3 to be rewritten as

$$\Pi_2 = \mathbf{R} = \rho\sqrt{g_n}\,H^{3/2}/\mu \qquad \text{and} \qquad \Pi_3 = \mathbf{W} = \rho g_n H^2/\sigma$$

The relationship among the four dimensionless groups is typically written as

$$\frac{q}{\sqrt{g_n}H^{3/2}} = f\left(\mathbf{R,\ W,\ }\frac{P}{H}\right)$$

in which the Froude number on the left-hand side is a direct measure of C_w; dimensional analysis thus shows the dependence of the weir coefficient on $\mathbf{R}$, $\mathbf{W}$, and P/H. Of these numbers P/H has been found to be the most important in determining the magnitude of C_w—as would be expected since this ratio, more than any of the others, has the greatest influence on the shape of the flowfield. The effect of $\mathbf{W}$ is negligible except at low heads where surface-tension effects may be large; the effect of $\mathbf{R}$ is small (except at low heads) since water is usually involved in weir flows, $\mathbf{R}$ is high, and viscous action small.

The experimental work of Rehbock[39] led to an empirical formula for the coefficient of well-ventilated, sharp-crested rectangular weirs for water measurement, which has an accuracy of better than 1 percent if care is taken with the details of installation. Rehbock's formula for C_w is

$$C_w = 0.605 + 0.08\frac{H}{P} + \frac{1}{1\ 000H} \tag{11.21}$$

which shows the strong influence of P/H, except where H (in m) is small and $1/1\ 000H$ of some significance. In this term H is (for a liquid of constant physical properties) a direct measure of $\mathbf{R}$ and $\mathbf{W}$(see the second forms of Π_2 and Π_3 above); this implies (as expected) that the influences of viscosity and surface tension on weir flow are strong only when H is small.

Weirs are reliable measuring devices only at heads above the range of strong action by viscosity and surface tension; in this range the formula shows that the coefficient can be expected to increase with increasing head and decreasing weir height. Although it is desirable to calibrate a new weir installation *in place,* this is frequently not possible and a formula must be selected for the weir coefficient. The Rehbock formula can be expected to give good results if such important details as adequate ventilation, stilling devices, crest sharpness, and smoothness of upstream face are not overlooked.

Broad-crested weirs (Fig 11.36) have been shown (Section 10.8) to be critical-

---

[39] Th. Rehbock, "Wassermessung mit scharfkantigen Uberfallwehren," *Zeitschrift des V.d.I.,* vol. 73, no. 24, June 15, 1929, and C. E. Kindsvater and R. W. Carter, "Discharge Characteristics of Rectangular Thin Plate Weirs," *Trans. A.S.C.E.,* vol. 124, 1959. See P. Ackers, W. R. White, J. A. Perkins, and A. J. M. Harrison, *Weirs and flumes for flow measurement,* John Wiley & Sons, 1978, for a complete discussion of the many formulas proposed.

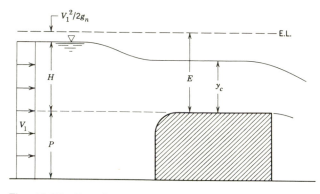

Fig. **11.36** Broad-crested weir.

depth meters; here, for ideal flow,

$$q = \sqrt{g_n \left(\frac{2E}{3}\right)^3} = \left(\frac{2}{3}\right)^{3/2} \sqrt{g_n} E^{3/2} \qquad (11.22)$$

The weir coefficient for the ideal broad-crested weir may be calculated by equating equations 11.20 and 11.22 to yield

$$C_w = \frac{1}{\sqrt{3}} \left(\frac{E}{H}\right)^{3/2}$$

For a very high weir $P/H \rightarrow \infty$, $E \rightarrow H$, $E/H \rightarrow 1$, and $C_w \rightarrow 1/\sqrt{3} = 0.577$ as shown in Section 10.8; for a lower weir $P/H < \infty$, $E > H$, $E/H > 1$, and $C_w > 0.577$. Hence the weir coefficient increases with decreasing P/H and thus exhibits the same trend as the Rehbock formula. Experimental measurements also substantiate this variation of C_w with P/H, but the values of C_w obtained from experiment are a few percent lower than those of ideal flow because of head loss accompanied by a falling energy line.

For small flowrates, *notch weirs* are widely used as measuring devices; of these the most popular is the triangular weir or V-notch (Fig. 11.37). A simplified analysis similar to that used on the rectangular weir (but neglecting velocity of approach) yields (after inserting the experimental coefficient) the fundamental formula

$$Q = C_w \tfrac{8}{15} \tan \alpha \sqrt{2g_n} H^{5/2} \qquad (11.23)$$

Triangular weirs of 90° notch angle (2α) have coefficients (for water) near 0.59, but these are affected by viscosity, surface tension, and weir plate roughness; increases of any one of these tend to increase the coefficient. A comprehensive study of triangular weir flow has been made by Lenz,[40] who used many liquids in order to discover the effects of viscosity and surface tension on weir coefficients, thus extending the utility

---

[40] A. T. Lenz, "Viscosity and Surface-Tension Effects on V-Notch Weir Coefficients," *Trans. A.S.C.E.*, vol. 108, 1943.

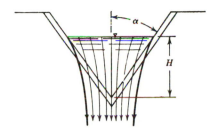

Fig. **11.37** Triangular weir.

of the triangular weir as a reliable measuring device. For notch angles of 90° Lenz proposed[41]

$$C_w = 0.56 + \frac{0.70}{\mathbf{R}^{0.165}\mathbf{W}^{0.170}}$$

as applying to all liquids providing that the falling sheet of liquid does not cling to the weir plate and that $H > 0.06$ m, $\mathbf{R} > 300$, and $\mathbf{W} > 300$. The work of Lenz has not only broadened the field of application of the weir as a measuring device but has also documented the increase of coefficient with decreasing $\mathbf{R}$ and $\mathbf{W}$, a characteristic of the coefficients for all sharp-crested weirs.

The crest of a *spillway structure* is shown in Fig 11.38. Major considerations in the design of such a spillway are structural stability against hydrostatic pressure and other loads, and prevention of separation (Section 7.5) and reduced pressures on the surface of the structure. The rectangular weir equation may be applied to the spillway, the coefficient C_w ranging from 0.60 to 0.75. The relatively high value of C_w may be explained by a comparison of a sharp-crested weir (Fig. 11.34) and a spillway crest designed exactly to fit the curvature of the lower side of the nappe of this weir for a certain design head, H_D. In spite of the greater friction of the spillway, with a fixed

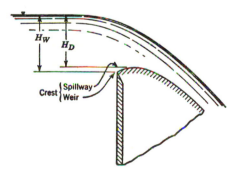

Fig. **11.38**

$\mathbf{R} = H\sqrt{g_n H}/\nu, \quad \mathbf{W} = \rho g_n \mathbf{H}^2/\sigma$

reservoir surface the flow over the two structures must be approximately the same, but the heads for each structure will be measured from their respective crests and will, therefore, be quite different, the head on the weir being greater than the head on the spillway. Since for (about) the same flowrate the smaller head must be associated with a larger coefficient, the spillway coefficient is seen to be larger than that of the sharp-crested weir.

A spillway profile which will fit the flow of a sharp-crested weir and thus prevent harmful discontinuities of pressure is shown with its coefficients in Fig 11.39. The results are both useful and instructive, since in defining the coefficients the head is taken as the vertical distance between spillway crest and energy line upstream from the structure and thus contains the velocity head of the approaching flow. After the design head and height of structure have been determined, the profile of the structure may be laid out and discharge coefficients accurately predicted for heads between 40% and 130% of the design head. For any weir height, P, a steady increase of C_w with H is as noted for the sharp-crested weir (equation 11.21). For the sharp-crested weir this trend was due to change in the overall shape of the flow picture; for the spillway crest it is due mostly to the steady decline in pressure over the crest with increasing head, which increases the effective head and is reflected in an increase of coefficient. This trend is beneficial up to the point where cavitation or separation occurs, inasmuch as an increase in C_w may be interpreted as an increase in the efficiency of the spillway.

The effects of submergence (Fig. 11.40) on weir flow are of some theoretical and practical interest. Sharp-crested weirs cannot be considered to be precise measuring devices when operating submerged, but the effect of submergence on broad-crested weirs is surprisingly small, making this type of weir a reliable measuring device even for high submergence. This reliability is due primarily to the straighter streamlines and essentially hydrostatic pressure distribution on the crest of the broad-crested weir which

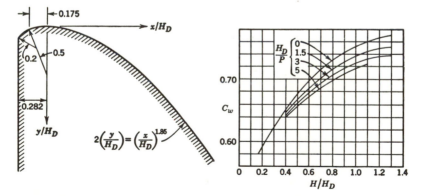

Fig. **11.39**   Spillway profiles and coefficients.[42]

[42] F. R. Brown, "Hydraulic Models as an Aid to the Development of Design Criteria," *Waterways Expt. Sta., Corps of Engrs., Bull.* 37, Vicksburg, Miss., June, 1951.

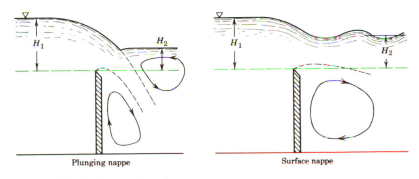

Fig. **11.40** Submerged weirs.

change little as the downstream water level rises above the crest of the weir. For the sharp-crested weir discharging freely the pressure distribution through the nappe is featured by zero pressure on top and bottom, and thus is far from hydrostatic; raising the downstream water level above the crest of the weir will drastically change this pressure distribution and immediately affect the whole flow picture. In all submerged weir problems the two flow situations (Fig. 11.40) of *plunging nappe* at low submergence and *surface nappe* at higher submergence are observed. Approximate results[43] of investigations on submerged rectangular weirs may be seen on the sketch of Fig. 11.41 in which submergence ratio H_2/H_1 is plotted against the ratio of measured flowrate (Q_S) to that which would have existed with free flow (Q_F) for a head H_1.

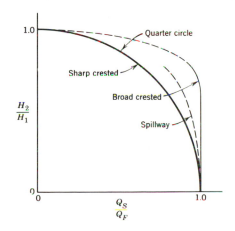

Fig. **11.41** Approximate effects of submergence.

[43] For accurate and detailed informaton see J. G. Woodburn, "Tests of Broad Crested Weirs," *Trans. A.S.C.E.*, vol. 96, 1932; J. K. Vennard and R. F. Weston, "Submergence Effect on Sharp-Crested Weirs," *Eng. News-Record,* June 3, 1943; J. R. Villemonte, "Submerged Weir Discharge Studies," *Eng. News-Record,* December 25, 1947.

Engineers require accurate information on weir submergence when flowrate measurements are needed in times of flood, in spillway design, and in situations where adequate vertical drop is unavailable for use of a (more precise) freely flowing weir.

11.18 Current-Meter Measurements

The construction of a weir for measuring the flowrate in large canals, streams, or rivers is impractical for many obvious reasons, but existing spillways whose coefficients are known may frequently serve as measuring devices. The standard method of river flow measurement is to measure the velocity by means of a current meter (Section 11.7) and integrate the results to obtain the flowrate.

Fundamental to the use of a current meter is some knowledge of the general properties of velocity distribution in open flow. As in pipes, the velocities are reduced at the banks and bed of the channel, but it must be realized that in open flow the roughnesses and turbulences are of such great and irregular magnitudes that the velocity distribution problem cannot be placed on the precise basis which it enjoys in pipe flow. However, from long experience and thousands of measurements, the United States Geological Survey has established certain average characteristics of velocity distribution in streams and rivers which serve as a basis for current-meter measurements. These characteristics of velocity distribution in a vertical are shown in Fig. 11.42 and may be amplified by the following statements: (1) the curve may be assumed parabolic;[44] (2) the location of the maximum velocity is from $0.05y$ to $0.25y$ below the water surface; (3) the mean velocity occurs at approximately $0.6y$ below the water surface; (4) the mean velocity is approximately 85 percent of the surface velocity; (5) a more accurate and reliable method of obtaining the mean velocity is to take a numerical average of the velocities at $0.2y$ and $0.8y$ below the water surface. These average values will, obviously, not apply perfectly to a particular stream or river, but numerous measurements with the current meter will tend toward accurate results since deviations from

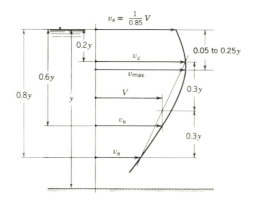

Fig. **11.42** Standard velocity distribution in a vertical in open flow.

[44] This is merely a convenient approximation; it does not imply laminar flow.

the average values will tend to compensate, thus giving a greater accuracy than can be obtained from individual measurements.

Current-meter measurements for calculation of flowrate may be taken in the following manner. A reach of river having a fairly regular cross section is selected. This cross section is measured accurately by soundings. It is then divided into vertical strips (Fig. 11.43), the current meter is suspended, and velocities are measured at the two-tenths and eight-tenths points in each vertical (1, 2, 3, etc., Fig. 11.43). From these measurements the mean velocities (V_1, V_2, V_3, etc.) in each vertical may be calculated. The mean velocity through each vertical strip is taken as the average of the mean velocities in the two verticals which bound the strip, and thus the rates of flow (Q_{12}, Q_{23}, etc.) through the strips may be calculated from

$$Q_{12} = b_{12}\left(\frac{y_1 + y_2}{2}\right)\left(\frac{V_1 + V_2}{2}\right)$$

$$Q_{23} = b_{23}\left(\frac{y_2 + y_3}{2}\right)\left(\frac{V_2 + V_3}{2}\right)$$

and the flowrate in the stream may be calculated by totaling the flowrates through the various strips.

FLOW VISUALIZATION

There are a number of visual methods of gaining data about fluid flows. Often these methods have the advantages of not requiring disturbance of the flow by instruments or of giving direct graphic evidence of essential flow patterns not obtainable by point measurements. Among the many visual methods only two are discussed briefly here. For more complete coverage the reader should consult the References at the end of the chapter.

11.19 Optical Methods

These methods are used for gas flows and employ the properties of light. No probes need to be inserted in the flow, but channel boundaries must be transparent to light. Three methods are in common use, that is, the shadowgraph, the Schlieren, and the

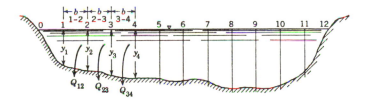

Fig. **11.43** Division of river cross section for current-meter measurements.

Mach-Zehender interferometer. All three utilize the refraction of light by a medium of changing density.

The simplest method is the shadowgraph. If parallel light rays pass through a test section, those encountering a density gradient are refracted (deflected) while the rest pass straight through. The resulting light pattern is observed on a screen where the light intensity depends on the second derivative of the density along the flow directions parallel to the screen (two-dimensional flow is assumed).

The Schlieren and interferometer techniques require more complex optical equipment, but measure, as the shadowgraph does, functions related to the gas density. The Schlieren results give a measure of density gradient, while the interferometer gives a direct measure of density variations in a test section. All three methods are very useful in high-speed compressible flows where density changes are significant. With these optical methods one obtains both density data and visual display of the flow structures such as shock waves.

11.20   The Hydrogen-Bubble Method

For low-speed water flows, the hydrogen-bubble method provides graphic flow visualization as well as a means for obtaining accurate data on velocities, streamline and

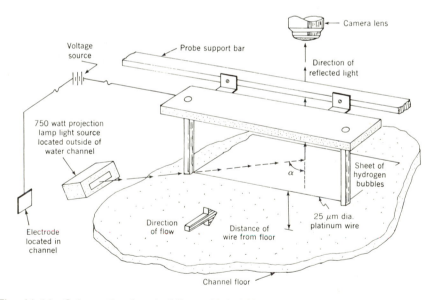

Fig. **11.44**   Schematic of typical flow with bubble wire and lighting arrangement.[45]

[45] Courtesy of Thermosciences Division, Department of Mechanical Engineering, Stanford University

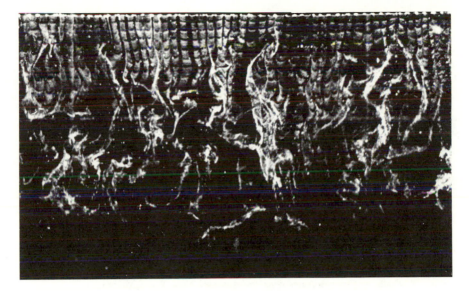

Fig. **11.45** Eddy structure of turbulent boundary layer on a flat plate ($u_o \approx 0.15$ m/s, $v_* y / \nu = 50.7$) observed with time-space markers (120 Hz).[45]

pathline configurations, and so forth, in steady and unsteady flows. The technique employs a fine wire, as the negative electrode of a D.C. electric circuit, to electrolyze water flowing in a channel. Very small hydrogen bubbles can be generated. They are swept from the wire and become markers for fluid particles. By illuminating the test section, it is possible to track and photograph the patterns. Figure 11.44 shows a schematic of a typical flow and lighting arrangement.

The hydrogen-bubble method is very flexible so a number of wire configurations and bubble-generating circuits can be used. One of the most interesting is the use of short insulated sections on the wire to create space marks (no bubbles) combined with a regular pulsing of the wire voltage to produce time marks (no bubbles). Figure 11.45 shows how this procedure leads to visualization of the eddy structure in a turbulent flow.

References

Ackers, P., White, W. R., Perkins, J. A., and Harrison, A. J. M. 1978. *Weirs and flumes for flow measurement*. New York: Wiley.

Asanuma, T. Ed. 1979. *Flow visualization*. New York: Hemisphere Pub. Corp.

A.S.M.E. 1931. Density determinations. Part 16 of *A.S.M.E. power test codes on instruments and apparatus*.

A.S.M.E. 1931. Determination of the viscosity of liquids. Part 17 of *A.S.M.E. power test codes on instruments and apparatus*.

A.S.M.E. 1933. Head measuring apparatus. Part 4 of *A.S.M.E. power test codes on instruments and apparatus*.

A.S.M.E. 1938–1942. Pressure measurement. Part 2 of *A.S.M.E. power test codes on instruments and apparatus*.

A.S.M.E. 1959. Flow measurement by means of thin plate orifices, flow nozzles, and Venturi tubes. Chapter 4 of Part 5 of *A.S.M.E. power test codes supplement on instruments and apparatus*.

A.S.M.E. 1961. *Flowmeter computation handbook*.

Bean, H. S. Ed. 1971. *Fluid meters—their theory and application*. 6th ed. A.S.M.E.

Benedict, R. P. 1977. *Fundamentals of temperature, pressure, and flow measurements*. 2nd ed. New York: Wiley.

Benedict, R. P. 1980. *Fundamentals of pipe flow*. New York: Wiley.

Buchhave, W. K., et al. 1979. The measurement of turbulence with the laser-Doppler anemometer. *Annual Review of Fluid Mechanics* 11: 443–504.

Cheremisinoff, N. P. 1979. *Applied fluid flow measurement*. New York: Marcel Dekker.

Comte-Bellot, G. 1976. Hot-wire anemometry. *Annual Review of Fluid Mechanics* 8: 209–232.

Corbett, D. M., et al. 1943. Stream gaging procedure. *U.S. Geological Survey Water Supply Paper* 888.

Coxon, W. F. 1959. *Flow measurement and control*. New York: Macmillan.

Durst, F., Melling, A., and Whitelaw, J. H. 1976. *Principles and practice of laser-Doppler anemometry*. New York: Academic Press.

Hayward, A. T. J. 1979. *Flowmeters*. New York: Wiley.

Holman, J. P. 1978. *Experimental methods for engineers*. 3rd ed. New York: McGraw-Hill.

Hydraulic Institute. 1979. Viscosity. Section IIC of *Engineering data book*. 1st ed.
    Illustrated experiments in fluid mechanics. 1972. MIT Press.

Liepmann, H. W., and Roshko, A. 1957. *Elements of gasdynamics*. New York: Wiley, Chapter 6.

Linford, A. 1961. *Flow measurement & meters*. 2nd ed. London: Spon.

Lowell, F. C., Jr., and Hirschfeld, F. 1979. Acoustic flowmeters for pipelines. *Mechanical Engineering* 101: 29–35.

Ower, E., and Pankhurst, R. C. 1977. *The measurement of air flow*. 5th ed. New York: Pergamon Press.

Sandborn, V. A. 1972. *Resistance temperature transducers*. Fort Collins: Metrology Press.

Schraub, F. A., et al. 1965. Use of hydrogen bubbles for quantitative determination of time-dependent velocity fields in low-speed water flows. *Journ. Basic Engineering, Trans. A.S.M.E.* 87, Ser. D, 2: 429–44.

Spink, L. K. 1967. *Principles and practice of flow meter engineering*. 9th ed. Foxboro: Foxboro.

U.S. Bureau of Reclamation. 1953. *Water measurement manual*.

Wilmarth, W. W. 1971. Unsteady force and pressure measurements. *Annual Review of Fluid Mechanics* 3: 147–70.

Films

Abernathy, F. H. Fundamentals of boundary layers. NCFMF/EDC Film No. 21623, Encyclopaedia Britannica Educ. Corp.

Coles, D. Channel flow of a compressible fluid. NCFMF/EDC Film No. 21616, Encyclopaedia Britannica Educ. Corp.

Kline, S. J. Flow visualization. NCFMF/EDC Film No. 21607, Encyclopaedia Britannica Educ. Corp.

Shell Oil Co. Schlieren. Shell Film Library, 1433 Sadlier Circle, West Drive, Indianapolis, Ind. 46239.

Problems

11.1. A pycnometer weighs 0.220 7 lb when empty and 0.925 6 lb when filled with liquid. If its volume is 0.007 639 ft^3, calculate the specific gravity of the liquid.

11.2. A plummet weighs 4.00 N in air and 2.97 N in a liquid. If the volume of liquid displaced by the plummet is 1.29×10^{-4} m^3, what is the specific gravity of the liquid?

11.3. A cylindrical plummet weighing 0.44 N, of 25 mm diameter, and having a specific gravity of 7.70, is suspended in a liquid from the end of a balance arm, 150 mm from the knife edge. The arm is balanced by a weight of 0.53 N, 100 mm from the knife edge. What is the specific gravity of the liquid? Neglect the weight of the balance arm.

11.4. A crude hydrometer consists of a cylinder of $\frac{1}{2}$ in. diameter and 2 in. length, surmounted by a cylindrical tube of $\frac{1}{8}$ in. diameter and 8 in. long. Lead shot in the cylinder brings the hydrometer's total weight to 0.30 oz. What range of specific gravities can be measured by this hydrometer?

11.5. To what depth will the bottom of the hydrometer of the preceding problem sink in a liquid of specific gravity 1.10?

11.6. A hydrometer weighing 1.1 N and having a volume of 99 100 mm^3 is placed in a liquid of specific gravity 1.60. What percent of the volume remains above the liquid surface?

11.7. If the torque required to rotate the inner cylinder of problem 1.62 at a constant speed of 4 r/min is 2.7 N·m, calculate the approximate viscosity of the oil.

11.8. A Stormer-type viscometer consists of two cylinders, one of 75 mm outside diameter, the other of 77.5 mm inside diameter; both are 250 mm high. A 4.45 N weight falls 1.5 m in 10 s, its supporting wire unwinding from a spool of 50 mm diameter on the main shaft of the viscometer. If the space between the cylinders is filled with oil to a depth of 200 mm and the space between the ends of the cylinders is 1.25 mm, calculate the viscosity of the oil.

11.9. A Saybolt Universal viscometer has tube diameter and length of 1.76 mm and 12.25 mm, respectively. The internal diameter of the cylindrical reservoir is 30 mm, and the height from the tube outlet to rim of reservoir is 125 mm. Assuming that the loss of head may be taken as the average of the total heads on the tube outlet at the beginning and end of the run, derive an approximate relationship between ν (m^2/s) and t (Saybolt seconds), and compare with the exact equation relating these quantities. Volume collected is 60 000 mm^3.

11.10. If in 150 s a Saybolt viscometer discharges a standard volume of one oil, how long will it take to discharge the same volume of a second oil which is 10% denser and has a 10% smaller absolute viscosity than the first?

11.11. This viscometer is to be used for liquids of kinematic viscosity between 0.000 02 and 0.000 5 ft^2/s. Calculate the minimum tube length which will hold all Reynolds numbers below 1 500. Consider tube friction only.

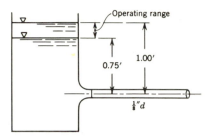

Problem **11.11**

11.12. In this viscosity test the time from start (S) to finish (F) is 200 s. The oil has s.g. 0.87, the pipe is 2.1 m long and of 12.5 mm diameter, K_L for the entrance is 1.2, and for the exit 1.0. Calculate the approximate viscosity of the oil.

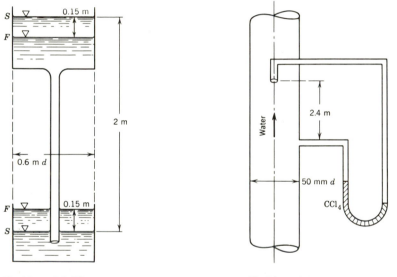

Problem **11.12**   Problem **11.18**

11.13. The disk of Fig. 11.4a and a pitot tube are placed in an airstream aligned properly with the flow, and connected to a U-tube containing water. If the difference of water elevation in the legs of the manometer is 100 mm, calculate the air velocity, assuming a specific weight of 12.0 N/m³.

11.14. A pitot-static tube installed in an airduct shows a differential manometer reading of 2 in. of water. If the pressure and temperature in the duct are 15.0 psia and 80°F, respectively, what velocity is indicated? Neglect compressibility effects.

11.15. A differential manometer containing water is attached to the pitot-static tube on an airplane flying close to the ground through the U.S. Standard Atmosphere (Appendix 4) at

150 mph (67 m/s). What manometer reading can be expected? What velocity would be indicated if this manometer reading occurred at altitude 20 000 ft (6.1 km)?

11.16. A pitot-static tube is placed at the center of a 6 in. smooth pipe in which there is an established flow of carbon tetrachloride at 68°F. The attached differential manometer containing mercury and carbon tetrachloride shows a difference of 3 in. What flowrate exists in the line?

11.17. In a laminar flow in a 0.3 m pipe two pitot tubes are installed, one on the centerline, the other 75 mm from the centerline. If the specific weight of the liquid flowing is 7.86 kN/m^3 and the flowrate is 0.28 m^3/s, calculate the reading of a differential manometer connected to the two tubes when there is mercury in the bottom of the manometer.

11.18. The friction factor of this pipe is 0.025. Calculate the manometer reading when the mean velocity is 3 m/s.

11.19. If the mean velocity is 8.02 ft/s and the fluid flowing has specific gravity and kinematic viscosity of 0.80 and 0.004 01 ft^2/s, respectively, calculate the manometer reading.

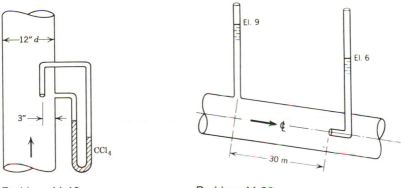

Problem **11.19**   Problem **11.20**

11.20. A liquid flows from left to right in this 150 mm diameter clean cast iron pipe at **R** > 10^6. Calculate the head lost per 30 m of pipe.

11.21. If the sensitivity of a diaphragm pressure transducer is 400 V/psi and the output voltmeter can be read to ±0.001 volts, how accurately can the pressure be measured?

11.22. A diaphragm pressure transducer has a linear relation between output voltage and pressure. If a pair of readings are (1.0 V, 1.38 Pa) and (2.0 V, 2.76 Pa), what is the slope of the calibration curve?

11.23. Calculate the velocity at a point in an airduct if the stagnation temperature is 150°F and the separate stagnation and static pressure manometers show readings of 8 in. and 15 in. of mercury vacuum, respectively. The barometric pressure is 28 in. of mercury.

11.24. Carbon dioxide flows in a pipe. At a point in the flow the stagnation temperature is 40°C, and the absolute static pressure 96.5 kPa. The differential manometer attached to the pitot-static tube shows a reading of 762 mm of mercury. Calculate the velocity at this point.

11.25. A pitot-static tube and temperature probe are installed in a duct where nitrogen is flowing. The stagnation pressure and temperature are 381 mm of mercury gage and 93.3°C, respectively. The static pressure is 75 mm of mercury vacuum. If the barometric pressure is 762 mm of mercury, what velocity is indicated?

11.26. Calculate the pressure and temperature on the nose of a projectile moving near the ground at 2 000 ft/s (610 m/s) through the U.S. Standard Atmosphere. (Appendix 4.)

11.27. The hot-film sensor of Fig. 11.18 produces an output voltage of 4.2 V; what is the indicated velocity? What increase in voltage will occur if the velocity is doubled?

11.28. For the sensor of Fig. 11.18 find the velocity-heat-transfer equation 11.10, that is, determine A and B. Assume room temperature is 20°C.

11.29. Prove that an X-array hot-wire can be used to obtain the turbulent components v_x and v_y in a two-dimensional flow. Assume the wires are inclined at $\pm45°$ to the mean flow direction.

11.30. A Stanton tube of height h of 0.015 in. is installed in the wall of a 12 in. smooth pipe in which a fluid is flowing with a mean velocity of 10 ft/s. The fluid has viscosity and density of 0.002 lb · s/ft^2 and 2 slugs/ft^3, respectively. What pressure difference is to be expected? If this pressure difference is interpreted (as for a pitot-static tube) as $\rho v^2/2$, how far from the wall will this velocity be located?

11.31. A laminar flow occurs in a 300 mm smooth pipe at a Reynolds number of 1 000. In the pipe wall a Stanton tube having $h = 1$ mm is installed. Calculate the ratio between the expected pressure difference and the wall shear. If this pressure difference is presumed to equal $\rho v^2/2$, how far from the wall will this velocty be located?

11.32. A Preston tube of 0.5 in. (12.7 mm) outside diameter is attached to the hull of a ship to measure the local shear. When the ship moves through freshwater (68°F or 20°C) the pressure difference is found to be 75 psf (3.6 kPa). Calculate the local shear.

11.33. Water flows in a horizontal 0.3 m diameter pipe. A force of 0.2 N is exerted on a 50 mm × 50 mm square shear plate by the flow. What is the pressure gradient in the pipe?

11.34. A 12 in. by 6 in. Venturi meter is installed in a horizontal waterline. The pressure gages read 30 and 20 psi. Calculate the flowrate if the water temperature is 68°F. Calculate the head lost between the base and throat of the meter. Calculate the total head lost by the meter if the diffuser tube has cone angle 7°. Calculate the flowrate if the pipe is vertical and the throat of the meter 2 ft below the base.

11.35. Crude oil flows through a horizontal 150 mm by 75 mm Venturi meter. What is the difference in pressure head between the base and throat of the meter when 7.6 litres/s flow at (a) 27°C and (b) 49°C? What is the head loss for each?

11.36. The maximum flowrate in a 250 mm waterline is expected to be 142 litres/s. To the Venturi meter is attached a mercury-under-water manometer 0.91 m long. Calculate the minimum throat diameter which should be specified?

11.37. If the head lost between base and throat of a 3 in by 6 in. Venturi meter is neglected, what coefficient of velocity would be required to allow for the change in velocity distribution between these points if α_1 and α_2 are 1.06 and 1.00, respectively (see Fig. 7.20)?

11.38. A pitot tube is installed at the center of the base of a 100 mm by 50 mm Venturi meter through which 28.1 litres/s of water is flowing. The pitot tube is connected to one side of a differential manometer (containing mercury and water); the other side of this manometer is connected to the throat piezometer ring. Calculate the manometer reading if V/v_c at sections 1 and 2 are 0.82 and 1.00, respectively. Let $C_v = 0.98$.

11.39. Air ($\gamma = 0.08$ lb/ft^3) flows through a 4 in. by 2 in. frictionless Venturi meter. The pressures in the 4 in. and 2 in. sections are 0.25 psi and 0.300 in. of mercury vacuum, respectively. Calculate the flowrate, neglecting compressibility of the air.

11.40. Carbon dioxide flows through a 150 mm by 75 mm Venturi meter. Gages at the base and throat read 138 kPa and 96.5 kPa and temperature in the fluid at the base of the meter is 26.7°C. Calculate the weight flowrate, assuming standard barometer and $C_v = 0.99$.

11.41. An A.S.M.E. flow nozzle of 75 mm diameter is installed in a 150 mm water (20°C) line. The attached manometer contains mercury and water and registers a difference of 381 mm. Calculate the flowrate through the nozzle. Calculate the head lost by the nozzle installation.

11.42. A 3 in. nozzle is installed at the end of a 6 in. airduct in which the specific weight is 0.076 3 lb/ft^3. A differential manometer connected to a piezometer opening 6 in. upstream from the base of the nozzle and to a pitot tube in the jet shows a reading of 0.25 in. of water. Calculate the flowrate, assuming uniform velocity distribution in the jet and C_v of 0.97. Assume the air incompressible.

11.43. If air flows through the pipe and nozzle of problem 11.41 and if open mercury manometers at points 1 and 2 show positive gage pressures of 762 mm and 508 mm, and if the temperature of the air at point 1 is 15.6°C, calculate the weight flowrate, assuming standard barometric pressure.

11.44. A 1 in. nozzle has C_v of 0.98 and is attached to a 3 in. hose. What flowrate (water) will occur through the nozzle when the pressure in the hose is 60 psi? What is the velocity of the jet at the nozzle tip? How much head is lost through the nozzle? To what maximum height will this jet rise (neglect air friction)?

11.45. A sharp-edged orifice with conventional pressure connections is to be installed in a 300 mm waterline. For a flowrate of 0.28 m^3/s the maximum allowable head loss is 7.6 m. What is the smallest orifice that may be used? Since calculations are approximate, asume $C_v = 1$.

11.46. A 4 in. orifice at the end of a 6 in. line discharges 5.30 cfs of water. A pressure gage upstream from the orifice reads 58.0 psi and a gage connected to a pitot tube in the vena contracta reads 60.0 psi. Calculate C_c and C_v for this orifice assuming $\alpha_2 = 1.00$.

11.47. A 150 mm flow nozzle is installed in a 300 mm waterline. An orifice of what diameter will produce the same head loss as the nozzle? Assume C_v the same for nozzle and orifice.

11.48. If the coefficient of an orifice of 3 in. diameter installed in a 6 in. line is approximately 0.65 for the conventional piezometer connections of Fig. 11.25, what approximate coefficient can be expected if the downstream connection is made at a point where the expanding jet has a 4 in. diameter?

11.49. Predict the location of the water surface in the middle piezometer tube relative to one of the other water surfaces; C_v is 0.97.

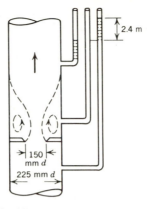

Problem **11.49**

11.50. Find the ratio of the manometer readings for upward and downward flow of the same flowrate. The manometer liquids are the same and each downstream pressure connection is opposite the vena contracta.

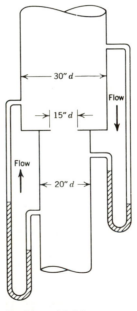

Problem **11.50**

11.51. A 100 mm sharp-edged orifice at the end of a 150 mm waterline has C_v of 0.97. Calculate the flowrate when the pressure in the line is 275 kPa.

11.52. A conical nozzle of 50 mm tip diameter and having C_c of 0.85 and C_v of 0.97 is attached to the end of a 100 mm waterline. A manometer (containing carbon tetrachloride and water) is connected to a pitot tube in the vena contracta and to a piezometer ring at the base of the nozzle. Taking $\alpha_2 = 1.0$, calculate the flowrate and the pressure at the base of the nozzle. The manometer reads 610 mm.

11.53. A 2 in. conical nozzle having C_v of 0.98 and C_c of 0.80 is attached to a 4 in. pipeline and delivers water to an impulse turbine. The pipeline is 1 000 ft long and leaves a reservoir of surface elevation 450 at elevation 420. The nozzle is at elevation 25. Assuming a square-edged pipe entrance and a friction factor of 0.02, calculate (a) the flowrate through the pipe and nozzle, (b) the horsepower of the nozzle stream, and (c) the horsepower lost in line and nozzle.

11.54. This *inlet orifice* is used to meter the flow of air into the pipe. Assuming the downstream pressure connection to be in the plane of the vena contracta, predict the flowrate. Consider the air incompressible.

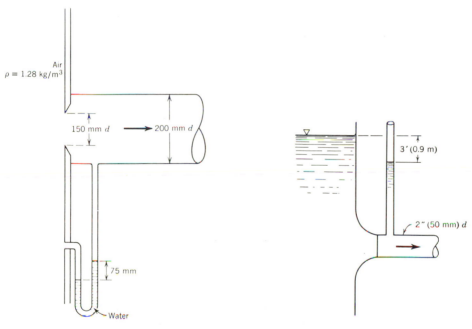

Problem **11.54**

Problem **11.55**

11.55. Calculate the flowrate if C_v for this entrance nozzle is 0.96.

11.56. Water (20°C or 68°F) discharges into the atmosphere from a 37.5 mm or 1.5 in. sharp-edged orifice under a 1.5 m or 5 ft head. Calculate the flowrate. Repeat the calculation for crude oil at 20°C or 68°F.

11.57. Water flows from one tank to an adjacent one through a 75 mm sharp-edged orifice. The head of water on one side of the orifice is 1.8 m and on the other 0.6 m. Taking C_c as 0.62 and C_v as 0.95, calculate the flowrate.

11.58. A jet discharges vertically upward from a 2 in. sharp-edged orifice located in a horizontal plane. If the head (on the vena contracta) is 20 ft and the jet rises to a height of 19 ft above the vena contracta, what is the flowrate if the diameter of the vena contracta is 1.6 in.? Air friction on the jet may be neglected.

11.59. A conventional sharp-edged orifice of 50 mm diameter discharges into the atmosphere from a large tank. At a point in the jet the height of the energy line is measured by pitot tube and found to be 0.09 m below the free surface level in the tank. Calculate the flowrate and the head on the orifice. Air friction on the jet may be neglected.

11.60. A 3 in. (75 mm) sharp-edged orifice discharges vertically upward. At a point 10 ft (3 m) above the vena contracta, the diameter of the jet is 3 in. (75 mm). Under what head is the orifice discharging?

11.61. The flowrate is 5.4 l/s, and the head lost in the diffuser is 0.15 m. Predict the flowrate when the diffuser is removed.

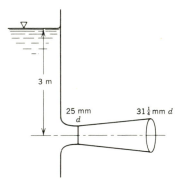

3 m

25 mm d

$31\frac{1}{4}$ mm d

Problem 11.61

11.62. Under a 4.42 ft head, 0.056 cfs of water discharges from a 1 in. sharp-edged orifice in a vertical plane; 3.30 ft outward horizontally from the vena contracta the jet has dropped 0.65 ft below the centerline of the orifice.Calculate C, C_v, and C_c.

11.63. A 2 in. (50 mm) sharp-edged orifice discharges with a 20 ft (6 m) head on its vena contracta. To what height will the jet rise (above the vena contracta) if the jet discharges (a) vertically upward, and (b) upward at an angle of 45°? Neglect air friction on the jet.

11.64. Water discharges through a 1 in. diameter sharp-edged orifice under a 3 ft head. At what head will the same flowrate occur through a horizontal pipe 1 in. in diameter, 12 in. long (friction factor 0.020), and having a square-edged entrance?

11.65. A short tube of 25 mm diameter and 37.5 mm length may flow full or not full at its exit. Calculate the approximate ratio between the respective flowrates for the same head.

11.66. Predict the discharge coefficients of the standard short tube and re-entrant short tube from the loss coefficients of Fig. 9.17. Assume that the tubes are 4 diameters in length with friction factors of 0.020.

11.67. Predict the coefficient of velocity for this short tube with restricted entrance if the friction factor for the tube is 0.020.

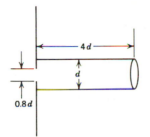

Problem **11.67**

11.68. A sluice gate 4 ft (1.2 m) wide is open 3 ft (0.9 m) and discharges onto a horizontal surface. If the coefficient of contraction is 0.80 and the coefficient of velocity 0.90, calculate the flowrate if the upstream water surface is 4 ft (1.2 m) above the top of the gate opening.

11.69. This sluice gate extends the full width of a rectangular channel 5 ft wide. Assuming C_v is 0.96 and C_c is 0.75, estimate the flowrate, neglecting the dynamics of the roller.

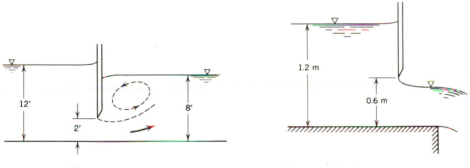

Problem **11.69**   Problem **11.70**

11.70. When this sluice gate is open 0.6 m, its vena contracta is in the plane of the brink of the outfall. If its coefficient of contraction is 0.75, channel and gate widths 2.4 m, and the flowrate 4.95 m³/s, what is its coefficient of velocity?

11.71. An elbow meter of 100 mm diameter has a coefficient of 0.815. What flowrate of water occurs through this meter when the attached manometer (containing mercury and water) shows a difference of 250 mm?

11.72. The flow in a brook is measured by the salt-dilution method; $\frac{3}{4}$ l/min of salt solution having a concentration of 24 N of salt per litre is introduced and mixes with the flow. A sample extracted below the mixing point shows a concentration of 9.4×10^{-5} N/l. Calculate the flowrate in the brook.

11.73. The salt-velocity method is to be used in a 24 in. pipeline, and electrodes are installed 100 ft apart. The time between deflection of the ammeter needles is 12.0 s. Calculate the flowrate in the line.

11.74. The salt-velocity method is to be used in a 1 m pipeline. What is the minimum spacing of electrodes if the time of passage must be greater than 10 seconds for measuring accuracy and the maximum flowrate is 1 m³/s.

11.75. Carry out the dimensional analysis to show that the weir equation can be written as $q/(\sqrt{g_n}H^{3/2}) = f(\mathbf{R}, \mathbf{W}, P/H)$.

11.76. The head on a sharp-crested rectangular weir 1.2 m long and 0.9 m high is 100 mm. Calculate flowrate and velocity of approach. Repeat the calculation for a weir of 0.3 m height.

11.77. A certain flowrate passes over a sharp-crested rectangular weir 0.6 m (2 ft) high under a head of 0.3 m (1 ft). Calculate the head on a similar weir 0.3 m (1 ft) high for the same flowrate.

11.78. What depth of water must exist behind a rectangular sharp-crested weir 1.5 m long and 1.2 m high, when a flow of 0.28 m³/s passes over it? What is the velocity of approach?

11.79. A rectangular channel 5.4 m (18 ft) wide carries a flowrate of 1.4 m³/s (50 cfs). A rectangular sharp-crested weir is to be installed near the end of the channel to create a depth of 0.9 m (3 ft) upstream from the weir. Calculate the necessary weir height.

11.80. Across one end of a rectangular tank 0.9 m wide is a sharp-crested weir 1.2 m high. In the bottom of the tank is a sharp-edged orifice of 75 mm diameter. If 57 1/s flow into the tank, what depth of water will be attained?

11.81. Treating the upper edge of the pipe as a sharp weir crest, estimate the flowrate when the water depth in the basin is 0.75 m.

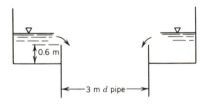

Problem **11.81**

11.82. Derive the theoretical flow equation for the triangular weir.

11.83. Calculate the flowrate of water (68°F or 20°C) over a smooth sharp-crested triangular weir of 90° notch angle when operating under a head of 6 in. or 150 mm. Repeat the calculation for crude oil at the same temperature.

11.84. A triangular weir of 90° notch angle is to be used for measuring water flowrates up to 1.5 cfs or 42.5 1/s. What is the minimum depth of notch which will pass this flowrate?

11.85. A 90° triangular weir discharges water at a head of 0.5 ft (0.15 m) into a tank with a 2.5 in. (62.5 mm) sharp-edged orifice in the bottom. Predict the depth of water in the tank.

11.86. Calculate the approximate flowrate to be expected through this sharp-edged opening. Assume a weir coefficient of 0.62.

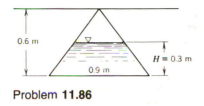

Problem **11.86**

11.87. The depth of water behind the weir plate is 4.8 ft. Predict the flowrate over the weir.

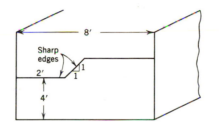

Problem **11.87**

11.88. Estimate the flowrate over this sharp-crested weir, assuming a coefficient of 0.62.

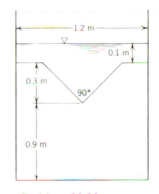

Problem **11.88**

11.89. A rectangular channel 6 m wide carries 2.8 m³/s at a depth of 0.9 m. What height of broad-crested rectangular weir must be installed to double the depth? Assume a weir coefficient of 0.56.

11.90. A broad-crested weir 0.9 m high has a flat crest and a coefficient of 0.55. If this weir is 6 m long and the head on it is 0.46 m, what flowrate will occur over it? What flowrate could be expected if the flow were frictionless?

11.91. A frictionless broad-crested weir 1.2 m high is built across a channel 2.4 m wide. If the energy line is 0.9 m above the weir crest, calculate the head, flowrate, and weir coefficient.

11.92. Using the specific energy and critical depth principles of Section 10.6, derive an expression for C_w as a function of P/H for the frictionless broad-crested weir of Fig. 11.36.

11.93. In order to justify the form of the relation between C_w and P/H of the Rehbock formula, calculate C_w for frictionless broad-crested weirs having values of P/H of 1, 5, and 9. Compare these with values obtained from the Rehbock formula, disregarding $1/1\,000\,H$.

11.94. In a semi-empirical analysis of this square-edged broad-crested weir it is usually *assumed* that $y_2 = H/2$ and the pressure distribution on the weir face is hydrostatic as shown. Using these assumptions derive an expression for the weir coefficient as a function of P/H.

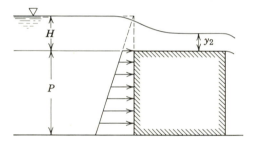

Problem **11.94**

11.95. What flowrate will occur over a spillway of 500 ft (150 m) length when the head there is 4 ft (1.2 m), if the coefficient (referenced to the static head) of the spillway is 0.72?

11.96. A spillway 300 m long is found by model experiments to have a coefficient (referenced to the static head) of 0.68. It has a crest elevation of 30.0. When a flood flow of 1 400 m³/s passes over the spillway, what is the elevation of the water surface in the reservoir just upstream from the spillway?

11.97. A spillway structure 20 ft high is designed by the methods of Fig. 11.39 for a head of 10 ft. Calculate the coefficients and flowrates for heads of 4, 10, and 12 ft. Calculate the corresponding coefficient for the design head when this is referenced to water surface rather than energy line.

11.98. Upstream and downstream from a sharp-crested rectangular weir 0.9 m high the water depths are 1.2 m and 1.05 m, respectively. Calculate the approximate flowrate.

11.99. The drop in the water surface in passing a submerged sharp-crested rectangular weir 1.2 m high is 76 mm. Calculate the approximate flowrate if the depth upstream from the weir is 1.5 m.

11.100. The following data are collected in a current-meter measurement at the river cross section of Fig. 11.43, which is 60 ft (18 m) wide at the water surface. Assume $V = 2.22 \times (r/s)$

[ft/s] or 0.677 × (r/s) [m/s], and calculate the flowrate in the river.

Station	0	1	2	3	4	5	6	7	8	9	10	11	12
Depth													
ft	0.0	3.0	3.2	3.5	3.6	3.7	3.9	4.0	4.4	4.4	4.2	3.5	0.0
m	0.0	0.9	0.96	1.05	1.08	1.11	1.17	1.20	1.32	1.32	1.26	1.05	0.0

rpm (r/min) of Rotating Element

$0.2y$	—	40.0	53.5	58.6	63.0	66.7	61.5	56.3	54.0	52.6	50.0	45.0	—
$0.8y$	—	30.7	42.8	50.0	54.2	58.8	53.3	49.4	46.5	43.2	40.1	32.5	—

12
ELEMENTARY HYDRODYNAMICS

Although the word *hydrodynamics* means (literally) *water-motion* to the layperson, the early scientists and mathematicians used it in a more definite way to define the study of the flowfields of any ideal fluid by mathematical methods—and thus gave it the specific meaning used here. In this sense the flow of an ideal fluid about an airfoil would be considered a problem in hydrodynamics, whereas the established flow of a real fluid in a pipe would not. Although modern hydrodynamics now includes problems with viscous action, the word is used in the classical sense described above in this chapter, except in the case of the flow of fluids in a porous medium.

The aim of this chapter is to provide a modest introduction to classical hydro-dynamics and its possibilities and limitations. To confine the treatment within realistic limits it is restricted to the incompressible ideal fluid, steady flow, and two-dimensional flowfields mostly in the horizontal plane.

In hydrodynamics an ideal fluid is assumed as in Chapters 4 and 5, accompanied by the same limitations discussed there. Since then the reader has learned that real fluids flow as ideal ones if outside the zone of viscous influence, or may be considered ideal as an approximation when viscous action is small. The flowfields about a submerged object or in a short smooth nozzle or pipe entrance are examples of this. Other problems in which viscous action predominates or triggers the separation of fluid from boundary obviously cannot be treated with the ideal fluid; however, hydrodynamical methods can be employed in the study of flowfields in which separation points are not determined by viscous action. For example, the prediction of the complete flowfield about an elliptical object in a free stream would be impossible with an ideal fluid because of the separation caused by viscous action in the boundary layer, but for a disk normal to the flow, where the separation at the edges is independent of viscous action, mathematical methods and the ideal fluid may be effectively used. Other examples are two real fluid flows whose analysis requires specific use of the hydrodynamic concepts and methods discussed below. The first flow is one involving a boundary layer, while the second is laminar flow in a porous medium.

In many flows past streamlined shapes, the frictional aspects of the flow are confined to a thin, boundary layer (see Section 7.4 and Figs. 7.8 and 7.13 in particular). Outside the boundary layer the fluid motion is accurately described by ideal fluid theory. This main ideal flow acts as an "outer" flow which establishes both the velocity at the edge of the "inner" flow or boundary layer and the pressure distribution along the

566

body.[1] An engineering approach to solution of such a flow problem is to solve first the "outer" problem of ideal fluid motion about the body, ignoring viscous effects entirely. Then, using the "outer" solution values of velocity and pressure at the surface of the body as approximate values for the edge of the boundary layer, the "inner" viscous flow problem is solved. Experiments have demonstrated that this is often an effective and accurate process. For streamlined shapes, this procedure gives, from the "outer" solution, the pressure distribution (including an accurate estimate of the lift force, if any) and, from the "inner" solution, an estimate of the friction force or drag on the shape.

In many common flows of liquids and gases through porous media—for example, sand or broken rock—the flow is laminar. Although the microscopic details of the flow in the interstices between the randomly arranged solids of the medium are impossible to define, the average flow or the apparent macroscopic velocity are both definable and obey a linear resistance law. In Section 12.8, these real fluid flows are shown to satisfy the equations of ideal fluid flow. Thus, all the mathematical rigor, elegance, and power of hydrodynamic theory can be applied directly to porous-media flow problems.

12.1 The Stream Function

Definition of the stream function, a concept based on the continuity principle and the properties of the streamline, provides a mathematical means of plotting and interpreting flowfields. Consider the streamline A of Fig. 12.1. By definition, no flow crosses it and thus the flowrate ψ across all lines OA is the same.[2] Accordingly, ψ is a constant of the streamline and, if ψ can be found as a function of x and y, the streamline can be plotted. Similarly, the flowrate between O and a closely adjacent streamline (B) will be $\psi + d\psi$, and the flowrate between the streamlines (i.e., in the streamtube) will be

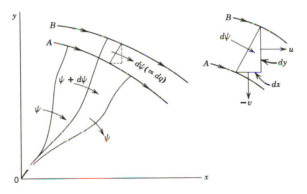

Fig. **12.1** Definition of the stream function.

[1] K. Karamcheti, *Principles of Ideal-Fluid Aerodynamics*, Wiley, 1966, pp. 52–55.
[2] Point O need not be the origin of coordinates.

$d\psi$. As the flowrates into and out of the elemental triangle are equal from continuity considerations,

$$d\psi = -v \, dx + u \, dy$$

However, if $\psi = \psi(x, y)$, the total derivative is (see Appendix 5)

$$d\psi = \frac{\partial \psi}{\partial x} \, dx + \frac{\partial \psi}{\partial y} \, dy \qquad (12.1)$$

and comparison of these equations yields[3]

$$u = \frac{\partial \psi}{\partial y} \quad \text{and} \quad v = -\frac{\partial \psi}{\partial x} \qquad (12.2)$$

Thus, if ψ is a known function of x and y, the components of velocity at any point may be obtained by taking the appropriate partial derivative of ψ. Conversely, if u and v are known as functions of x and y, ψ may be obtained by integration of equation 12.1, yielding

$$\psi = \int \left(\frac{\partial \psi}{\partial x} \right) dx + \int \left(\frac{\partial \psi}{\partial y} \right) dy + C$$

The equation of continuity,

$$\frac{\partial u}{\partial x} + \frac{\partial v}{\partial y} = 0 \qquad (3.18)$$

can be easily (and usefully) expressed in terms of ψ by substituting the relations of equations 12.2; this leads to the identity

$$\frac{\partial}{\partial x}\left(\frac{\partial \psi}{\partial y} \right) = \frac{\partial}{\partial y}\left(\frac{\partial \psi}{\partial x} \right) \quad \text{or} \quad \frac{\partial^2 \psi}{\partial x \, \partial y} = \frac{\partial^2 \psi}{\partial y \, \partial x}$$

which expresses the fact that if $\psi = \psi(x, y)$ the derivatives taken in either order yield the same result and that a flow described by a stream function satisfies the continuity equation automatically.

The equation for vorticity,

$$\xi = \frac{\partial v}{\partial x} - \frac{\partial u}{\partial y} \qquad (3.22)$$

can also be expressed in terms of ψ by similar substitutions:

[3] Analogous relations for polar coordinates are

$$v_r = \frac{\partial \psi}{r \, \partial \theta} \quad \text{and} \quad v_t = -\frac{\partial \psi}{\partial r}$$

in which v_r is positive radially outward from the origin; v_t and θ are positive counterclockwise.

$$\xi = -\frac{\partial^2 \psi}{\partial x^2} - \frac{\partial^2 \psi}{\partial y^2}$$

However, for irrotational flows, $\xi = 0$, and the classic Laplace equation,

$$\frac{\partial^2 \psi}{\partial x^2} + \frac{\partial^2 \psi}{\partial y^2} = \nabla^2 \psi = 0$$

results. This means that the stream functions of all irrotational flows must satisfy the Laplace equation and that such flows may be identified in this manner; conversely, flows whose ψ does not satisfy the Laplace equation are rotational ones. Since both rotational and irrotational flowfields are physically possible, the satisfaction of the Laplace equation is no criterion of the physical existence of a flowfield.

—————————————— **Illustrative Problem** ——————————————

A flowfield is described by the equation $\psi = y - x^2$. Sketch the streamlines $\psi = 0$, $\psi = 1$, and $\psi = 2$. Derive an expression for the velocity V at any point in the flowfield. Calculate the vorticity of this flow.

Relevant Equations and Given Data

$$\xi = \frac{\partial v}{\partial x} - \frac{\partial u}{\partial y} \tag{3.22}$$

$$u = \frac{\partial \psi}{\partial y} \quad \text{and} \quad v = -\frac{\partial \psi}{\partial x} \tag{12.2}$$

$$\psi = y - x^2$$

Solution. From the equation for ψ, the flowfield is a family of parabolas symmetrical about the y-axis with the streamline $\psi = 0$ passing through the origin of coordinates.

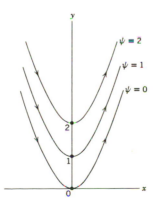

$$u = \frac{\partial}{\partial y}(y - x^2) = 1 - 0 = 1 \tag{12.2}$$

$$v = -\frac{\partial}{\partial x}(y - x^2) = 0 + 2x = 2x \tag{12.2}$$

which allows the directional arrows to be placed on the streamlines as shown. The magnitude V of the velocity is calculated from

$$V = \sqrt{u^2 + v^2} = \sqrt{1 + 4x^2} \blacktriangleleft$$

and the vorticity from

$$\xi = \frac{\partial}{\partial x}(2x) - \frac{\partial}{\partial y}(1) = 2\,\mathrm{s}^{-1} \qquad \text{(counterclockwise)} \blacktriangleleft \tag{3.22}$$

Since $\xi \neq 0$, this flowfield is rotational.

12.2 Basic Flowfields

(a) Rectilinear Flow

In dealing with the flow about solid objects immersed in a stream, the approaching flow is frequently of practically infinite extent and possesses straight and parallel streamlines, and uniform velocity distribution at a great distance from the object. Such a horizontal flow of velocity U is shown in Fig. 12.2. Clearly, $u = U$ and $v = 0$ for this flowfield; its stream function may be obtained from

$$\psi = \int U\, dy + \int (0)\, dx = Uy + C$$

If the streamline coincident with the x-axis is designated by $\psi = 0$, the constant

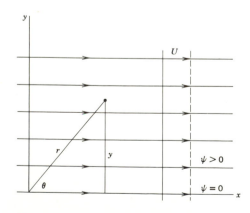

Fig. **12.2** Horizontal rectilinear flow.

vanishes and the stream function of the flowfield is

$$\psi = Uy \qquad (12.3)$$

or, expressed in polar coordinates (where $y = r \sin \theta$),

$$\psi = Ur \sin \theta \qquad (12.4)$$

(b) Source and Sink

Imagine now a symmetrical flowfield consisting of radial streamlines directed outward from a common point where fluid is supplied at a constant rate (Fig. 12.3). Continuity considerations show that the velocities diminish as the streamlines spread, and symmetry requires that all velocities be the same at the same radial distance from the origin. Across all circles of radius r will pass the same flowrate q, and thus the velocity at any point in the flowfield may be determined from $v_r = q/2\pi r$ and $v_t = 0$. Here a mathematical singularity occurs at the center of the source, since, as $r \to 0$, $v_r \to \infty$ because the flowrate is presumed constant; this singularity is of no importance in practical problems which are always concerned with the flowfield away from the center of the source.

The stream function for a source flow can be easily found from

$$\psi = \int \left(\frac{q}{2\pi r}\right) r \, d\theta + \int (0) \, dr = \frac{q\theta}{2\pi} + C$$

and if the streamline $\psi = 0$ is chosen along the radial line $\theta = 0$ the constant vanishes and

$$\psi = \frac{q\theta}{2\pi} \qquad (12.5)$$

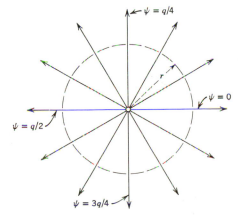

Fig. **12.3** Source flow.

The exact opposite of the *outward* flow from a source is the *inward* flow to a *sink*, whose stream function is (without separate proof)

$$\psi = -\frac{q\theta}{2\pi} \tag{12.6}$$

(c) Free Vortex

Another useful basic flow is the free vortex described by concentric circular streamlines (Fig. 12.4) and a velocity distribution such that the flowfield is irrotational; in this flowfield the radial component of velocity is everywhere zero. To find the velocity distribution and stream function for this flowfield, select a small convenient differential element and calculate the circulation $d\Gamma$ around it, the circulation to be zero if the flow is irrotational. Proceed around the element in counterclockwise direction from A to B, B to C, C to D, and D to A, writing the components $d\Gamma$ in that order:

$$d\Gamma = 0 = (v_t + dv_t)(r + dr)\, d\theta + 0 - v_t r\, d\theta + 0$$

whence (canceling $d\theta$ and neglecting $dv_t\, dr$)

$$v_t\, dr + r\, dv_t = d(rv_t) = 0$$

Thus the velocity distribution is characterized by $v_t r = $ constant, showing that with $r \to 0$, $v_t \to \infty$, and a mathematical singularity is to be expected at the origin.

The circulation Γ along any closed curve coincident with any streamline may be calculated as

$$\Gamma = (2\pi r)v_t$$

in which $v_t r = $ constant. Thus the circulation Γ is a constant of the vortex, being the

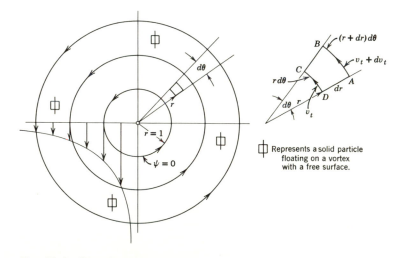

Represents a solid particle floating on a vortex with a free surface.

Fig. **12.4**   Free (irrotational) vortex.

same along all streamlines. However, if the circulation is finite along a streamline at an infinitesimal distance from the origin, the vorticity within the infinitesimal area enclosed by this streamline cannot be zero. Thus the free vortex is a flowfield which is everywhere irrotational except at the singularity in the core of the vortex. In practical problems, however, this paradox causes no difficulties since such problems always involve the irrotational flowfield away from the vortex core.

The irrotationality of a free vortex may be easily visualized by imagining such a vortex with a free surface on which small particles of solid material are floating. As these particles move with the fluid they will be found not to rotate about their own axes, a directional line on a particle remaining parallel to itself during the motion (see Fig. 12.4). This may be convincingly demonstrated in the laboratory with the "drain hole vortex" (Section 4.8) which closely approximates the free vortex in spite of viscous action and radial component of velocity.

The stream function for the vortex may be found from

$$v_t = \frac{\Gamma}{2\pi r} = -\frac{\partial \psi}{\partial r} \qquad \text{and} \qquad v_r = 0$$

and

$$\psi = \int \left(-\frac{\Gamma}{2\pi r}\right) dr + \int (0) r \, d\theta + C$$

which gives $\psi = -(\Gamma/2\pi) \ln r + C$ and, with ψ taken as 0 for the streamline passing through $r = 1$, the constant of integration vanishes and

$$\psi = -\frac{\Gamma}{2\pi} \ln r \qquad (12.7)$$

in which Γ is known as the *vortex strength*. From the sense of the foregoing derivation it can be seen without proof that the stream function of a *clockwise* vortex of strength Γ is given by

$$\psi = +\frac{\Gamma}{2\pi} \ln r \qquad (12.8)$$

(d) Forced Vortex

The so-called forced vortex is defined by the velocity distribution $v_t = \omega r$ and is thus another name for the problem (of rotation of a body of fluid about an axis) discussed in Section 2.7. Although the forced vortex has fewer applications in engineering than the free vortex, it is of considerable interest to study it here for further understanding of vorticity.

The stream function for this clockwise vortex (Fig. 12.5) may be found from $v_t = -\omega r$, $v_r = 0$, and

$$\psi = \int (\omega r) \, dr + \int (0) r \, d\theta + C$$

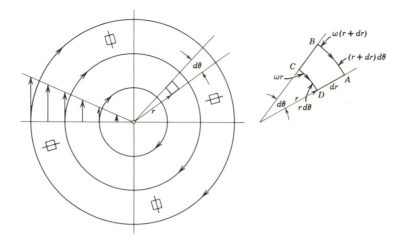

Fig. **12.5** Forced (rotational) vortex.

which yields $\psi = \omega r^2/2 + C$; if ψ is then taken to be zero where $r = 0$, the constant disappears and $\psi = \omega r^2/2$. This could also have been obtained directly from the original definition of the stream function (Fig. 12.1) as a flowrate between origin and (circular) streamline by writing

$$\psi = q = \text{(mean velocity)(area)} = \left(\frac{\omega r}{2}\right) r = \frac{\omega r^2}{2} \qquad (12.9)$$

Consider now the differential circulation around any differential element of fluid in the forced vortex. Starting from A and proceeding in a counterclockwise direction,

$$d\Gamma = -\omega(r + dr)(r + dr)\, d\theta + 0 + \omega r(r\, d\theta) + 0 = -2\omega r\, dr\, d\theta$$

However, the vorticity ξ is defined by $d\Gamma/dA$, in which $dA = r\, d\theta\, dr$, and thus $\xi = -2\omega$, which shows the vorticity to be directly related to the angular velocity of the fluid mass. From this it may be concluded that fluid elements in the forced vortex rotate about their own axes, whereas in the free (irrotational) vortex they do not. This may be generalized to conclude that vorticity is a measure of the rotational aspects of the fluid particles as they move through the flowfield.

12.3 Combining Flows by Superposition

By combination of rectilinear, vortex, source, and sink flows, flowfields of considerable engineering importance can be developed. At a point P (see Fig. 12.6) in a flowfield ψ_A there will be some velocity V_A tangent to the streamline passing through this point; *at the same point* in a flowfield ψ_B there will be some velocity V_B. When these flows are superposed the resulting velocity V at the point must be the vector sum (resultant) of the two velocities V_A and V_B. At other points the same vector addition of velocities occurs and a third flowfield ψ results. Although the principle of superposing

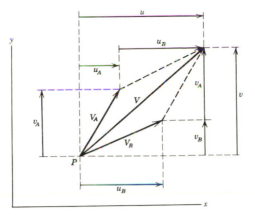

Fig. **12.6**

flows is basically a simple process, its application on a point-by-point basis over a flowfield would be excessively tedious and time-consuming. A simpler procedure may be deduced from Fig. 12.6, where it is easily seen that

$$u = u_A + u_B$$
$$v = v_A + v_B$$

and, using the relations of equations 12.2, that

$$\frac{\partial \psi}{\partial y} = \frac{\partial \psi_A}{\partial y} + \frac{\partial \psi_B}{\partial y} = \frac{\partial}{\partial y}(\psi_A + \psi_B)$$

$$\frac{\partial \psi}{\partial x} = \frac{\partial \psi_A}{\partial x} + \frac{\partial \psi_B}{\partial x} = \frac{\partial}{\partial x}(\psi_A + \psi_B)$$

Similarly, because the Laplace equation is linear,

$$\nabla^2 \psi_A + \nabla^2 \psi_B = \nabla^2(\psi_A + \psi_B)$$

It follows that

$$\psi = \psi_A + \psi_B$$

Thus, obtaining a flowfield by superposition of other flowfields merely involves algebraic addition of the stream functions of the latter; the sum of these stream functions is the stream function of the resultant (combined) flowfield.

12.4 Some Useful Combined Flowfields

(a) Source in a Rectilinear Flow

Superposition of a source and rectilinear flow yields, if the source is at the origin of coordinates, the stream function

$$\psi = Ur \sin \theta + \frac{q\theta}{2\pi} \qquad (12.10)$$

from which the streamlines may be plotted, using polar coordinates. The result (Fig. 12.7) is the streamline picture about the *nose of a solid object* (of special form) in a free stream. Although the reader may easily confirm this flowfield in a mathematical plotting exercise, certain physical features are worthy of discussion and necessary to a full understanding of the problem.

Near the source at O the velocities are very high, but they vary inversely with the radial distance from the source. Accordingly, the effect of the source is nil at large (theoretically infinite) distances from O and there the free stream velocity is found to be unchanged by the presence of the source. However, at some point (on the x-axis) to the left of O, the velocity U of the free stream must exactly equal and cancel the source velocity, yielding a net velocity of zero for the combined flowfield. This condition defines the stagnation point s (where the velocity is zero) whose distance x_s from the origin may be computed from $U = q/2\pi x_s$, showing x_s to be dependent on source strength q and free stream velocity U. Fluid particles issuing from the source and moving to S cannot proceed further to the left, so they must move up or down from S and be carried to the right along a streamline which separates the source flow from that of the free stream. This streamline may be considered to be the contour of a solid body around which the original free stream flow is forced to pass. The equation of the body contour can be completely determined once the value of ψ for its (coincident) streamline is found. Using the coordinates of the stagnation point, which is a known point on the body contour,

$$\psi = U\left(\frac{q}{2\pi U}\right) \sin \pi + \frac{q\pi}{2\pi} = \frac{q}{2}$$

and the body contour is thus described by the equation

$$Ur \sin \theta + \frac{q\theta}{2\pi} = \frac{q}{2}$$

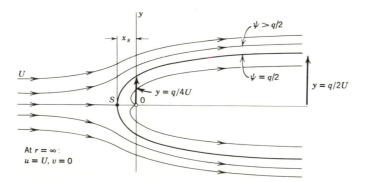

Fig. **12.7** Half-body: source in horizontal rectilinear flow.

Other principal dimensions of the body contour are found by setting $\theta = \pi/2$ and $\theta = 0$ or 2π. For $\theta = \pi/2$, $r = y = q/4U$; for $\theta = 0$, r is infinite but $r \sin \theta = y = q/2U$ (for $\theta = 2\pi$, $y = -q/2U$), showing that the body contour approaches these lines as asymptotes, that is, the body has a finite width.

Anywhere in the flowfield the velocities v_r and v_t are

$$v_r = \frac{\partial \psi}{r \partial \theta} = \frac{1}{r} \frac{\partial}{\partial \theta}\left(Ur \sin \theta + \frac{q\theta}{2\pi} \right) = U \cos \theta + \frac{q}{2\pi r}$$

$$v_t = -\frac{\partial \psi}{\partial r} = -\frac{\partial}{\partial r}\left(Ur \sin \theta + \frac{q\theta}{2\pi} \right) = -U \sin \theta$$

and the resultant velocity V is $V^2 = v_t^2 + v_r^2$. From the definition of velocity in terms of stream function derivative, the reader can easily deduce that the velocity normal to a streamline is zero, that is, $v_n = -\partial \psi/\partial s = 0$, where n and s are the normal and tangential coordinates at any point on a streamline. This then is the boundary condition at a solid body in a fluid flow.

The body contour formed by a source in a rectilinear flow is unique in that it has a head (or nose) but no tail; for this reason it is known as a *half-body*. Half-bodies are used effectively in studying the pressure and velocity distributions near the upstream end of symmetrical objects such as aircraft stabilizers, struts, and bridge piers, or torpedo and aircraft fuselages for the axisymmetric three-dimensional case. A generalization of this scheme is the use of many sources of different strengths arrayed along the x-axis to produce a boundary streamline coincident with a prescribed body contour; by this method, the velocity and pressure field about the upstream end of an object may be accurately predicted without costly experimentation.

(b) Source and Sink of Equal Strength

In the flowfield produced by a source and sink of equal strength the total flowrate passes from one to the other and thus features a family of streamlines originating at the source and ending in the sink. A source and sink on the x-axis are shown in Fig. 12.8. The stream function of the flowfield is

$$\psi = \frac{q\theta_1}{2\pi} - \frac{q\theta_2}{2\pi} = \frac{q}{2\pi}(\theta_1 - \theta_2) \tag{12.11}$$

Consider any point P in the flowfield; from geometrical considerations the angle between the radii r_1 and r_2 is given by $\alpha = \theta_2 - \theta_1$. Thus the equation of any streamline is

$$\psi = -\frac{q}{2\pi}\alpha \tag{12.12}$$

and, for ψ constant along any streamline, α is constant, showing (again from geometry) that all streamlines of the flowfield are circles which pass through source and sink. The similarity of this flowfield to (1) the flux lines between the poles of a magnet and (2) the flow of electrical current through a homogeneous conductor between points of high and low potential can be observed.

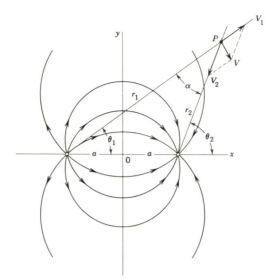

Fig. **12.8** Source and sink of equal strength.

In Cartesian coordinates the stream function for this flowfield may be written by replacing the θ's with the appropriate trigonometric function. This yields

$$\psi = \frac{q}{2\pi}\left(\arctan \frac{y}{x+a} - \arctan \frac{y}{x-a}\right) \tag{12.13}$$

No simple expressions for u and v may be derived from this, but the velocity at any point P is the vector sum of those (V_1 and V_2) caused separately by the source and sink; these are easily calculated from $q/2\pi r_1$ and $q/2\pi r_2$.

(c) Source and Sink of Equal Strength in a Rectilinear Flow

To close the downstream end of the half body cited in part (a) of this section, a source and sink of equal strength may be used, resulting in the classic *Rankine oval* of Fig. 12.9. Let the source and sink be on the x-axis with the origin of coordinates midway between them. The stream function for the combination of source, sink, and rectilinear flow is then

$$\psi = Uy + \frac{q}{2\pi}\left(\arctan \frac{y}{x+a} - \arctan \frac{y}{x-a}\right) \tag{12.14}$$

from which any streamline in the flowfield may be plotted. The body contour is of particular interest, and its principal dimensions ($l/2$ and $b/2$) may be obtained as in part (a). Stagnation points (S) are to be expected to the left of the source and to the right of the sink. Here the resultant velocity from source, sink, and rectilinear flow is zero, so

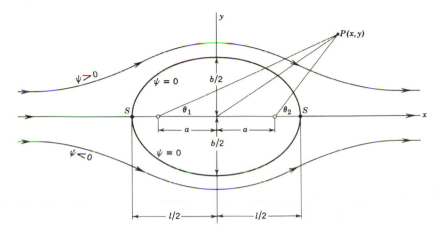

Fig. **12.9** Rankine oval: source and sink (of equal strength) in horizontal rectilinear flow.

$$0 = -\frac{q}{2\pi(l/2 - a)} + \frac{q}{2\pi(l/2 + a)} + U$$

and so

$$\frac{l}{2} = a\sqrt{1 + \frac{q}{a\pi U}}$$

The stagnation points are at $y = 0$ and $x \neq a$, so ψ (from equation 12.14) for the body contour streamline will be zero, from which the body contour may be plotted. The other principal dimension $b/2$ may be obtained by substituting $x = 0$, $y = b/2$, and $\psi = 0$ into equation 12.14, which reduces to

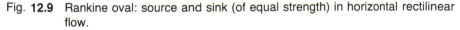

$$0 = U\frac{b}{2} + \frac{q}{\pi} \arctan \frac{b}{2a} - \frac{q}{2}$$

and from which $b/2$ is obtained by trial.

The flowfields about certain oval bodies may thus be predicted and the pressure distributions on the body contour deduced from them.[4] To the engineer the results will be somewhat unrealistic downstream from the midsection of the oval because of separation and wake formation in real fluid flow (see Figs. 7.10, 7.14, 13.11, and 13.12). However, on the upstream end of the object, where boundary layers are thin and flow accelerating, the theoretical predictions will be found to agree closely with experimental measurements.

[4] The top half of such a flow picture is a mathematical representation of the flow of the wind over an oval building (such as a dirigible hangar or quonset hut); the wind pressure on such structures may be studied (very approximately) in this manner.

A generalization of this basic scheme (but with considerably greater mathematical difficulties) is the distribution of sources and sinks of *zero net strength* along the x-axis. In combination with a rectilinear flow a closed body contour is to be expected, but the form of the body may be made by this technique into a more useful streamlined one than that of the Rankine oval.

(d) The Doublet

The flowfield for a *doublet* is produced by reducing the distance a of Fig. 12.8 to zero, that is, superposing a source and sink of equal strength. However, if this operation were carried out without some restriction the result would be no flowfield at all, since source and sink would exactly cancel each other! To obtain a useful flowfield a certain strategem must be employed to prevent this cancellation. Consider now (on Fig. 12.10) the same arrangement of source, sink, and point P as that of Fig. 12.8. The common side AB of the two triangles ABP and ABC is seen to be

$$AB = r_2 \sin \alpha = 2a \sin \theta_1$$

Now as $a \to 0$: $\alpha \to 0$, $\sin \alpha \to \alpha$, $r_2 \to r_1 \to r$, and $\theta_1 \to \theta_2 \to \theta$. The foregoing equation thus reduces to (for $a \to 0$)

$$r\alpha = 2a \sin \theta$$

Substituting this in equation 12.12, the stream function for the doublet is obtained as

$$\psi = -\frac{q}{2\pi}\left(\frac{2a \sin \theta}{r}\right)$$

This form of the equation is quite useless since, with $a = 0$, $\psi = 0$. Suppose, however, that as a approaches zero, q is increased in such a manner that $2qa$ remains a constant, m, known as the *strength* of the doublet. Then the stream function of the doublet becomes

$$\psi = -\frac{m \sin \theta}{2\pi r} \tag{12.15}$$

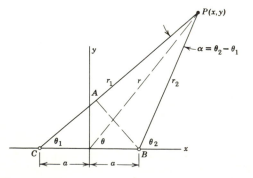

Fig. **12.10**

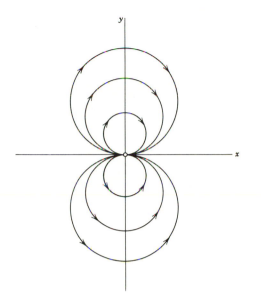

Fig. **12.11** Doublet.

For any streamline, ψ is a constant and $r = C \sin \theta$, showing the streamlines to be circles tangent to the x-axis at the origin. The flowfield produced by a doublet is therefore that of Fig. 12.11.

(e) Doublet in Rectilinear Flow

When the doublet is combined with the rectilinear flow, a limiting case of the Rankine oval results. From experience with the latter a closed body contour will be expected and from the stream function

$$\psi = Ur \sin \theta - \frac{m \sin \theta}{2\pi r} \qquad (12.16)$$

the form of the body may be deduced once the value of ψ for the body streamline is determined. From experience with the Rankine oval, the value of ψ for the body contour is zero and, with $\psi = 0$ in equation 12.16, the dimension R of the body may be obtained as $R = \sqrt{m/2\pi U}$. This shows R to be a constant and the body contour a circle (Fig. 12.12).

The stream function for this flowfield may now be expressed in terms of R instead of the doublet strength by substitution of $m = 2\pi UR^2$ into equation 12.16. This gives

$$\psi = U\left(r - \frac{R^2}{r}\right) \sin \theta \qquad (12.17)$$

and simple expressions for radial and tangential velocity components anywhere in the flowfield are obtained; these are

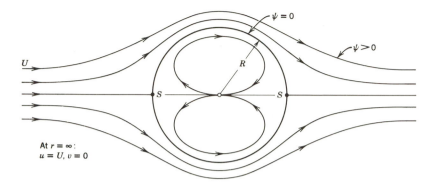

Fig. **12.12**   Doublet in horizontal rectilinear flow.

$$v_r = \frac{\partial \psi}{r\, \partial \theta} = U\left(1 - \frac{R^2}{r^2}\right)\cos\theta \qquad (12.18)$$

$$v_t = -\frac{\partial \psi}{\partial r} = -U\left(1 + \frac{R^2}{r^2}\right)\sin\theta \qquad (12.19)$$

Since the radial velocity component on the body contour (where $r = R$) is zero, $V = v_t = -2U\sin\theta$, from which the pressure variation over the body may be deduced. Of particular interest are the facts that (1) the velocity on the surface of the body at its midsection is exactly twice[5] that of the free stream and (2) on the body surface, at 30° from the x-axis,[5] the velocity and the pressure are exactly equal to those in the undisturbed free stream (see the next to last Illustrative Problem in Section 4.8).

(f) Doublet in Rectilinear Flow with Circulation

Another useful flowfield may be constructed by superposing[6] free vortex, doublet, and rectilinear flowfields. The stream function is (for the clockwise vortex)

$$\psi = U\left(r - \frac{R^2}{r}\right)\sin\theta + \frac{\Gamma}{2\pi}\ln r \qquad (12.20)$$

The use of the first term, representing the doublet in the rectilinear flow, implies that the body contour, a circle of radius R, is unchanged by the superposition of the vortex flow; this is easily justified by recalling that the velocities induced by the vortex are wholly tangential, therefore possess no component normal to the body, and thus cannot change its form. When the flowfield is plotted it will be found (for small Γ) to be as shown in Fig. 12.13 with the stagnation points at angle α below the horizontal diameter of the circle. The velocity components anywhere in the flowfield are

---

[5] For the corresponding three-dimensional flowfield about a sphere these values are $3U/2$ and 42°, respectively.

[6] With vortex and doublet both at the origin of coordinates.

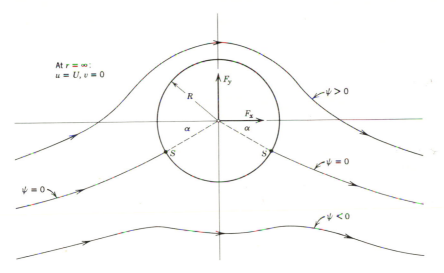

Fig. 12.13 Doublet and free vortex in horizontal rectilinear flow.

$$v_r = \frac{\partial \psi}{r \partial \theta} = U\left(1 - \frac{R^2}{r^2}\right) \cos \theta$$

$$v_t = -\frac{\partial \psi}{\partial r} = -U\left(1 + \frac{R^2}{r^2}\right) \sin \theta - \frac{\Gamma}{2\pi r}$$

On the body contour, $r = R$, $v_r = 0$, and $v_t = -2U \sin \theta - \Gamma/2\pi R$, from which the stagnation points may be located. At the stagnation points, $v_t = 0$ and $\theta = -\alpha$; therefore $\sin \alpha = \Gamma/4\pi UR$. For the largest value of α, which is seen to be $\pi/2$, the two stagnation points will merge. Here $\sin \alpha = 1$, $\Gamma/4\pi UR = 1$, and the flowfield is that of Fig. 12.14a. For $\Gamma/4\pi UR > 1$, the stagnation point is found at some point below the circle and on its vertical axis as in Fig. 12.14b.

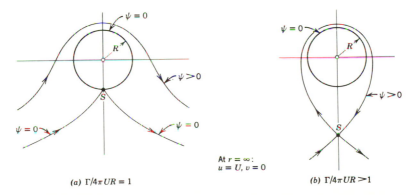

(a) $\Gamma/4\pi UR = 1$ (b) $\Gamma/4\pi UR > 1$

Fig. 12.14

For the flowfield of Fig. 12.13 the forces F_x and F_y exerted by fluid on the circle may be computed by integrating the appropriate differential force components over the perimeter of the circle. Applying the Bernoulli equation between a point in the free stream (where velocity is U and pressure p_o) and another point on the surface of the body where pressure is p and velocity is $(-2U \sin \theta - \Gamma/2\pi R)$,

$$p = p_o + \tfrac{1}{2}\rho U^2 - \tfrac{1}{2}\rho \left(-2U \sin \theta - \frac{\Gamma}{2\pi R} \right)^2 \tag{12.21}$$

The pressure p (see Fig 12.15) will be normal to the perimeter of the circle, will act on a differential area $R \, d\theta$ at the point (R, θ), and will produce a differential force $pR \, d\theta$ with horizontal and vertical components of $pR \cos \theta \, d\theta$ and $pR \sin \theta \, d\theta$, respectively. Evidently, then,

$$-F_x = \int_0^{2\pi} pR \cos \theta \, d\theta$$

$$-F_y = \int_0^{2\pi} pR \sin \theta \, d\theta$$

The integrations may be carried out after inserting the expression for p given by equation 12.21. The results are

$$F_x = 0 \quad \text{and} \quad F_y = \rho U \Gamma \tag{12.22}$$

The first of these might have been anticipated from symmetry of the flowfield, and it shows that the drag force exerted by an ideal fluid on the body contour is zero. This may be generalized to apply to bodies of any shape and is known as the *d'Alembert paradox.*[7] The expression for F_y is of more far-reaching consequence in modern fluid

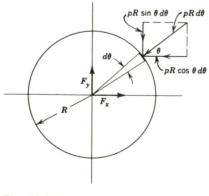

Fig. 12.15

---

[7] In d'Alembert's time the discrepancy between theory and observation was considered to be a contradiction; today it is completely explained by the action of viscosity.

mechanics as it shows that a circulation Γ and a velocity U are both necessary to the existence of a transverse (*lift*) force. The foregoing derivation, when generalized to apply to a body contour of any shape, is called the *Kutta-Joukowsky theorem* after those who first developed it. It serves to explain certain familiar phenomena in which bodies spinning in real fluids create their own circulation and when exposed to a rectilinear flow are acted on by a transverse force; some examples are the forces on the rotating cylinders of a "rotorship" and the transverse force which causes a pitched baseball to curve. However, its most important engineering application is in the theory of the lift force exerted on airfoils, hydrofoils, and the blades of turbines, pumps, propellers, windmills, and so forth. The theory does not, of course, explain the origin of circulation about a nonspinning object but does demonstrate that a circulation is required for the generation of a transverse force.

By the use of the mathematics of complex variables, *conformal transformations* of these flows may be constructed. Using the Joukowsky transformation, Fig. 12.12 may be distorted into an airfoil or hydrofoil in a flowfield as shown in Fig. 12.16a, which yields no transverse force and (the unrealistic) flow around the trailing edge of the foil. However, the transformation of Fig. 12.13 yields the more realistic flow picture of Fig. 12.16b; here, with circulation, and a flow tangent to the surfaces at the trailing edge, a lift is produced and experimental results are closely approximated.[8]

(g) Vortex in Rectilinear Flow

The stream function of a clockwise vortex of strength Γ in a rectilinear flow of velocity U is

$$\psi = Ur \sin \theta + \frac{\Gamma}{2\pi} \ln r \qquad (12.23)$$

from which

$$v_r = \frac{\partial \psi}{r\partial \theta} = U \cos \theta$$

$$v_t = -\frac{\partial \psi}{\partial r} = -U \sin \theta - \frac{\Gamma}{2\pi r}$$

The resulting flowfield will appear as in Fig. 12.17, resembling that of Fig. 12.14b.

(a)   (b)

Fig. **12.16**

[8] As noted earlier viscous effects in the boundary layer must be accounted for to reconcile theoretical and experimental drag results.

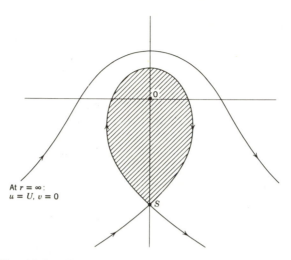

Fig. **12.17**   Free vortex in horizontal rectilinear flow.

The closed streamline may be considered to be a body contour on which an upward force (by the generalized Kutta-Joukowsky theorem) $\rho U \Gamma$ may be expected. Such a flowfield, therefore, becomes a mathematical representation of the large flowfield about a small wing or blade element. When used in company with another vortex (of the same sense) the flowfield may be made to represent that about a biplane; an infinite number of vortices in line and uniformly spaced will yield the flowfield near an infinite series of blades such as encountered in propeller or turbine theory.

(h) Vortex Pair

Two vortices of equal strength but opposite sense in close proximity are known as a *vortex pair*. If these vortices are on the x-axis (Fig 12.18) the stream function of the

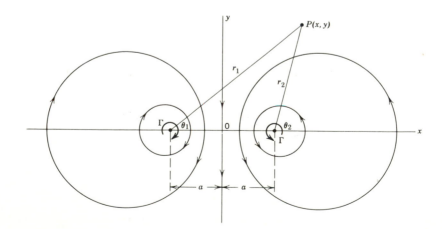

Fig. **12.18**   Vortex pair.

combined flow will be

$$\psi = \frac{\Gamma}{2\pi} \ln r_1 - \frac{\Gamma}{2\pi} \ln r_2 = \frac{\Gamma}{2\pi} \ln \frac{r_1}{r_2} \qquad (12.24)$$

and, when plotted, a symmetrical flowfield will result as shown. Since the velocities induced by the vortices in the zone between $-a$ and $+a$ are larger than elsewhere in the flowfield, a region of lower pressure is to be expected there and such vortices have a tendency to move toward each other.

(i) Vortex Pair in Rectilinear Flow

When combined with a rectilinear flow the vortex pair will form a closed body contour (Fig. 12.19) analogous to the Rankine oval of Fig. 12.9. If the vortex pair is on the x-axis and the rectilinear flow upward and parallel to the y-axis, the stream function of the flowfield will be

$$\psi = \frac{\Gamma}{2\pi} \ln \frac{r_1}{r_2} - Vx \qquad (12.25)$$

From symmetry of the flow, two stagnation points are expected on the y-axis and may be located by setting the resultant velocity equal to zero:

$$2\left(\frac{\Gamma}{2\pi r_s}\right) \cos \alpha - V = 0$$

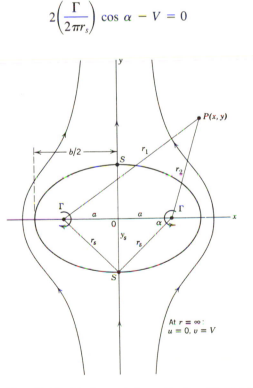

Fig. **12.19** Vortex pair in vertical rectilinear flow.

and with $r_s^2 = y_s^2 + a^2$ and $\cos \alpha = a/r_s$,

$$y_s^2 = \frac{\Gamma a}{\pi V} - a^2$$

from which y_s may be obtained.[9] The equation for the body contour may be obtained by noting that, at the stagnation point, $r_1 = r_2$ and $x = 0$ and therefore $\psi = 0$. The equation of the body contour (see Fig. 12.19) is then (from equation 12.25)

$$x = \frac{\Gamma}{4\pi V} \ln \frac{(x + a)^2 + y^2}{(x - a)^2 + y^2}$$

from which it may be plotted; the half-width $b/2$ may be obtained by solving (by trial) for x with $y = 0$.

(j) Source and Vortex

Superposition of a source and vortex produces a flowfield of some utility in the analysis of flow through fluid machinery. The stream function of this flow is (for a clockwise vortex)

$$\psi = \frac{q\theta}{2\pi} + \frac{\Gamma}{2\pi} \ln r \qquad (12.26)$$

The equation of the streamline, $\psi = 0$, is

$$r = e^{-q\theta/\Gamma}$$

showing the flowfield to be a family of logarithmic spirals (Fig 12.20). The center of the source-vortex combination contains the double singularity of infinite tangential and radial velocity components; this point is to be avoided in practical calculations. However, the flow between circles 1 and 2 is of considerable engineering interest. Through the zone between circles 1 and 2 the velocity diminishes from V_1 to V_2, showing the zone to be a *diffuser* through which a rise of pressure is to be expected. The velocity components anywhere in the flowfield are

$$v_r = \frac{\partial \psi}{r \partial \theta} = \frac{q}{2\pi r} \quad \text{and} \quad v_t = -\frac{\partial \psi}{\partial r} = -\frac{\Gamma}{2\pi r}$$

With Γ constant,

$$(v_t r)_1 = (v_t r)_2$$

which means that the moments of momentum (see Section 6.12) at sections 1 and 2 are the same; since no torque is exerted on the fluid between sections 1 and 2, this is to be expected. Also it may be noted (from the velocity triangle) that $\tan \alpha$ is a constant throughout the flowfield, confirming a well-known property of the logarithmic spiral. Frequently in engineering this flow is encountered in the annular passage (called *vortex chamber or vaneless diffuser*) just downstream from the rotating impeller of a centrif-

---

[9] For the existence of stagnation points on the y-axis, $\Gamma a/\pi V$ must be larger than a^2.

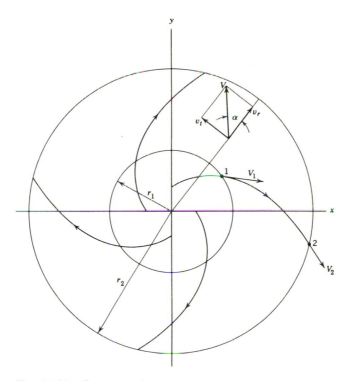

Fig. **12.20** Source and vortex.

ugal pump or supercharger. Similarly a sink-vortex combination is a mathematical representation of the flowfield (if two-dimensional) between the guide vanes and runner of a hydraulic turbine. (See Fig. 6.24.)

(k) The Effect of a Wall; Images

Figure 12.21 shows a collection of the bodies represented above by a combination of sources, sinks, or vortices with a rectilinear flow. However, now the dividing, central streamline has been interpreted as a wall outside the body (see footnote 4). In the first three cases the sources were placed along the wall to create the hump or body. However, in the case of the vortex pair, they are located at equal distances to each side of the wall, so that one vortex can be imagined to be the "mirror image" of the other.

What then is the effect of the presence of a wall on flow in a vortex in a rectilinear flow? In general, the line $x = 0$ is not a streamline in the presence of a vortex (Fig 12.4) located a distance $x = a$ above $x = 0$ (see Fig. 12.18). However, it is clear that the flow of a vortex near a wall is simulated by locating an *image* vortex of opposite sign at $x = -a$. Then $x = 0$ is a streamline (compare Fig 12.4), and the apparent effect of a wall is to produce the cellular flow in Fig. 12.18.

A number of useful flowfields can be created by beginning with basic source, sink, vortex, or other flowfields, and then locating equivalent flowfields of the same or

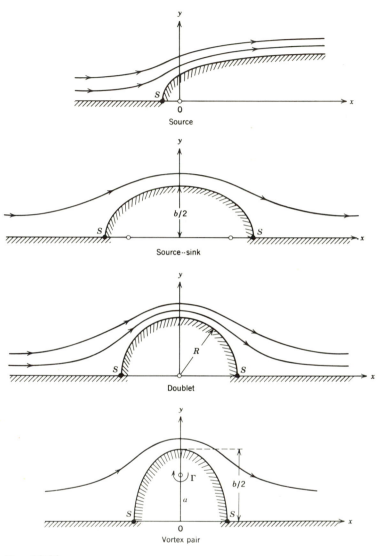

Fig. **12.21**

opposite sign on the opposite side of a boundary that is to be simulated. An example is given in Section 12.8 where sources and sinks have practical interpretations as recharge or withdrawal (supply) wells in groundwater flow.

12.5   The Velocity Potential

Suppose now that another function $\phi(x, y)$ is defined such that the negative of its derivative with respect to distance in any direction yields the velocity in that direction.

For example,

$$\mathbf{V} = -\text{grad } \phi = -\left[\frac{\partial \phi}{\partial x}\mathbf{e}_x + \frac{\partial \phi}{\partial y}\mathbf{e}_y\right] = -\nabla\phi$$

so in Cartesian coordinates

$$u = -\frac{\partial \phi}{\partial x} \qquad \text{and} \qquad v = -\frac{\partial \phi}{\partial y} \tag{12.27}$$

or, in polar coordinates,

$$v_r = -\frac{\partial \phi}{\partial r} \qquad \text{and} \qquad v_t = -\frac{\partial \phi}{r\,\partial\theta} \tag{12.28}$$

The function ϕ is known as the *velocity potential,* and it has some significant properties which may now be examined in general terms.

The continuity equation

$$\frac{\partial u}{\partial x} + \frac{\partial v}{\partial y} = 0 \tag{3.18}$$

may be written in terms of ϕ by substitution of the above definitions, to yield the Laplacian differential equation,

$$\frac{\partial^2 \phi}{\partial x^2} + \frac{\partial^2 \phi}{\partial y^2} = 0 \tag{12.29}$$

Thus all practical flows (which must conform to the continuity principle) must satisfy the Laplacian equation in terms of ϕ.

Similarly the equation for vorticity,

$$\xi = \frac{\partial v}{\partial x} - \frac{\partial u}{\partial y} \tag{3.22}$$

may be put in terms of ϕ to give

$$\xi = \frac{\partial}{\partial x}\left(-\frac{\partial \phi}{\partial y}\right) - \frac{\partial}{\partial y}\left(-\frac{\partial \phi}{\partial x}\right) = -\frac{\partial^2 \phi}{\partial x\,\partial y} + \frac{\partial^2 \phi}{\partial y\,\partial x}$$

from which a valuable conclusion may be drawn: Since $\partial^2\phi/\partial x\,\partial y = \partial^2\phi/\partial y\,\partial x$, *the vorticity must be zero for the existence of a velocity potential.* From this it may be deduced that only irrotational ($\xi = 0$) flowfields can be characterized by a velocity potential ϕ; for this reason *irrotational* flows are also known as *potential* flows.

—————————————— **Illustrative Problem** ——————————————

Calculate the velocity potential ϕ and sketch the equipotential lines of the flowfield produced by a doublet and rectilinear flow.

Solution. From part (*e*) of Section 12.4,

$$v_r = U\left(1 - \frac{R^2}{r^2}\right)\cos\theta \qquad v_t = -U\left(1 + \frac{R^2}{r^2}\right)\sin\theta \qquad \text{(12.18 and 12.19)}$$

but

$$v_r = -\frac{\partial\phi}{\partial r} \qquad \text{and} \qquad v_t = -\frac{\partial\phi}{r\,\partial\theta} \qquad\qquad \text{(12.28)}$$

and

$$\phi = \int\left(\frac{\partial\phi}{\partial r}\right)dr + \int\left(\frac{\partial\phi}{r\partial\theta}\right)r\,d\theta + C$$

Therefore

$$\phi = \int -U\left(1 - \frac{R^2}{r^2}\right)\cos\theta\,dr + \int U\sin\theta\left(1 + \frac{R^2}{r^2}\right)r\,d\theta + C$$

Expanding and rearranging,

$$\phi = -U\left[\int\int(\cos\theta)\,dr - \int(r\sin\theta)\,d\theta\right]$$

$$+ UR^2\left[\int\int\left(\frac{\cos\theta}{r^2}\right)dr + \int\left(\frac{\sin\theta}{r}\right)d\theta\right] + C$$

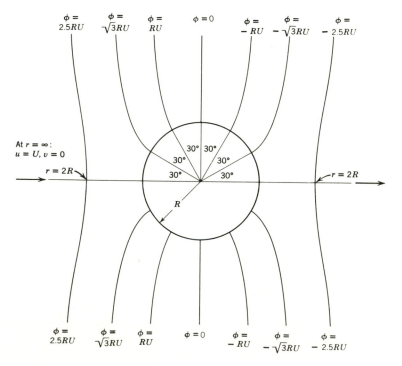

The bracketed quantities may be shown to be (respectively) equivalent to the integrals indicated in

$$\phi = -U \int d(r \cos \theta) + UR^2 \int d\left(\frac{-\cos \theta}{r}\right) + C$$

and thus

$$\phi = -U\left(r + \frac{R^2}{r}\right) \cos \theta + C \blacktriangleleft$$

Assuming for convenience that the constant of integration is zero, it is seen that $\phi = 0$ when $\theta = \pi/2$ or $3\pi/2$; thus the line $\phi = 0$ coincides with y-axis. For $\theta \to \pi$ and $r \to \infty$, $\phi \to +\infty$, and, for $\theta \to 0$ and $r \to \infty$, $\phi \to -\infty$. For $\theta = \pi/3$ and $r = R$, $\phi = -RU$, and when plotted or sketched the equipotential lines must appear as shown. It should be noted that these lines appear to be perpendicular (orthogonal) to the streamlines of Fig. 12.12.

12.6   Relation Between Stream Function and Velocity Potential

A geometric relationship between streamlines and equipotential lines may be derived from the foregoing equations and restatement of certain mathematical definitions; the latter are (with definitions of u and v inserted)

$$d\psi = \frac{\partial \psi}{\partial x} dx + \frac{\partial \psi}{\partial y} dy = -v \, dx + u \, dy$$

$$d\phi = \frac{\partial \phi}{\partial x} dx + \frac{\partial \phi}{\partial y} dy = -u \, dx - v \, dy$$

However, along any streamline ψ is constant and $d\psi = 0$, so $dy/dx = v/u$ along a streamline; also along any equipotential line ϕ is constant and $d\phi = 0$, so $dy/dx = -u/v$ along an equipotential line. The geometric significance of this is seen in Fig. 12.22 (and has been suggested in the preceding Illustrative Problem): *The equipotential lines are normal to the streamlines.* Thus the streamlines and equipotential lines (for an irrotational flow)[10] form a net, called a *flownet*, of mutually perpendicular families of lines, a fact of great significance for the study of flowfields where formal mathematical expressions of ϕ and ψ are unobtainable. Another feature of the velocity potential which may be deduced from Fig. 12.22 and equations 12.27 is that the value of ϕ drops *along the direction of the flow;* that is, $\phi_3 < \phi_2 < \phi_1$.

---

[10] There can be no equipotential lines for rotational flows; see Section 12.5.

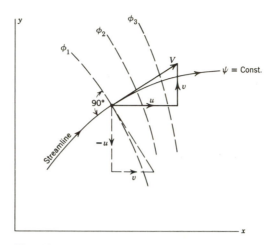

Fig. **12.22**

12.7   The Flownet

The fact that streamlines and equipotential lines are orthogonal offers a graphical technique for the solution of two-dimensional irrotational flow problems.[11] In Fig. 12.23, it is seen that, with $\Delta s = \Delta n$ and both of these dimensions approaching zero, the figures approach tiny squares; if streamlines and equipotential lines can be constructed to satisfy this requirement, the only possible flowfield may be found, the velocities determined by the streamline spacing, and the pressures from the Bernoulli equation. Considerable patience is needed to apply the technique because of trial and adjustment of streamline and equipotential line positions, but results are invariably obtained more quickly and less expensively than from an experimental program. The accuracy of such results depends on the fineness of the flownet, and it is obvious at once that greater accuracy will require more time and patience since there are many more

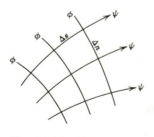

Fig. **12.23**   Flownet.

[11] It is possible also to formulate mathematical models using the equations and boundary conditions for ϕ and ψ. Few of these may be solved analytically, but a range of numerical techniques can be used. See J. M. Robertson, *Hydrodynamcis in Theory and Application,* Prentice-Hall, 1965.

lines to adjust; however, the engineer will usually have no trouble reaching a suitable compromise between available time and desired accuracy of results.

Illustrative Problems

Draw the flownet for rectilinear flow of unit velocity past a circular cylinder of unit radius.

Solution. The expressions for ϕ and ψ are known for this flow, that is, for $U = 1$,

$$\phi = -(r + r^{-1}) \cos \theta$$

from the previous Illustrative Problem and

$$\psi = (r - r^{-1}) \sin \theta \tag{12.17}$$

If these are plotted for values of ϕ and ψ in increments ± 0.5, the result is as shown. It is evident that the flownet is approximated by squares and that the streamlines and equipotential lines are perpendicular. (How do the velocities estimated from this computer output compare with those found in Section 12.4e?)

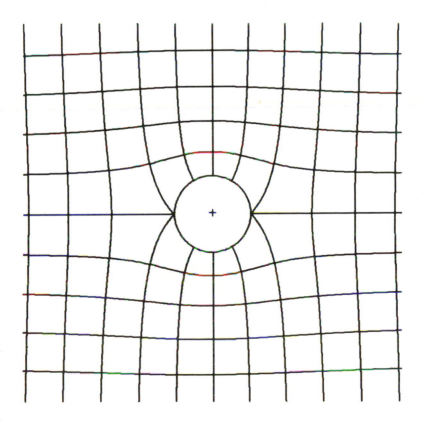

Predict the flowrate q from the large reservoir over this outflow structure for a given head H.

Solution.   First sketch a plausible flow picture about as shown. From the position of free streamlines *AB* and *CD,* the velocity at any point on them may be determined by the vertical distance from the energy line; this distance is $v^2/2g_n$. At a section downstream from *BD* the sheet of liquid will be freely falling and a velocity distribution and total flowrate may be estimated. At this section the streamlines may be spaced in such a manner that the flowrate is the same in all the streamtubes and the streamlines may be sketched back to the reservoir. The flownet is then drawn by trial and error to satisfy the condition of orthogonality of streamlines and equipotential lines. Once the flownet is established the velocities obtained from streamline spacing must conform to those obtained from the energy line. If they do not, a new flow picture must be drawn and the process repeated until this conformance is obtained; when it is, the total flowrate *q* will be the sum of the flowrates in the separate streamtubes.

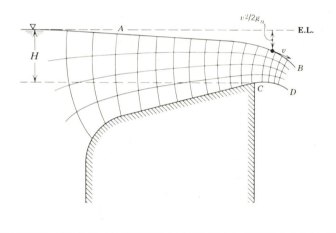

12.8   Flow in a Porous Medium

The flownet and flowfield superposition techniques may also be applied to the flow of real fluids under some restrictions which (although severe) are frequently encountered in engineering practice. Consider the one-dimensional flow of an incompressible real fluid in a streamtube. The Bernoulli equation written in differential form is

$$d\left(\frac{p}{\gamma} + \frac{V^2}{2g_n} + z\right) = -dh_L$$

Suppose now that *V* is small [so that $d(V^2/2g_n)$ may be neglected] and the head loss dh_L given by

$$dh_L = \frac{1}{K}V\,dl \tag{12.30}$$

in which *dl* is the differential length along the streamtube and *K* is a constant. The

Bernoulli equation above then reduces to

$$d\left(\frac{p}{\gamma} + z\right) = -\frac{1}{K}V\,dl \qquad \text{or} \qquad V = -\frac{d}{dl}K\left(\frac{p}{\gamma} + z\right)$$

and, if this may be extended[12] to the two-dimensional case,

$$u = -\frac{\partial}{\partial x}K\left(\frac{p}{\gamma} + z\right) = -\frac{\partial\phi}{\partial x} \qquad (12.31)$$

$$v = -\frac{\partial}{\partial y}K\left(\frac{p}{\gamma} + z\right) = -\frac{\partial\phi}{\partial y} \qquad (12.32)$$

and $K(p/\gamma + z)$ *is seen to be the velocity potential of such a flowfield.*

The conditions of the foregoing hypothetical problem are satisfied when fluid flows in a laminar condition through a homogeneous porous medium. The media of interest are those having a set of interconnected pores that will pass a significant volume of fluid, for example, sand, certain rock formations, some ceramics used in heat exchangers, and the human liver. The head-loss law 12.30 is usually written as

$$V = K\frac{dh_L}{dl} = -K\frac{dh}{dl}$$

(where $h = p/\gamma + z$) and is an experimental relation called Darcy's law; K is known as the coefficient of permeability, has the dimensions of a velocity, and ranges in value from 3×10^{-11} m/s (10^{-10} ft/s) for clay to 0.3 m/s (1 ft/s) for gravel.

A Reynolds number is defined for porous media flow as $\mathbf{R} = Vd/\nu$, where V is the *apparent velocity* or *specific discharge* (Q/A) and d is a characteristic length of the medium, for example, the effective or median grain size in sand. When $\mathbf{R} < 1$, the flow is surely laminar and Darcy's linear law is valid. If $\mathbf{R} \gg 1$, it is likely that the flow is turbulent, that $V^2/2g_n$ is not negligible, and equation 12.30 is not valid. Note that V is *not* the actual velocity in the pores, but is the velocity obtained by measuring the discharge Q through an area A. The *average* velocity in the pores is $V_p = V/n$ where n is the porosity of the medium;

$$n = \frac{\text{Volume of voids}}{\text{Volume of solids plus voids}}$$

Even though the actual fluid flow in the porous medium is viscous-dominated and rotational, the "apparent flow," represented by V and the velocities u and v (equations 12.31 and 12.32), is irrotational. Thus, all the previous analyses of this chapter can be applied. In particular both the flownet and superposition of flowfield concepts can be used. The flownet is very useful in obtaining engineering information for the "seepage flow" of water through or under structures, to wells and under-drains, or for the flow of petroleum through the porous materials of subsurface "reservoirs." Flowfield super-

---

[12] Validation of this extension will be found in any treatise on flow through porous media.

position is most useful in defining the flow pattern in groundwater aquifers under the action of recharge and withdrawal wells.

It is common for groundwater flow to occur through a confined porous medium (aquifer), that is, bounded above and below by impermeable rock or other structures. In this case the flow is often in a horizontal plane under the action of some distant supply. The vertical variation of flow in an aquifer of thickness T is negligible under these circumstances, and the flow is representable as a uniform rectilinear flow in a plane. Clearly, when the flow in the porous medium satisfies Darcy's law, ψ and ϕ satisfy the Laplace equation, and a well that penetrates the entire aquifer will input or withdraw fluid equally at each level, thus appearing as a source or sink in the flow. Figure 12.7 accordingly represents a plan view of the effect of a recharge well (source) on a uniform flow in a confined aquifer. The consequence of putting a recharge well upstream of a supply (withdrawal) well is graphically illustrated in Fig. 12.9, that is, the supply well simply recollects the recharged fluid if both pump at the same rate.

--------------------------------- **Illustrative Problems** ---------------------------------

Fine-grained uniform sand is packed in this vertical permeameter tube. Calculate the coefficient of permeability, K, if the flowrate is 2.8 ml/s.

Given Data

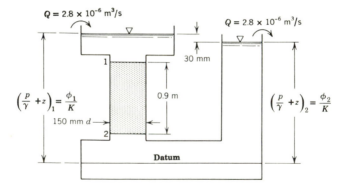

Solution.   Neglecting the velocities in the reservoirs, $(p/\gamma + z)$ at 1 and 2 are as shown for the top and bottom of the sand column, respectively. Therefore,

$$\frac{\Delta\phi}{\Delta\ell} = \frac{K \times 0.03}{0.9} = V = \frac{2.8 \times 10^{-6}}{(\pi/4)(0.15)^2} \qquad K = 4.8 \times 10^{-3} \text{ m/s} \blacktriangleleft$$

However, the foregoing calculation is valid only if the flow is laminar, that is, only if $Vd/\nu < 1$. From this experiment it is not hard to infer that K can depend on the fluid properties (μ and ρ) and the medium properties (d, particle shape, particle arrangement, and porosity n).

Show how to calculate the seepage flowrate from reservoir to reservoir beneath the sheet piling.

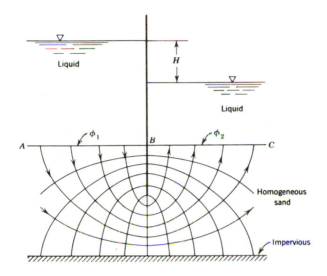

Solution. Line AB is an equipotential line having $\phi_1 = K(p/\gamma + z)_1$; likewise for line BC, $\phi_2 = K(p/\gamma + z)_2$. Sketch streamlines and equipotential lines about as shown; note that the former must all be normal to the top of the sand bed and the latter normal to the sheet piling and impervious layer. Refine these into a net of squares as accurately as possible. Assign values of ϕ to all the equipotential lines so that $\Delta\phi$ is the same between any consecutive pair of lines. Divide $(\phi_1 - \phi_2)$ by the number of spaces between the equipotential lines to obtain $\Delta\phi$, and assign values to the equipotential lines $\phi_1 - \Delta\phi$, $\phi_1 - 2\Delta\phi$, $\phi_1 - 3\Delta\phi$, and so forth, in order downstream. The flowrate q through any square (of side $\Delta\ell$) may be computed from $\Delta q = V\Delta\ell = -(\Delta\phi/\Delta\ell)\,\Delta\ell = -\Delta\phi$, which establishes the flowrate through the streamtube containing that square. Use the same procedure for squares in the other streamtubes and obtain the total flowrate from $q = \sum(\Delta q)$. If N_s is the number of streamtubes, N_ϕ is the number of spaces between equipotential lines, and H is the difference in water levels on each side of the sheet pile, it follows that

$$q = \frac{N_s}{N_\phi} KH \blacktriangleleft$$

because

$$\phi_1 - \phi_2 = KH$$

Find the potential for flow through a confined aquifer of thickness T from the river to the withdrawal well shown.

Given Data

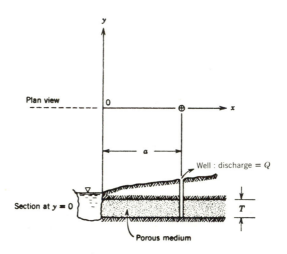

Plan view

Well : discharge = Q

Section at y = 0

Porous medium

Solution. The potential for a sink located at $(a, 0)$ is obtained by use of the radial velocity for a sink from Section 12.2b; that is,

$$v_r = -q/2\pi r = -Q/2\pi rT \qquad (q = Q/T = [\text{ft}^3/\text{s/ft}])$$

and equation 12.28, that is,

$$v_r = -\partial\phi/\partial r$$

Thus, by equating these equations and integrating, we obtain

$$\phi_S = \int \left(\frac{Q}{2\pi rT}\right) dr + C_1$$

$$= \frac{Q}{2\pi T} \ln r + C_1$$

where Q is the well discharge rate and $r^2 = (x - a)^2 + y^2$. At the river $(x = 0)$ the head $H_R = (p/\gamma + z)_R$ is constant, so $\phi(0, y) = KH_R = $ constant; that is, $x = 0$ is an equipotential line for the pattern sought, but surely is not for the single sink flow.

Placement of an image source outside the flow area under consideration, that is, at $x = -a$, yields the combined flow

$$\phi_c = \frac{Q}{2\pi T} \ln [(x - a)^2 + y^2]^{1/2} - \frac{Q}{2\pi T} \ln [(x + a)^2 + y^2]^{1/2} + C_1 - C_2$$

$$= \frac{Q}{2\pi T} \ln \left[\frac{(x - a)^2 + y^2}{(x + a)^2 + y^2}\right]^{1/2} + C_1 - C_2$$

$$= \frac{Q}{4\pi T} \ln \left[\frac{(x - a)^2 + y^2}{(x + a)^2 + y^2}\right] + C_1 - C_2$$

At $x = 0$, $\phi_c(0, y) = C_1 - C_2$ for all y; that is, $x = 0$ is an equipotential line for the combined flow. The integration constants $C_1 - C_2$ can be set equal to KH_R, so the solution for ϕ is

$$\phi(x, y) = \frac{Q}{4\pi T}\ln\left[\frac{(x - a)^2 + y^2}{(x + a)^2 + y^2}\right] + KH_R \blacktriangleleft$$

from which the velocities at any place in the flow can be found. Because this flow is the superposition of a source flow and a sink flow of equal strength, the region $x > 0$ in Fig. 12.8 gives a picture of the streamlines for flow from the river to the well.

References

Batchelor, G. K. 1967. *An introduction to fluid dynamics.* Cambridge: Cambridge University Press.

Bear, J. 1979. *Hydraulics of groundwater.* New York: McGraw-Hill.

DeWiest, R. J. M. 1965. *Geohydrology.* New York: Wiley.

Freeze, R. A., and Cherry, J. A. 1979. *Groundwater.* Englewood Cliffs, N.J.: Prentice-Hall, Inc.

Karamcheti, K. 1966. *Principles of ideal-fluid aerodynamics.* New York: Wiley.

Lamb, H. 1945. *Hydrodynamics.* 6th ed. New York: Dover Publications.

Milne-Thomson, L. M. 1968. *Theoretical hydrodynamics.* 5th ed. New York: Macmillan.

Robertson, J. M. 1965. *Hydrodynamics in theory and application.* Englewood Cliffs, N. J.: Prentice Hall.

Todd, D. K. 1980. *Groundwater Hydrology.* 2nd ed. New York: Wiley.

Vallentine, H. R. 1967. *Applied hydrodynamics.* 2nd ed. London: Butterworths Scientific Publications.

Film Loops

Hele-Shaw analog to potential flows. Part I. Sources and Sinks in Uniform Flow, Loop No. S-FM080. Hele-Shaw analog to potential flows. Part II. Sources and Sinks, Loop No. S-FM081. Encyclopaedia Britannica Educ. Corp.

Problems

12.1. Determine the stream functions for the flowfields of problem 3.3 and plot the streamline $\psi = 2$.

12.2. Determine the stream function for a parabolic velocity profile (as in laminar flow) between parallel plates separated by a distance b. Take the origin of coordinates midway between the plates.

12.3. A flowfield is characterized by the stream function $\psi = 3x^2y - y^3$. Is this flow irrotational? If not, calculate the vorticity. Show that the magnitude of velocity at any point in the flowfield depends only on its distance from the origin of coordinates. Plot the streamline $\psi = 2$.

12.4. A flowfield is characterized by the stream function $\psi = xy$. Is this flow irrotational? Plot sufficient streamlines to determine the flow pattern. Identify at least two possible physical interpretations of this flow.

12.5. Determine the stream function for a uniform rectilinear flow at an arbitrary angle α to the x-axis and with velocity magnitude U.

12.6. Plot the body contour formed by a source (at the origin) of 40π m^3/s · m in a uniform horizontal stream (from left to right) of velocity 10 m/s. Calculate the velocities at the body contour for values of θ of 180°, 150°, 120°, 90°, and 60°. Calculate the velocities at radial distance of 2, 3, and 4 m for $\theta = 180°$. Calculate the pressures at the above eight points if water is flowing and the pressure in the undisturbed free stream is 0 Pa. At what point on the body contour will the velocity be a maximum?

12.7. The nose of a two-dimensional strut having a 150 mm (6 in.) thickness is made in the shape of a half-body. If the strut is designed to operate at a speed of 9 m/s (30 ft/s), find the source strength required to simulate the nose of the strut, the distance between source and stagnation point, and the pressure on the body at a point 90° from the stagnation point if water is flowing.

12.8. A bluff overlooking a plain approximates a section of a half-body as shown. For use as a guide in soaring, derive an equation for lines of constant vertical component of velocity for a wind velocity of 48 km/h.

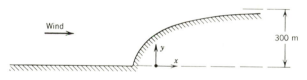

Problem **12.8**

12.9. Devise a graphical scheme for plotting the contour of a half-body.

12.10. A sink of strength 100 m^2/s (or ft^2/s) located at the origin of coordinates is combined with a horizontal rectilinear flow (from left to right) having a velocity of 10 m/s (or ft/s). Determine the stream function of the combined flowfield. Locate any stagnation points. Derive the equation for the line of separation between the flow entering the sink and that passing it by.

12.11. Two sources of equal strength are located on the x-axis at a distance a from the origin of coordinates. Determine the stream function of the flowfield and locate any stagnation points. If this flowfield is combined with one of stream function Uy, sketch the flowfield and locate the stagnation points.

12.12. Two sources of strength 100 m^2/s (or ft^2/s) are located on the y-axis; the first one metre (foot) above the origin, the second one metre (foot) below. Their flowfields are combined with one of stream function $10y$. Determine the stream function of the combined flowfield. Sketch the resulting body contour. Determine the location of any stagnation points. Find the point where the body contour intersects the y-axis and calculate the velocity there. Sketch this flowfield when the uniform velocity is 30 m/s (or ft/s).

12.13. A source and sink each of strength 60 m^2/s are located on the x-axis (see Fig. 12.8) with $a = 3$ m. Calculate the velocity at the origin of coordinates. Calculate the value of ψ for the streamline passing through the point (0, 4) and calculate the velocity there.

12.14. If the sink of the preceding problem is replaced by a source, calculate the stream function for the flowfield and the other quantities requested.

12.15. A source of strength 100 m²/s (ft²/s) at $(-4, 0)$ and a sink of strength 50 m²/s (ft²/s) at $(4, 0)$ are located in a horizontal rectilinear flow (to the right) of velocity 10 m/s (ft/s). Determine the locations of any stagnation points. Sketch the flowfield carefully. Find the points of intersection of any body contours with the y-axis.

12.16. A Rankine oval 8 ft (or m) long and 4 ft (or m) wide is in a uniform rectilinear flow of velocity 10 ft/s (or m/s). Calculate the velocity along the body contour at the midsection.

12.17. A source and a sink of equal strengths 6.28 m²/s are placed in a uniform stream moving vertically upward. The source is at $(-1, 0)$ and the sink at $(1, 0)$. Derive the equation for the stream function and sketch the flowfield. Locate the stagnation points if any. Calculate the pressure at the origin relative to that in the undisturbed uniform stream, which has a velocity of 1 m/s.

12.18. What strength of doublet at the origin of coordinates will be needed to produce a velocity of 10 at the point $(0, 5)$? What is the value of ψ for the streamline passing through this point?

12.19. Calculate the velocities and pressures on the surface of a cylinder of 4 m (or ft) diameter in a horizontal uniform flow of velocity 10 m/s (or ft/s) at values of θ of 165°, 140°, 115°, and 90°. Also calculate the velocities (at the foregoing angles) along the streamline $\psi = 5$. Assume water flowing and pressure zero in the free stream.

12.20. Estimate the resultant dynamic force on the upstream half of a cylindrical bridge pier 5 m (or ft) in diameter when the water approaching it is 10 m (or ft) deep and moving at a mean velocity of 3 m/s (or ft/s).

12.21. For a doublet in a rectilinear flow there will be a line in the flowfield at all points of which the velocity and pressure are the same as those of the undisturbed free stream. Show that the equation of this line is $(R/r)^2 = 2 \cos 2\theta$.

12.22. A horizontal rectilinear flow (to the right) and a clockwise free vortex and a source (both at the origin of coordinates) are all superposed. Determine the components of velocity in Cartesian and polar coordinates. How many stagnation points are to be expected and what are their locations? Derive expression for their coordinates. Sketch the flowfield carefully. Is a "body contour" to be expected?

12.23. A flowfield is defined by the stream function

$$\psi = Uy + \frac{q}{2\pi}\left(\arctan \frac{y - a}{x} - \arctan \frac{y + a}{x}\right)$$

Derive a general expression for the locations of all stagnation points. Sketch the flowfields implied by this stream function.

12.24. For the stream function

$$\psi = \frac{q}{2\pi}\left(\arctan \frac{y - a}{x} - \arctan \frac{y + a}{x}\right) - \frac{\Gamma}{2\pi} \ln \sqrt{x^2 + y^2}$$

locate any stagnation points and sketch the flowfield. Derive an expression for the velocity at $(a, 0)$.

12.25.   A flowfield is defined by the stream function

$$\psi = 12\left(\arctan\frac{y-2}{x-2} + \arctan\frac{y+2}{x+2}\right) + \frac{3}{4\sqrt{2}}(x^2 + y^2)$$

Identify the components of this flowfield. Is the flowfield rotational or irrotational? Why? Are there any stagnation points? How many? Where are they located? What is the velocity at (4, 4)? Sketch the flowfield.

12.26.   Calculate the velocities and pressures at 90° and 270° on the surface of the cylinder of problem 12.19 after a clockwise vortex of strength 150 ft²/s (or m²/s) has been imposed on the flowfield at the center of the cylinder. Also calculate these velocities and pressures for a vortex strength of 300 ft²/s (or m²/s).

12.27.   A rotorship has two cylindrical rotors, each of 10 m height and 3 m diameter. The ship moves due north at 30 km/h when a 50 km/h east wind is blowing. If the circulation induced by rotating each cylinder is 300 m²/s, calculate the propulsive force on the ship. Assume that the specific weight of the air is 12.0 N/m³.

12.28.   A uniform horizontal flow (from left to right) of velocity 10 is combined with a clockwise vortex whose center is at the origin of coordinates. The stagnation point is located at (0, −5). Plot the streamline passing through the stagnation point in the region between $x = -5$ and $x = 5$. What is the velocity in this flowfield at (0, 5)?

12.29.   Clockwise vortices of equal strength are located on the y-axis at points $(0, a)$ and $(0, -a)$. Derive the stream function of the resulting flowfield and sketch the streamlines. Calculate the velocity at the origin of coordinates.

12.30.   Solve the preceding problem if the upper vortex is clockwise and the lower one counterclockwise.

12.31.   Sketch the flowfield suggested by Fig. 12.19 when: (a) $\Gamma a/\pi V = a^2$ and (b) $\Gamma a/\pi V < a^2$. Locate the stagnation points for (a) and for $\Gamma a/\pi V = a^2/4$.

12.32.   A flowfield is defined by the stream function

$$\psi = 10r \sin \theta - 50 \ln r - (40 \sin \theta)/r$$

Locate any stagnation points and sketch the flowfield. Calculate the velocities at $r = 2$ and $\theta = 0°$, 90°, 180°, and 270°.

12.33.   The two-dimensional vaneless diffuser of a centrifugal pump has entrance and exit diameters of 0.6 m (2 ft) and 0.9 m (3 ft), respectively. Water enters at a velocity of 12 m/s (40 ft/s) and at an angle of 60° with a radial line. Calculate the stream function for this flowfield and the expected pressure rise through the diffuser.

12.34.   The exit circle for the guide vanes of a hydraulic turbine (see Fig. 6.24) is of 4.25 m diameter. The entrance to the runner blades is of 3 m diameter. If water leaves the guide vanes at a velocity of 24 m/s and at an angle of 75° with a radial line, calculate the velocity at the runner entrance, the stream function for the flowfield, and the pressure drop between guide vanes and runner.

12.35.   A source is located a distance of 2 units above the x-axis in a horizontal rectilinear flow with $U = 5$; the source strength $q = 5$. Plot the shape of the resulting half-body. Now introduce a wall along the x-axis by use of an image source. Plot the new shape of the half-body (in the region $x > 0$ above the wall) to see the "ground effect."

12.36. Determine a general expression for ψ for a source in the presence of a wall. Plot the streamlines for a typical case.

12.37. Determine the effect of a wall on the flow shown in Fig 12.4 by an image method.

12.38. Determine (if possible) the velocity potentials of the flowfields of problem 3.3.

12.39. Determine (if possible) the velocity potentials of the four basic flowfields of Section 12.2.

12.40. Determine the velocity potentials of the flowfields of: (a) Fig. 12.7, (b) Fig. 12.12, (c) Fig. 12.17, and (d) Fig. 12.20.

12.41. Detemine the velocity potential ϕ for (a) the flow in problem 12.3 and (b) the flow in problem 12.4.

12.42. Plot representative ϕ and ψ lines for the flow of problem 12.4.

12.43. For a frictionless flow of water through the U-bend find the pressures at points A and B by the flownet method. The two-dimensional flowrate is 4 m³/s · m, the bend is in a horizontal plane, and the pressure in the straight passage is 70 kPa. Draw the bend to a scale of 50 mm = 1 m and divide the flow into 10 streamtubes.

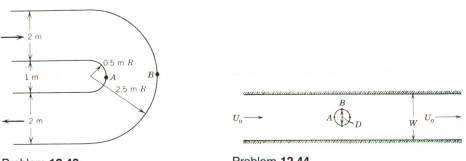

Problem **12.43**   Problem **12.44**

12.44. Construct a flownet for ideal flow past a cylinder between confining walls in a water tunnel. Compute the pressure at points A and B according to velocities estimated from the flownet. Compare these with values obtained analytically for the case where the walls are absent. Let $U_o = 10$ m/s, $D = 0.3$ m, $W = 1$ m, and the pressure far from the body $p_o = 100$ kPa.

12.45. Construct a flownet for ideal flow in a contraction and obtain the pressure differences between points A–B, A–C, A–D, and D–E.

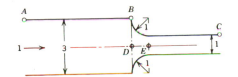

Problem **12.45**

12.46. Predict the seepage flowrate under this dam if the coefficient of permeability of the previous layer is 50 ft/day. Use a scale of 1 in. = 4 ft and divide the flow into 10 streamtubes.

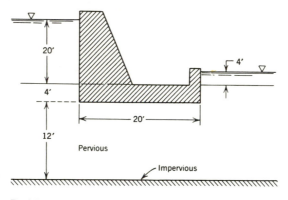

Problem **12.46**

12.47. Predict the seepage flowrate through this dike on an impervious foundation if the dike material is homogeneous and has a coefficient of permeability of 12 m/day. Draw the dike to a scale of 75 mm = 1 m and use 10 streamtubes.

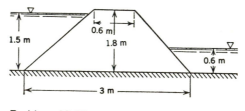

Problem **12.47**

12.48. Determine ψ for the third Illustrative Problem of Section 12.8. Plot curves of representative values of ϕ and ψ in the region $x > 0$ to determine flow pattern. Estimate the velocity at several points by use of the plotted flownet and compare to analytic results.

12.49. Determine ϕ and ψ for a well penetrating a confined horizontal aquifer of thickness T near an impermeable rock barrier. Plot streamlines for a typical case.

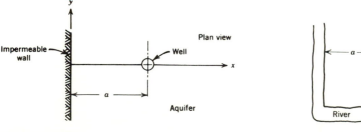

Problem **12.49**   Problem **12.50**

12.50. Determine the stream function ψ and velocity potential ϕ for flow through a confined aquifer of thickness T to a supply well located near a bend in a river. Plot streamlines for a typical case.

12.51. Determine the direction of flow, the head, and the pressure distributions in the medium for each case. Assume that the flow is one-dimensional and that fluid is added to or removed from the reservoirs at each end of the medium as needed to maintain steady conditions. What is the role of the pressure gradient or the slope of the medium in these cases? What determines the flowrate?

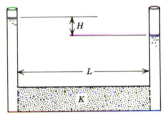

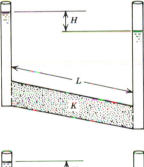

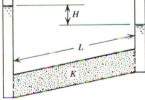

Problem **12.51**

FLUID FLOW ABOUT IMMERSED OBJECTS

Problems involving the forces exerted on a solid body when fluid flows by it (external flows) no longer belong exclusively to the naval architect and aeronautical engineer. For the effective and safe design of buildings, bridges, automobiles, and trains, a knowledge of fluid resistance or drag is of great importance; the fluid principles describing lift are applied in the design of propellers, turbines, and pumps. Because these principles find wide application, it is necessary for all engineers to be familiar with the fundamental mechanics of the fluid motion. It is the purpose of this chapter to outline the fundamental and elementary aspects of external flow for both incompressible and compressible fluids.

13.1 Fundamentals and Definitions

In general, when flow occurs about an object which is either asymmetrical or whose axis is not aligned with the flow, the flowfield will be asymmetrical, the local velocities and pressures on either side of the object will be different, and a force normal to the oncoming flow will be exerted. Accompanying this, the action of frictional stress in the boundary layers over the surface of the object will produce a force along the direction of the oncoming flow. These forces are known (from their aeronautical backgrounds) as *lift* and *drag,* respectively. The classic and most useful example is the airfoil (or hydrofoil) of Fig. 13.1, on which the lift $L,$ drag $D,$ resultant force $F,$ angle of attack $\alpha,$ and chord c are shown; the length of the foil perpendicular to the plane of the paper is termed the *span.* The force $F,$ which is seen to be the resultant of L and $D,$ is also the resultant of all forces of pressure and friction exerted by fluid on foil. However, often the contribution of the frictional stresses to the lift may be neglected when such stresses are small compared to the pressure and act in a direction roughly normal to $L.$ When permissible this is an important simplification of the problem in that it allows L to be considered to be the result of pressure variation alone and thus permits the use of the ideal fluid and the methods of the preceding chapter for analytical predictions of lift.

Prediction of the drag force on immersed objects is much more difficult than that of lift, since usually no simplifications are possible and both pressure and frictional forces must be considered. However, both lift and drag may be predicted from experimental measurements on small models in wind tunnels, water tunnels, or towing basins. In addition, a combination of hydrodynamical and boundary-layer theory can often be used (see the introduction to Chapter 12).

608

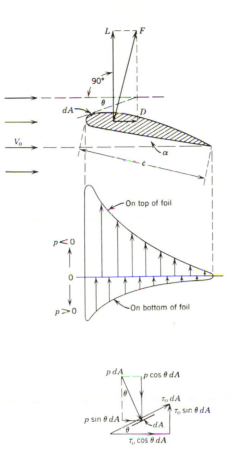

Fig. **13.1**

The foregoing preliminary discussion becomes more meaningful through formal application to a surface element of an immersed body, followed by appropriate integration. Consider the element dA of Fig. 13.1 on which a pressure p and frictional stress τ_o act. The differential drag and lift on this element are seen to be

$$dD = p\, dA\, \sin\theta + \tau_o\, dA\, \cos\theta$$

$$dL = -p\, dA\, \cos\theta + \tau_o\, dA\, \sin\theta$$

which may be integrated to yield

$$D = \int^s p\, dA\, \sin\theta + \int^s \tau_o\, dA\, \cos\theta$$

$$L = -\int^s p\, dA\, \cos\theta + \int^s \tau_o\, dA\, \sin\theta$$

in which $\int^s$ designates the *integral over the surface of the object*. Under the assumption that the second integral of the lift expression is often negligible, the lift is expressed by

$$L = -\int^s p \, dA \, \cos \theta$$

The integrals in the drag equation are equally important; the first one is called the *pressure drag* D_p, and the second the *frictional drag* D_f. The former will depend on the form of the object and flow separation, the latter on the extent and character of the boundary layer. Although the prediction of separation points is generally a very complex problem dependent on both body form and boundary-layer properties, the breakdown of total drag into pressure drag and frictional drag proves of great value in studying these separately in further detail. Frictional drag may be isolated by considering the flow past a thin flat plate parallel to the oncoming flow (Fig. 13.2a); here $\sin \theta = 0$, $\cos \theta = 1$, $D_p = 0$, and $D = D_f = \int^s \tau_o \, dA$. Pressure drag may be isolated by studying the flow about a flat plate normal to the oncoming flow (Fig. 13.2b); here $\cos \theta = 0$, $\sin \theta = 1$, $D_f = 0$, and $D = D_p = \int^s p \, dA$. Although these problems (and others) will be examined more intensively in Sections 13.3 through 13.6, a descriptive preview will prove useful in setting up certain guideposts.

Consider the circular disk, sphere, and streamlined body of Fig. 13.3, all of which have the same cross-sectional area, and are immersed in the same turbulent flow; all these flows will feature a stagnation point on the upstream side of the object and a maximum local pressure there.

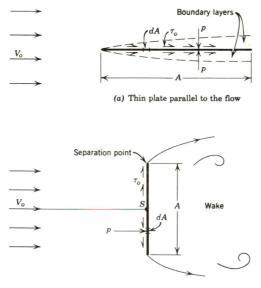

(a) Thin plate parallel to the flow

(b) Thin plate normal to the flow

Fig. **13.2**

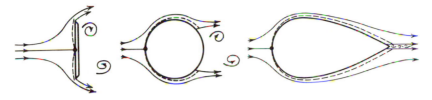

Fig. **13.3**

For the disk, separation will be expected at the edges with high local velocity and low pressure. This reduced pressure, being adjacent to the wake, will be transmitted into it, causing the downstream side of the disk to be exposed to a mean pressure considerably below that on the upstream side. The result is a large drag force caused wholly by pressure since none of the shear forces on the disk has components in the original direction of motion.

For the sphere, the wake is smaller than that of the disk and (from the streamline picture) will contain a somewhat higher pressure, leading to the expectation that the drag of the sphere is considerably smaller than that of the disk.[1] For the sphere, the frictional drag is not zero since all shear stresses acting on the sphere will have components parallel to the oncoming flow. These shear stresses are extremely difficult to calculate, but they are small and result in a frictional drag which is negligible compared to the pressure drag for spheres and other objects of similar (blunt) form.

For the well-streamlined body, the wake may be extremely small, being only the width of the boundary layer at the tail of the object. The pressure in such a wake is comparatively large, since the gentle contour of the body allows deceleration of the flow and consequent regain of pressure without incurring separation. Thus the pressure drag of such objects is a very small fraction of that of the disk. However, the frictional drag of streamlined bodies is considerably larger than that of the sphere, since stream-lining has brought more surface area in contact with the flow. For well-streamlined objects, frictional drag is usually larger than pressure drag but both are so small that their total is only about one-fortieth that of the disk.

The foregoing examples illustrate the fact that the viscosity property of a fluid is the root of the drag problem. Viscosity has been seen to cause drag either by frictional effects on the surface of an object or through pressure drag by causing separation and the creation of a low-pressure wake behind the object. By streamlining an object, the size of its wake is decreased and a reduction in pressure drag is accomplished, but in general an increase in frictional drag is incurred.

For an ideal fluid in which there is no viscosity and thus no cause for frictional effects or wake formation regardless of the shape of the object about which flow is occurring, it is evident that the drag of the object is zero. Two centuries ago, d'Alembert's observation that all objects in an ideal fluid exhibit no drag was a fundamental

---

[1] Experiments show the total drag of the sphere to be about one-third that of the disk.

and disturbing paradox; today this fact is a logical consequence of the fundamental reasoning presented above.

13.2 Dimensional Analysis of Drag and Lift

The general aspects of drag and lift forces on completely[2] immersed bodies may be examined advantageously by dimensional analysis before further consideration of the physical details of the problem.

The smooth object of Fig. 13.4 having area[3] A moves through a fluid of density ρ, viscosity μ, and modulus of elasticity E, with a velocity V_o. If the drag force exerted on the body is D,

$$D = f_1(A, \rho, \mu, V_o, E)$$

and similarly

$$L = f_2(A, \rho, \mu, V_o, E)$$

The Buckingham Π-method of dimensional analysis (Section 8.2) shows that, in each case, three distinct nondimensional groups can be formed. Choosing a length ($\sqrt{A}$), ρ, and V_o as the repeating variables (as in Section 8.2) leads to the following Π-terms:

$$\Pi_1 = \rho\sqrt{A}V_o/\mu = \mathbf{R}$$

$$\Pi_2 = \rho V_o^2/E = V_o^2/a^2 = \mathbf{M}^2$$

(because $a = \sqrt{E/\rho}$) which are common to both the lift and drag cases plus

$$\Pi_3 = D/A\rho V_o^2$$

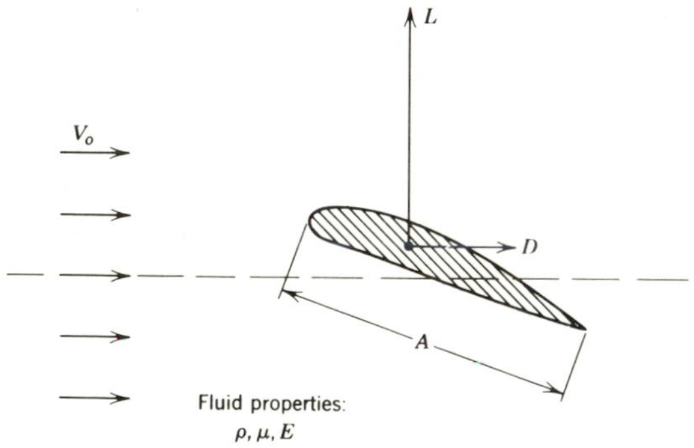

V_o

Fluid properties:
ρ, μ, E

Fig. **13.4**

---

[2] The drag of a surface vessel, a partially submerged body, was analyzed in Sections 8.1 and 8.2.
[3] A convenient significant area may be selected, for example, the product of maximum chord and span for a wing.

and

$$\Pi_4 = L/A\rho V_o^2$$

Accordingly, the very general results are

$$D = \frac{f_3(\mathbf{R}, \mathbf{M})A\rho V_o^2}{2} \quad \text{and} \quad L = \frac{f_4(\mathbf{R}, \mathbf{M})A\rho V_o^2}{2}$$

If drag and lift coefficients, C_D and C_L, respectively, are defined by[4]

$$C_D = \frac{D}{\frac{1}{2}A\rho V_o^2} \quad \text{and} \quad C_L = \frac{L}{\frac{1}{2}A\rho V_o^2} \tag{13.1}$$

it follows that

$$C_D = f_3(\mathbf{R}, \mathbf{M}) \quad \text{and} \quad C_L = f_4(\mathbf{R}, \mathbf{M})$$

These equations indicate: (1) that bodies having the same shape and the same alignment with the flow (i.e., models of each other) possess the same drag and lift coefficients if their Reynolds numbers and Mach numbers are the same, or (2) that the drag and lift coefficients of bodies of given shape and alignment may be expected to depend on their Reynolds and Mach numbers only. Thus dimensional analysis has, as in previous problems (ship resistance and pipe friction), opened the way to a comprehensive treatment of the resistance of immersed bodies by indicating the dimensionless combinations of variables on which the drag coefficient depends. An example of this is given in Fig. 13.5, where experimentally determined drag coefficients for spheres are plotted as contours on a Reynolds number-Mach number graph.

From the foregoing dimensional analysis alone no conclusion can be reached on the quantitative effects of $\mathbf{R}$ and $\mathbf{M}$ on the drag and lift coefficients, but theory and experiment both show that $\mathbf{R}$ is predominant when the fluid may be considered incompressible and $\mathbf{M}$ predominant when compressibility effects must be considered. Usually this means that over a wide range of $\mathbf{R}$, where $\mathbf{M}$ is small and velocities subsonic, the fluid may be considered incompressible and C_D and C_L functions of $\mathbf{R}$ only. On the other hand, when $\mathbf{M}$ approaches or exceeds unity and velocities approach or exceed that of sound, C_D and C_L are functions of $\mathbf{M}$ only, whatever the magnitude of $\mathbf{R}$. Although such a division of flow problems is useful and convenient, it is somewhat arbitrary as there is no definite point at which the effects of compressibility begin, such effects being present at all velocities; therefore it is to be expected that there will be exceptions to this convenient division of flow problems and situations encountered where the effects of $\mathbf{R}$ and $\mathbf{M}$ are of the same order and in which neither may be ignored. Such problems are highly complex and far beyond the scope of an elementary text, but the beginner should be aware of their existence and not expect them to be classified by usual and convenient methods. Reynolds and Mach numbers have been shown (Section 8.1) to be, respectively, ratios of inertia-to-viscous and inertia-to-elastic forces; from this it

---

[4] L/A or D/A is a force per unit area, while $\frac{1}{2}\rho V_o^2$ is the dynamic pressure.

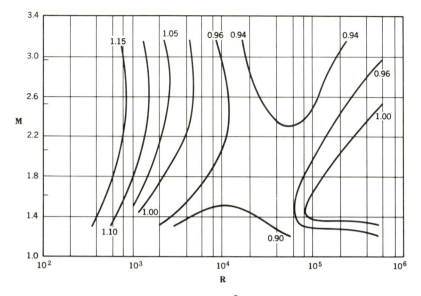

Fig. **13.5** Drag coefficients for spheres.[5]

may be concluded directly that the flow phenomena, when governed by Reynolds number, will result from viscous action (boundary layers, etc.) but, when governed by Mach number, from elastic phenomena (shock waves, etc.). Examples of problems dealing with the first situation are the settling of solid particles through a fluid, the forces of wind on structures, and most aerodynamic problems in the field of commercial aviation. Examples of problems concerned with the second situation are the motions of missiles, rockets, propeller tips, supersonic aircraft, and elements of gas turbines and high-speed compressors.

After an introductory discussion of boundary layers, the remainder of the chapter follows the pattern above, drag and lift on objects in incompressible fluids being treated first, followed by a parallel, but very brief, treatment for compressible fluids.

Illustrative Problem

The lift and drag coefficients of an approximately rectangular airfoil of 36 m span and 7.5 m chord are 0.6 and 0.05, respectively, when at an angle of attack of 7°. Calculate the power required to drive this airfoil (in horizontal flight) at 600 km/h through still, standard air at altitude 4 km. What lift force is obtained when this power is expended? Also calculate the Reynolds and Mach numbers.

[5] A. May, "Supersonic Drag of Spheres at Low Reynolds Numbers in Free Flight," *Jl. Appl. Physics*, vol. 28, 1957, pp. 910–912.

Relevant Equations and Given Data

$$\mathbf{R} = V_o l \rho / \mu \qquad \mathbf{M} = V_o / a$$

$$a = \sqrt{kRT} \tag{1.11}$$

$$C_D = \frac{D}{\frac{1}{2}A\rho V_o^2} \quad \text{and} \quad C_L = \frac{L}{\frac{1}{2}A\rho V_o^2} \tag{13.1}$$

$C_D = 0.05 \qquad C_L = 0.6 \qquad c = 7.5 \text{ m} \qquad s = 36 \text{ m} \qquad A = 7.5 \times 36 \text{ m}^2$

$V_o = 600 \text{ km/h} \qquad \text{alt} = 4 \text{ km} \qquad R = 286.8 \text{ J/kg} \cdot \text{K} \qquad k = 1.4$

Solution. From Appendix 4,

$$\rho = 0.909 \text{ kg/m}^3$$

Converting, 600 km/h yields 166.7 m/s.

$$D = \frac{0.05 \times (36 \times 7.5) \times 0.909(166.7)^2}{2} = 170.5 \text{ kN} \tag{13.1}$$

From mechanics,

$$\text{Power} = 170.5 \times 166.7 = 28.4 \text{ MW}$$

$$L = \frac{0.6 \times (36 \times 7.5) \times 0.909(166.7)^2}{2} = 2\ 046 \text{ kN} \blacktriangleleft \tag{13.1}$$

From Appendix 4,

$$T = -4.5°C$$

and

$$\mu = 1.661 \times 10^{-5} \text{ Pa} \cdot \text{s}$$

$$R = \frac{166.7 \times 7.5 \times 0.909}{1.661 \times 10^{-5}} = 68.4 \times 10^6 \blacktriangleleft$$

$$a = \sqrt{1.4 \times 286.8(-4.5 + 273.2)} = 328.5 \text{ m/s} \tag{1.11}$$

$$M = \frac{166.7}{328.5} = 0.508 \blacktriangleleft$$

Note that the Boeing 747B has engines whose thrust is up to 209 kN each and has flown at a gross weight of 3 650 kN.

THE BOUNDARY LAYER

13.3 Characteristics of the Boundary Layer

As real fluid flows past a smooth stationary immersed object, the effects of viscosity produce velocity profiles featured by no velocity at the solid surface, high shear and velocity gradient, and the velocity of ideal flow at a relatively small distance from the surface. The zone in which the velocity profile is governed by frictional action is known as the *boundary layer* (see Section 7.4). If this layer is thin and there is no flow separation, the frictional effects are confined to the boundary layer and the flow outside the boundary layer may be considered ideal (irrotational). Thus, it may be treated by the methods of mathematical hydrodynamics (Chapter 12).

Boundary-layer phenomena may be most easily visualized on a smooth flat plate parallel to the oncoming flow (Fig. 13.6). Assuming the leading edge of the plate to be smooth and even, a *laminar boundary layer* is to be expected adjacent to the upstream portions of the plate since there the boundary layer is thin, viscous action intense, and turbulence impossible.[6] The laminar nature of the boundary layer may be characterized in a Reynolds number, $\mathbf{R}_\delta = V_o\delta/\nu$, which experiments have shown to be less than a critical value of about 3 900. The boundary layer must start from no thickness at the leading edge of the plate where viscous action begins and steadily increases in thickness as increased viscous action extends into the flow and slows down more and more fluid; in terms of Reynolds number, $\mathbf{R}_\delta$ will vary over the length of the laminar boundary layer from zero at the leading edge of the plate to the critical value of about 3 900 at the downstream end[7] of the layer. At this point instabilities in the boundary layer produce breakdown of the laminar structure of the flow and cause turbulence to begin. After the onset of turbulence the (transition) boundary layer thickens rapidly, developing into a turbulent boundary layer having many of the

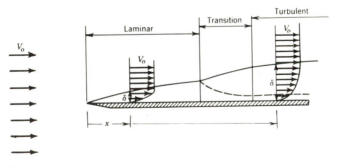

Fig. **13.6**   Boundary layers on a flat plate.

characteristics of turbulent pipe flow; as in pipe flow, a thin *viscous sublayer* will exist between the solid surface and the turbulence of the boundary layer.

Another important characteristic of boundary layers can be deduced[8] from their relative thinness; that is, the tangential component of velocity u is much greater than the normal component v and the gradients $\partial u / \partial x$, $\partial v / \partial x$, and $\partial v / \partial y$ are negligible compared to $\partial u / \partial y$. In the general case noted here, the coordinate directions x and y are taken to be locally tangent and normal to each point along the body on which the boundary layer is growing. As a consequence of the velocity behavior in the boundary layer,[8] the pressure change across it is negligible, so $\partial p / \partial y \approx 0$ and $\partial p / \partial x \approx dp / dx$.

While solution of the flow in a boundary layer is difficult, often requiring sophisticated mathematical and/or computer methods, the so-called *von Kármán momentum integral equation* of the boundary layer is simple to derive and very useful in the analysis of boundary layer behavior. The derivation leads in turn to natural definitions of useful quantities such as the boundary layer, displacement, and momentum thicknesses, as well as to determination of the surface shear stress, τ_o, by analysis or experiment.

The development of a boundary layer in a compressible, two-dimensional flow over a streamlined shape is shown in Fig. 13.7. The integral equation is developed by analysis of a control volume $ABCD$ of height h (that is greater than the boundary layer thickness) and differential length dx. The velocity, $V_o(x)$, at the edge of the boundary layer, the pressure, $p_o(x)$, and density, $\rho_o(x)$, are presumed known from analysis of the outer ideal flow (see the introduction to Chapter 12). Neither the pressure nor the velocity of the ideal flow is expected to change significantly over y-distances of the order of the boundary-layer thickness, δ. In addition the curvature of the body has a

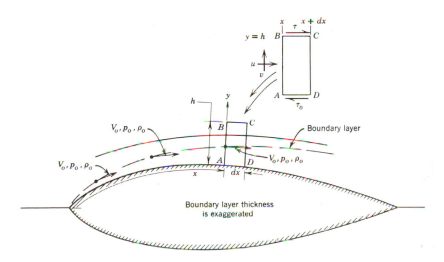

Fig. **13.7**

[8] H. Schlichting, *Boundary-Layer Theory*, 7th ed., McGraw-Hill, 1979, Chap. VII.

negligible influence on the dynamics of the boundary layer, provided that the curvature is not appreciable, that is, that the local radius of curvature is very large compared to δ. Accordingly the continuity equation 3.16 and the momentum equation 6.4 can be applied directly to the control volume $ABCD$.

From equation 3.16

$$\oint_{ABCD} \rho \mathbf{V} \cdot \mathbf{n}\, dA = \left[\int_0^h \rho u\, dy \right]_{at\, x+dx} + [\rho v]_{y=h}\, dx - \left[\int_0^h \rho u\, dy \right]_{at\, x} = 0$$

$$(3.16)$$

Thus, by rewriting the terms evaluated at $x + dx$ and x as a derivative, we find that

$$\frac{d}{dx}\left(\int_0^h \rho u\, dy \right) dx + [\rho v]_{y=h}\, dx = 0$$

The second term represents the mass transport through the top of the control volume resulting from the fact that the streamlines are not parallel to the body surface because of the growing boundary layer. From equation 6.4,

$$\sum F_x = [-p_o]_{at\, x+dx}h + [p_o]_{at\, x}h - \tau_o\, dx + [\tau]_{y=h}\, dx$$

and

$$\oint_{ABCD} \mathbf{V}_x(\rho \mathbf{V} \cdot \mathbf{n}\, dA) = \left[\int_0^h \rho u^2\, dy \right]_{at\, x+dx} + [\rho u v]_{y=h}\, dx - \left[\int_0^h \rho u^2\, dy \right]_{at\, x}$$

Hence,

$$-\frac{dp_o}{dx} h\, dx - \tau_o\, dx + [\tau]_{y=h}\, dx = \frac{d}{dx}\left(\int_0^h \rho u^2\, dy \right) dx + [\rho u v]_{y=h}\, dx$$

Because $h > \delta$, $[\tau]_{y=h} = 0$ and other terms evaluated at $y = h$ have free stream or ideal flow values. Thus, the above equations become

$$\rho_o v_o = -\frac{d}{dx}\left(\int_0^h \rho u\, dy \right)$$

$$(13.2)$$

where v_o is the small vertical velocity component at the edge of the boundary layer. Letting[9] $u_o = u_{y=h} = V_o$,

$$-\frac{dp_o}{dx} h - \tau_o = \frac{d}{dx}\left(\int_0^h \rho u^2\, dy \right) + \rho_o V_o v_o$$

$$(13.3)$$

Introducing equation 13.2 into equation 13.3,

---

[9] Strictly speaking, the magnitude of the velocity at the edge of the boundary layer $V_o = (u_o^2 + v_o^2)^{1/2}$. However, usually $u_o \gg v_o$, so $u_o \approx V_o$.

$$-\frac{dp_o}{dx}h - \tau_o = \frac{d}{dx}\left(\int_0^h \rho u^2 \, dy\right) - V_o\frac{d}{dx}\left(\int_0^h \rho u \, dy\right) \tag{13.4}$$

However, p_o and V_o are related in the ideal, outer flow by the Euler equation (Section 4.1). Neglecting gravity effects,

$$\frac{dp_o}{\rho_o} + V_o dV_o = 0$$

Hence,

$$\frac{dp_o}{dx} = -\rho_o V_o \frac{dV_o}{dx}$$

and equation 13.4 becomes

$$-\tau_o = \frac{d}{dx}\left(\int_0^h \rho u^2 \, dy\right) - V_o\frac{d}{dx}\left(\int_0^h \rho u \, dy\right) - \rho_o V_o\frac{dV_o}{dx}h \tag{13.5}$$

If the density variation and velocity profile in the boundary layer are known or measured, along with p_o, ρ_o, and V_o, equation 13.5 allows direct evaluation of the shear stress along the body. In using equation 13.5 it is convenient to define certain physically based terms to simplify the equation.

First, δ must be defined; this is difficult because there is no sharp demarcation between the boundary layer and the ideal fluid zones. Usually δ is defined as the distance from the solid boundary at which the velocity u reaches 99% of the free stream value V_o.

Second, the term $\int_0^\delta \rho u \, dy$ is the mass flowrate through the boundary layer. In the absence of the boundary layer, the mass flowrate would be $\rho_o V_o \delta > \int_0^\delta \rho u \, dy$ because there is a *flow defect* in the boundary layer caused by retardation of the fluid. A *displacement thickness*, δ_1, is defined to represent the thickness of an imaginary layer needed to carry the defect flow; then,

$$\rho_o V_o \delta_1 = \int_0^\delta (\rho_o V_o - \rho u) \, dy = \int_0^h (\rho_o V_o - \rho u) \, dy$$

because $\rho_o V_o - \rho u \equiv 0$ for $h \geq y \geq \delta$. Thus,

$$\delta_1 = \int_0^{\delta \text{ or } h} \left(1 - \frac{\rho u}{\rho_o V_o}\right) dy \tag{13.6}$$

and δ_1 is a measure of the outward displacement of the ideal flow streamlines caused by the boundary layer.

Third, the retardation of the flow in the boundary layer also causes a momentum flux defect in comparison to the ideal fluid flow momentum flux. Let δ_2 be the *momentum thickness* and represent the thickness of an imaginary layer of fluid of velocity V_o that carries a momentum flux equal to the defect caused by the boundary-layer profile. It follows that (remember to account for flux into the top of the control volume)

$$\rho_o V_o^2 \delta_2 = \int_0^\delta \rho u (V_o - u) \, dy = \int_0^h \rho u (V_o - u) \, dy$$

or

$$\delta_2 = \int_0^{\delta \text{ or } h} \frac{\rho u}{\rho_o V_o} \left(1 - \frac{u}{V_o}\right) dy \qquad (13.7)$$

Equation 13.5 can now be recast in terms of δ_1 and δ_2 and values in the free stream only. From equations 13.6 and 13.7,

$$\int_0^h \rho u \, dy = -\rho_o V_o \delta_1 + \rho_o V_o h$$

$$\int_0^h \rho u^2 \, dy = V_o(-\rho_o V_o \delta_1 + \rho_o V_o h) - \rho_o V_o^2 \delta_2$$

Introducing these results and making a number of rearrangements in equation 13.5 leads to

$$\frac{\tau_o}{\rho_o V_o^2} = \frac{d\delta_2}{dx} + \delta_2 \left[\left(2 + \frac{\delta_1}{\delta_2}\right)\frac{dV_o/dx}{V_o} + \frac{d\rho_o/dx}{\rho_o}\right] \qquad (13.8)$$

It is customary to express τ_o in terms of a dimensionless *local friction coefficient* c_f in the form $\tau_o = c_f \rho_o V_o^2 / 2$. Thus,

$$\frac{c_f}{2} = \frac{d\delta_2}{dx} + \delta_2 \left[\left(2 + \frac{\delta_1}{\delta_2}\right)\frac{dV_o/dx}{V_o} + \frac{d\rho_o/dx}{\rho_o}\right] \qquad (13.9)$$

For a constant density flow, $d\rho_o/dx = 0$ and

$$\frac{c_f}{2} = \frac{d\delta_2}{dx} + \delta_2 \left[\left(2 + \frac{\delta_1}{\delta_2}\right)\frac{dV_o/dx}{V_o}\right] \qquad (13.10)$$

For ρ_o = constant and in the absence of a pressure gradient,

$$\frac{c_f}{2} = \frac{d\delta_2}{dx} \qquad (13.11)$$

that is, the shear stress is proportional to the rate of change of the momentum defect along the surface.

13.4   The Laminar Boundary Layer—Incompressible Flow

An approximate analysis of the laminar boundary layer developing on a flat plate in a zero pressure gradient (Fig. 13.6) can be carried out by assuming the velocity profile in the layer to be parabolic.[10] For a parabolic profile

$$u = \frac{V_o(2\,\delta y - y^2)}{\delta^2}$$

Hence, from equation 13.7,

$$\delta_2 \cong \int_0^\delta \frac{1}{\delta^2}(2\,\delta y - y_2)\left(1 - \frac{2\,\delta y - y^2}{\delta^2}\right) dy = \frac{2}{15}\delta \tag{13.7}$$

From equation 13.11 it follows that (with $\rho_o = \rho =$ constant here)

$$\frac{\tau_o}{\rho V_o^2} = \frac{c_f}{2} \cong \frac{2}{15}\frac{d\delta}{dx} \tag{13.12}$$

and

$$\tau_o \cong \frac{2}{15}\rho V_o^2\frac{d\delta}{dx} \cong \mu\left(\frac{2V_o}{\delta}\right) \tag{13.13}^{11}$$

because $\tau_o = \mu(du/dy)_{y=0}$ and $(du/dy)_{y=0} = 2V_o/\delta$. The variables of this equation may be separated and the equation integrated:

$$\int_0^\delta \delta\,d\delta \cong \frac{15\mu}{\rho V_o}\int_0^x dx \qquad \frac{\delta^2}{2} \cong \frac{15\mu x}{\rho V_o} \tag{13.14}^{11}$$

giving the desired relationship between δ and x, which is conveniently expressed in terms of another Reynolds number, $\mathbf{R}_x = V_o x/\nu$.

$$\frac{\delta}{x} \cong \sqrt{\frac{30}{V_o x/\nu}} \cong \sqrt{\frac{30}{\mathbf{R}_x}} \tag{13.15}^{11}$$

Information on shear stress and its variation along the plate can now be obtained from equation 13.13, with $d\delta/dx$ obtained from equation 13.15. The result is

$$\tau_o \cong \frac{\rho V_o^2}{2}\sqrt{\frac{8}{15\mathbf{R}_x}} \tag{13.16}^{11}$$

This equation shows τ_o to vary *inversely* with $\sqrt{x}$; this means that the frictional stress

[10]The actual velocity profile is nearly parabolic but not precisely so, and this fact will necessitate subsequent adjustments in the derived equations. A comprehensive analytical solution of this boundary-layer problem was accomplished by H. Blasius (*Zeit. Math. Physik*, vol. 56, p. 1, 1908), and his predicted velocity distributions have been confirmed experimentally many times. The advanced mathematical methods used by Blasius preclude development of his exact solution in an elementary text.

[11] For agreement with the exact results of Blasius' analysis and also with experimental values, the following changes should be made in the coefficients of the equations: 2 to 1.723 and 2/15 to 0.1278 in equation 13.13, 2 × 15 to 27 in equation 13.14, 30 to 27 in equations 13.15 and 13.19, $\sqrt{8/15}$ to 0.664 in equations 13.16 and 13.17, $\sqrt{32/15}$ to 1.328 in equation 13.18.

on the upstream portions of the plate will be larger than that on the portions down-stream. From equation 13.12,

$$c_f \cong \sqrt{\frac{8}{15\mathbf{R}_x}} \tag{13.17[11]}$$

If equation 13.17 is integrated along the plate, the drag coefficient,[12] C_f, is obtained

$$C_f = \frac{1}{A} \int_0^x c_f \, dx = \frac{1}{x} \int_0^x c_f \, dx = \sqrt{\frac{32}{15\mathbf{R}_x}} \tag{13.18[11]}$$

which shows the drag coefficient to be a function of Reynolds number (for incompressible flow) as predicted by the dimensional analysis of Section 13.2.

Other useful relationships may be derived from the foregoing equations. The relation between $\mathbf{R}_x$ and $\mathbf{R}_\delta$ may be obtained by mere rearrangement of equation 13.15; the result is

$$\mathbf{R}_x \cong \frac{\mathbf{R}_\delta^2}{30} \tag{13.19[11]}$$

When the critical value of 3 900 is substituted for $\mathbf{R}_\delta$, $\mathbf{R}_x \cong 500\,000$, showing that, in terms of plate length x, a laminar boundary layer is not to be expected beyond this critical value of $\mathbf{R}_x$.

Further insight into boundary layer properties may be obtained by comparing (see Fig. 13.8) the assumed distributions of velocity and shear with those derived by the exact analysis of Blasius (for which the velocity profile has been accurately confirmed by experiment). Experience with pipe flow has shown (Section 9.2) that a parabolic velocity profile in laminar flow is accompanied by a linear shear stress distribution. However, the exact analysis shows that the velocity profile is not precisely parabolic nor the stress profile precisely linear, the most important property of both of these profiles being their asymptotic character at the outer edge of the boundary layer. From equation 13.6 and the parabolic profile, the displacement thickness $\delta_1 = \delta/3$, which is very close to the value obtained for the exact analysis. This quantity is important in precise calculations since it in effect augments the thickness of plate or body and thus alters the pressure distributions on them; in the case of the flat plate, for example, the deflection of the streamlines causes a slight increase of free stream velocity along the plate accompanied by a slight drop of pressure, and a small favorable pressure gradient. Even though these features have been ignored in Blasius' exact analysis of the laminar boundary layer, there is no appreciable effect on the final results, all of which have been confirmed by experiment.

---

[12] The subscript f is used here to emphasize the wholly frictional nature of the coefficient. Because the flow is two-dimensional, C_f is the coefficient per unit width; that is, $A = 1 \cdot x$.

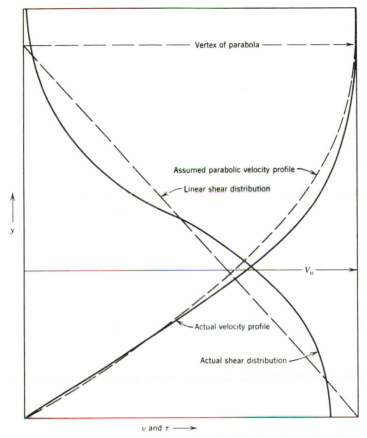

Fig. **13.8**   Shear and velocity profiles in a laminar boundary layer.

13.5   The Turbulent Boundary Layer—Incompressible Flow

An approximate analysis of the turbulent boundary layer developing on a flat plate in a zero pressure gradient (Fig. 13.6) can be made by assuming that the seventh-root velocity profile and the accompanying shear stress expression developed by Blasius for established turbulent pipe flow are applicable. For the boundary layer, equations 9.27 and 9.28 (Section 9.3) are (with $\rho_o = \rho$ here)

$$u = V_o \left(\frac{y}{\delta}\right)^{1/7} \tag{9.27}$$

$$\tau_o = 0.046\ 4 \left(\frac{\nu}{V_o \delta}\right)^{1/4} \frac{\rho V_o^2}{2} \tag{9.28}$$

Hence, from equation 13.7,

$$\delta_2 \cong \int_0^{\delta} \left(\frac{y}{\delta}\right)^{1/7} \left[1 - \left(\frac{y}{\delta}\right)^{1/7}\right] dy = \frac{7}{72}\delta \tag{13.20}$$

From equation 13.11, it follows that

$$\frac{\tau_o}{\rho V_o^2} = \frac{c_f}{2} \cong \frac{7}{72}\frac{d\delta}{dx} \tag{13.21}$$

and using equation 9.28

$$\tau_o \cong \frac{7}{72}\rho V_o^2 \frac{d\delta}{dx} \cong 0.046\ 4 \left(\frac{\nu}{V_o\delta}\right)^{1/4}\frac{\rho V_o^2}{2}$$

Separating variables and integrating,[13]

$$\int_0^{\delta} \delta^{1/4}\ d\delta \cong 0.238\ 5 \left(\frac{\nu}{V_o}\right)^{1/4}\int_0^x dx$$

Performing the integration and substituting $\mathbf{R}_x$ for $V_o x/\nu$ gives

$$\frac{\delta}{x} \cong \left(\frac{0.007\ 9}{\mathbf{R}_x}\right)^{1/5} = \frac{0.38}{\mathbf{R}_x^{0.2}} \tag{13.22}$$

allowing the approximate shape and size of the turbulent boundary layer to be predicted. Substitution of δ from equation 13.22 into equation 13.21 yields

$$\frac{\tau_o}{\rho V_o^2} = \frac{c_f}{2} \cong \frac{0.30}{\mathbf{R}_x^{0.2}} \tag{13.23}$$

It follows that the drag coefficient is

$$C_f \cong \frac{0.074}{\mathbf{R}_x^{0.2}} \tag{13.24}$$

and again the drag coefficient is seen to be a function of Reynolds number only. The seventh-root law in boundary layers, as in pipe flow, can be expected to be adequate only over a limited range of Reynolds numbers which must be determined by experiment; comparison of these on Fig. 13.9 shows this range of $\mathbf{R}_x$ to be from 10^5 to about 10^8 and implies that computations for boundary-layer thickness may be made with equation 13.22 in this range.

A more refined analysis of the turbulent boundary layer has been made by von Kármán,[14] who used the logarithmic law of velocity distribution which leads, with some adjustment of coefficients, to

$$\frac{1}{\sqrt{C_f}} = 1.70 + 4.15 \log C_f \mathbf{R}_x$$

[13] It is assumed for this integration and the definition of $\mathbf{R}_x$ that the boundary layer is turbulent all the way to the leading edge of the plate.
[14] Th. von Kármán, "Turbulence and Skin Friction," *Jour. Aero. Sci.*, vol. 1, p. 1, 1934.

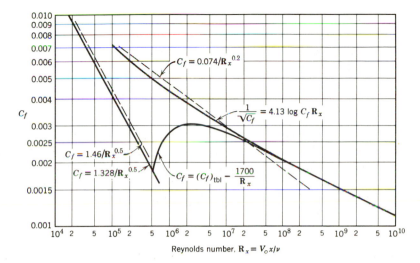

Fig. **13.9** Drag coefficients for smooth, flat plates.

However, Schoenherr[15] found that further adjustment of the constants gave an equation which more accurately represented the results of a wide range of experiments; his equation, now used as a standard in American practice, is

$$\frac{1}{\sqrt{C_f}} = 4.13 \log C_f \mathbf{R}_x \qquad (13.25)$$

A plot of C_f against $\mathbf{R}_x$ for smooth flat plates (Fig. 13.9) with laminar and turbulent boundary layers bears a striking resemblance to the Stanton diagram for smooth circular pipes (Fig. 9.9). However, the critical Reynolds number is not so well defined as its counterpart in pipe flow because of flow conditions which are not as well controlled. With increased initial turbulence in the approaching flow, earlier breakdown of the laminar boundary layer occurs, thus reducing the critical Reynolds number; roughening the leading edge of the plate has also been found to decrease the critical Reynolds number by decreasing the flow stability and causing earlier breakdown of the laminar layer. On the other hand, a decrease in pressure along the flow (*favorable pressure gradient*) produced by curvature of the boundary surface or at a pipe inlet (Fig. 7.16) has been found to delay the breakdown of the laminar boundary layer; application of this principle has produced the so-called *laminar flow wing* for aircraft which exhibits a drag well below that of ordinary wings. The sizable reduction in drag obtained by maintenance of a laminar boundary layer is obvious at once from Fig. 13.9; in aerodynamic practice, suction slots and porous materials, along with smooth leading edges, and appropriately shaped wing profiles are used to accomplish this reduction.

[15] K. E. Schoenherr, "Resistance of Flat Plates Moving through a Fluid," *Trans. Soc. Nav. Arch. and Marine Engrs.*, vol. 40, p. 279, 1932.

When smooth plates feature a laminar boundary layer followed by a turbulent one (Fig. 13.6), experimental drag coefficients fall between the laminar and turbulent lines of Fig. 13.9, leaving the former abruptly and approaching the latter asymptotically. Prandtl expressed this fact mathematically by

$$C_f = (C_f)_{\text{tbl}} - \frac{1\,700}{\mathbf{R}_x} \tag{13.26}$$

in which $(C_f)_{\text{tbl}}$ would be obtained from the Schoenherr equation 13.25; equation 13.26 is also plotted on Fig. 13.9, giving a critical Reynolds number of 537 000. This value, however, should be taken only as a typical or nominal one; the range of critical Reynolds numbers determined experimentally without a favorable pressure gradient range from 100 000 to 600 000. For favorable pressure gradients they may be considerably larger than 600 000.

_____ **Illustrative Problem** _____

A ship model 1.5 m long with draft of 0.15 m is towed at a velocity of 0.3 m/s in a basin containing water at 16°C. Assuming that one side of the immersed portion of the hull may be approximated by a smooth flat plate (1.5 m × 0.15 m), estimate the frictional drag of the hull and the thickness of the boundary layer at the stern of the model if the boundary layer is (a) laminar and (b) turbulent. If the measured total drag of the model is 0.178 N, estimate the total drag of the prototype if the model scale is 1:64.

Relevant Equations and Given Data

$$\mathbf{R}_x = V_o x / \nu$$

$$\mathbf{F}_p = \mathbf{F}_m \qquad \left(\frac{V_o^2}{lg_n}\right)_p = \left(\frac{V_o^2}{lg_n}\right)_m \tag{8.7}$$

$$\mathbf{R} = V_o l / \nu \tag{8.11}$$

$$C_f = C_D = \frac{D_x}{\frac{1}{2}A\rho V_o^2} \tag{13.1}$$

$$\frac{\delta}{x} \cong \sqrt{\frac{30}{V_o x / \nu}} \cong \sqrt{\frac{30}{\mathbf{R}_x}} \tag{13.15}$$

$$\frac{\delta}{x} \cong \left(\frac{0.0079}{\mathbf{R}_x}\right)^{1/5} \cong \frac{0.38}{\mathbf{R}_x^{0.2}} \tag{13.22}$$

$(V_o)_m = 0.3$ m/s $l_m = 1.5$ m $w_m = 0.15$ m $A_m = 1.5 \times 0.15$ m²

$C_f(2 \text{ sides})_m = 0.178$ N $l_m / l_p = 1/64$ $T_m = T_p = 16°C$

Solution. Obtaining ν from Appendix 2,

$$\mathbf{R}_x = \frac{0.3 \times 1.5}{1.13 \times 10^{-6}} = 4 \times 10^5$$

From Fig. 13.9, (a) $C_f = 0.002\ 0$ and (b) $C_f = 0.005\ 2$. For the laminar boundary layer, the drag

$$D_x(2 \text{ sides of plate}) = \frac{2 \times 0.002 \times (1.5 \times 0.15)998.8 \times (0.3)^2}{2} = 0.04 \text{ N} \blacktriangleleft$$

$$(13.1)$$

Similarly, for the turbulent boundary layer,

$$D_x(2 \text{ sides}) = 0.105 \text{ N} \blacktriangleleft$$

For thickness of the laminar boundary layer,

$$\delta = \frac{1.5 \times 5.20}{\sqrt{4 \times 10^5}} = 0.012 \text{ m} \blacktriangleleft \qquad (13.15)$$

For the turbulent boundary layer,

$$\delta = \frac{1.5 \times 0.38}{(4 \times 10^5)^{0.2}} = 0.043 \text{ m} \blacktriangleleft \qquad (13.22)$$

Restudying Section 8.1, the corresponding speed of the prototype is obtained from the Froude law of similitude:

$$\frac{(0.3)^2}{1.5 \times g_n} = \frac{V_o^2}{(1.5 \times 64)g_n} \qquad V_o = 2.4 \text{ m/s} \qquad (8.7)$$

The Reynolds number for the prototype is therefore

$$\mathbf{R}_x = \frac{2.4 \times 1.5 \times 64}{1.13 \times 10^{-6}} = 2.04 \times 10^8$$

giving C_f (from Fig. 13.9) 0.001 85 and allowing the frictional drag of the prototype to be estimated:

$$D_x(2 \text{ sides}) = \frac{2 \times 0.001\ 85(1.5 \times 0.15 \times 64^2)998.8 \times (2.4)^2}{2} = 9\ 809 \text{ N} \blacktriangleleft \qquad (13.1)$$

With the boundary layer turbulent for the prototype, the turbulent boundary-layer assumption for the model must be used to estimate the wave drag, D_w; this is the difference between total and frictional drags:

$$D_w = 0.178 - 0.105 = 0.073 \text{ N}$$

Since the wave drag is modelled by the Froude law and the Froude numbers are the same in model and prototype, D_w will vary with $A\rho V_o^2$; hence D_w for the prototype may be calculated from the proportion

$$\frac{D_w}{0.073} = \frac{(64^2 \times 0.15 \times 1.5)2.4^2}{(0.15 \times 1.5)(0.3)^2} \qquad D_w = 19\ 136 \text{ N}$$

Finally, the total drag of the prototype (at a speed of 2.4 m/s) is estimated to be

$$D = 19\ 136 + 9\ 809 = 28\ 945 \text{ N} \blacktriangleleft$$

DRAG AND LIFT—INCOMPRESSIBLE FLOW

13.6   Profile Drag

Although the total drag force on any immersed object is always the sum of frictional and pressure drag, it will be seen later that this breakdown of the drag is inconvenient for objects (such as airfoils) on which a transverse (lift) force is exerted. Here the total drag is considered to be the sum of (1) that which would be developed if the airfoil had no ends (i.e., two-dimensional flow), and (2) that produced by any end effects. Since the former depends only on the shape (profile) and orientation of the airfoil, it is called *profile drag,* whereas the latter, which depends on the airfoil plan form and is *induced* by the lift force, is termed *induced drag.* Evidently, for objects which exhibit no lift, the induced drag will be zero and the profile drag equal to the total drag.

Pressure drag has been shown (Section 13.1) to be that part of total drag resulting from pressure variation over the surface of an object and to be dependent on wake formation downstream from the object. In general, when wakes are large, pressure drag is large and, when wake width is reduced by streamlining, pressure drag is reduced also; pressure drag is thus critically dependent on the existence and position of flow separation which in turn depends on shape of object and structure of the boundary layer. Consider first the case of a well-streamlined object (Fig. 13.10 and Fig. 7.8); here there is no separation, and the flowfield for ideal or real fluid flow is essentially the same except for boundary-layer growth and lack of trailing edge stagnation point in the latter. In this case a good approximation to the pressure distribution on the object may be made by neglecting the boundary layer thickness and applying the methods of mathematical hydrodynamics (Chapter 12); such computations may be refined (with considerable difficulty) through altering the shape of the object by the displacement thickness of the boundary layer. These methods cannot, of course, yield the pressure at the trailing edge of the body, which for ideal flow is the stagnation pressure but for real flow is considerably less than this. However, the pressure may be estimated (by extrapolating

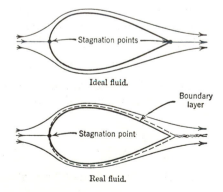

Ideal fluid.

Real fluid.

Fig. **13.10**

the pressure distribution over the body) with sufficient accuracy that the pressure variation over the body may be appropriately integrated to allow a reasonable estimate of the pressure drag. More accurate values of pressure drag may be obtained from experimentally determined pressure distributions in the same manner.

For a blunt object (Fig. 13.11) there is a drastic difference between the ideal and real flowfields caused by flow separation and wake formation in the latter. Here the critical feature either for analytical calculations or for understanding of experimental results is the position of the separation points. Lack of resistance and energy dissipation in the ideal flow will allow fluid particles adjacent to the object to move between stagnation points, accelerating over the upstream end of the object in a favorable pressure gradient and decelerating over the downstream end through an unfavorable pressure gradient. For the real fluid, boundary-layer growth will begin at the stagnation point and energy will be dissipated in overcoming resistance caused by shear stresses in the boundary layer. The momentum of fluid particles in the boundary layer will thus be considerably less than those at corresponding positions in the ideal flowfield; the momentum of such particles will be further reduced by the unfavorable pressure gradient until at some point they will come to rest, accumulate, and be given a rotary motion by the surrounding flow; separation of the live flow from the object then results as the eddy increases in size. This description of the separation process applies at the inception of flow and at the beginning of wake formation. Once separation has occurred, a new flowfield is established and there is no reason to expect reattachment[16] of the live stream to the object.

With separation dependent on boundary-layer growth, it may be expected that the laminar or turbulent character of the boundary layer will be of critical importance in determining the position of separation. A simple comparison of laminar and turbulent layers of the same ρ, V_o, and δ will show their momentum fluxes to be $8\rho V_o^2 \delta/15$ and

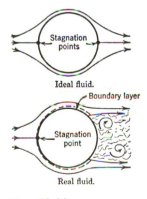

Fig. **13.11**

---

[16] However, this may occur under special conditions for well-streamlined objects.

$7\rho V_o^2\delta/9$, respectively—the momentum flux of the turbulent layer being nearly 50% greater[17] than that of the laminar one. Thus the turbulent boundary layer may be considered the "stronger" of the two and better able to survive an unfavorable pressure gradient; accordingly, it may be expected that the separation point for a turbulent boundary layer will be found farther downstream than that for a laminar boundary layer.

Quantitative aspects of profile drag may be obtained from a study of the experimentally determined drag coefficients of various objects. First, consider the basic sphere, which has been exhaustively studied. At low Reynolds number (Fig. 13.12*a*) the flow will close behind the sphere and no wake will form; under these conditions profile drag is composed almost entirely of frictional drag. Stokes[18] has shown analytically that, in laminar flow at very low Reynolds numbers, where inertia forces may be neglected and those of viscosity alone considered, the drag of a sphere of diameter

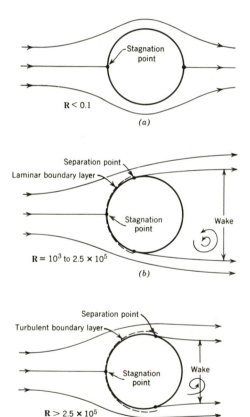

Fig. **13.12**   Flow about a sphere at various Reynolds numbers.

---

[17] A better comparison of laminar and turbulent boundary layers on the same object shows an even larger difference.

[18] G. G. Stokes, *Mathematical and Physical Papers*, vol. III, p. 55, Cambridge University Press, 1901.

d, moving at a velocity V_o through a fluid of viscosity μ, is given by

$$D = 3\pi\mu V_o d \qquad (13.27)$$

and this equation has been confirmed by many experiments. The drag coefficient, C_D, for the sphere under these conditions may be found by equating the preceding expression to equation 13.1:

$$\frac{C_D A \rho V_o^2}{2} = 3\pi\mu V_o d$$

Taking A to be the area of the projection of the object on a plane, normal to the direction of V_o, $A = \pi d^2/4$, and substituting this above gives

$$C_D = \frac{24\mu}{V_o d\rho} = \frac{24}{\mathbf{R}}$$

Thus the drag coefficients of spheres at low velocities are dependent only on the Reynolds number—another confirmation of the results of the dimensional analysis of Section 13.2.

As the Reynolds number increases, the drag coefficients of spheres continue to depend only on the size of this number, and a plot of experimental results over a large range of Reynolds numbers for spheres of many sizes, tested in many fluids, gives the single curve of Fig. 13.13.

Up to a Reynolds number of 0.1, the Stokes equation applies accurately and the

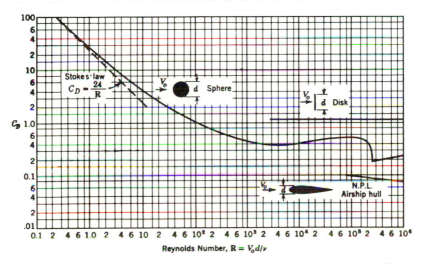

Fig. **13.13** Drag coefficients for sphere, disk, and streamlined body.[19]

[19] Data from L. Prandtl, *Ergebnisse der aerodynamischen Versuchsanstalt zu Göttingen*, vol. II, p. 29, R. Oldenbourg, 1923; and G. J. Higgens, "Tests of the N. P. L. Airship Models in the Variable Density Wind Tunnel," *N.A.C.A. Tech. Note* 264, 1927.

drag coefficient results from frictional effects. As the Reynolds number is increased to about 10, separation and weak eddies begin to form, enlarging into a fully developed wake near a Reynolds number of 1 000; in this range the drag coefficient results from a combination of pressure and frictional drag, the latter becoming about 5% of total drag as a Reynolds number of 1 000 is reached (see Fig. 13.12). Above this figure the effects of friction become even smaller and the drag problem becomes primarily one of pressure drag.

The drag coefficient of the sphere ranges (approximately) between 0.4 and 0.5 from $\mathbf{R} \cong 1\ 000$ to $\mathbf{R} \cong 250\ 000$, at which point it suddenly drops more than 50% and then increases gradually with further increase in the Reynolds number (see Figs. 13.12 and 13.13). In this range of Reynolds numbers, experiments have shown the separation point to be upstream from the midsection of the sphere, resulting in a relatively wide turbulent wake; the boundary layer on the surface of the sphere from stagnation point to separation point has been found to be laminar up to $\mathbf{R} \cong 250\ 000$. With further increase in $\mathbf{R}$ the length of the laminar boundary layer decreases, the boundary layer flow past the separation point becomes turbulent, and the separation point moves to a position downstream from the center of the sphere (Fig. 13.12c), causing a decrease in the width of the wake and consequent decrease in the drag coefficient.

The change from laminar to turbulent boundary layer on a flat plate has been seen (Section 13.3) to occur at a critical Reynolds number dependent on the turbulence of the approaching flow. It also occurs with a sphere, and with increased turbulence in the approaching flow the sudden drop in the drag coefficient curve occurs at lower Reynolds number. Thus a sphere may be used as a relative measure of turbulence by noting the Reynolds number at which a drag coefficient of 0.30 (see Fig. 13.13) is obtained. Before the development of the hot-wire anemometer (Section 11.8), this method was used to compare the turbulence characteristics of different wind tunnels.

The drag coefficient of a thin circular disk placed normal to the flow shows practically no variation with the Reynolds number, since the separation point is fixed at the edge of the disk and cannot shift from this point, regardless of the condition of the boundary layer. Thus the width of the wake remains essentially constant, as does the drag coefficient. This idea may be usefully generalized and applied to all brusque or very rough objects in a fluid flow; experiments indicate that such objects have drag coefficients which are essentially constant in the range of high Reynolds numbers.[20]

The drag coefficients of circular cylinders placed normal to the flow show characteristics similar to those of spheres. The coefficients shown in Fig. 13.14 are for infinitely long cylinders. The drag coefficients of streamlined struts[21] and flat plates of infinite length are also shown for comparison. The total drags of the flat plate and cylinder contain negligible frictional drag at ordinary velocities, whereas the stream-

---

[20] Compare this with the relation of the friction factor f, and Reynolds number, $\mathbf{R}$, for rough pipes, Figs. 9.9 and 9.10, and also the fact that the minor loss coefficients of pipe flow show little variation with the Reynolds number.

[21] The area to be used in the drag equation is the projection of the body on a plane normal to the direction of flow.

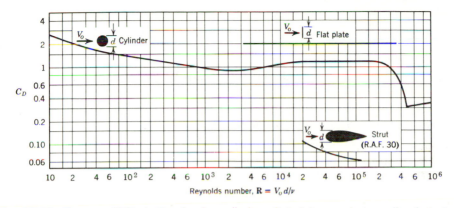

Fig. **13.14** Drag coefficients for circular cylinders, flat plates, and streamlined struts of infinite length.[22]

lined strut, because of its small turbulent wake, possesses little pressure drag. The curves are typical of those resulting from tests of brusque and streamlined objects.

Long blunt objects such as cylinders, when placed crosswise to a fluid flow, sometimes exhibit the property of shedding large eddies regularly and alternately from opposite sides (Fig. 13.15).[23] Because of von Kármán's studies of the stability of these regular vortex patterns they are generally known as *Kármán vortex streets*. Experiments have shown that eddies will be shed regularly (with frequency n) from a circular cylinder for Reynolds numbers between 60 and 5 000 (see Fig. 13.14) and that the nondimensional frequency of the phenomenon, the so-called *Strouhal number, S* = nd/V_o, is equal to approximately 0.21 over much of this range. The regular periodic nature of the eddy formation produces transverse forces on the cylinder that are also periodic, and thus tend to produce transverse oscillations. These considerations are vital to the design of elastic structures such as tall chimneys and suspension bridges which

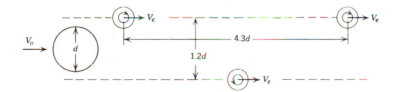

Fig. **13.15**   Schematic of typical vortex street.

[22] Data from L. Prandtl, *Ergebnisse der aerodynamischen Versuchsanstalt zu Göttingen,* vol. II, p. 24, R. Oldenbourg, 1923; and B. A. Bakhmeteff, *Mechanics of Fluids,* Part II, p. 44, Columbia University Press, 1933.

[23] See Film Loop No. S-FM012, "Flow Separation and Vortex Shedding," Encyclopaedia Britannica Educational Corp.

are exposed to the wind; they become critical if the natural frequency of vibration of the structure is close to the frequency of the eddy formation.

_____   **Illustrative Problem**   _____

Calculate the ratio between the drag forces on the same sphere in the same fluid stream at Reynolds numbers 400 000 and 200 000.

Solution.   From Fig. 13.13 the drag coefficients at these Reynolds numbers are 0.20 and 0.43, respectively. With sphere size and fluid the same, Reynolds number will vary directly with velocity, and drag with $C_D\mathbf{R}^2$. Writing this as a ratio,

$$\frac{D_4}{D_2} = \frac{0.2(4)^2}{0.43(2)^2} = 1.86 \blacktriangleleft$$

Frequently for small changes of Reynolds number such calculations are made neglecting the change of drag coefficient. Had this been done here, the error would have been 115%.

13.7   Lift and Circulation

The Kutta-Joukowsky theorem (Section 12.4f) has demonstrated that circulation about an object is one of the requirements for the existence of a lift force on the object. Although it is not difficult to imagine a rotating body in a viscous fluid inducing its own circulation, to explain the origin of circulation about an airfoil, or an element of a propeller or turbine blade, requires knowledge of other principles.

   Consider the flow conditions about a typical airfoil as it starts to move. Before motion begins, the circulation about the foil is obviously zero (Fig. 13.16a). As motion starts, the circulation about the airfoil tends to remain zero, and the potential flow of Fig. 13.16b tends to be set up, but such a flow, which includes a stagnation point near the rear of the airfoil and flow around its sharp trailing edge, cannot be maintained in a real fluid because of separation. This momentary potential flow gives way immediately to the flow of Fig. 13.16c, and in the process a circulation Γ develops about the airfoil and a vortex, the *starting vortex* (Fig. 13.16d), is shed from the airfoil. During the creation of this vortex, however, the circulation around a closed curve, including and at some distance from the airfoil, is not changed and must still be zero; thus, from the properties of circulation, the circulation about, or the strength of, the starting vortex must be equal and opposite to that about the airfoil; the existence of circulation about an airfoil is thus dependent on the creation of a starting vortex.[24]

_____

[24] See Film Loop No. S-FM010, "Generation of Circulation and Lift for an Airfoil," Encyclopaedia Britannica Educational Corp.

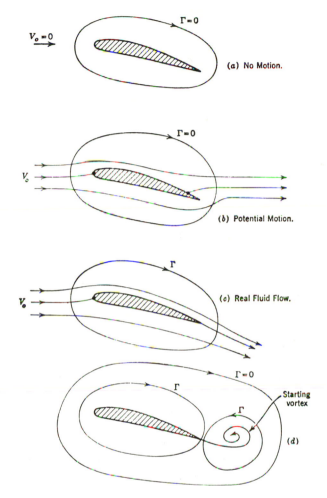

Fig. 13.16 Development of circulation about an airfoil.

13.8 Airfoils of Finite Length

When fluid flows about airfoils of finite length, flow phenomena result which affect both lift and drag of the airfoil; these phenomena may be understood by further investigation and application of the foregoing circulation theory of lift.

Since pressure on the bottom of an airfoil is greater than that on the top, flow will escape from below the airfoil at the ends and flow toward the top, thus distorting the general flow about the airfoil, causing fluid to move inward over the top of the airfoil and outward over the bottom (Fig. 13.17). As the fluid merges at the trailing edge of the airfoil, a surface of discontinuity is set up, and flows above and below this surface are, respectively, inward and outward as shown. The tendency for vortices to form from

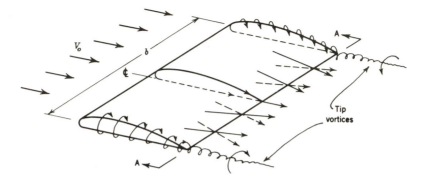

Fig. **13.17** Airfoil of finite length.

these velocity components is apparent, and, in fact, this surface of discontinuity is a sheet of vortices. However, such a vortex sheet is unstable, and the rotary motions contained therein combine to form two large vortices trailing from the tips of the airfoil (Fig. 13.17); these are called tip vortices and are often visible when an airfoil passes through dust-laden air or as vapor trails produced by condensation of atmospheric moisture. (See Fig. A.9 for an illustration of tip vortices made visible by cavitation on a ship propeller.)

Since the pressure difference between top and bottom of an airfoil must reduce to zero at the tips, it is evident that the lift per unit length of span varies over the span (Fig. 13.18), being maximum at the center and reducing to zero at the tips. The total lift of the airfoil is, of course, the total resulting from this lift diagram. Since lift per unit length of span varies directly with circulation ($L = \Gamma\rho V_o$), a diagram showing distribution of circulation over the span has the same shape as a diagram of lift distribution. The variation of lift and circulation over the span of an airfoil cannot, of course, be disregarded in a rigorous treatment of the subject, but such treatment leads to mathematical and physical complexities which are beyond the scope of this volume. A simple physical picture may be obtained, however, from the following analysis in which lift and circulation will be assumed to be distributed uniformly over the span (Fig. 13.18).

One of the properties of vortices is that their axes can end only at solid boundaries. Since there is no solid boundary at the end of the airfoil, the circulation Γ cannot stop here, but must continue to exist about the axes of the tip vortices (Fig. 13.19). The axes

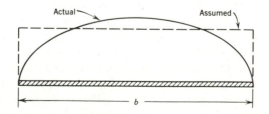

Fig. **13.18** Distribution of lift and circulation over an airfoil of finite length.

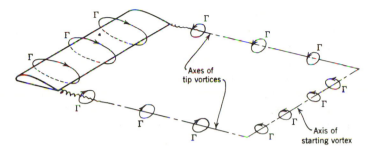

Fig. **13.19** Circulation about an airfoil of finite length.

of the tip vortices extend rearward to the axis of the starting vortex; thus, according to the theory, the axis of the vortex having circulation Γ does not end, but is a closed curve composed of the axes of the airfoil, tip vortices, and starting vortex. In the real fluid the circulation persists only about the airfoil and portions of the tip vortices close to the airfoil; the starting vortex and remainder of the tip vortices are extinguished by viscous action.

The circulations about the tip vortices induce a downward motion in the fluid passing over an airfoil of finite length and in so doing affect both lift and drag *by changing the effective angle of attack.* The strength of this induced motion will, obviously, depend on the proximity of the tip vortices and, thus, upon the span of the airfoil which may be expressed in terms of the *aspect ratio* b^2/A.

An airfoil of finite span is shown at angle of attack α in the horizontal flow of Fig. 13.20. The vertical velocity induced near the wing by the tip vortices decreases the angle of attack by a small angle α_i, making the effective angle of attack $(\alpha - \alpha_i)$. This effective angle of attack is that for no induced velocity or, in other words, it is the angle of attack which would be obtained if the foil had infinite span and aspect ratio. Calling this angle of attack α_o,

$$\alpha_o = \alpha - \alpha_i \tag{13.28}$$

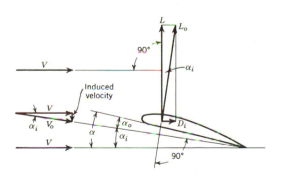

Fig. **13.20**

Now, treating the airfoil as one of infinite span at an angle of attack α_o, the lift L_o exerted on such an airfoil is by definition normal to the direction of flow in which it is placed; therefore L_o is normal to the effective velocity V_o and at an angle α_i with the vertical. The lift, L, on the airfoil of finite span is normal to the approaching horizontal velocity V and is the vertical component of L_o. But L_o also has a component in the direction of the original velocity V, which is a drag force, D_i, called the *induced drag* because its existence depends on the downward velocity induced by the tip vortices. Thus an additional drag force, D_i, must be added to profile drag in computing the total drag of a body of finite length about which a circulation exists. Calling the profile drag D_o, since it is the drag of an airfoil of infinite span (which has no end effects and, therefore, no induced drag) the total drag, D, of an airfoil of finite length is given by

$$D = D_o + D_i$$

which, by dividing by $A\rho V^2/2$, may be expressed in terms of dimensionless drag coefficients as

$$C_D = C_{D_o} + C_{D_i} \tag{13.29}$$

From the foregoing statements and Fig. 13.20, it is evident that induced drag, D_i, is related to lift L, angle α, and aspect ratio b^2/A; the equations relating these variables are of great practical importance. Since α_i is small,

$$L = L_o \qquad V = V_o \qquad D_i = L\alpha_i$$

If the distribution of lift over a wing of finite span is taken to be a semiellipse[25] (see Fig. 13.18), it may be shown that

$$\alpha_i = \frac{C_L}{\pi b^2/A} \tag{13.30}$$

With drag and lift proportional to their respective coefficients, $D_i = L\alpha_i$ may be written $\alpha_i = C_{D_i}/C_L$, and by substitution of this into equation 13.30 there results

$$C_{D_i} = \frac{C_L^2}{\pi b^2/A} \tag{13.31}$$

which relates lift and induced drag through their dimensionless coefficients and shows that induced drag is inversely proportional to aspect ratio, becoming zero at infinite aspect ratio (infinite span) and increasing as aspect ratio and span decrease—thus offering mathematical proof of the foregoing statements on the effect of span, aspect ratio, and proximity of tip vortices on induced downward velocity and induced drag.

The use of the derived expressions for α_i and C_{D_i} in equations 13.30 and 13.31 allows airfoil data obtained at one aspect ratio to be converted into corresponding

---

[25] An assumption which gives minimum induced drag and conforms well (but not perfectly) with fact.

conditions at infinite aspect ratio, and these data, in turn, to be reconverted to airfoils of any aspect ratio; thus extensive testing of airfoils of the same profile at various aspect ratios becomes unnecessary.

────────────────── **Illustrative Problem** ──────────────────

A rectangular airfoil of 6 ft (1.83 m) chord and 36 ft (11 m) span (aspect ratio 6) has a drag coefficient of 0.054 3 and lift coefficient of 0.960 at an angle of attack of 7.2°. What are the corresponding lift and drag coefficients and angle of attack for a wing of the same profile and aspect ratio 8?

Relevant Equations and Given Data

$$\alpha_o = \alpha - \alpha_i \tag{13.28}$$

$$C_D = C_{D_o} + C_{D_i} \tag{13.29}$$

$$\alpha_i = \frac{C_L}{\pi b^2 / A} \tag{13.30}$$

$$C_{D_i} = \frac{C_L^2}{\pi b^2 / A} \tag{13.31}$$

$c = 6 \text{ ft}/1.83 \text{ m}$ $b = 36 \text{ ft}/11 \text{ m}$ $A = 216 \text{ ft}^2/20 \text{ m}^2$

$C_D = 0.054\ 3$ $C_L = 0.960$ at $\alpha = 7.2°$ $b^2/A = 6$

Solution.   For aspect ratio 8:

$$C_L = 0.960 \text{ (negligible change of lift coefficient)} \blacktriangleleft$$

For aspect ratio 6:

$$C_{D_i} = \frac{(0.960)^2}{6\pi} = 0.048\ 9 \tag{13.31}$$

For aspect ratio ∞:

$$C_{D_o} = 0.054\ 3 - 0.048\ 9 = 0.005\ 4 \tag{13.29}$$

For aspect ratio 8:

$$C_D = 0.005\ 4 + \frac{(0.960)^2}{8\pi} = 0.042\ 1 \blacktriangleleft \tag{13.29}$$

For aspect ratio 6:

$$\alpha_i = \frac{0.960}{6\pi} = 0.050\ 9 \text{ radian} = 2.9° \tag{13.30}$$

For aspect ratio ∞:

$$\alpha_o = 7.2 - 2.9 = 4.3° \tag{13.28}$$

For aspect ratio 8:

$$\alpha = 4.3 + \left(\frac{0.960}{8\pi}\right)\left(\frac{360}{2\pi}\right) = 6.5° \blacktriangleleft \qquad (13.28 \ \& \ 13.30)$$

13.9 Lift and Drag Diagrams

The relation between lift and induced drag coefficients suggests plotting lift coefficient against drag coefficient and gives the so-called *polar diagram* of Fig. 13.21, which is

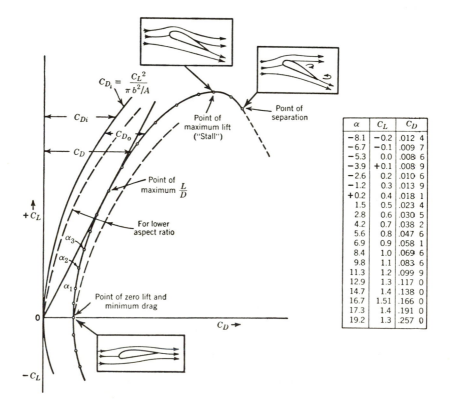

α	c_L	c_D
−8.1	−0.2	.012 4
−6.7	−0.1	.009 7
−5.3	0.0	.008 6
−3.9	+0.1	.008 9
−2.6	0.2	.010 6
−1.2	0.3	.013 9
+0.2	0.4	.018 1
1.5	0.5	.023 4
2.8	0.6	.030 5
4.2	0.7	.038 2
5.6	0.8	.047 6
6.9	0.9	.058 1
8.4	1.0	.069 6
9.8	1.1	.083 6
11.3	1.2	.099 9
12.9	1.3	.117 0
14.7	1.4	.138 0
16.7	1.51	.166 0
17.3	1.4	.191 0
19.2	1.3	.257 0

Fig. 13.21 Polar diagram for a typical airfoil and numerical data for the Clark-Y airfoil.[26]

---

[26] A 14.6 m × 2.4 m (48 ft × 8 ft) rectangular airfoil tested at **R** ~ 6 000 000. Data from A. Silverstein, "Scale Effect on Clark-Y Airfoil Characteristics from N.A.C.A. Full-Scale Wind-Tunnel Tests," *N.A.C.A. Rept.* 502, 1934.

used extensively in airplane design. On this diagram equation 13.31 appears as a parabola passing through the origin and symmetrical about the C_D-axis, the position of the parabola depending on the aspect ratio. Since the two curves are for airfoils of the same aspect ratio, the horizontal distance between them is the profile drag coefficient C_D. However, the diagram shows much more than this. The important ratio of lift to drag is the slope of a straight line drawn between origin and the point for which this ratio is to be found; the maximum value of this ratio is the slope of a straight line tangent to the curve and passing through the origin; on the diagram are also easily seen the points of zero lift and minimum drag, and the point of maximum lift or *stall*,[27] which determines stalling angle above which lift no longer continues to increase with angle of attack; the end of the upper solid portion of the curve is the point at which the flow separates completely from the upper side of the wing, forming a wake which increases the profile drag (and, therefore, the drag coefficient) and is accompanied by a large drop in lift and lift coefficient because of increased pressure on the upper side of the wing. Curves for other aspect ratios may be obtained[28] by the methods of the preceding Illustrative Problem and will be found to follow the trend indicated. Worthy of note are the equal horizontal distances C_{D_o}, between corresponding curves and the decrease of L/D ratio with decreasing aspect ratio.

Another method of presenting airfoil data is to plot C_L, C_D, and L/D against angle of attack. Such a plot for the Clark-Y data of Fig. 13.21 is shown on Fig. 13.22. Of significance are the slope of the straight portion of the C_L curve, the location of the point of maximum L/D, the overall shape of the C_D curve, and the change in the position of the curves with change of aspect ratio.

The data of Fig. 13.21 were obtained at Reynolds numbers around 6×10^6, and from many foregoing statements it should be expected that the data will change with changing Reynolds number. The following trends, which are confirmed by experiment, are of some interest in the light of foregoing principles. With increasing Reynolds number the drag coefficient at low angles of attack decreases; here the drag coefficient contains predominantly frictional effects and its variation with Reynolds number is similar to that of the flat plate (Fig. 13.9). With increased turbulence, due either to increased initial turbulence or increased Reynolds number, the maximum lift coefficient usually increases; in other words, higher angles of attack can be attained without causing separation. Here the momentum of the turbulent boundary layer delays separation, allowing high-velocity flow to cling to the upper side of the airfoil, causing lower pressures and greater lift.

---

[27] The mechanism of stall has received some intensive study. See H. W. Emmons, R. E. Kronauer, and J. A. Rockett, "A Survey of Stall Propagation—Experiment and Theory," and S. J. Kline, "On the Nature of Stall." Both papers in *Trans. A.S.M.E. (Series D)*, vol. 81, 1959.

[28] Because of deviations from the assumed semielliptical lift distribution which led to equations 13.30 and 13.31, the latter must be corrected by experimental coefficients which depend on the plan form of the wing. The size of the corrections increases with the aspect ratio, but the order of these is 5% on C_{D_i} and 15% on α_i for rectangular wings.

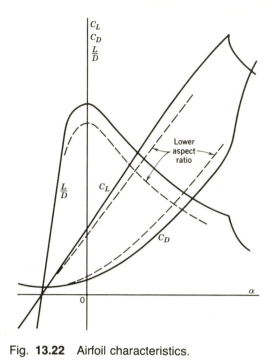

Fig. **13.22** Airfoil characteristics.

DRAG AND LIFT—COMPRESSIBLE FLOW

The problems of flow of compressible fluids about solid objects cover a vast field and are highly complex in their physical and mathematical aspects. Analytical solutions are available for only a small number of the less difficult problems, but steady progress is being made by theoretical research and through physical and numerical experiments. In view of the complexity and advanced nature of such problems, only the most rudimentary ideas and results can be presented in the following elementary treatment.

13.10 The Mach Wave

Consider the motion of a tiny source of disturbance such as a needle point or razor edge moving through a fluid (Fig. 13.23). If the fluid were truly incompressible its modulus of elasticity would be infinite, the velocity of propagation (equation 1.9, 1.10, or 1.11) of the disturbance through the fluid would also be infinite, and the Mach number always zero. The effects of the disturbance would be felt instantaneously through the flowfield, and the fluid in front of the object would "know" of the presence of the object and adjust itself accordingly as the object approached. This situation is closely approximated in real fluid when the flow velocities are small compared with the sonic velocity (which

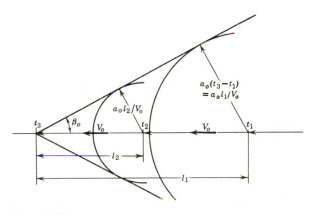

Fig. **13.23**

is the velocity of propagation of the disturbance) and produces the conventional flowfields with which the reader is now familiar.

For the opposite case in which the velocity of motion of the disturbances is appreciably greater than the sonic velocity, a very different situation develops in which fluid upstream from the object is "unaware" of its existence; as the object encounters this fluid, the fluid must suddenly change its direction, producing the sharp discontinuities known as *shock waves* (Section 6.8). To investigate this, assume the point of Fig. 13.23 to move at a (supersonic) velocity V_o, and let it occupy positions 1, 2, 3, at times t_1, t_2, and t_3. At time t_1 the disturbance sends out an elastic wave from point 1 with a celerity a_o, and after the time $(t_3 - t_1)$ has elapsed the distance covered by the wave is $a_o(t_3 - t_1)$. In this elapsed time, however, the source of disturbance has moved to a point 3, and $l_1 = V_o(t_3 - t_1)$; eliminating $(t_3 - t_1)$, the distance covered by the wave is $a_o l_1/V_o$. Similarly the wave which started at point 2 has (when the disturbance reaches point 3) covered a distance of $a_o l_2/V_o$, and many other waves have done likewise from numerous intermediate positions. The result is a wave front (or *oblique shock wave*) which represents the line of advance of the elastic waves; the fluid to the left of this is at rest and "unaware" of the presence of the disturbance. This wave front, known as a *Mach wave*, will be conical (three-dimensional) if produced by the needle point, and wedge-shaped (two-dimensional) if produced by the razor edge. The *Mach angle*, β_o, may be seen from the figure to be defined by $\sin \beta_o = a_o/V_o$. This is useful in estimating from photographs of such wave fronts the velocity, V_o, by measuring β_o on the photograph and computing a_o from the fluid properties.

The same wave picture is, of course, produced when the fluid moves with velocity V_o past a source of disturbance at rest. On such a picture the streamlines may also be conveniently shown, and for a vanishingly small disturbance these will be straight parallel lines. For a disturbance of finite magnitude (Fig. 13.24) the velocity of propagation will be greater than a_o and the angle of the wave, therefore, greater than the Mach angle. Through a finite wave the streamlines will be deflected away from the object, velocities will diminish, and pressures increase (see Section 6.8), these changes

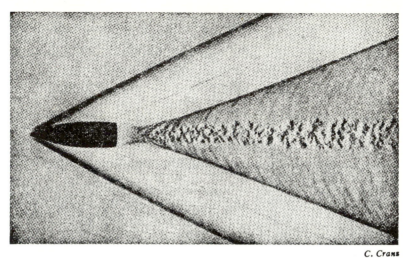

C. *Cranz*

Fig. **13.24** Small-bore bullet in flight.[29]

being comparable in sense (not in magnitude)[30] to those occurring through the normal shock wave (Section 6.7). All these changes are very rapid as the shock wave is exceedingly thin; for most practical purposes it may be considered an abrupt discontinuity.

13.11 Phenomena and Definitions

To visualize the flow phenomena in compressible flow about an object, consider the chain of events which occurs as the free stream velocity is increased from low subsonic to high subsonic to supersonic. At low subsonic speeds the fluid may be considered incompressible, resulting in a conventional streamline picture featuring a stagnation point, S, on the nose of the object and a point, m, of maximum velocity and minimum pressure on the upper side (Fig. 13.25a). Increasing the velocity of the free stream will, of course, raise the stagnation pressure, increase the maximum velocity, and lower the minimum pressure, even if compressibility of the fluid is neglected; inclusion of compressibility exaggerates these changes, however, and they continue until the velocity past m becomes equal to the *local sonic velocity*[31]; this means that the *local Mach number* at point m is now unity, although the free stream Mach number is considerably less than this. In other words, the body is moving at subsonic speed, but sonic phenomena are beginning to appear at a point on its surface. This situation marks the

[29] From C. Cranz, *Lehrbuch der Ballistik*, vol. I, B. G. Teubner, Leipzig, 1917.

[30] The changes are largest in the case of the normal shock wave through which the velocity decreases from supersonic to subsonic; the velocities upstream and downstream from the oblique shock wave are both supersonic except in the case of relatively large β, where the oblique shock approaches the normal shock.

[31] It is very important to note that the pressure, density, and temperature at m are considerably less than those properties in the free stream and that the local sonic velocity (equation 1.11) is thus smaller than free stream sonic velocity.

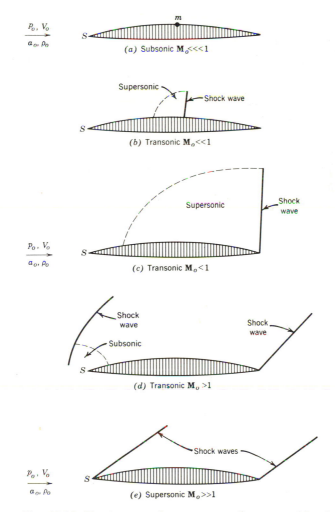

Fig. **13.25** Shock-wave phenomena on the upper side of an airfoil.

end of the *subsonic* flow region (where all velocities in the flowfield are subsonic) and the beginning of the *transonic* regime (in which some velocities are subsonic and others supersonic); the transonic regime ends and the supersonic begins when all velocities in the field are supersonic.

A simplified[32] picture of a transonic flow is shown in Fig. 13.25*b*. Of particular

---

[32] The complex problem of shock wave-boundary layer interaction is omitted from these simplified sketches. In the boundary layer the velocities near the surface are subsonic while the flow outside the boundary layer and upstream from the shock wave is supersonic; the subsonic region allows transmission upstream of the adverse pressure gradient produced by the shock and thus spreads out the region of shock phenomena at the boundary surface.

interest is the large region of supersonic flow and the presence of the shock wave through which the velocity decreases from supersonic to subsonic. Through the shock wave the sudden jump in pressure produces an adverse pressure gradient in the boundary layer, which promotes separation and the usual effects on lift and drag. As the free stream velocity is further increased (but V_o maintained less than a_o) the shock wave is forced rearward (Fig. 13.25c) and the major portion of the flow immediately above the surface is supersonic although the free stream Mach number $\mathbf{M}_o$ is still less than 1.

Increasing the free stream velocity V_o to slightly more than a_o will produce the wave arrangement of Fig. 13.25d, featuring a new shock wave upstream from the body and the flowfield completely supersonic except in a small region[33] between this new shock wave and the nose of the object. The supersonic regime (Fig. 13.25e) will exist at higher free stream Mach numbers when the subsonic zone becomes of negligible size or vanishes entirely, as it will for a sharp-nosed object, to which the upstream shock wave will attach itself.

13.12 Drag

In compressible fluid motion, drag results from energy dissipated in shock waves as well as from the skin friction and separation effects discussed in Sections 13.3 through 13.6. Skin friction drag may be computed approximately by the methods of Sections 13.3 through 13.5 up to free stream Mach numbers around 2, but boundary-layer thicknesses are much greater in compressible fluid motion and stability is greatly affected by the transfer of heat between solid surface and boundary layer. At higher Mach numbers, frictional heating in the boundary layer and adjacent surface becomes a serious problem. The approach to compressible fluid motion which assumes Mach number effects predominant and Reynolds number effects negligible implies that frictional effects and boundary layers are of little importance compared to shock phenomena. This should be recognized as an adequate working assumption but not an absolute fact, and many exceptions to this convenient rule are to be expected. For example, if methods are found for minimizing shock wave effects and preventing separation, a very large portion of total drag force will be composed of frictional drag, and Reynolds number effects will then by no means be negligible.

The variation of drag coefficients with Mach number (Fig. 13.26) is central to drag problems in compressible flows and may be examined fruitfully by considering the extreme cases of a streamlined airfoil or body of revolution and a blunt body such as a flat-nosed projectile. For the latter, separation of fluid from body is fixed in position by the geometry of the body, skin friction is small, and a steady increase in drag coefficient with Mach number results from the predominant compressibility effects on or near the flat nose of the projectile. Streamlining the tail of such an object will increase rather than reduce the drag coefficient, but pointing the nose, and thus reducing

---

[33] Evidently the extent of this region depends on the nose geometry of the body since considerations of Section 13.10 have shown it to be nonexistent if the nose is sharply edged or pointed.

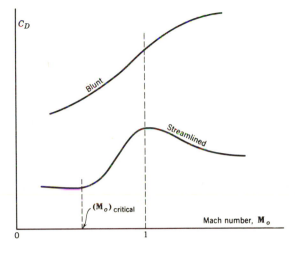

Fig. **13.26**

the frontal area near the stagnation point, will materially reduce the drag coefficient (see Fig. 13.27).

For a streamlined object the variation of drag coefficient is more interesting because of the changing position of the separation point which acccompanies the formation of shock wave phenomena. Through the subsonic range (Fig. 13.25*a*) the slight decrease in drag coefficient with Reynolds number will appear on the plot (Fig. 13.26) of drag coefficient against Mach number since both of these numbers are directly proportional to V_o. As the transonic range is entered (Fig. 13.25*b* and accompanying discussion), shock phenomena and separation appear in the vicinity of point *m* and drag coefficient begins to increase[34] rapidly. Toward the middle of the transonic range (Fig. 13.25*c*) the separation point has moved well to the rear, reducing the size of the wake, but shock phenomena have intensified and drag coefficient continues to rise but more slowly, eventually reaching a maximum value as these opposite effects cancel each other. In the supersonic range (Fig. 13.25*e*), the drag coefficient depends primarily on the energy dissipation through the inclined shock waves and decreases steadily with further increase in Mach number.

Of great practical interest is the conversion of airfoil data obtained for incompressible flow to flow at higher velocities where the effects of compressibility are significant. The appearance of shock phenomena prevents such conversion above the critical Mach number, but below this there is a considerable range of engineering interest, covering free stream velocities between 20 and (roughly) 80% of the velocity of sound. The higher figure is the critical Mach number and has been found to depend on the thickness of the airfoil, being larger for thinner airfoils. The method of con-

---

[34] Calculations show the major part of this increase to be due more to the separation than to the energy dissipated in the shock waves.

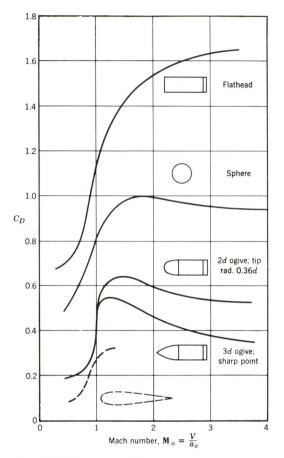

Fig. **13.27** Drag coefficients for bodies of revolution.[35]

version depends only on the Mach number and is known as the *Prandtl–Glauert rule*; the critical Mach number (the limit of application of the rule) may be established fairly reliably by application of elementary principles of compressible flow.

Pressure distribution data obtained for an airfoil at a certain angle of attack in incompressible flow are usually presented in dimensionless fashion as a plot of *pressure coefficient*, C_p, against the chord of the airfoil. The pressure coefficient is defined by

$$C_p = \frac{p - p_o}{\rho_o V_o^2/2} \tag{13.32}$$

[35] Data from: F. R. W. Hunt, *The Mechanical Properties of Fluids*, p. 341, Blackie and Son, 1925; and A. C. Charters and R. N. Thomas, "The Aerodynamic Performance of Small Spheres from Subsonic to High Supersonic Velocities," *Jour. Aero. Sci.*, vol. 12, no. 4, October, 1945.

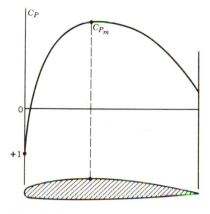

Fig. **13.28**

in which p_o, V_o, and ρ_o are at free stream conditions and p is the pressure at any point on the airfoil. A typical plot of pressure coefficient is shown on Fig. 13.28, on which it is customary to plot C_p positive downward. The Prandtl–Glauert rule states[36] that such data may be corrected for use at higher Mach numbers by simply dividing C_p by $\sqrt{1 - \mathbf{M}_o^2}$, providing the critical Mach number for the wing is not exceeded. Space does not permit proof of the Prandtl–Glauert rule here, but its limit of applicability, the critical Mach number, may be easily found. Referring to the discussion of Fig. 13.25a, ignoring the boundary layer, and using equation 5.14 for an ideal compressible fluid between the free stream and point m, which is the point of minimum pressure, p_m, and minimum pressure coefficient, C_{p_m},

$$\frac{V_m^2 - V_o^2}{2} = \frac{p_o}{\rho_o}\frac{k}{k - 1}\left[1 - \left(\frac{p_m}{p_o}\right)^{(k-1)/k}\right] \tag{5.14}$$

Shock phenomena begin to appear at m when V_m attains the *local sonic velocity* $\sqrt{kp_m/\rho_m}$. Substituting this for V_m, and a_o for $\sqrt{kp_o/\rho_o}$, this equation may be reduced to

$$\frac{p_m\rho_o}{\rho_m p_o} - \mathbf{M}_o^2 = \frac{2}{k - 1}\left[1 - \left(\frac{p_m}{p_o}\right)^{(k-1)/k}\right]$$

Substituting $(p_m/p_o)^{1/k}$ for ρ_o/ρ_m, and solving for p_m/p_o yields

$$\frac{p_m}{p_o} = \left[\frac{2 + (k - 1)\mathbf{M}_o^2}{k + 1}\right]^{k/(k-1)} \tag{13.33}$$

By dividing the numerator and denominator on the right-hand side of equation 13.32 by p_o and k, C_{p_m} becomes

[36] A more refined (and more complicated) rule is that of von Kármán and Tsien.

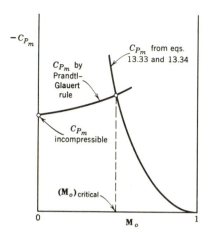

Fig. **13.29**

$$C_{Pm} = \frac{p_m - p_o}{\rho_o V_o^2/2} = \frac{2(p_m/p_o - 1)}{k\rho_o V_o^2/p_o k} = \frac{2(p_m/p_o - 1)}{k M_o^2} \qquad (13.34)$$

into which p_m/p_o from equation 13.33 may be inserted to yield the limiting C_p below which values obtained by the Prandtl–Glauert rule are invalid. The solution for C_{Pm} and the resulting (critical) M_o may be done by trial or by plotting, as indicated on Fig. 13.29.

13.13 Lift

Because of the steady decrease in pressure over the major portion of the upper side of an airfoil with increase of Mach number in the subsonic range (Fig. 13.25a), the lift coefficient at a given angle of attack will increase through this range. The Prandtl–Glauert rule shows that this may be computed approximately by dividing the lift coefficient of incompressible flow by $\sqrt{1 - M_o^2}$. The result may be seen on Fig. 13.30 and applies only up to the point of stall.

With increase of Mach number into the transonic range the lift coefficient continues to increase (for the same angle of attack) until the shock phenomena produce separation (Fig. 13.25b). The presence of the shock wave (through which there is a sudden rise in pressure) and the resulting separation produce a region of increased pressure on the upper side of the airfoil which will cause the lift coefficient to drop sharply. This so-called *shock-stall* usually occurs at Mach numbers slightly above the critical (Fig. 13.31).

The variation of the lift coefficient with Mach number through the transonic and into the supersonic range is pictured in Fig. 13.32 and related to the simplified flow pictures of Fig. 13.25. The first drop in lift coefficient brought about by the shock-stall is arrested by the formation of a shock wave and increase of pressure on the lower side

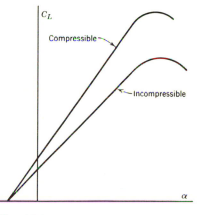

Fig. **13.30**

of the airfoil and by the change in position of the shock wave on the upper side, which reduces the region of separation and high pressure there. The lift coefficient then increases with Mach number to another maximum until intense shock wave phenomena become predominant, after which the trend is steadily downward with increasing Mach number.

Although analytical methods are unavailable for treating problems in the transonic range, simple expressions for C_L have been worked out for the subsonic and supersonic flows about thin symmetrical airfoils of infinite span and small angle of attack, which have been confirmed by experiment and may be used by engineers as rough guides. These expressions are

$$C_L = \frac{2\pi\alpha}{\sqrt{1 - \mathbf{M}_o^2}} \quad \text{and} \quad C_L = \frac{4\alpha}{\sqrt{\mathbf{M}_o^2 - 1}}$$

for the subsonic and supersonic cases, respectively. In spite of the tremendous advances

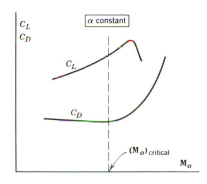

Fig. **13.31**

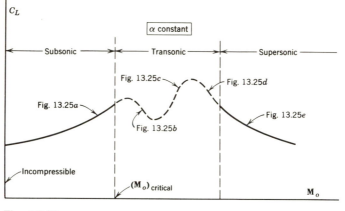

Fig. **13.32**

in theory, however, reliable results for airfoils of finite length and arbitrary profile cannot yet be predicted, and experimental data are considered the only safe basis for design.

References

Carafoli, E. 1956. *High-speed aerodynamics*. New York: Pergamon Press.

Durand, W. F., ed. 1963. *Aerodynamic theory*. New York: Dover Publications. 6 Vols.

Glauert, H. 1932. *Aerofoil and airscrew theory*. Cambridge: Cambridge Univ. Press.

Hayes, W. D., and Probstein, R. F. 1959. *Hypersonic flow theory*. New York: Academic Press.

Hoerner, S. F. 1965. *Fluid-dynamic drag*. Published by the author.

Kovasznay, L. S. G. 1970. The turbulent boundary layer. *Annual Review of Fluid Mechanics* 2: 95–112.

Kuethe, A. M., and Schetzer, J. D. 1959. *Foundations of aerodynamics*. 2nd ed. New York: Wiley.

Liepmann, H. W., and Roshko, A. 1957. *Elements of gasdynamics*. New York: Wiley.

Loitsianskii, L. G. 1970. The development of boundary-layer theory in the U.S.S.R. *Annual Review of Fluid Mechanics* 2: 1–14.

Nieuwland, G. Y., and Spee, B. M. 1973. Transonic airfoils: recent developments in theory, experiment, and design. *Annual Review of Fluid Mechanics* 5: 119–150.

Prandtl, L., and Tietjens, O. G. 1934. *Fundamentals of hydro- and aeromechanics; Applied hydro- and aeromechanics*. New York: McGraw-Hill.

Schlichting, H. 1979. *Boundary-layer theory*. 7th ed. New York: McGraw-Hill.

Shapiro, A. H. 1953. *The dynamics and thermodynamics of compressible fluid flow*. Vol. II. New York: Ronald Press.

Sutton, O. G. 1949. *The science of flight*. New York: Penguin Books.

Thwaites, B., ed. 1960. *Incompressible aerodynamics*. New York: Oxford Univ. Press.

Films

Abernathy, F. H. Fundamentals of boundary layers. NCFMF/EDC Film No. 21623, Encyclopaedia Britannica Educ. Corp.

Hazen, D. Boundary layer control. NCFMF/EDC Film No. 21614, Encyclopaedia Britannica Educ. Corp.

Shapiro, A. H. The fluid dynamics of drag. NCFMF/EDC Film Nos. 21601 through 21604, Encyclopaedia Britannica Educ. Corp.

Taylor, G. Low-Reynolds-number flows. NCFMF/EDC Film No. 21617, Encyclopaedia Britannica Educ. Corp.

Problems

13.1. A rectangular airfoil of 40 ft span and 6 ft chord has lift and drag coefficients of 0.5 and 0.04, respectively, at an angle of attack of 6°. Calculate the drag and horsepower necessary to drive this airfoil at 50, 100, and 150 mph horizontally through still air (40°F and 13.5 psia). What lift forces are obtained at these speeds?

13.2. A rectangular airfoil of 9 m span and 1.8 m chord moves horizontally at a certain angle of attack through still air at 240 km/h. Calculate the lift and drag, and the power necessary to drive the airfoil at this speed through air of (a) 101.3 kPa and 15°C, and (b) 79.3 kPa and −18°C. $C_D = 0.035$; $C_L = 0.46$. Calculate the speed and power required for condition (b) to obtain the lift of condition (a).

13.3. If $C_L = 1.0$ and $C_D = 0.05$ for an airfoil, then find the span needed for a rectangular wing of 30 ft (10 m) chord to lift 800 000 lb (3 560 kN) at a take-off speed of 175 mph (282 km/h). What is the wing drag at take-off?

13.4. The drag coefficient of a circular disk when placed normal to the flow is 1.12. Calculate the force and power necessary to drive a 12 in. (0.3 m) disk at 30 mph (48 km/h) through (a) standard air at sea level (Appendix 4), and (b) water.

13.5. The drag coefficient of a blimp is 0.04 when the area used in the drag formula is the two-thirds power of the volume. Calculate the drag of a blimp of this shape having a volume of 14 000m³ when moving at 100 km/h through still standard air at sea level (Appendix 4).

13.6. A wing model of 5 in. chord and 2.5 ft span is tested at a certain angle of attack in a wind tunnel at 60 mph using air at 14.5 psia and 70°F. The lift and drag are found to be 6.0 lb and 0.4 lb, respectively. Calculate the lift and drag coefficient for the model at this angle of attack.

13.7. A wing of 46.5 m² plan form area is to produce a lift of 44.5 kN in level flight through standard air at sea level (Appendix 4) at 402 km/h by expending 450 kW. What C_L and C_D are required?

13.8. A steel sphere of 0.25 in. diameter is fired at a velocity of 2 000 ft/s at an altitude of 30 000 ft in the U.S. Standard Atmosphere (Appendix 4). Calculate the drag force on this sphere.

13.9. In an incompressible boundary-layer flow, the velocity profile is found to be $u/V_o = (y/\delta)^{1/5}$. Find δ_1 and δ_2 as functions of δ.

13.10.   Prove that, if $u/V_o = (y/\delta)^{1/m}$ in incompressible flow in a boundary layer, then $\delta_2/\delta = m/[(m + 1)(m + 2)]$; find δ_1/δ also.

13.11.   For flow of problem 13.10, find an expression for c_f in the zero pressure-gradient case.

13.12.   Use the momentum integral equation to derive expressions for δ_1, δ_2, and c_f for flow in an incompressible boundary layer with the linear velocity profile $u/V_o = y/\delta$. How do the results differ from those obtained in Section 13.4?

13.13.   Repeat problem 13.12 for a velocity profile $u/V_o = \sin(\pi y/2\delta)$.

13.14.   A smooth plate 3 m long and 0.9 m wide moves through still sea level air (Appendix 4) at 4.5 m/s. Assuming the boundary layer to be wholly laminar, calculate (a) the thickness of the layer at 0.5, 1.0, 1.5, 2.0, 2.5, and 3.0 m from the leading edge of the plate; (b) the shear stress, τ_o, at those points; and (c) the total drag force on one side of the plate. (d) Calculate the thickness at the above points if the layer is turbulent. (e) Calculate the total drag for the turbulent boundary layer. (f) What percentage saving in drag is effected by a laminar boundary layer?

13.15.   A smooth flat plate 2.4 m long and 0.6 m wide is placed in an airstream (101.3 kPa and 15°C) of velocity 9 m/s. Calculate the total drag force on this plate (2 sides) if the boundary layer at the trailing edge is (a) laminar, (b) transition, and (c) turbulent.

13.16.   A flat-bottomed barge having a 150 ft by 20 ft bottom is towed through still water (60°F) at 10 mph. What is the frictional drag force exerted by the water on the bottom of the barge? How long could the laminar portion of the boundary layer be, using a critical Reynolds number of 537 000? What is the thickness of the laminar layer at its downstream end? What is the approximate thickness of the boundary layer at the rear end of the bottom of the barge?

13.17.   A streamlined train 400 ft (120 m) long is to travel at 90 mph (145 km/h). Treating the sides and top of the train as a smooth flat plate 30 ft (9 m) wide, calculate the total drag on these surfaces when the train moves through still air at sea level (Appendix 4). Calculate the possible length of the laminar boundary layer and the thickness of this layer at its downstream end. What is the thickness of the boundary layer at the rear end of the train? What power must be expended to overcome this resistance?

13.18.   Estimate the frictional drag encountered by the hull of a submarine when traveling deeply submerged at a speed of 33 km/h. The length of the hull is 60 m and its surface area 1 700 m². Assume water density 1 026 kg/m³ and kinematic viscosity 1.2×10^{-6} m²/s.

13.19.   The two rectangular smooth flat plates are to have the same drag in the same fluid stream. Calculate the required x. If the two plates are combined into the T-shape indicated, what ratio exists between the drag of the combination and that of either one? Assume laminar boundary layers in all calculations.

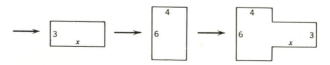

Problem **13.19**

13.20. If at the trailing edge of a smooth flat plate 10 ft (3 m) long, covered by a laminar boundary layer, the shearing stress is 0.000 01 lb/ft^2 (0.000 48 Pa), what is the total drag force on one side of the plate? What is the shearing stress at a point halfway between leading and trailing edges? What are the drag forces on the front and rear halves of one side of the plate?

13.21. A smooth flat plate 2.5 ft (0.76 m) long is immersed in water at 68°F (20°C) flowing at an undisturbed velocity of 2 ft/s (0.61 m/s). Calculate the shear stress at the center of the plate, assuming a laminar boundary layer over the whole plate.

13.22. A fluid stream of uniform velocity 3 m/s approaches a flat plate 30 m long which is parallel to the oncoming flow. The Reynolds number calculated with the full length of the plate is 400 000. How much space (m^2) is occupied by the boundary layer?

13.23. A boundary layer on a flat plate immersed in water increases in thickness from 0.152 m to 0.155 m in a distance of 0.3 m. The velocity of the undisturbed flow is 7.6 m/s. The flow in the boundary layer is not laminar. If there is a (fictitious) linear velocity distribution in the boundary layer, calculate the mean local shearing stress on this section of the plate.

13.24. The velocity profile through the boundary layer at the downstream end of a flat plate is found to conform to the equation $v/V_o = (y/\delta)^{1/8}$ in which V_o is 20 ft/s (6.1 m/s) and δ is 1 ft (0.3 m). Calculate the drag force exerted on (one side of) this plate if the fluid density is 2.0 slugs/ft^3 (1 031 kg/m^3).

13.25. The velocity distributions upstream and downstream from the transition region of a flat plate boundary layer are as indicated. Calculate the total drag force on the plate length AB.

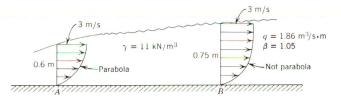

Problem **13.25**

13.26. Show that for a laminar boundary layer on a flat plate the kinetic energy lost between free stream and any point in the boundary layer is 31.5% of the total kinetic energy of a portion of the free stream containing the flowrate in the boundary layer at the above point. How may this loss of kinetic energy be accounted for?

13.27. Accurate values of C_f from Fig 13.9 are 0.001 85, 0.001 81, and 0.001 99 for Reynolds numbers 514 000, 537 000, and 560 000, respectively, which seem to indicate that the local wall shear is larger at the beginning of transition than at the end of the laminar boundary layer. From these figures what percentage increase in mean wall shear is indicated for the reaches implied by the Reynolds numbers?

13.28. A smooth flat plate 20 ft long moves through a fluid of kinematic viscosity 0.000 1 ft^2/s and s.g. 0.80. When $\mathbf{R}_x$ is 537 000 the shear stress on the plate is 0.067 lb/ft^2. Estimate the total drag (per foot of width) on one side of the plate.

13.29.   A javelin is about 2.6 m long and 50 mm in diameter. By treating its surface as a flat plate of length 2.6 m and $50\,\pi$ mm width, estimate the friction drag as a function of speed in sea level air. Plot the result. Up to what speed does the boundary layer remain laminar? What is the drag at 15 m/s?

13.30.   A steel sphere (s.g. = 7.8) of 13 mm diameter falls at a constant velocity of 0.06 m/s through an oil (s.g. = 0.90). Calculate the viscosity of the oil, assuming that the fall occurs in a large tank.

13.31.   What constant speed will be attained by a lead (s.g. = 11.4) sphere of 0.5 in. diameter falling freely through an oil of kinematic viscosity 0.12 ft^2/s and s.g. 0.95, if the fall occurs in a large tank?

13.32.   Assuming a critical Reynolds number of 0.1, calculate the approximate diameter of the largest air bubble which will obey Stokes' law while rising through a large tank of oil of viscosity 0.004 lb · s/ft^2 (0.19 Pa · s) and s.g. 0.90.

13.33.   Glass spheres of 0.1 in. diameter fall at constant velocities of 0.1 and 0.05 ft/s through two different oils (of the same specific gravity) in very large tanks. If the viscosity of the first oil is 0.002 lb · s/ft^2, what is the viscosity of the second?

13.34.   Calculate the drag of a smooth sphere of 12 in. or 0.3 m diameter in a stream of standard sea level air (Appendix 4) at Reynolds numbers of 1, 10, 100, and 1 000.

13.35.   Calculate the drag of a smooth sphere of 0.5 m diameter when placed in an airstream (15°C and 101.3 kPa) if the velocity is (*a*) 6 m/s, and (*b*) 8.4 m/s. At what velocity will the sphere attain the same drag which it had at a velocity of 6 m/s?

13.36.   Estimate the drag on a model of an N.P.L. airship hull of 0.15 m diameter which is to be tested in an airstream (101.3 kPa and 15°C) at 27 m/s.

13.37.   A modern submarine is 252 ft (77 m) long. It is shaped much like a blimp, with a circular cross section and a length-diameter ratio of 7.6 to 1. Compute and plot a curve of power versus speed in seawater (s.g. = 1.03) if $C_D \approx 0.08$.

13.38.   A sphere of 10 in. diameter is tested in a wind tunnel with standard sea level air (Appendix 4) at 80 mph. At what speed must a 2 in. sphere be towed in water (68°F) for these spheres to have the same drag coefficients? What are the drag forces on these two spheres?

13.39.   A steel sphere (s.g. 7.82) of 51 mm diameter is released in a large tank of oil (s.g. 0.82, viscosity 0.96 Pa · s). Calculate the terminal velocity of this sphere.

13.40.   A cylindrical chimney 0.9 m in diameter and 22.5 m high is exposed to a 56 km/h wind (15°C and 101.3 kPa); estimate the bending moment at the bottom of the chimney. Neglect end effects.

13.41.   The drag force exerted on an object (having a volume of 0.028 m^3 and s.g. of 2.80) is 445 N when moving at 6 m/s through oil (s.g. 0.90). If this object is allowed to fall through the same oil, is its terminal velocity larger or smaller than 6 m/s?

13.42.   A standard marine torpedo is 0.533 m in diameter and about 7.2 m long. Make an engineering estimate of the power required to drive this torpedo at 80 km/h through freshwater at 20°C. Assume hemispherical nose, cylindrical body, and flat tail. C_D for a solid hemisphere (flat side downstream) is about 0.35.

13.43. This thin smooth wing lands at a speed of 200 km/h in air of weight density 12.0 N/m³ and kinematic viscosity 1.4×10^{-5} m²/s. A braking parachute is released to slow it down. Calculate the approximate diameter of the parachute required to produce an extra drag equal to the wing drag at this speed. Assume flow about the wing two-dimensional.

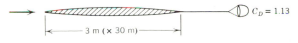

$C_D = 1.13$

3 m (× 30 m)

Problem **13.43**

13.44. A thin circular disk is placed (normal to the flow) in an airstream of velocity 152 m/s, absolute pressure 138 kPa, and density 1.29 kg/m³. If the pressure on the upstream side of the disk may be assumed to vary elliptically from stagnation pressure at the center to p_x at the edges, estimate the magnitude of p_x if this pressure is also exerted over the downstream side of the disk.

13.45. A large truck has an essentially boxlike body that causes flow separation at the front edges of the cab at any speed. The drag is mostly profile drag and $C_D = 0.75$. If the projected frontal area of the truck is 6 m² (65 ft²), determine and plot as a function of speed between zero and the legal limit the power that must be delivered to the road to propel the truck.

13.46. A guy wire on a 300 m high television transmitting antenna has a 25 mm diameter. In a wind of 3 m/s, determine the drag load on the 400 m long wire and the frequency of vortex shedding. The natural frequency of vibration of a wire varies according to Mersenne's law: $n = (T/\rho)^{1/2}/2l$ where T is the wire tension, $\rho = 4$ kg/m, and l is the wire length. At what value of T is the natural vibration of the wire in resonance with the vortex shedding? Let $\nu_{air} = 1.6 \times 10^{-5}$ m²/s. If the allowable stress in the steel wire is 575 000 kPa, can T ever be great enough to produce resonance?

13.47. Determine above what wind velocity the disturbance in the air induced by vortex shedding from telephone wires of 2 mm (0.08 in.) diameter can be heard. Humans can receive sound in the range of 20 to 20 000 Hz. What is the highest frequency tone generated within the range of **R** for regular vortex shedding?

13.48. If Γ is the mean circulation about a wing per foot or metre of span, calculate the circulation about the wing at midpoint and quarter-points of the span, assuming a semielliptical lift distribution.

13.49. Derive a general expression for lift coefficient in terms of circulation.

13.50. An airfoil of 1.5 m chord and 9 m span develops a lift of 14 kN when moving through air of specific weight 12.0 N/m³ at a velocity of 160 km/h. What is the mean circulation about the wing?

13.51. If the mean velocity adjacent to the top of a wing 1.8 m chord is 40 m/s and that adjacent to the bottom of the wing 31 m/s when the wing moves through still air (11.8 N/m³) at 33.5 m/s, estimate the lift per metre of span.

13.52. A model wing of 5 in. chord and 3 ft span is tested in a wind tunnel (60°F and 14.5 psia) at 60 mph, and the lift and drag are found to be 9.00 and 0.460 lb, respectively, at an angle of attack of 6.7°. Assuming a semielliptical lift distribution, calculate (a) the lift and drag coefficients, (b) C_{D_i}, (c) C_{D_o}, (d) the corresponding angle of attack for an airfoil of infinite span, (e) the corresponding angle of attack for a foil of this type with aspect ratio 5, and (f) the lift and drag coefficients at this aspect ratio.

13.53. An airfoil of infinite span has lift and drag coefficients of 1.31 and 0.062, respectively, at an angle of attack of 7.3°. Assuming semielliptical lift distribution, what will be the corresponding coefficients for an airfoil of the same profile but aspect ratio 6? What will be the corresponding angle of attack?

13.54. From the data of Fig. 13.21, calculate the lift and drag coefficients for a Clark-Y airfoil of aspect ratio 8, and plot the polar diagram for this airfoil.

13.55. The Clark-Y airfoil of Fig. 13.21 is to move at 180 mph (290 km/h) through standard sea level air (Appendix 4). Determine the minimum drag, drag at optimum L/D, and drag at point of maximum lift. Calculate the lift at these points and the power that must be expended to obtain these lifts.

13.56. Using Fig. 13.9 and assuming a turbulent boundary layer, what approximate percentage of the total drag (at zero lift) can be attributed to skin friction for the Clark-Y airfoil of Fig. 13.21?

13.57. A jet passes overhead at an altitude of 1 000 ft (300 m) but the sound is not heard until 0.5 s later. What is the jet's Mach number?

13.58. A supersonic airliner (1 800 mph or 2 900 km/h) passes a jumbo jet going in the opposite direction at 400 mph or 645km/h at a distance of 1 mile or 1.61 km. When can the jumbo jet be expected to feel the shock wave?

13.59. If the pointed artillery projectile of Fig. 13.27 is of 12 in. (0.3 m) diameter and is moving at 2 000 ft/s (600 m/s) through standard sea level air (Appendix 4), what drag force is exerted on it?

13.60. What is the drag of the blunt-nosed projectile of Fig. 13.27 (if its diameter is 3 in.) when it travels at (a) 700 mph, and (b) 800 mph, through standard sea level air (Appendix 4)?

13.61. An airfoil travels through standard sea level air (Appendix 4) at 500 mph (805 km/h). Calculate the local pressure and velocity on the airfoil at which shock wave phenomena will be expected.

13.62. The minimum pressure coefficient for an airfoil at a certain angle of attack in incompressible flow is −0.7. Predict the critical Mach number ($\mathbf{M}_o$) for this airfoil at this angle of attack.

13.63. Plot lift coefficient against angle of attack for the Clark-Y airfoil of Fig. 13.21 for free stream Mach numbers of 0 and 0.4.

13.64. A thin airfoil at a small angle of attack produces a certain lift force at a free stream Mach number of 0.90. At what Mach number(s) will the same airfoil at the same angle of attack in the same air stream produce the same lift?

Appendix 1
SYMBOLS, UNITS AND DIMENSIONS

Units and Dimensions

Symbol	Quantity	SI Units	FSS Units	Mass-Length-Time Dimensions[a]
A	Area	m^2	ft^2	L^2
a	Wave velocity	m/s	ft/s	L/t
a	Linear acceleration	m/s^2	ft/s^2	L/t^2
B	Bottom width (open channel)	m	ft	L
b	Surface width (open channel)	m	ft	L
b	Span of airfoil	m	ft	L
C	Chezy coefficient	$m^{1/2}/s$	$ft^{1/2}/s$	$L^{1/2}/t$
c_p	Specific heat at constant pressure	$J/kg \cdot K$	$ft \cdot lb/slug \cdot °R$	L^2/Tt^2
c_v	Specific heat at constant volume	$J/kg \cdot K$	$ft \cdot lb/slug \cdot °R$	L^2/Tt^2
D	Drag force	N	lb	ML/t^2
d	Diameter	m	ft	L
E	Modulus of elasticity	$Pa(N/m^2)$	lb/ft^2	M/Lt^2
E	Unit energy	J/N	$ft \cdot lb/lb$	L
e	Size of roughness	m	ft	L
F	Force	$N(kg \cdot m/s^2)$	$lb(slug \cdot ft/s^2)$	ML/t^2
G	Weight flowrate	N/s	lb/s	ML/t^3
g_n	Gravitational acceleration	m/s^2	ft/s^2	L/t^2
H	Energy per unit mass	m^2/s^2	ft^2/s^2	L^2/t^2
H	Head on weir	m	ft	L
H	Total head	m	ft	L
h	Enthalpy	J/kg	$ft \cdot lb/slug$	L^2/t^2
h	Head or height	m	ft	L
I	Second moment of an area	m^4	ft^4	L^4
ie	Internal energy	J/kg	$ft \cdot lb/slug$	L^2/t^2
K	Coefficient of permeability	m/s	ft/s	L/t
L	Lift force	N	lb	ML/t^2
ℓ	Length	m	ft	L

Symbol	Quantity	SI Units	FSS Units	Mass-Length-Time Dimensions[a]
M	Mass	kg	slug	M
M	Moment	N·m	ft·lb	ML^2/t^2
$\dot{m}$	Mass flowrate	kg/s	slug/s	M/t
P	Power	W(J/s)	ft·lb/s	ML^2/t^3
P	Perimeter	m	ft	L
P	Weir height	m	ft	L
p	Pressure	Pa(N/m²)	lb/ft²(psf)	M/Lt^2
Q	Flowrate	m³/s	ft³/s(cfs)	L^3/t
q	Two-dimensional flowrate	m²/s	ft²/s	L^2/t
R	Radius	m	ft	L
R	Engineering gas constant	J/kg·K	ft·lb/slug·°R	L^2/Tt^2
R_h	Hydraulic radius	m	ft	L
r	Radial distance	m	ft	L
T	Torque	N·m	ft·lb	ML^2/t^2
T	Absolute temperature	K	°R	T
T	Temperature	°C	°F	T
t	Time (seconds)	s	s	t
u	Velocity	m/s	ft/s	L/t
V	Velocity	m/s	ft/s	L/t
V	Volume	m³	ft³	L^3
v	Velocity	m/s	ft/s	L/t
W	Weight	N	lb	ML/t^2
w	Velocity	m/s	ft/s	L/t
y	Depth (open channel)	m	ft	L
y	Distance from solid boundary	m	ft	L
z	Height above datum	m	ft	L
Γ	Circulation	m²/s	ft²/s	L^2/t
γ	Specific weight	N/m³	lb/ft³	M/L^2t^2
δ	Boundary layer thickness	m	ft	L
δ_1	Displacement thickness	m	ft	L
δ_2	Momentum thickness	m	ft	L
δ_v	Viscous sublayer thickness	m	ft	L
ϵ	Eddy viscosity	Pa·s(N·s/m²)	lb·s/ft²	M/Lt
μ	Viscosity	Pa·s(N·s/m²)	lb·s/ft²	M/Lt
ν	Kinematic viscosity	m²/s	ft²/s	L^2/t
ξ	Vorticity	s⁻¹	s⁻¹	$1/t$
ρ	Density	kg/m³	slug/m³	M/L^3
σ	Surface tension	N/m	lb/ft	M/t^2
τ	Shear stress	Pa	lb/ft²	M/Lt^2
ϕ	Velocity potential	m²/s	ft²/s	L^2/t
ψ	Stream function	m²/s	ft²/s	L^2/t
ω	Angular velocity	s⁻¹	s⁻¹	$1/t$

[a] Mass (M), length (L), time (t), and thermodynamic temperature (T).

Symbols for Dimensionless Quantities

Symbol	Quantity	Symbol	Quantity
C	Cauchy number	n	Polytropic exponent; Manning's
C_c	Coefficient of contraction		coefficient
C_D	Drag coefficient	**R**	Reynolds number
C_f	Frictional drag coefficient	S	Slope of energy line
C_L	Lift coefficient	S_o	Bottom slope (open channel)
C_p	Pressure coefficient	S_c	Critical slope
C_v	Coefficient of velocity	**W**	Weber number
C_w	Weir coefficient	Y	Expansion factor
E	Euler number	α	Kinetic energy correction factor
F	Froude number	β	Momentum correction factor
f	Friction factor	η	Efficiency
K_L	Loss coefficient	κ	von Kármán's turbulence
k	Adiabatic exponent		"constant"
M	Mach number	Π	Dimensionless group

Conversion Factor Table

Abbreviations used:

BTU = British Thermal Unit
cfs = cubic feet per second
ft/s = feet per second
ft = foot
gpm = gallons per minute
hp = horsepower
h = hour
Hz = hertz
in. = inch
J = joule = N·m
kg = kilogram = 10^3 gram
lb = pound force

m = metre (SI) = mile (FSS)
mb = millibar = 10^{-3} bar
mm = millimeter = 10^{-3} meter
mm^2 = square millimeter
mph = miles per hour
m/s = metres per second
N = newton
Pa = pascal = N/m^2
psi = pound per square inch
s = second
W = watt = J/s

Absolute viscosity: 1 Pa·s = 10 poises = 0.020 89 lb·s/ft^2
Acceleration due to gravity: 9.806 65 m/s^2 = 32.174 ft/s^2
Area: 1m^2 = 10.76 ft^2; 1 mm^2 = 0.001 55 in.^{2}
Density: 1 kg/m^3 = 0.001 94 slug/ft^3
Energy: 1 N·m = 1 J = 0.737 5 ft·lb
Flowrate: 1 m^3/s = 35.31 ft^3/s; 1 litre/s = 10^{-3} m^3/s = 0.022 83 mgd
Force: 1 N = 0.224 8 lb
Kinematic viscosity: 1m^2/s = 10^4 Stokes = 10.76 ft^2/s
Length: 1 mm = 0.039 4 in.; 1 m = 3.281 ft; 1 km = 0.622 miles
Mass: 1 kg = 0.068 5 slug
Power: 1 W = 1 J/s = 0.737 5 ft·lb/s; 1 kW = 1.341 hp = 737.5 ft·lb/s

Pressure: 1 kN/m^2 = 1 kPa = 0.145 psi; 1 mm Hg = 0.039 4 in. Hg = 133.3 Pa;
   1 mm H$_2$O = 9.807 Pa; 101.325 kPa = 760 mm Hg = 29.92 in. Hg = 14.70 psi;
   1 bar = 100 kPa = 14.504 psi
Specific Heat; Engineering Gas Constant: 1 J/kg·K = 5.98 ft·lb/slug·°R
Specific Weight: 1 N/m^3 = 0.006 365 lb/ft^3
Temperature: 1°C = 1 K = 1.8°F = 1.8°R (see section 1.3)
Velocity: 1 m/s = 3.281 ft/s = 3.60 km/h = 2.28 mph; 1 knot = 0.515 5 m/s
Volume: 1 m^3 = 10^3 litres = 35.31 ft^3; 1 U.S. gallon = 3.785 litres;
   1 U.K. gallon = 4.546 litres

Appendix 2
PHYSICAL PROPERTIES OF WATER

FSS Units[f]

Temperature, °F	Specific Weight,[a] γ, lb/ft^3	Density,[a] ρ, slug/ft^3	Modulus of Elasticity,[b,c] $E/10^3$, psi	Viscosity,[a] $\mu \times 10^5$, lb·s/ft^2	Kinematic Viscosity,[a] $\nu \times 10^5$, ft^2/s	Surface Tension,[a,d] σ, lb/ft	Vapor Pressure,[e] p_v, psia
32	62.42	1.940	287	3.746	1.931	0.005 18	0.09
40	62.43	1.940	296	3.229	1.664	0.006 14	0.12
50	62.41	1.940	305	2.735	1.410	0.005 09	0.18
60	62.37	1.938	313	2.359	1.217	0.005 04	0.26
70	62.30	1.936	319	2.050	1.059	0.004 98	0.36
80	62.22	1.934	324	1.799	0.930	0.004 92	0.51
90	62.11	1.931	328	1.595	0.826	0.004 86	0.70
100	62.00	1.927	331	1.424	0.739	0.004 80	0.95
110	61.86	1.923	332	1.284	0.667	0.004 73	1.27
120	61.71	1.918	332	1.168	0.609	0.004 67	1.69
130	61.55	1.913	331	1.069	0.558	0.004 60	2.22
140	61.38	1.908	330	0.981	0.514	0.004 54	2.89
150	61.20	1.902	328	0.905	0.476	0.004 47	3.72
160	61.00	1.896	326	0.838	0.442	0.004 41	4.74
170	60.80	1.890	322	0.780	0.413	0.004 34	5.99
180	60.58	1.883	318	0.726	0.385	0.004 27	7.51
190	60.36	1.876	313	0.678	0.362	0.004 20	9.34
200	60.12	1.868	308	0.637	0.341	0.004 13	11.52
212	59.83	1.860	300	0.593	0.319	0.004 04	14.70

[a] From "Hydraulic Models," *A.S.C.E. Manual of Engineering Practice*, No. 25, A.S.C.E., 1942. See footnote 1.
[b] Approximate values averaged from many sources.
[c] At atmospheric pressure. See footnote 1.
[d] In contact with air.
[e] From J. H. Keenan and F. G. Keyes, *Thermodynamic Properties of Steam*, John Wiley & Sons, 1936.
[f] Compiled from many sources including those indicated, *Handbook of Chemistry and Physics*, 54th Ed., The CRC Press, 1973, and *Handbook of Tables for Applied Engineering Science*, The Chemical Rubber Co., 1970.

[1] Here, if $E/10^3 = 287$, then $E = 287 \times 10^3$ psi; while if $\mu \times 10^5 = 3.746$, then $\mu = 3.746 \times 10^{-5}$ lb·s/ft^2, and so on.

S.I. Units[f]

Tem-pera-ture, $^\circ$C	Specific Weight,[a] γ, kN/m^3	Density,[a] ρ, kg/m^3	Modulus of Elasticity,[b,c] $E/10^6$, kPa	Viscosity,[a] $\mu \times 10^3$, Pa·s	Kinematic Viscosity,[a] $\nu \times 10^6$, m^2/s	Surface Tension,[a,d] σ, N/m	Vapor Pres-sure,[e] p_v, kPa
0	9.805	999.8	1.98	1.781	1.785	0.075 6	0.61
5	9.807	1 000.0	2.05	1.518	1.518	0.074 9	0.87
10	9.804	999.7	2.10	1.307	1.306	0.074 2	1.23
15	9.798	999.1	2.15	1.139	1.139	0.073 5	1.70
20	9.789	998.2	2.17	1.002	1.003	0.072 8	2.34
25	9.777	997.0	2.22	0.890	0.893	0.072 0	3.17
30	9.764	995.7	2.25	0.798	0.800	0.071 2	4.24
40	9.730	992.2	2.28	0.653	0.658	0.069 6	7.38
50	9.689	988.0	2.29	0.547	0.553	0.067 9	12.33
60	9.642	983.2	2.28	0.466	0.474	0.066 2	19.92
70	9.589	977.8	2.25	0.404	0.413	0.064 4	31.16
80	9.530	971.8	2.20	0.354	0.364	0.062 6	47.34
90	9.466	965.3	2.14	0.315	0.326	0.060 8	70.10
100	9.399	958.4	2.07	0.282	0.294	0.058 9	101.33

[a] From "Hydraulic Models," *A.S.C.E. Manual of Engineering Practice,* No. 25, A.S.C.E., 1942. See footnote 2.

[b] Approximate values averaged from many sources.

[c] At atmospheric pressure. See footnote 2.

[d] In contact with air.

[e] From J. H. Keenan and F. G. Keyes, *Thermodynamic Properties of Steam,* John Wiley & Sons, 1936.

[f] Compiled from many sources including those indicated, *Handbook of Chemistry and Physics,* 54th Ed., The CRC Press, 1973, and *Handbook of Tables for Applied Engineering Science,* The Chemical Rubber Co., 1970.

[2] Here, if $E/10^6 = 1.98$, then $E = 1.98 \times 10^6$ kPa, while if $\mu \times 10^3 = 1.781$, then $\mu = 1.781 \times 10^{-3}$ Pa·s, and so on.

Appendix 3
WAVE VELOCITIES

The velocity (or celerity) of small waves through fluids may be predicted by application of the continuity and impulse-momentum principles. For the incompressible (inelastic) fluid as in a rigid (inelastic) body, such waves would be expected to travel at infinite speeds as disturbances are transmitted instantaneously from point to point. For a fluid assumed incompressible, such as a liquid with a free surface, a small wave appears as a slight localized rise in the liquid surface; this type of wave is called a *gravity wave*. For a compressible fluid (liquid or gas), a small wave is featured by a slight localized rise in pressure and density of the fluid; such a wave is called a *wave of compression*. These waves are depicted in Fig. A.1 moving at velocity a on or through fluids at rest. To a stationary observer the fluid motion is unsteady as the wave passes, but if the observer moves with the wave at velocity a he sees a steady flow which may be easily analyzed with elementary principles. Taking V to be the velocity at the centers of the waves and applying the continuity principle (equations 3.14 and 3.8),

$$\text{Gravity wave:} \qquad ay = V(y + dy)$$

$$\text{Compression wave:} \qquad Aa\rho = AV(\rho + d\rho)$$

Applying the impulse-momentum principle (equation 6.4),

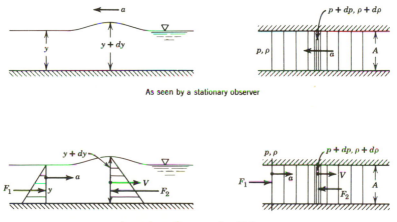

Fig. **A.1** Motion of small waves.

665

Gravity wave: $\quad \dfrac{\gamma}{2}y^2 - \dfrac{\gamma}{2}(y + dy)^2 = ay\dfrac{\gamma}{g_n}(V - a)$

Compression wave: $\quad pA - A(p + dp) = Aa\rho(V - a)$

Simultaneous solution of the appropriate equations with the elimination of V yields

Gravity wave: $\quad a^2 = g_n(y + dy)$

Compression wave: $\quad a^2 = \dfrac{dp}{d\rho}\left(\dfrac{\rho + d\rho}{\rho}\right)$

As dy and $d\rho$ approach zero (the basic condition for a vanishingly small wave) these equations reduce to

Gravity wave: $\quad a = \sqrt{g_n y}$

Compression wave: $\quad a = \sqrt{\dfrac{dp}{d\rho}}$

Appendix 4
THE U.S. STANDARD ATMOSPHERE

FSS Units[a]

Altitude, ft	Temperature, °F	Absolute Pressure, psia	Specific Weight, lb/ft³	Density, slug/ft³	Viscosity × 10⁷, lb · s/ft²
0	59.00	14.696	0.076 47	0.002 377	3.737
5 000	41.17	12.243	0.065 87	0.002 048	3.637
10 000	23.36	10.108	0.056 43	0.001 756	3.534
15 000	5.55	8.297	0.048 07	0.001 496	3.430
20 000	−12.26	6.759	0.040 69	0.001 267	3.325
25 000	−30.05	5.461	0.034 18	0.001 066	3.217
30 000	−47.83	4.373	0.028 57	0.000 891	3.107
35 000	−65.61	3.468	0.023 67	0.000 738	2.995
40 000	−69.70	2.730	0.018 82	0.000 587	2.969
45 000	−69.70	2.149	0.014 81	0.000 462	2.969
50 000	−69.70	1.690	0.011 65	0.000 364	2.969
55 000	−69.70	1.331	0.009 17	0.000 287	2.969
60 000	−69.70	1.049	0.007 22	0.000 226	2.969
65 000	−69.70	0.826	0.005 68	0.000 178	2.969
70 000	−67.42	0.651	0.004 45	0.000 139	2.984
75 000	−64.70	0.514	0.003 49	0.000 109	3.001
80 000	−61.98	0.404	0.002 63	0.000 086	3.018
85 000	−59.26	0.322	0.002 15	0.000 067	3.035
90 000	−56.54	0.255	0.001 70	0.000 053	3.052
95 000	−53.82	0.203	0.001 34	0.000 042	3.070
100 000	−51.10	0.162	0.001 06	0.000 033	3.087

[a] Data from *U.S. Standard Atmosphere, 1962,* U.S. Government Printing Office, 1962. Data agrees with ICAO standard atmosphere to 20 km and with ICAO proposed extension to 30 km. For atmospheric tables depicting conditions other than mid-latitude mean represented by standard atmosphere, see *U.S. Standard Atmosphere Supplements, 1966,* U.S. Government Printing Office, 1966.

667

S.I. Units[a]

Altitude, km	Temperature, °C	Absolute Pressure, kPa	Specific Weight, N/m³	Density, kg/m³	Viscosity $\times 10^5$, Pa·s
0	15.00	101.33	12.01	1.225	1.789
2	2.00	79.50	9.86	1.007	1.726
4	−4.49	70.12	8.02	0.909	1.661
6	−23.96	47.22	6.46	0.660	1.595
8	−36.94	35.65	5.14	0.526	1.527
10	−49.90	26.50	4.04	0.414	1.458
12	−56.50	19.40	3.05	0.312	1.422
14	−56.50	14.17	2.22	0.228	1.422
16	−56.50	10.35	1.62	0.166	1.422
18	−56.50	7.57	1.19	0.122	1.422
20	−56.50	5.53	0.87	0.089	1.422
22	−54.58	4.05	0.63	0.065	1.432
24	−52.59	2.97	0.46	0.047	1.443
26	−50.61	2.19	0.33	0.034	1.454
28	−48.62	1.62	0.24	0.025	1.465
30	−46.64	1.20	0.18	0.018	1.475

[a] Data from *U.S. Standard Atmosphere, 1962,* U.S. Government Printing Office, 1962. Data agrees with ICAO standard atmosphere to 20 km and with ICAO proposed extension to 30 km. For atmospheric tables depicting conditions other than mid-latitude mean represented by standard atmosphere, see *U.S. Standard Atmosphere Supplements, 1966,* U.S. Government Printing Office, 1966.

Appendix 5
BASIC MATHEMATICAL OPERATIONS

If F is some function of x, y, and z, written $F(x, y, z)$, the partial derivative $\partial F / \partial x$ is obtained by differentiating F with respect to x, while y and z are held constant. To obtain $\partial F / \partial y$, the differentiation is performed with respect to y with x and z held constant, and so forth. Thus, partial derivatives give the rates of change of F in each of the coordinate directions. The *gradient* of F gives the maximum rate of change of F and has a direction perpendicular to lines of equal value of F, that is, perpendicular to contours of F drawn in space.

Example

$$\text{Let } F = 2x^2 + 3xy + 4y^2 + 5z^2 + C$$

$$\frac{\partial F}{\partial x} = 4x + 3y$$

$$\frac{\partial F}{\partial y} = 3x + 8y$$

$$\frac{\partial F}{\partial z} = 10z$$

$$\frac{\partial^2 F}{\partial x \, \partial y} = 3 \qquad \frac{\partial^2 F}{\partial z^2} = 10$$

and so on.

A *scalar quantity* or *scalar* is a quantity that can be described by a single number. A scalar has only a *magnitude* (size). A *vector* is a directed quantity that has both a *magnitude* and a *direction*. Pressure, density, and temperature are scalars. Velocity and force are vectors. A vector $\mathbf{F}$ can be expressed in terms of its components along the axes of the Cartesian system (x, y, z), that is,

$$\mathbf{F} = F_x \mathbf{e}_x + F_y \mathbf{e}_y + F_z \mathbf{e}_z$$

where F_x, F_y, F_z are the projections of the magnitude of $\mathbf{F}$ on the x, y, z axes, respectively. The unit vectors (their magnitude is unity) $\mathbf{e}_x$, $\mathbf{e}_y$, $\mathbf{e}_z$ are directed along the mutually perpendicular coordinate axes (see Fig. A.2). The magnitude F of $\mathbf{F}$ is

$$F = |\mathbf{F}| = (F_x^2 + F_y^2 + F_z^2)^{1/2}$$

There are three important vector operations that may be new to the reader. The first

669

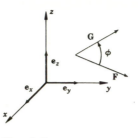

Fig. **A.2**

is the *scalar or dot product*

$$S = \mathbf{F} \cdot \mathbf{G} = |\mathbf{F}|\,|\mathbf{G}| \cos \phi$$

where ϕ is the angle between the vectors $\mathbf{F}$ and $\mathbf{G}$ and S is a scalar. If $\mathbf{n}$ is a unit vector normal (perpendicular) to some surface or contour, the component of $\mathbf{F}$ along $\mathbf{n}$ is

$$F_n = \mathbf{F} \cdot \mathbf{n} = |\mathbf{F}| \cos \phi$$

that is, the scalar or dot product gives the projection of $\mathbf{F}$ on $\mathbf{n}$ (recall $|\mathbf{n}| = 1$). (See Fig. A.3.)

The second operation is the *vector product*

$$\mathbf{V} = \mathbf{F} \times \mathbf{G}$$

where the magnitude of $\mathbf{V}$ is

$$|\mathbf{V}| = |\mathbf{F}|\,|\mathbf{G}| \sin \phi$$

and the direction of $\mathbf{V}$ is perpendicular to the plane of $\mathbf{F}$ and $\mathbf{G}$ and in accordance with the usual right-hand rule. (See Fig. A.4.)

Example

Calculate the moment $\mathbf{T}$ arising about 0 from the action of a force $\mathbf{F} = F\mathbf{e}_\theta$ at a point P shown in Fig. A.5. In polar cylindrical coordinates the unit vectors are $\mathbf{e}_r$, $\mathbf{e}_\theta$, $\mathbf{e}_z$. Thus,

$$\mathbf{T} = \mathbf{r} \times \mathbf{F} = r\mathbf{e}_r \times F\mathbf{e}_\theta = rF\mathbf{e}_r \times \mathbf{e}_\theta$$

but $\mathbf{e}_r \times \mathbf{e}_\theta = \mathbf{e}_z$ according to the definition of the vector product. Therefore,

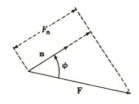

Fig. **A.3**

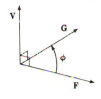

Fig. **A.4**

$$\mathbf{T} = T\mathbf{e}_z = rF\,\mathbf{e}_z$$

where $T = rF$ is the magnitude of $\mathbf{T}$.

The third operation is the derivative of vectors. For example,

$$\frac{\partial \mathbf{F}}{\partial s} = \frac{\partial F_x}{\partial s}\,\mathbf{e}_x + \frac{\partial F_y}{\partial s}\,\mathbf{e}_y + \frac{\partial F_z}{\partial s}\,\mathbf{e}_z$$

(The Cartesian system's unit vectors have fixed magnitude and direction.) The gradient of F is

$$\mathrm{grad}\ F = \frac{\partial F}{\partial x}\,\mathbf{e}_x + \frac{\partial F}{\partial y}\,\mathbf{e}_y + \frac{\partial F}{\partial z}\,\mathbf{e}_z = \nabla F$$

for a scalar $F(x, y, z)$. The divergence of $\mathbf{F}$ is defined as

$$\mathrm{div}\ \mathbf{F} = \nabla \cdot \mathbf{F} = \left(\frac{\partial}{\partial x}\,\mathbf{e}_x + \frac{\partial}{\partial y}\,\mathbf{e}_y + \frac{\partial}{\partial z}\,\mathbf{e}_z \right) \cdot \mathbf{F}$$

$$= \frac{\partial F_x}{\partial x}\,\mathbf{e}_x + \frac{\partial F_y}{\partial y}\,\mathbf{e}_y + \frac{\partial F_z}{\partial z}\,\mathbf{e}_z$$

It is easy to show (why not try it?) that

$$\mathrm{div}\ (\mathrm{grad}\ F) = \nabla \cdot \nabla F = \nabla^2 F$$

$$= \frac{\partial^2 F}{\partial x^2} + \frac{\partial^2 F}{\partial y^2} + \frac{\partial^2 F}{\partial z^2}$$

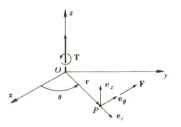

Fig. **A.5**

for the scalar $F(x, y, z)$. The Laplace equation $\nabla^2 F = 0$ is accordingly a partial differential equation that is important in analyses of ideal fluid flows.

Example

With

$$F = 2x^2 + 3xy + 4y^2 + 5z^2 + C$$

as before,

$$\text{div (grad } F) = 4 + 8 + 10 = 22 = \nabla^2 F$$

The total derivative of $F(x, y, z)$ is

$$dF = \frac{\partial F}{\partial x} dx + \frac{\partial F}{\partial y} dy + \frac{\partial F}{\partial z} dz$$

Example

Suppose the fluid density is given by $\rho(x, y, t)$, that is, is a function of space and time in a compressible flow. Then,

$$d\rho = \frac{\partial \rho}{\partial x} dx + \frac{\partial \rho}{\partial y} dy + \frac{\partial \rho}{\partial t} dt$$

and

$$\frac{d\rho}{dt} = \frac{\partial \rho}{\partial x}\frac{dx}{dt} + \frac{\partial \rho}{\partial y}\frac{dy}{dt} + \frac{\partial \rho}{\partial t}\frac{dt}{dt}$$

Because $u = dx/dt$ and $v = dy/dt$,

$$\frac{d\rho}{dt} = u\frac{\partial \rho}{\partial x} + v\frac{\partial \rho}{\partial y} + \frac{\partial \rho}{\partial t}$$

Often derivatives are known and the function is sought. A typical example (in polar coordinates) is to find $F(r, \theta)$ when

$$\frac{\partial F}{\partial r} = -\frac{\sin \theta}{r^2} \qquad \text{and} \qquad \frac{\partial F}{r\,\partial \theta} = \frac{\cos \theta}{r^2}$$

Here

$$dF = \left(\frac{\partial F}{r\,\partial \theta}\right) r\, d\theta + \left(\frac{\partial F}{\partial r}\right) dr$$

and the integral is

$$F = \int \frac{\cos \theta}{r^2} r\, d\theta - \int \frac{\sin \theta}{r^2} dr + C$$

However, $\cos \theta\, d\theta/r - \sin \theta\, dr/r^2$ may be recognized as $d[(\sin \theta)/r]$, so the solution is

$$F = \int d\left(\frac{\sin\theta}{r}\right) + C = \frac{\sin\theta}{r} + C$$

References

Kreyszig, E. 1979. *Advanced engineering mathematics*. 4th ed. New York: Wiley.
McQuistan, R. B. 1965. *Scalar and vector fields*. New York: Wiley.

Appendix 6
PROPERTIES OF AREAS AND VOLUMES

	Sketch	Area or Volume	Location of Centroid	I or I_c
Rectangle		bh	$y_c = \dfrac{h}{2}$	$I_c = \dfrac{bh^3}{12}$
Triangle		$\dfrac{bh}{2}$	$y_c = \dfrac{h}{3}$	$I_c = \dfrac{bh^3}{36}$
Circle		$\dfrac{\pi d^2}{4}$	$y_c = \dfrac{d}{2}$	$I_c = \dfrac{\pi d^4}{64}$
Semicircle [1]		$\dfrac{\pi d^2}{8}$	$y_c = \dfrac{4r}{3\pi}$	$I = \dfrac{\pi d^4}{128}$
Ellipse		$\dfrac{\pi bh}{4}$	$y_c = \dfrac{h}{2}$	$I_c = \dfrac{\pi bh^3}{64}$
Semiellipse		$\dfrac{\pi bh}{4}$	$y_c = \dfrac{4h}{3\pi}$	$I = \dfrac{\pi bh^3}{16}$
Parabola		$\tfrac{2}{3}bh$	$y_c = \dfrac{3h}{5}$ $x_c = \dfrac{3b}{8}$	$I = \dfrac{2bh^3}{7}$
Cylinder		$\dfrac{\pi d^2 h}{4}$	$y_c = \dfrac{h}{2}$	

	Sketch	Area or Volume	Location of Centroid	I or I_c
Cone		$\dfrac{1}{3}\left(\dfrac{\pi d^2 h}{4}\right)$	$y_c = \dfrac{h}{4}$	
Paraboloid of revolution		$\dfrac{1}{2}\left(\dfrac{\pi d^2 h}{4}\right)$	$y_c = \dfrac{h}{3}$	
Sphere		$\dfrac{\pi d^3}{6}$	$y_c = \dfrac{d}{2}$	
Hemisphere		$\dfrac{\pi d^3}{12}$	$y_c = \dfrac{3r}{8}$	

[1] For the quarter-circle, the respective values are $\pi d^2/16$, $4r/3\pi$, and $\pi d^4/256$.

Appendix 7
CAVITATION

The phenomenon of cavitation is of great importance in the design of high-speed hydraulic machinery such as turbines, pumps, and marine propellers, in the overflow and underflow structures of high dams, and in the high-speed motion of underwater bodies (such as submarines and hydrofoils). Cavitation also may be of critical significance in pipeline design and in certain problems of fluid metering. Typically considered a "water-problem," cavitation has, in recent years, become a concern in a range of liquid flows, including liquid metal and cryogenic fluid pumps, and in a variety of areas, including medicine and industrial cleaning.

Cavitation may be expected in a flowing liquid[1] wherever the local pressure falls to the vapor pressure of the liquid. Local vaporization of the liquid will then result, causing development of a vapor-filled bubble or cavity in the flow. Cavitation is often accompanied by erosion (pitting) of solid boundary surfaces in machines or on hydrofoils, losses of efficiency, and serious vibration problems.

The nature of cavitation may be most easily observed by study of the ideal flow of a liquid through a constriction in a passage (Fig. A.6). With the valve partially open, the variation of pressure head through passage and constriction is given by hydraulic grade line A, the point of lowest pressure occurring at the minimum area, where the velocity is highest. Increase of valve opening (causing larger flowrate) produces hydraulic grade line B, for which the absolute pressure in the throat of the constriction falls to the vapor pressure of the liquid, causing the *inception* of cavitation. Further opening of the valve *does not increase the flowrate* but serves to extend the zone of vapor pressure downstream from the throat of the constriction; here the live stream of liquid separates from the boundary walls, producing a cavity in which the mean pressure is the vapor pressure of the liquid. The cavity contains a swirling mass of droplets and vapor and, although appearing steady to the naked eye, actually forms and reforms many times a second. The formation and disappearance of a single cavity are shown schematically in Fig. A.7, and the disappearance of the cavity is the clue to the destructive action caused by cavitation. The low-pressure cavity is swept swiftly downstream into a region of high pressure where it collapses suddenly, the surrounding liquid rushing in to fill the void. At the point of disappearance of the cavity the inrushing liquid comes together, momentarily raising the local pressure within the liquid to a very high value. If the point of collapse of the cavity is in contact with the boundary wall, the wall receives a blow as from a tiny hammer, and its surface may be stressed locally beyond its elastic limit, resulting in fatigue, probably enhanced chemical corrosion, and eventual destruction of the wall material (Fig. A.8).

---

[1] Cavitation is not possible in a gas because of its capacity for expansion.

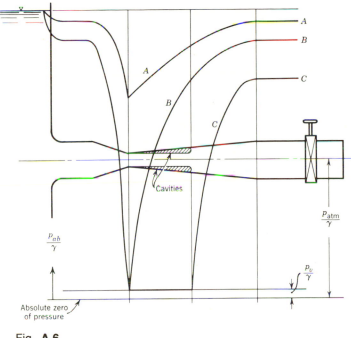

Fig. **A.6**

Another form of cavitation is the *steady-state cavity* frequently observed in the tip vortices (Section 13.8) of marine propellers or surrounding high-speed hydrofoils. In the model test in a water tunnel shown in Fig. A.9, the cavitation number (cf., equation 8.16)

$$\sigma = \frac{p_o - p_v}{\rho V_o^2}$$

is determined by the pressure p_o in the flow upstream from the propeller (to the right in the figure), the vapor pressure p_v existing in the cavities on the blades, the liquid density ρ, and the speed of advance V_o of the propeller. Because the propeller is fixed in the tunnel, V_o is the speed of the water in the tunnel upstream from the propeller. In the test shown, the propeller's effective-axial-speed V_o and r/min were held constant while p_o was reduced to increase cavitation. In Fig. A.9a, σ is relatively large and only a small amount of cavitation occurs on the low pressure side of the blades and in the tip vortices. In Figs. A.9a and A.9b, the blade cavities collapse directly on the blade.

In Fig. A.9c, the propeller is operating close to its design condition as a fully cavitating or supercavitating propeller, and σ is very small. Although the downstream ends of such cavities exhibit certain unsteady phenomena, the large portion of the cavity is steady with its outer boundary acting as a streamline of the flowfield. Such a streamline is a *free streamline* (Section 4.8), because the pressure along it will be that in the cavity and equal to the vapor pressure of the liquid. Here the engineering problem is the prediction of cavity form and location of the separation points; the latter may be

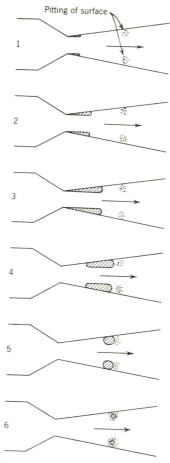

Fig. **A.7**

Before

After

M. I. T.

Fig. **A.8** Pitting of brass plate after 5 hours' exposure to cavitation (magnification 10×).

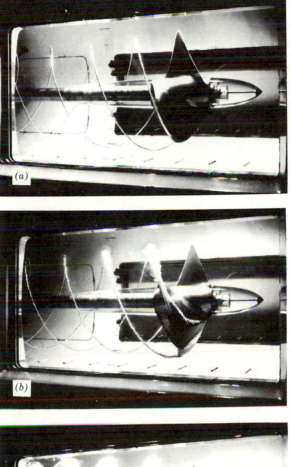

Fig. **A.9**   Cavitating propeller in a water tunnel. (Courtesy of National Physical Laboratory, Teddington, Middlesex, England; Crown Copyright reserved.)

critically dependent on boundary-layer growth (Section 13.6) and on the fine details of vaporization in the flow as the pressure falls to the vapor pressure of the liquid.

One of the usual objectives in the design of hydraulic machinery and structures is the prevention of cavitation, which the designer accomplishes by improved forms of boundary surfaces and by setting limits beyond which the machine or structure should not be operated. In the case of high-speed ships and modern, high-performance pumps, there is little hope of preventing large-scale cavitation on propellers and pump impellers. Then, it is the objective of the design to predict the exact nature of the cavitation and to control its extent so, for example, large steady-state cavities exist that do not damage the machinery. As a result fully cavitating propellers and pumps can be made to have predictable and reliable performance.

References

A.S.M.E. 1965. Symposium, *Cavitation in fluid machinery*.

Eisenberg, P., and Tulin, M. P. 1961. Cavitation. Sec. 12, *Handbook of fluid dynamics*. New York: McGraw-Hill.

IAHR, Institute of High Speed Mechanics (Sendai, Japan). 1962. Symposium, *Cavitation and hydraulic machinery*.

Knapp, R. T., Daily, J. W., and Hammitt, F. G. 1970. *Cavitation*. New York: McGraw-Hill.

Thomas, H. A., and Schuleen, E. P. 1942. Cavitation in outlet conduits of high dams. *Trans. A.S.C.E.* 107.

Film

Eisenberg, P. Cavitation. NCFMF/EDC Film No. 21620, Encyclopaedia Britannica Educ. Corp.

Appendix 8
THE EXPANSION FACTOR,[1] Y

$\frac{A_2}{A_1}$	k	$\frac{p_2}{p_1}$	0.95	0.90	0.85	0.80	0.75
0	1.40		0.973	0.945	0.916	0.886	0.856
	1.30		0.971	0.941	0.910	0.878	0.846
	1.20		0.968	0.936	0.903	0.869	0.834
0.2	1.40		0.971	0.942	0.912	0.881	0.850
	1.30		0.969	0.938	0.906	0.873	0.839
	1.20		0.967	0.933	0.899	0.863	0.827
0.3	1.40		0.969	0.938	0.907	0.875	0.842
	1.30		0.967	0.934	0.900	0.866	0.831
	1.20		0.965	0.929	0.862	0.856	0.819
0.4	1.40		0.966	0.932	0.899	0.864	0.830
	1.30		0.964	0.928	0.891	0.855	0.818
	1.20		0.961	0.922	0.886	0.844	0.805
0.5	1.40		0.962	0.923	0.886	0.848	0.811
	1.30		0.959	0.918	0.878	0.839	0.799
	1.20		0.955	0.912	0.869	0.827	0.786
0.6	1.40		0.954	0.910	0.867	0.825	0.785
	1.30		0.951	0.904	0.858	0.814	0.772
	1.20		0.947	0.896	0.848	0.801	0.757

$$Y = \sqrt{\frac{1 - \left(\frac{A_2}{A_1}\right)^2}{1 - \left(\frac{A_2}{A_1}\right)^2 \left(\frac{p_2}{p_1}\right)^{2/k}} \cdot \frac{\frac{k}{k-1}\left(\frac{p_2}{p_1}\right)^2\left[1 - \left(\frac{p_2}{p_1}\right)^{(k-1)/k}\right]}{1 - \frac{p_2}{p_1}}}$$

[1] The tabulated values were computed by J. P. Robb and R. E. Royer at the University of Wyoming.

INDEX

683